MATHEMATICAL MODELING

MATHEMATICAL MODELING

Applications with GeoGebra™

JONAS HALL AND THOMAS LINGEFJÄRD

Published by John Wiley & Sons, Inc., Hoboken, New Jersey
Published simultaneously in Canada

For general information on our other products and services or for technical support, please contact our Customer Care Department within the United States at (800) 762-2974, outside the United States at (317) 572-3993 or fax (317) 572-4002.

Wiley also publishes its books in a variety of electronic formats. Some content that appears in print may not be available in electronic formats. For more information about Wiley products, visit our web site at www.wiley.com.

Library of Congress Cataloging-in-Publication Data

Names: Hall, Jonas, 1963– | Lingefjärd, Thomas, 1952–
Title: Mathematical modeling : applications with GeoGebra / Jonas Hall, Thomas Lingefjärd.
Description: Hoboken, New Jersey : John Wiley & Sons, Inc., [2017] | Includes index.
Identifiers: LCCN 2016009974 (print) | LCCN 2016019552 (ebook) | ISBN 9781119102724 (cloth) | ISBN 9781119102694 (pdf) | ISBN 9781119102847 (epub)
Subjects: LCSH: Mathematical analysis–Data processing. | Mathematical models–Data processing.
Classification: LCC QA300 .H353 2017 (print) | LCC QA300 (ebook) | DDC 003–dc23
LC record available at https://lccn.loc.gov/2016009974

Set in 10/12pt Times by SPi Global, Pondicherry, India

CONTENTS

PREFACE

Welcome to the world of GeoGebra! GeoGebra is a mathematical environment that will allow you to work with graphing, dynamic 2D and 3D geometry, dynamic and symbolic algebra, spreadsheets, probability, complex numbers, differential equations, dynamic text, fitting functions of any kind to data, and so forth. All representations of mathematical objects are linked and allow you to view, experiment with, and analyze problems and situations in a laboratory-like setting. You can build a geometrical model in one window, animate it, and collect data to a spreadsheet or directly to a graph. GeoGebra works on computers, tablets, and iPhones, under multiple operating systems and is free to use for noncommercial purposes. It can be used in over 65 different languages delivered from a menu option.

Welcome also to the world of mathematical modeling in high school (grades 10–12). With computer tools such as GeoGebra and Wolfram Alpha you can now expand the relatively modest modeling normally done in class to include more real-life situations and solve more interesting and difficult problems. While the computer does the necessary calculations and graphing, with GeoGebra your students will learn the process of translating problem situations to the mathematical language that the computer needs to be able to work efficiently. They will also learn to interpret the results that the computer gives them and draw sensible conclusions from them.

These competencies—translating problem situations to mathematical language suitable for computer processing and analyzing, reasoning about and presenting results—are more important in a modern world than performing specific algorithmic calculations. After all, no one today complains about the fact that we no longer teach the manual algorithm for calculating square roots.

This book came about because the authors strongly believe that modeling lies at the heart of mathematics. And in order to model interesting problems, you need powerful tools. The strong user community and fast development of GeoGebra made it the tool of choice. With it, we were able to explore a multitude of modeling situations, some quite simple and some involving systems of differential equations, something not normally taught in high school but rendered possible with the use of computers.

In Sweden, modeling is one of seven competencies the students are to develop in mathematics education. On the cover of this book, we have symbolically placed the modeling competence at the center of the other competencies, indicating our belief that modeling is absolutely central in mathematics.

In this English translation of the book, we have restructured our material so that it now comes in order of mathematical content. Thus we explore linear models, non-linear models, models requiring calculus and differential equations, and discrete and geometrical models. It is then up to you to decide what problems in this book might be suitable for what courses and what students you teach in your school.

This book should be useful for both experienced teachers and for those students who are in teacher education programs. In addition, students of mathematical modeling at starting level university courses may find this book useful, as may anyone wishing to learn GeoGebra really well. Indeed, the book is intended to be a handy reference on both modeling and GeoGebra, for the reader to keep and return to look up the details of problem solving, modeling, and GeoGebra techniques. If you are new to GeoGebra, we suggest that you read Appendix A, *An Introduction to GeoGebra*, first before continuing with the rest of the book.

Several digital resources are available with this book. On the book's website you will find a collection of all the GeoGebra files used in the book's problems, a list of clickable links, and some screencasts showing basic techniques.

Last we wish to acknowledge the continued support we have received from the GeoGebra community, which has encouraged us to write this book, and we wish to thank our families for being, above all else, patient.

Sweden, May 2016

JONAS HALL AND
THOMAS LINGEFJÄRD

INTRODUCTION

ABOUT THIS BOOK

This book is written primarily for teachers of mathematical modeling in upper secondary schools or in high schools. Students in a teacher training program at a university or studying mathematical modeling in an introductory course at the university may also want to explore the possibilities that GeoGebra can afford. The book was conceived from the standpoint of the Swedish curriculum, which regards mathematical modeling competence to be one of seven competencies that should be taught and assessed in upper secondary school.

As a school subject, mathematics is no longer only about calculation. Some parts of mathematics, of course, relate strongly to procedures and counting, but altogether this part of the curriculum has less emphasis today than it used to have. Today, mathematics is treated as a tool, as an aid, as a language, and as logic. The curriculum in many countries is nowadays expressed in terms of competency objectives. The competencies are general and not related to a specific mathematical content. Yet, the competencies are developed in levels by students' processing specific content. The modeling competency is one of these competencies that draw heavily on functions and differential equations.

Mathematical models and other mathematical representations such as diagrams, histograms, functions, graphs, tables, and symbols normally make it easier for abstract mathematical concepts to be understood and for other phenomena to be described in mathematical terms. Educators today are facing a world that is shaped by increasingly complex, dynamic, and powerful systems of information that are meet through various media. Being able to interpret, understand, and work with mathematical models and other complex systems involves important mathematical processes that become discernible and obvious when teaching mathematical modeling.

In mathematics education, as seen from the K–12 perspective, teachers work with different representations in order to help students understand mathematical objects and concepts. Models such as geometrical constructions, graphs of functions, and a variety of diagrams are used to introduce new concepts and to show relationships, dependency, and change. Mathematical models, structures, and constructions are also used in different scientific fields, such as in physics and the social sciences. To be able to construct, interpret, and understand mathematical models is becoming increasingly important for students all over the world.

Our main academic position is that once modeling competency is acquired in the classroom, all other competencies will be addressed automatically. With training in mathematical modeling, instead of always asking "Why are we doing this?" students will find classroom work to be interesting and related to reality, and then concepts, procedures, problem solving, reasoning, communication, and relevance will follow without much effort. If you, the teacher, try to do it the other way around, you may soon discover that in sticking with too many routine calculations you will end up without time to address the modeling and reasoning competencies.

There were some basic considerations that we needed to address in writing this text on mathematical modeling. We could have chosen to only focus on the process of constructing and developing models or instead on the evaluation of already produced mathematical models. We decided to try and address both situations in this book. However, for those of you teaching mathematical modeling in upper secondary school, it may be a good idea to start with existing and well-developed models. Then, as students become familiar with the mathematical modeling concept, they could be started on constructing their own mathematical models.

To place mathematical modeling into a particular branch of mathematics, one could consider it as applied problem solving using data that have already been gathered in some way. We try to address the many different data that can be used in our selection of modeling examples in order to show how mathematical models are applied everywhere in our society. In some instances, however, we investigate purely geometrical models.

In today's schools, teachers have the possibility to allow every student to use powerful mathematical instruments that help them learn and do mathematics in a way that humans once only could dream about. Students can tackle difficult problems a lot earlier with these tools, so they can connect concepts and procedures to more realistic situations and open up their minds to a more nuanced communications.

In this book we decided to mainly work with GeoGebra, but other tools, primarily Wolfram Alpha, can be used as well. GeoGebra was created in 2001 by Marcus Hohenwarter, and as a tool, it could be considered a mathematical laboratory, or even an environment. GeoGebra is free and platform independent, and it handles algebra, plane geometry, 3D geometry, functions, statistics, spreadsheet calculations, and symbolic algebra. GeoGebra has been translated to over 50 languages and is used all over the world. In this book we show how to use GeoGebra for mathematical modeling as well as how to apply it to teaching mathematics in general.

We have organized the mathematical modeling examples in the following order:

Chapter 1: Some Introductory Problems
Chapter 2: Linear Models
Chapter 3: Nonlinear Empirical Models I
Chapter 4: Nonlinear Empirical Models II
Chapter 5: Modeling with Calculus
Chapter 6: Using Differential Equations
Chapter 7: Geometrical Models
Chapter 8: Discrete Models

Then we have added four more chapters on the teaching and assessing of mathematical modeling, in accord with the methodology of the teaching profession:

Chapter 9: Modeling in the Classroom
Chapter 10: Assessing Modeling
Chapter 11: Assessing Models
Chapter 12: Interpreting Models

For those of you who are new to GeoGebra, we have added an introduction to this interactive, dynamic platform. We have further added a function library that can be browsed for different functions to fit data.

Appendix A: Introduction to GeoGebra
Appendix B: Function Library

In trying to model different phenomena, you will soon discover that you need different prerequisites in mathematics. We address this issue with different mathematical modeling examples at different levels of learning. In this regard the mathematical hierarchy we present is probably much the same in your school system as it is around the world. If you have previous experience with GeoGebra, we recommend that you study the chapters in sequence. If you have no previous experience at all, we suggest that you study Appendix A first, and thereafter the chapters in order.

Chapter 1–8 contain the modeling tasks. In each of these chapters there are a number of solved modeling tasks with at least one, sometimes several, thorough solution suggestions. The solutions are very detailed, both mathematically and technically. By reading—and doing—these solutions, your students will learn mathematics, mathematical modeling, and GeoGebra techniques, and so become good modelers and problem solvers. Each chapter also has a number of unsolved tasks at the end.

Each task may be varied in a number of ways. Sometimes it may be that only a value should be calculated, sometimes several different models could be created, sometimes an error estimate could be included, and sometimes it may be better to write a report. These different ways to work out a solution can be applied to all tasks.

We have varied the tasks somewhat randomly and encourage you to adapt them to your students' needs and current levels.

The modeling tasks, both those we have solved and those left to be solved are often quite comprehensive and cannot be fitted in just one lesson. We think of them as requiring students to have at least a few lessons before starting the assignment or a week to work on the assignment. You, the teacher, may choose to do nothing else for a week or so, but we prefer to think about the tasks as parallel assignments that allow the student to work on them for long time periods while learning new concepts, asking questions, discussing the tasks with classmates, and so forth. This way they will learn and progress in reasoning, communication, conceptual growth, and more.

We also believe that students need to see many examples of written mathematical reasoning in order for them to be able to start producing written mathematical reports. The typical syllabus mentions detailed and nuanced reasoning, and this competency needs to be discussed so that students know what is expected of them. It is our hope that the many solved examples in this book will be one such source of wisdom for students and serve as an inspiration for teachers to help students develop their competencies.

Chapter 9 addresses how to organize everyday work in the classroom, and it gives some examples of different approaches to that task. Teachers can easily turn into coaches when introducing and maintaining modeling processes, so they do need to assume that students have certain inner motivation and experience together with a desire to learn. The learning that takes place is built on the assumption that the students get continual opportunities to test, validate, and rebuild their previous knowledge. Moreover in the classroom students will learn from each other, so their social interactions are important. Lecturing should be seen as a compliment, and not as the main core of teaching in the classroom.

Chapters 10 and 11 are about assessing and evaluation. **Chapter 10** is on evaluating students' work, and **Chapter 11** is more technical and discusses how to evaluate models and do basic mathematical error analysis.

It is obviously a challenging task to evaluate mathematical models as well as mathematical modeling. In play here is not only the pure mathematical ability to analyze and select the "best" model. Anyone who does mathematical modeling also needs to attain a certain level of technical knowledge in order to do the modeling. Students who have gone through the modeling process must be given time and space to explain and discuss their work. As they fulfill the discussion and presentation requirements, their competencies will become quite apparent.

In **Chapter 10** we also describe what forms the students' mathematical modeling presentations might take.

In **Chapter 11** we look at different ways to evaluate models. Even a model that follows Newton's cooling law describing the cooling of coffee in a cup or a linear model for pole vault results can be expected to have some pervasive errors. That means that mathematical models seldom are correct in the same sense as $2+2=4$ is correct.

When you compare a model's conditions with reality, you might find good conformity. The populations of many animals and plant species, both globally and in different countries, are decreasing exponentially. These species will, if nothing is done, go extinct. But reality is always more complex than the model. Many species can be saved. One such example is the Swedish peregrine, which decreased in number from the 1950s to the 1970s when large rescue efforts were made. That work changed the decreasing trend of the population, and now the previous model no longer works.

You may find some populations that are growing exponentially, however. Bacteria may, under certain conditions, grow exponentially. This could also hold true for larger spices such as birds. The cormorant has increased exponentially since the mid-1950s. In other situations species numbers may oscillate, often due to deep and complex feedback loops. In **Chapter 11** we describe and illustrate some basic techniques for the evaluation and estimation of the errors inherent in all models.

Chapter 12 is about the different ways a student may interpret a modeling situation. You will see that there is a difference between interpreting the situation as an *event* or as a *process*.

In the **Appendixes** are technical descriptions of GeoGebra and math functions. **Appendix A** provides a basic introduction to GeoGebra for teachers not yet familiar with this excellent mathematical laboratory. **Appendix B** provides a function library showing how different, in upper secondary school more or less common, functions can be parameterized in order to be used in the mathematical modeling process.

This is not a textbook in mathematics. Our intention in writing this book was to provide an introduction to the subject of mathematical modeling, and not a comprehensive text on mathematical modeling. The book does present the many different ways that mathematics can be used and applied in mathematical modeling. We are convinced that anyone who learns mathematical modeling can also learn to use mathematics in a new way and therefore learn more mathematics. Through mathematical modeling we humans also learn to see the importance of mathematics in the world around us.

When should students be taught mathematical modeling during their school years? In certain countries mathematical modeling is already compulsory in primary schools, while in other countries students are taught modeling in upper secondary schools. Today, it appears that with the increasing availability of computers, teachers in many countries are becoming more and more interested in teaching mathematical modeling early in the educational system. Whatever the case is in your country, many students already think in terms of informal models when they compare different cell-phone plans. Even though the modeling exercises in this book relate to the upper secondary school level, it is up to the teacher using this book to introduce mathematical modeling when it suits the course's content.

As already mentioned, the main purpose of this book is to teach mathematical modeling but there are also other purposes:

- To serve as a guidebook for mathematical modeling with GeoGebra and thereby also as training material for mathematics teachers
- Together with additional material on the Internet (GeoGebra constructions and screencasts), to serve as a problem sets repository for teachers to use in working with mathematical modeling
- To provide solutions in such full detail as to assist students and teachers in preparing qualitative short mathematical reports

For the many teachers who have not yet worked either with mathematical models or with GeoGebra in this way before, the book may prove to be appropriate as course literature in teacher training programs or for theme-based teacher in-service training.

ABOUT MATHEMATICAL MODELING

The main strength of mathematical modeling is learning to make decisions. With the help of mathematical modeling, sometimes predict future trends can be predicted, decisions made on global environmental issues, and even the kind of resistor to select when building a hobby radio. Mathematical modeling can be used in artwork, in creating perspectives and in compositions. Mathematical modeling is used in packaging industry, in economics, in biological systems, in medical trials, and in computational physics. It is in fact difficult to find any area where humans work that has no use at all of mathematical models.

In today's competition-driven markets, the modern engineer needs to reduce start-up times and costly trial series productions and other construction costs. Thus, virtually all product development is now dependent on successful modeling and simulation, whether we are talking about cars, cellphones, computer parts, medical equipment, or more "invisible" things such as efficient queueing strategies for airports and trains or software development.

Obviously, the computer has given mathematical modeling a huge boost. Everybody with a modern mobile telephone and access to the Internet also has access to tools like Wolfram Alpha. Science is even changing with technology. Once we talked about theoretical and experimental physics. Today, we also talk about computational physics built on powerful computational models that we could not construct before the computer arrived. Once we talked about the dividing of the world in matter and energy. Today, we also talk about the importance of information for the structuring of matter and energy. Today, we also talk about artificial intelligence and the singularity date when computers become more intelligent than humankind. Mathematical modeling is an important part of all the information we need to handle.

Here are some further arguments as to why it is important to construct models:

- The real system is impossible to experiment with. It could be a living human being or a distant stellar cluster.
- The real system is too expensive to toy around with. It could be a space exploring satellite or the regulating system of a chemical production plant.
- The real system is too dangerous to experiment with. It could be a new aircraft model or finding the correct dose of medicine.
- Modeling gets you a better understanding of the original problem.
- Modeling makes visible a system that is not yet built or constructed.
- Modeling can help you foresee the future and make prognoses.

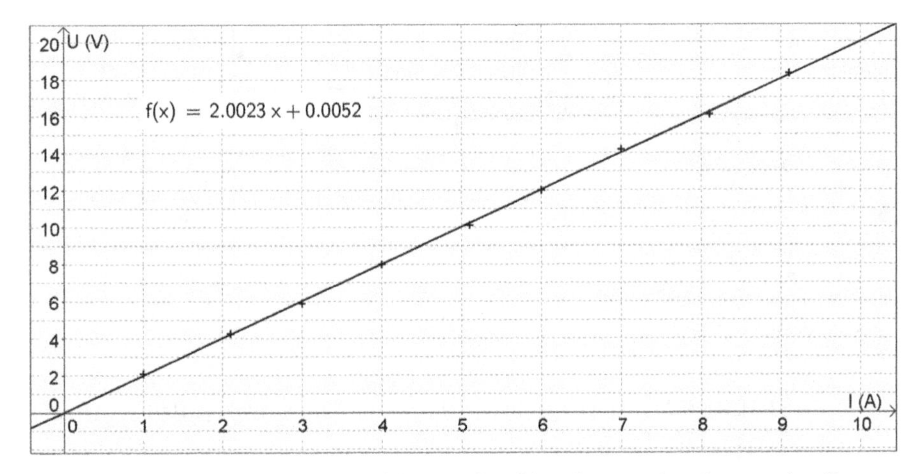

FIGURE 0.1 Example of a model presenting Ohms law as a (near) proportionality.

Students who enter upper secondary schools around the world have, of course, already met some simple mathematical models. One such example is Ohm's law from compulsory school, as shown in Figure 0.1.

Ohm's law states that the current through a conductor between two points is directly proportional to the potential difference across the two points. This introduces the constant of proportionality, the resistance, and the usual mathematical equation that describes this relationship $U = \mathrm{R} \cdot I$ is a proportionality between the variables I = current (Ampere), U = voltage (volt) and the parameter R = resistance (ohm).

When describing the mathematical modeling process in theoretical terms, researchers often talk about a cycle that humans go through when they are involved in the process of mathematical modeling. It could be understood in the following way:

The first step in a mathematical modeling process is to translate the situation or the phenomenon at hand into a mathematical problem using mathematical terms. This means that you sometimes need to define variables or know something about the relations that exist between the variables in the example above. The original problem will go through a mathematical process, an analysis, and then, perhaps it is solved. The results should be validated against theories or common knowledge and then perhaps tested in order to see if the solution works. Figure 0.2 shows one interpretation of this process.

An example that we think makes this mathematical modeling process more understandable is the *bear problem*, a generic problem that could be used in many mathematics courses around the world.

> The number of bears is increasing (in some region). In 2000 there were 2,500 bears and in 2010 there were 2,700 bears.
>
> a) How many bears will there be in 2025?
>
> b) When will there be 3,500 bears?

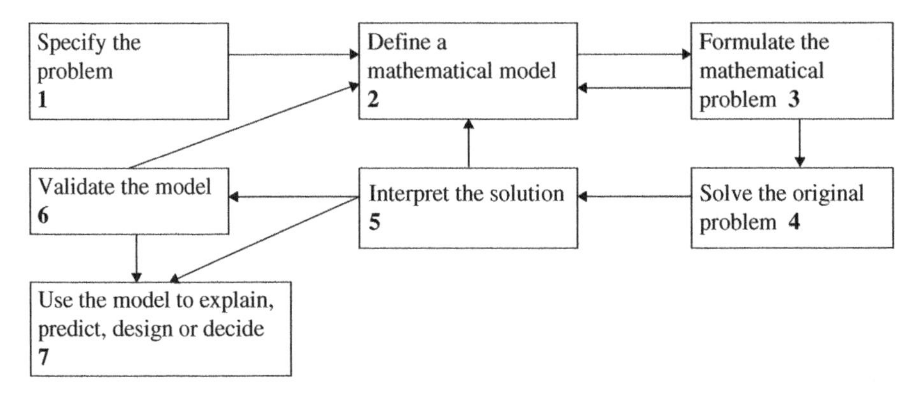

FIGURE 0.2 Main stages in modeling (from Mason, 1988, p. 209).

In our view, this is a good task to start with, but it has to be very detailed when presented in the classroom. It contains mathematical modeling, linear functions, and exponential functions; students will learn how to find the model's or the function's parameters and how to solve different types of equations. We will illustrate step by step how the *bear problem* can be solved by hand or with GeoGebra.

Step 1—Select and Organize a Model

To begin, you need to explain what kind of mathematical model you want the class to work with. If the text says: "The number of bears is increasing …

- … equally much/the same amount/by 20 bears/by the same number/linearly … every year (month, second, hour, …), then you should use a **linear** model: The number of bears $B(t) = \mathrm{a} \cdot t + \mathrm{b}$.
- … by the same rate/percentage/by 1.5 %/exponentially … every year (month, second, hour, …), then you should use an **exponential** model:

  The number of bears $B(t) = \mathrm{C} \cdot \mathrm{r}^t$.

Step 2—Define Your Variables

Clearly explain to the class what the variables stands for:

- t is the time in years starting at year 2000 (give the **unit** and what $t = 0$ corresponds to).
- B is the number of bears (or perhaps the number of **thousand** bears).

Step 3—Define the Parameters

If you decided to put the time as $t = 0$ equal to year 2000, then the intersection between the graph and the y-axis (b or C) will correspond to the first measurement. In this case you can directly write that: "I see that …

- b = 2,500 (**linear** model), or
- C = 2,500 (**exponential** model).

The second parameter requires a calculation:

- In the **linear** model you must calculate a (the slope), which you do like this:

$$a = \frac{\Delta B}{\Delta t} = \frac{(2700 - 2500)\,\text{bears}}{10\,\text{years}} = 20\,\text{bears / year}$$

  or you could solve the equation $B = a \cdot t + b$ and plug in $2{,}700 = a \cdot 10 + 2{,}500$. Note that this equation is about finding the **second measurement**. You used the **first** equation to find the value of m. The annual increase is 20 bears per year.
- Working with the exponential model in a similar way, you can solve an equation for the **second measurement**: $B = C \cdot r^t$, yielding $2{,}700 = 2{,}500 \cdot r^{10}$. Dividing both sides of the equation with 2,500, you get $1.08 = r^{10}$; leading to $r = 1.08^{1/10} = 1.0077258$, which means an annual increase of about 0.77%.

In GeoGebra the points can be defined as `A = (0, 2500)` and `B = (10, 2700)`. You get a linear model with the command `FitPoly[A, B, 1]` or an exponential model with `FitGrowth[A, B]`. You can find the value of the parameters from the algebra window.

Step 4—Write up Your Conclusions by Formulating Your Model

Write: "I have found that the number of bears can be predicted by …

- $B(t) = 20 \cdot t + 2500$ (**linear** model)," or
- $B(t) = 2500 \cdot 1.,0077258^t$ (**exponential** model)."

Step 5—Find the Number of Bears after a Certain Time

Since you now have the model, all you have to do is to enter the value of the time into your model. Be careful when you explain how you use the t-value. The year 2025 corresponds to $t = 25$ and so:

- $B(t) = 20 \cdot 25 + 2500 = 3000$ bears (**linear** model), or
- $B(t) = 2500 \cdot 1.0077258^{25} = 3030$ bears (**exponential** model).

In GeoGebra you type

```
x = 25
```

and can then decide where this line intersects with the graph by asking GeoGebra to construct an intersection with the Intersect tool.

Step 6—Find When the Bears Come to a Certain Number

In a similar way you can solve equations:

- $B(t)=20\cdot t+2500$ with $B=3500$, yielding the equation $3500=20\cdot t+2500$, which gives $t=1000/20=50$ years corresponding to the year 2050 (**linear model**), or
- $B(t)=2500\cdot 1.0077258^t$ with $B=3500$, yielding the equation $3500=2500\cdot 1.0077258^t$, where $1.4=1.0077258^t$ and $t=(\lg\ 1.4)/(\lg\ 1.0077258)=43.7$ years, corresponding to the year 2043 (**exponential** model).

In GeoGebra you type

```
y = 3500
```

and find the intersection point as before.

As you notice, all of the steps in the modeling cycle are not always present in all mathematical modeling tasks and all the practical steps do not always correlate one-to-one with the theoretical steps. However, as we will show further on in the book, when we get to more complex modeling exercises, we will often walk around the modeling cycle a few times

We will round up this excursion into practicality by looking at another phenomenon that actually affects our life regardless of where we live or of what we do. This phenomenon is usually called **price elasticity.** The basic question is the following:

How much will the demand for a product decrease if the price is increased?

We begin by making a basic assumption that there is a proportionality between the relative changes in price and demand. This is a reasonable assumption, at least when we are talking about small changes. Using algebra this situation can be defined as

$$\frac{q-q_0}{q_0}=\varepsilon\frac{p-p_0}{p_0}$$

where p is the price, q is the quantity of the product, and ε is the price elasticity. We can write q as a function of p and get $q=q(p)$, where p is the price q is the quantity of the product. If we only have two observations of price and quantity for a product, (p_1, q_1) and (p_2, q_2), then the average price elasticity can be defined as

$$\varepsilon=\frac{q_1-q_2}{p_1-p_2}\cdot\frac{p_1+p_2}{q_1+q_2}$$

So, in order to understand a model of price elasticity—the price that consumer are willing to pay for a product—you have to learn and understand quite a lot of mathematics such as rational expressions, linear approximation, proportionality, and negative numbers. In fact you will almost always end up in fairly complicated models by just asking apparently easy questions. Here are some examples:

- What happens in a lake when the temperature changes? How long a time does it take for the lake to turn into ice? How long will it take for the iced lake to turn into water again?
- Think of two or more intersecting roads. Should we have a traffic light system or a rotary? What would a model look like?
- How fast will a forest fire spread?
- How fast will a contaminated lake clean up?
- In games on computers—how will objects move? In what way do we want three-dimensional objects to appear in a two-dimensional monitor?
- In heating a house, it is well known that countries in the upper Northern Hemisphere have houses constructed with a short north wall containing few windows. But how short is the wall and how many windows are usually present?
- In our connected world, the placement of telecommunication masts is important. The economic principle is to cover as large area as possible with every mast. The result, when analyzing this phenomenon, will be a geometrical model. We will also have a choice of geometrical objects, such as regular hexagons or circles.

The types of questions asked are often categorized in four groups: understanding the system, predicting the system, regulating the system, and constructing the system. Understanding is the most basic category. Prediction is not possible without understanding, regulation is not possible without predictive power, and constructing something requires you to be able to control and regulate its behavior.

In teaching mathematical modeling, it is crucial that you get the students used to the different models and techniques. They should in fact go several times through the steps or stages in the modeling cycle.

Within different academic fields there is a need to study the relationships between measured variables. When you are constructing a mathematical model that describes the relationships between the variables mathematically, you are giving yourself the opportunity to predict and explain what will happen in the future. Therefore you, the teacher, should ensure that your students accomplish the following by working with mathematical modeling:

- Learn to problematize realistic situations
- Learn to fulfill assignments
- Learn about a variety of subject areas where people work with mathematics
- Learn to write mathematical reports with correct mathematical language
- Learn to use powerful tools and methods
- Learn to use powerful digital tools as GeoGebra and Wolfram Alpha
- Achieve different mathematical competencies from practicing mathematical modeling

All the different mathematical modeling tasks that we discuss in this book are solved for you, but the solutions are merely suggestions. Several of the problems can be solved in other ways.

ABOUT MATHEMATICAL MODELING TECHNIQUES

Among the different basic techniques within mathematical modeling, perhaps the most basic is to just fit a functional dependency to a given data set. That technique is sometimes called *empirical mathematical modeling*, and it is useful for *interpolation* (i.e., used to predict values *within* the range) but less useful for *extrapolating* (i.e. used to predict result *outside* the range). If there is an underlying theory, that can help make extrapolations work better.

Theoretical considerations make it possible to create mathematical models independent of measured data. Such a model often contains a few parameters that can be adjusted to get the best fit for the measurements you have done. One such common example is that coffee cools of exponentially down to room temperature, which gives you a model of the form $T=T_0 \cdot a^t+T_R$, where T_0 is the temperature difference against the room temperature at the time $t=0$, T_R is the room temperature, which you probably know, and a is a constant that directs the speed of the cooling. The parameters T_0, a, and eventually T_R can be varied to get the model to adjust as well as possible to the measured values. The process is often called "regression analysis" and GeoGebra have very strong tools for doing this.

If you do not have access to your own data, you may get data in the graph window by building representations of physical models that you can investigate and measure. These measurements give data that you can analyze in another window, as in Problem 8.6 *Inner Areas in a Triangle*. This is what is known as doing *simulations*. A specific sort of simulation is when you are working with pure mathematical objects, such as triangles and squares. In our case, we don't normally say that we are doing a simulation, but the technique is the same.

Another sort of simulations is called *stochastic*, or *Monte Carlo simulations*. These simulations are built on the concept of randomizing the control of certain variables. By doing the simulation many times, we can investigate a model in another way than a theoretical model would be able to do. It is possible to find quite many simulations on the Internet. Two useful websites with such collections are http://www.shodor.org/interactivate/activities/ and http://dmentrard.free.fr/geeogebra/index.htm.

Another sort of modeling, sometimes called *deterministic* in contrast to *stochastic*, is based on the creation of *differential equations* that you can solve and adapt to start values. In chapter 6 we will learn the powerful ways to solve complex problems. Without today's technology you would probably only be able to solve the simplest different equations, but with the aid of GeoGebra and Wolfram Alpha you have the capability to solve very advanced models and with solutions that you can find symbolically or numerically.

In some types of problems you will have systems of differential equations, as when studying how the numbers of predators and prey affect each other's living conditions. You can then do more analyzing with the help of phase diagram, where the number of predators set on one axis and the number of prey on the other axis.

There are, of course, other ways to do mathematical modeling, but in this book we have chosen the techniques described above because we believe these to be quite suitable for upper secondary schools.

ABOUT BEING A GOOD MODELER

What does it take to become a good mathematical modeler and how do you orchestrate your teaching so that your students become good at mathematical modeling? The students need to develop certain skills:

- Competence enough to reason about the real-life situation and thereby to connect to it and build a mathematical model that relates to reality. Students could, for instance, understand the relationship between the price of an ice-cream cone and the number of ice cream scoops you get, or for the higher level student, that the upper bridge span on Sydney Harbour bridge can be modeled by $y = 1/(x^2 + 1)$.
- Experience using different basic functions and how these functions are applied and behave. In particular, what straight lines look like, odd or even polynomials, rational functions, exponential functions, trigonometric functions, logistical and hyperbolic functions, the normal distribution function, and then again, the products of functions.
- Understanding as to how to move and stretch functions algebraically, for example by change x to $x - x_1$ when a function has to be moved x_1 steps to the right.
- Knowledge of suitable parametric transformations of functions, such as when cooling against the room temperature T_0 can be modeled by the function $y = C \cdot e^{k \cdot t} + T_0$.
- Digital competency using tools, for example, how to get GeoGebra to create points from your measured data and then fit a suitable function with the regressions tools. Or how to enter a differential equation with initial conditions into Wolfram Alpha.
- Competence to critically evaluate the mathematical modeling process, be open to changing the model, and have an understanding for the errors that can be carried by the model and transferred over to the responses from the model to the questions they were given at the start, for example, the way a water parabola is affected by an air stream and the fact that this could not be modeled sufficiently by a parabola.
- Competence to organize, follow, and document the mathematical modeling process from start to finish.
- Competence to select the most essential parts and ignore the rest. As Albert Einstein said: "Make it as simple as possible, but not simpler."

Clearly, this is not something that is easily developed over the course of a few weeks. Rather you have to embrace a *modeling mind set* where modeling is the natural starting point of many, if not all, modules, lessons and assignments. To develop all these modeling sub competencies requires that this mind set is be transferred to the students over time. As the student's experiences accumulate, it is your job as a teacher to forge these experiences together to a meaningful whole. If you hold this modeling mind set central you will find that competencies such as the ability to reason, communicate, and calculate will come naturally as consequences of working with modeling problems.

ABOUT THE MODELING PROBLEMS

Many of the problems we have selected might be experienced as difficult. Some of the problems are suitable for different mathematics courses, depending on your student's experiences and at the level of the mathematics that is taught. Remember, the aim of each problem is that it be solved over a long time so that the students will be able to understand the problem completely. The discussions the teacher will have with the students and the discussions the students have with each other as they tackle solutions to problems are generally very rewarding in that they help develop a student's language and conceptual understanding.

We have supplied most of the problems with extended solutions that might be seen as a way to write up a mathematical solution. Some of the problems can be used to show students how to write up the solution for the next problem.

The problems were selected to impart a certain breadth to the mathematical contents as well as to the problem solving techniques using GeoGebra. It means that it could be a good idea for you to first work through the solutions to several problems so that you will know how to help your students most effectively.

As authors, we like to share good mathematical modeling activities with our readers. If you have a suggestion for a good activity, preferably also with a solution, or a new way to solve one of our problems, please send it to us. You will receive a link to an Internet address where all contributed material will be published. It is our hope that in this way we will get a large inventory of suitable mathematical modeling activities that we all can use.

ABOUT GEOGEBRA

GeoGebra is a dynamical mathematical laboratory suitable for teaching or learning all kind of mathematics including mathematical modeling. GeoGebra runs on all platforms and can be downloaded from www.geogebra.org.

GeoGebra is specifically built to show many different representations of dynamic objects at once. It handles everything you are likely to meet in secondary mathematics, and then more. While it may stop short of professional programs like MS Excel, Mathematica, and MATLAB that do pure number crunching, GeoGebra combines every feature you could wish for in a package free for educational use.

Some procedures are done more often than others. You will often need to enter measured values into GeoGebra's spreadsheet and create points from these values. You can zoom in and change labels on axes and change the background color. You can even change the font sizes and the number of decimals. We have chosen to describe these procedures in the running text as well as in Appendix A, *Introduction to GeoGebra*. In this way we hope that it not only will be possible to read the book from beginning to end but also to access a collection of modeling activities that you can dip into at any point.

We have used the Windows version of GeoGebra 5.0, which was released in 2014 but constantly updates with new functions, as described in the *change log* found at http://wiki.geogebra.org/en/Reference:Changelog_5.0.

ABOUT WOLFRAM ALPHA

Wolfram Alpha is a powerful mathematical Internet based tool that combines a search engine and a computer algebra system. It can solve equations, differential equations, and sketch graphs, and further find all sorts of mathematical facts as well as information on physics, chemistry, biology, and more. It complements GeoGebra very well, and it is seldom that a mathematics teacher would need any other tools than these two except for personal preferences. The best way to get familiar with Wolfram Alpha is to have a look at some of the many examples at http://www.wolframalpha.com/examples/.

ABOUT TYPOGRAPHY

We have used the following conventions:

- Variables are written in *italic*:

$$x, y, p(x), t\ldots$$

- Parameters are written in normal text. We do this so that you can separate independent and dependent *variables* from the parameters we can control, change, and fit to data:

$$y=\mathrm{k}\cdot x+\mathrm{m},\ m(x)=\mathrm{a}/x+1, f(x)=\mathrm{a}\cdot x^2+\mathrm{b}\cdot x+\mathrm{c}\ldots$$

- Input commands in GeoGebra's input bar, spread sheet, or Wolfram Alpha are written in **Courier bold**:

`=A2/B2, Fit[list1,m], Max = Extremum[f]`...

- Keyboard shortcuts, menu alternatives, and commands that are discussed rather than typed are written in **bold**:

Ctrl-Shift-2, View – Spreadsheet…, the **Locus** command

- Tool names are written in normal text, but often followed by a tool image: Use the Point tool •A to create a point.

ABOUT THE DIGITAL MATERIAL FOR THE BOOK

If you have the paper version of this book, it is in black and white. Any color references in the book refer to the GeoGebra file that we have created for the solutions to the problems. Download the GeoGebra solutions to all problems from the book's website. There is a GeoGebra file for just about every figure in the book, numbered accordingly.

You are also welcome to send similar problems with solutions directly to the authors so that you can access the materials other teachers have submitted. In this way we hope to enrich the book by creating new useful modeling teaching materials in an online resource bank.

ABOUT THE AUTHORS

Jonas Hall is head of mathematics at Rodengymnasiet in Norrtälje, north of Stockholm, in Sweden, where he teaches mathematics and physics. He is a hobby mathematician with a special interest in problem solving, the aesthetics of mathematics, and teaching with technology. He is a multiple finalist in *Kappa*, a mathematical competition for mathematics teachers in Sweden held by the University of Stockholm, and he has dabbled in creating mathematical art, which can be found at http://www.geogebrainstitut.se/art/art.asp.

Thomas Lingefjärd is a professor of mathematics education at the University of Gothenburg. He started his teaching, however, as an upper secondary teacher in mathematics, physics, and ICT. Lingefjärd has for many years been interested in how the presence of modern technology can help students to learn and understand mathematics better. His main activities at the University of Gothenburg are to supervise PhD students, to teach courses in the mathematics teacher program, and to do research on the learning of mathematics.

Together, the authors are responsible for the Swedish GeoGebra institute, http://www.geogebrainstitut.se, one of over 100 regional and national GeoGebra institute over the world: http://www.geogebra.org/institutes. The aim of the institute is to spread the use of GeoGebra in Sweden. The authors have translated GeoGebra to Swedish and also written articles for teacher journals about GeoGebra. They also contribute to the international research around GeoGebra and its influence on mathematical teaching and learning in the Nordic–Baltic GeoGebra Network, http://nordic.geogebra.no/and run the Swedish Facebook support group.

REFERENCES

Arditi, R., & Ginzburg, L. R. (1989). Coupling in predator-prey dynamics: ratio dependence, *Journal of Theoretical Biology* 139: 311–326.

Chi, M. T. H., Roscoe, R., Slotta, J., Roy, M., & Chase, C. C. (2012). Misconceived causal explanations for emergent processes. *Cognitive Science*, 36, 1–61.

Duval, R. (2006). A cognitive analysis of problems of comprehension in a learning of mathematics. *Educational Studies in Mathematics*, 61, 103–131.

Elby, A. (2000). What students' learning of representations tells us about constructivism. *Journal of Mathematical Behavior*, 19 (4 4th quarter), 481–502.

Friel, S. N., Curcio, F. R., Bright, G. W. (2001) Making sense of graphs: critical factors influencing comprehension and instructional implications: *Journal for Research in Mathematics Education*, 32, 124–158.

Kreider, D., Lahr, D., and Diesel, S. (2005). *Principles of Calculus Modeling: An Interactive Approach*. Published on the Internet at https://math.dartmouth.edu/~klbooksite/

Lingefjärd, T. (2000). Mathematical modeling by prospective teachers using technology. PhD dissertation. University of Georgia, Athens, GA.

Lingefjärd, T. (2002a). Mathematical modeling for preservice teachers: a problem from anesthesiology. *The International Journal of Computers for Mathematical Learning* 7(2), 117–143.

Lingefjärd, T. (2002b). Teaching and assessing mathematical modeling. *Teaching mathematics and its Applications* 21(2), s. 75–83.

Lingefjärd, T. (2006). Faces of mathematical modeling. *Zentralblatt für Didaktik der Mathematik*, Volume 38, Number 2, s. 96–112.

Lingefjärd, T. (2012). Learning mathematics through mathematical modeling. *Journal of Mathematical Modeling and Application*, 1 (5), 41–49.

Lingefjärd, T. (2013). Connections between geometry and number theory. *At Right Angle*, 2 (2), 50–58.

Lingefjärd, T., & Holmquist, M. (2001). Mathematical modeling and technology in teacher education—Visions and reality. In: J. Matos, W. Blum, K. Houston, & S. Carreira (ed.), *Modeling and Mathematics Education ICTMA 9: Applications in Science and Technology*, 205–215. Horwood: Chichester.

Lingefjärd, T., & Holmquist, M. (2003). Learning mathematics using dynamic geometry tools. In: S. J. Lamon, W. A. Parker, & S. K. Houston (ed.), *Mathematical Modeling: A Way of Life. ICTMA 11*, 119–126. Horwood: Chichester.

Lingefjärd, T., & Holmquist, M. (2007). Model transitions in the real world: the Catwalk problem. In: C. Haines, P. Galbraith, W. Blum, & S. Khan (ed.), *Mathematical Modeling ICTMA 12 Education, Engineering and Economics*, 368–376. Horwood: Chichester.

Lingefjärd, T., & Holmquist, M. (2005). To assess students' attitudes, skills and competencies in mathematical modeling. *Teaching Mathematics and Its Applications* 24 (2–3), 123–133.

Lingefjärd, T., & Meier, S. (2010). Teachers as managers of the modeling process. *Mathematics Education Research Journal*, 22 (2), 92–107.

Mason, J. (1988). Modeling: what do we really want pupils to learn? In: D. Pimm (ed.), *Mathematics, Teachers and Children*,201–215. London: Hodder Stoughton.

http://www.geogebrainstitut.se/artiklar/artiklar.asp

http://www.ugrad.math.ubc.ca/coursedoc/math103/site2012/keshet.notes/Chapter9.pdf

http://energyfromthorium.com/tech/physics/decay2/

http://lwd.dol.state.nj.us/labor/lpa/census/1990/poptrd1.htm

http://www.infoplease.com/ipa/A0004986.html

http://www.nist.gov/data/PDFfiles/jpcrd615.pdf

http://jwilson.coe.uga.edu/Texts.Folder/SRR/DaD5.html

http://www.airspacemag.com/flight-today/falcon.html?c=y&page=1

http://en.wikipedia.org/wiki/Compartmental_models_in_epidemiology

http://www.geogebra.org

ABOUT THE COMPANION WEBSITE

This book is accompanied by a companion website:

www.wiley.com/go/Hall/MathematicalModeling

The website includes:

- Figures
- GeoGebra files
- Videos
- Web links

1

SOME INTRODUCTORY PROBLEMS

The introductory problems in this chapter focus on developing students' abilities to reason and produce a reasonable model from a real-world situation. Photographs are used in several examples and the student may easily expand the problem assignments by providing their own photographs and related questions. The problems are not necessarily "beginners" problems; they are selected because they are "nonstandard," and offer a glimpse into some of the different possibilities of modeling with GeoGebra. The problems presented in the next chapter are more standard and may for some beginners be a better choice.

Students should first try to solve some part of a problem "by head and hand," using a calculator and paper and pencil and then expand to the other parts using GeoGebra. To learn mathematics in the present digital age, we need to recognize that old school content and methods have been overtaken by a digital technology that has moved us into a mathematical golden age. To be effective in teaching mathematics with technology-assisted instruction, we suggest that you study the TPACK framework and work toward an integrated approach where technology, mathematical content, and pedagogy are all integrated seamlessly in every lesson.

We begin with a couple of different introductory problems. As students enter their upper secondary mathematical studies, the core content suddenly needs to be mastered in greater depth than previously. Some examples include, but are not limited, to the following:

- Understanding numbers, arithmetic, and algebra.
- Properties of a range of whole numbers, different number bases, and the concepts of prime numbers and divisibility.

Mathematical Modeling: Applications with GeoGebra™, First Edition. Jonas Hall and Thomas Lingefjärd.

Companion website: www.wiley.com/go/Hall/MathematicalModeling

- Calculations in everyday life and with real numbers, written in different forms, including powers with real exponents, together with strategies for calculations when using digital tools.
- Generalization of the rules of arithmetic to handle algebraic expressions. The concept of linear inequality.
- Algebraic and graphical methods for solving linear equations and inequalities, and exponential equations.
- Concepts of sine, cosine, and tangent, and methods of calculating angles and lengths of right-angled triangles. The concept of a vector and its representations, such as direction, length, and points in a coordinate system.
- Addition and subtraction with vectors and scalar multiplication to produce a vector.
- Mathematical reasoning using basic logic, including implication and equivalence, and comparisons with how to argue in everyday contexts and in science subjects.
- Illustration of the concepts of definition, theorem, and proof, such as the Pythagorean Theorem and the sum of the angles of a triangle.
- Advanced percentage concepts: ppm and percentage points.
- Concepts of rate of change and indexes, as well as methods for calculating interest and mortgage payments for different types of loans.
- Concept of a function, its domain and range of a definition, and also properties of linear functions, exponential, and polynomial functions.
- Representations of functions in the form of words, functional expressions, tables, and graphs.
- Differences in the concepts of equation, inequality, algebraic expression, and function.
- Appreciation of how statistical methods and results are used in society and in science.
- Concepts of dependent and independent events, as well as methods for calculating probabilities in multi-stage random trials, using examples from games and from risk and safety assessments.
- Strategies for mathematical problem solving including the use of digital tools.
- Mathematical problems relevant to personal finances, social life, and applications in other subjects.
- Mathematical problems related to the cultural history of mathematics.

As the teacher you need to be clear about what your goals are and how your methods will reach those goals. Your students need to know where you are going with your teaching and how you aim to get there. You also need to structure your time but leave ample room for flexibility. The modeling tasks you plan to use should be simple to begin with, so that they may be solved in one or two lessons. Use detailed instructions for your students, preferably together with screencasts to explain technical details in GeoGebra and other digital tools that your students can watch over and over again if necessary. You will also need detailed and clear summaries of your lessons to help the students who just want to keep on working without much reflection. It is beneficial to start with several short easy tasks in the beginning so that your students learn how to model, and when they have learned that, you can expand the mathematics in further modeling tasks.

1.1 TICKET PRICES

A travel agency is advertising a favorably priced sports holiday to the 2017 FIFA Confederation Cup in Russia. The transport to Russia from the United States will be by chartered flight with seating for 200 passengers. If a group intends to travel to the Confederation Cup fills the airplane, then the price for each person will be 400 USD. Understandably, the price for each sold ticket will rise for every empty seat on the airplane. The flight company owning the plane demands a minimum of 20,000 USD in rent for the plane taking this trip.

Make a model of the travel agency's total income possibilities for travel to this international football event. If the travel agency wants the same income for 199 tickets as they want for 200 passengers, what seat pricing would be reasonable then?

Instead of doing this calculation for every number of possible passengers and therefore presenting a table of possible prices for each customer, the travel agency chooses to simplify and add a standard fee for each empty seat. How will the model then describe the travel agency's total income for different numbers of passengers?

Also calculate the maximal price that a single passenger may need to pay. What is the least number of passengers that the travel agency must attract to avoid losing money on this trip?

Your model should clearly show the relation between the price of an air seat and the number of sold air seats as well as the relation between the total revenues and the number of sold seats.

Proposed Solution 1

With 200 air seats at 400 USD the total income is 80,000 USD. If the travel agency still wants the 80,000 USD even with just 199 sold air seats, the price needs to be 80,000 USD/199 air seats = 402 USD. It therefore seems reasonable to ask for 2 USD more from every passenger for every empty air seat.

The number of sold air seats x, and the price in USD y, should exhibit a linear relationship $y = a \cdot x + b$. If the number of unsold seats increases, the price will increase, and the price will be reduced if more seats are sold, which in turn means that the slope is negative. If the number of sold air seats increase with 1, the price will decrease with 2 USD, so $a = -2$. You also know that when the number of sold seats = 200, so the price is 400 USD. This gives $400 = -2 \cdot 400 + b$, and hence $b = 1200$. The equation becomes $y = 1200 - 2x$. This is represented in Table 1.1.

TABLE 1.1 Linear Relationship between the Number of Sold Air Seats and the Price for a Seat

Number of sold air seats	Price for one air seat (USD)
200	400
199	402
198	404
197	406
196	408
195	410
190	420
180	440
170	460
160	480
150	500
100	600
50	700
0	800

The maximal price, with just one passenger in the plane, will be 798 USD.

If you label the total revenues with R, you can write

$$R = \text{number of sold seats} \cdot \text{price per seat}$$

The most convenient way to handle this is to put the values into a spreadsheet, for example, in Excel or GeoGebra. To show the spreadsheet in GeoGebra, you would press **Ctrl-Shift-S** or select **View > Spreadsheet** in the menu. When you have entered the values in the first column you may type

```
=1200 - 2 A2
```

In cell B2 and press Enter. Observe that the space works as multiplication sign in GeoGebra. Then enter

```
=A2 B2
```

in cell **C2** and press **Enter**.

Now you may select cells **B2** and **C2** and then grab the small square in the lower right corner, the *fill handle*, dragging it downward, as indicated in Figure 1.1. This will copy the formulas downward. This is the standard procedure for copying formulas in spreadsheets.

It is very convenient to represent the relations graphically in GeoGebra. Select all the values in columns A and B (but not the titles) and right-click on the selection. Then select **Create... > List with points**, which will create both the data points and a list containing these points. You will not see them in the Graphic window yet, since the values are too large. Drag the background to pan the window. Hold down **Ctrl** and drag one axis at the time to zoom in or out until you see something similar to what you see in Figure 1.2.

Spreadsheet

B2 =1200 - 2A2

	A	B	C	D
1	Number of seats	Price per seat	Revenues	
2	200	800	160000	
3	199			
4	198			
5	197			
6	196			
7	195			
8	190			

FIGURE 1.1 How to copy a formula using the fill handle.

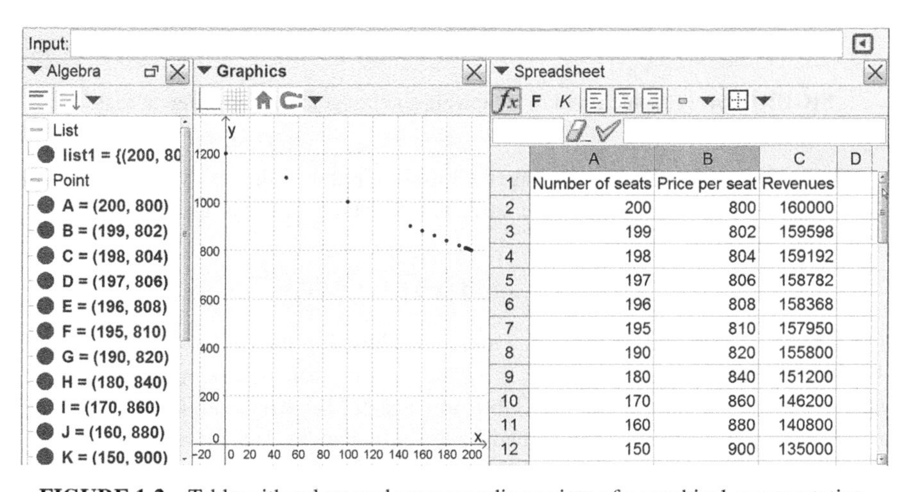

	A	B	C	D
1	Number of seats	Price per seat	Revenues	
2	200	800	160000	
3	199	802	159598	
4	198	804	159192	
5	197	806	158782	
6	196	808	158368	
7	195	810	157950	
8	190	820	155800	
9	180	840	151200	
10	170	860	146200	
11	160	880	140800	
12	150	900	135000	

FIGURE 1.2 Table with values and corresponding points of a graphical representation.

In the Figure 1.2 the point's labels (name) are not shown. This is because of the setting **Settings>Name on object>No new object**. If you want to change this afterward, just right-click on the word **Point** in the Algebra window and select **Show label** to toggle the visibility of labels.

We would also like to create a diagram that shows how the revenues will vary with the number of air seats. Do this by choosing **Show>Graphic 2** from the menu. It might be shown as a free-floating window without axes or a grid, but if you want to, you can dock this window to the main window by clicking on the dock button in the upper right corner of the window shown in Figure 1.3. Note that the axes and the grid have their own buttons in the upper left corner in the Style Bar.

FIGURE 1.3 The dock button that allows you to dock a free window.

	A	B	C	D
1	Number of seats	Price per seat	Revenues	
2	200	800	160000	(200, 160000)
3	199	802	159598	(199, 159598)
4	198	804	159192	(198, 159192)
5	197	806	158782	(197, 158782)
6	196	808	158368	(196, 158368)
7	195	810	157950	(195, 157950)
8	190	820	155800	(190, 155800)
9	180	840	151200	(180, 151200)
10	170	860	146200	(170, 146200)
11	160	880	140800	(160, 140800)

FIGURE 1.4 Graphic Window 2 showing another graph with a different scale.

When you have docked and arranged Graphic Window 2 the way you want it, you can create the points directly in the spreadsheet. Make sure that Graphic Area 2 is active and then type

```
=(A2,C2)
```

in cell **D2** and copy this formula downward by dragging the fill handle. Your result should be similar to Figure 1.4. This is a good way to create new points directly from values in the spreadsheet. The GeoGebra spreadsheet can handle all kinds of objects in its cells, while most other spreadsheets only handle real numbers, text, and formulas.

In order to find the number of passengers it takes to not lose money for the travel agency, you can do some investigations directly in the spreadsheet. Enter new values in columns A and B and look for the income that goes under USD. It takes at least 18 passengers to break even. They will then have to pay 1164 USD for their tickets.

Proposed Solution 2

There is, of course, nothing that can stop you from going directly to graphical solutions. When you have reached the relation $y = 1200 - 2x$, you can graph this directly in GeoGebra by entering $y = 1200 - 2x$ in the command line, and then pressing **Enter**. If you have already created the points, you could also create a line by using the Create New Line tool. Click once on the symbol and then click on two of

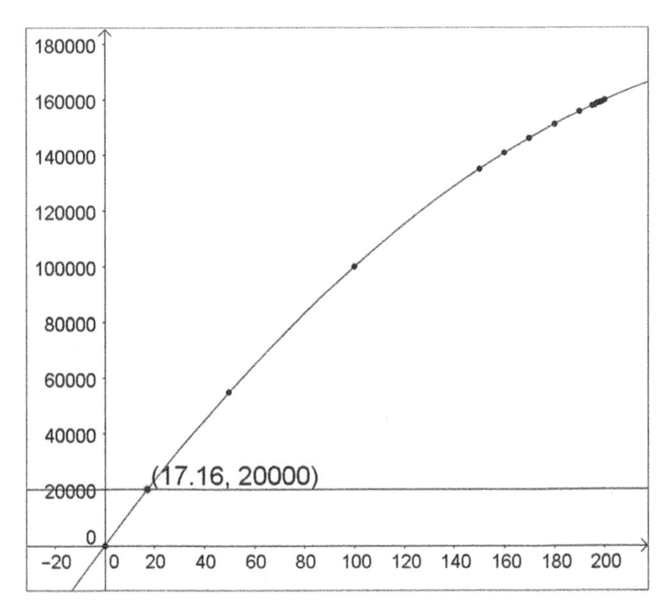

FIGURE 1.5 Income as it decreases dramatically after the first 50 empty seats.

the points. A line is created and you find the symbolic representation of it in the Algebra window.

By using substitution, you can write the relationship between income and sold air seats as $R(x)=x \cdot y=x(1200-2x)=1200x-2x^2$. This model can also be represented graphically. In order to find the number of sold air seats that will give 20,000 USD as income you could draw a line $y=20{,}000$ and observe where the line intersects the graph. The intersection point is found by clicking on the Intersect tool and then clicking on the graph and the line. You could also have written `Intersect[a,b]`, if a and b are names for the line and the graph. The coordinates for the intersection can be found in the Algebra window or if you right-click the point and select **Object Properties** and then on the **Basic tab**, **Show Label: Value**, as in Figure 1.5.

1.2 HOW LONG WILL THE PASTURE LAST IN A FIELD?

The pasture in a field like the one shown in Figure 1.6 has been known to last for 3 days for 6 cows and for 7 days for 3 cows.

Based on this experience, how long will the pasture likely last for a single cow?

Proposed Solution 1

Allow the students to think about a situation for a while, but in the end you will probably need to encourage the students to formulate the situation in a mathematical language in some way. In all events where you translate a situation into a mathematical language, you are in some sense always using a mathematical model. A very simple model would be a straight line between the two points that are given by (6 cows, 3 days) and (3 cows, 7 days), shown in Figure 1.7. It gives you the possibility to follow the line to the point where 1 cow is and see that it points at …

FIGURE 1.6 Cow in a dandelion field. Photo: Keith Weller/USDA (www.ars.usda.gov: Image Number K5176-3) [Public domain], via Wikimedia Commons. https://upload.wikimedia.org/wikipedia/commons/0/0c/Cow_female_black_white.jpg.

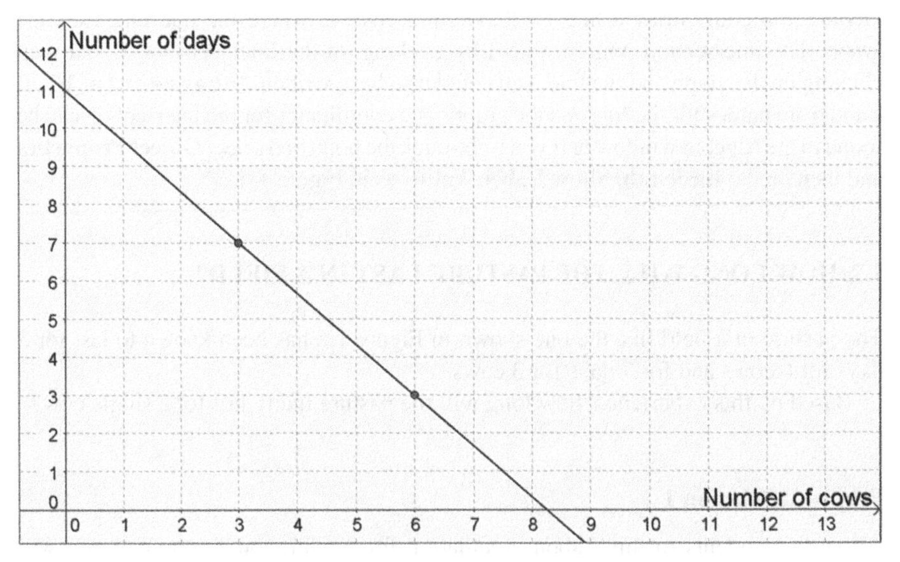

FIGURE 1.7 Linear model.

Proposed Solution 2

Another model could be found by reasoning around the concept of "cow days," where one cow days is equal to a certain amount of food. The two solutions will then deliver two different answers. That helps when thinking about the validity of the model, errors, how the model could become better, and so on.

We find that the first bit of information in the problem indicates that you have enough food for 18 cow days. The next bit of information indicates that you have food for 21 cow days. So you can assume that there is food for about 20 cow days (the differences may have something to do with rounding errors). In Figure 1.8 you can see a mathematical model where

$$Number\,of\,cows \cdot Number\,of\,days = 20 \Leftrightarrow Number\,of\,days = 20/Number\,of\,cows$$

We see that the prognosis in this case is remarkable higher. The food should be enough for approximately 20 days for a single cow, and possibly a lot longer if you factor in that grass is growing all the time.

Comments

This is a good example of a task that can be used even in lower secondary school to get students to think mathematically and see how the strength in using different representations might get a solution to reveal itself. The authors of this book, who teach mathematical modeling at an upper secondary school and at a university, usually start with a simple task, before they introduce more complex modeling activities. In this way students at all levels get a good introduction to the modeling process before they have to think through a complex, complicated, and mathematically more demanding situation.

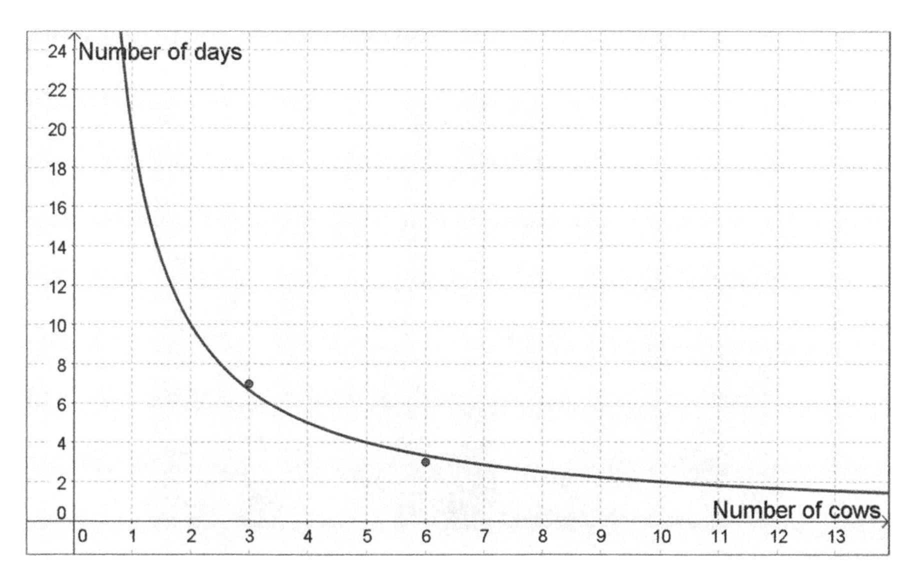

FIGURE 1.8 Nonlinear model based on the concept of cow days.

1.3 A BIT OF CHEMISTRY

A laboratory assistant has a large container with a 20% hydrochloric acid solution and a small container with 300 grams of 43% hydrochloric acid solution as shown in Figure 1.9. The assistant would like to dilute the strong solution until it is 28% acid. How much of the 20% hydrochloric acid should the laboratory assistant use?

Try to solve the problem in different ways and try to construct or derive a general solution model that shows how much of a solution you need to add to m grams of b% solution in order to get c% solution.

Proposed Solution 1

We start with an algebraic solution. In problems with a lot of information it is generally a good idea to organize the data. First, introduce a variable. Let x be the amount of added 20% hydrochloric acid solution, measured in grams. Always be clear about what your variables stand for and what units they are measured in.

Once you have done that, you can now represent the data as a bullet list. You have to determine how much acid and solution you have in the small container, before and after the dilution:

- Amount of hydrochloric acid before: 43% of 300 grams = 129 grams
- Amount hydrochloric acid after: 129 grams + 20% of x grams
- Total amount of solution before: 300 grams
- Total amount of solution after: 300 grams + x grams

These values are well prepared to be presented in a table, such as that shown in Table 1.2. The table setup will help you organize your data and your modeling

FIGURE 1.9 Solutions. Photo: Flikr, user usehung (CC BY 2.0) https://www.flickr.com/photos/13639809@N03/2295905317/sizes/o/.

TABLE 1.2 Organizing Your Data in a Table

Grams	Before	After
Pure hydrochloric acid	129	$129+0.2x$
Total amount of solution	300	$300+x$

strategies. The solution process becomes easier if you focus on one part of the problem at a time, and using a table to organize problems like these helps you in doing this.

Finally use the only bit of information that you haven't yet used to set up your equation. The new solution should be 28%; in other words, 28% of your new total acid should be hydrochloric acid. The equation that we use to solve the problem takes the form

$$0.28\cdot(300+x)=129+0.2x, \quad \text{which yields } x=562.5 \text{ grams}$$

Equations of this type are often viewed as difficult to set up as well as difficult to solve, but letting students practice on problems like these often gives a well-needed boost to their general problem-solving skills.

So 562.5 grams of the 20% solution need to be added to the small container in order to get 28% strength for the final solution.

To derive a general model, you need to substitute the parameters a, b, c, and m into suitable places in the equation above to get

$$\mathrm{c}\cdot(\mathrm{m}+x)=\mathrm{b}\cdot\mathrm{m}+\mathrm{a}\cdot x, \text{ which yields the solution}$$

$$x=\mathrm{m}\cdot\frac{\mathrm{b}-\mathrm{c}}{\mathrm{c}-\mathrm{a}}$$

Proposed Solution 2

Now you will make a dynamic model in GeoGebra where you will vary a slider until you get the right concentration for the solution. In order to make a general model, begin by creating five sliders: a, b, c, m, and x_0. If you wish, you can simply create these as numbers first, and then show them as sliders by clicking on the white bullet to the left of their algebraic representations in the Algebra window. Define a, b, and c to vary between 0 and 1 with a step of 0.01, while m and x_0 will vary between 0 and 600 with a step of 1. Then adjust the sliders to see the values of the current problem. All of these settings can be found in the properties of the numbers, on the **Slider** tab.

Next adjust the window settings so that the graphic area shows 0–1000 along the x-axis and 0–1 along the y-axis. Since every calculation of the amount of hydrochloric acid is a multiplication of a percentage and a total mass, you could represent the amount of hydrochloric acid as the area of a rectangle, with the percentage as the height along the y-axis and the total mass as the width on the x-axis. Use the following commands:

`O=(0,0)`	Origin
`A=(0,a)`	
`B=(0,b)`	A, B, and C represent percentages
`C=(0,c)`	
`M=(m,0)`	
`X=(x_0,0)`	M, T, and X represent
`T=(m+x_0,0)`	masses, in grams
`H_1=(x_0,a)`	
`H_2=(m,b)`	H1, H2, and H3 are
`H_3=(m+x_0,c)`	vertices of the rectangle

After that, create the rectangles with the following command:

```
Added=Polygon[O,A,H_1,X]
Existing=Polygon[O,B,H_2,M]
Total=Polygon[O,C,H_3,T]
```

The result can be seen in Figure 1.10. By changing the value of $x_{0,}$ you can search for the value when the sum of the Existing and the Added rectangles' area

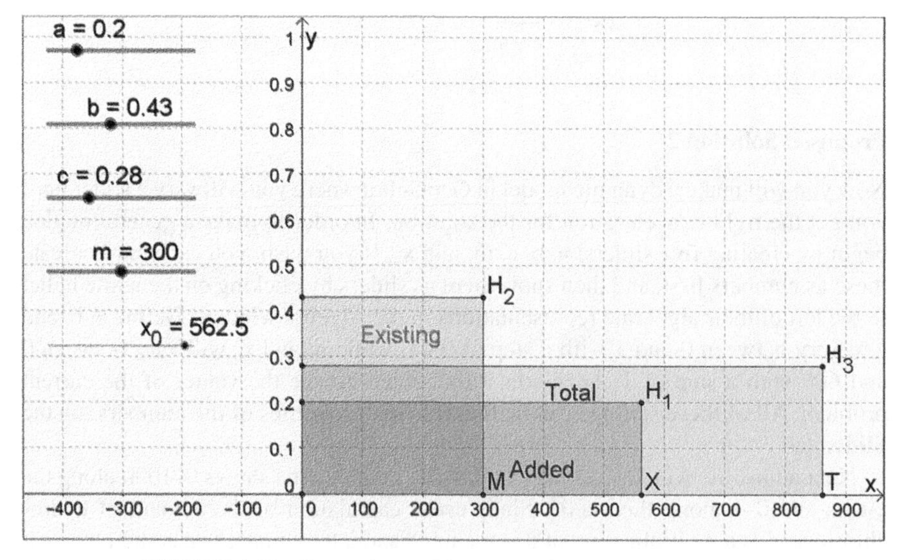

FIGURE 1.10 Dynamic graphic model of the chemical mixture.

is equal to the Total rectangles' area. The best way to do this is to first calculate the value

```
Tot = Existing + Added
```

Here, Tot is the actual total amount of hydrochloric acid while Total is 28% of the total amount of liquid. You could drag these values from the Algebra window to the graphics area so that you more easily can see and compare the values.

In order to achieve better precision when you adjust x_0, select x_0 and use the arrow keys to change the value one step at the time. For even better precision, hold down **Shift** while using the arrow keys. By adjusting the values for the other sliders, you could, in principle, even solve all similar problems.

After playing with this for a while, you may quickly realize that it is a bit cumbersome to adjust the value of x_0 manually. By typing

```
x_0=m (b-c)/(c-a)
```

you can get the value of x_0 to automatically adjust to the right value for every value for a, b, c, and m. The slider for x_0 will then disappear because it is no longer possible to adjust x_0 manually. Therefore drag instead the value from the Algebra window into the Graphic window as a text box as shown in Figure 1.11.

It may be an interesting challenge for your students to ask them what will happen to the model if you increase or decrease any of the four parameters.

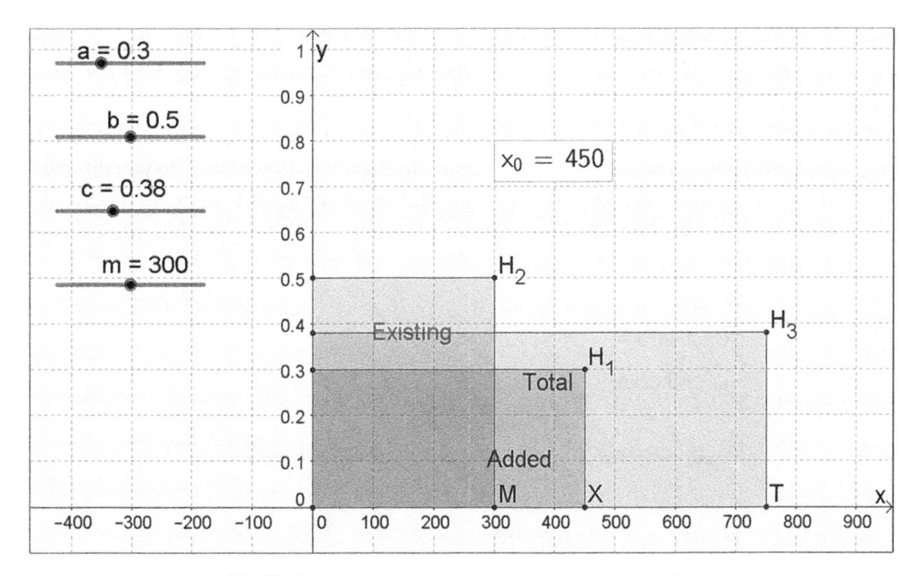

FIGURE 1.11 Solving a problem automatically.

Proposed Solution 3

The previous solution suggestion was visual and pedagogical but could just as easily be obtained on a calculator. Therefore you will now build a calculator in GeoGebra where you may enter values and receive answers.

Start a new session of GeoGebra by pressing **Ctrl-N**. Type

```
x_0 = m (b-c)/(c-a)
```

in the command field, and let GeoGebra create sliders for a, b, c, and m. Since it is hard to reach exact values with a slider, you will now represent the other parameters with input boxes instead. First you need to adjust some properties for a, b, c, and m. Right click on any of them in the Algebra window and select **Properties**, on the **Sliders** tab. In the left-hand pane you will find the properties dialogue with all objects represented in an object list. You should **Ctrl-Click** and select a number of different objects all at the same time. Enter `Min = 0` and `Max = 1000` for m and `Max = 1` for a, b, and c. The standard values `Min = -5` and `Max = 5` that GeoGebra uses for sliders are normally not ideal. So hide the sliders by clicking the blue buttons to the left of the numbers in the Algebra view.

You should create an input box for each of your four parameters by clicking on the Input Box tool and then in the Graphics area to show where you want them placed, often far to the left, one above the other. In the dialogue shown in Figure 1.12, you can enter a text label (normally the variable name) and the variable to be connected to the input field.

Adjust the boxes according to the grid after all boxes have been created. You can also chose to set the input field length to 5 in the **Input Box Length**, Fix Objects in the **Basic** tab and set text size to **Medium** in the **Text** tab in the properties dialogue. Last, drag x_0 from the Algebra window to the graphics area to create a text box. Select the text size to be **Medium**. This can be done in the style bar. Then hide the grid and the axis, and also, if you want to, close the Algebra window. Figure 1.13 shows a typical result.

This calculator calculates x_0 for any parametric value. But if you know how much you should add and want to calculate the concentration it will get, how do you do that?

Start by taking away the text box showing the value for x_0 and create an input field for x_0 instead, just like you did for the other parameters. Make sure that you change

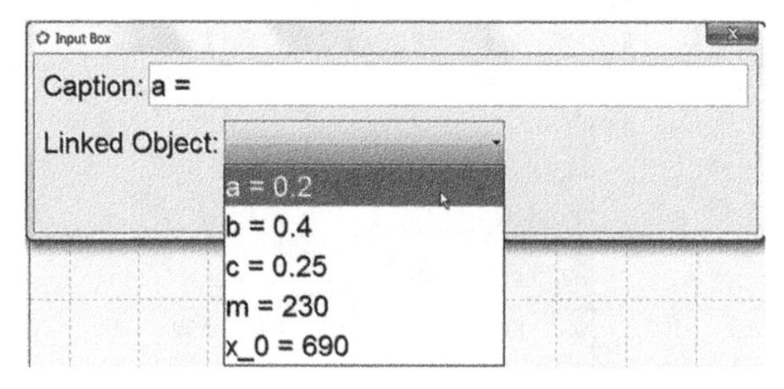

FIGURE 1.12 Input box dialogue.

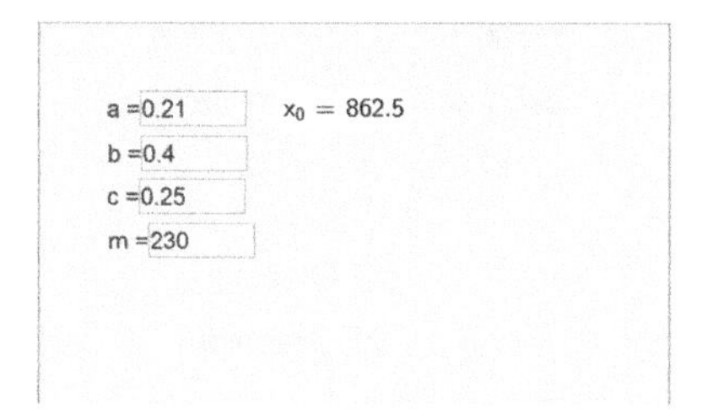

FIGURE 1.13 A simple calculator.

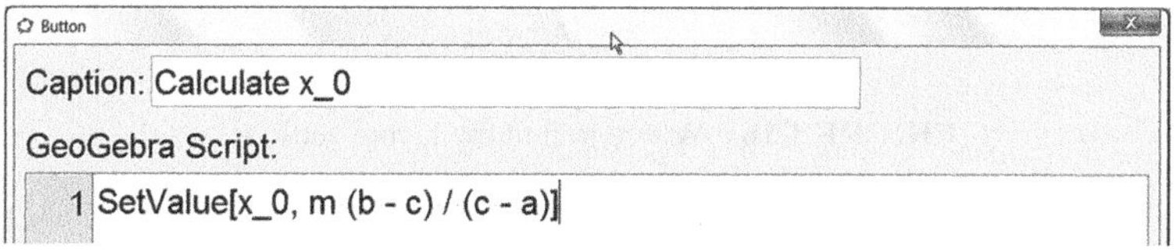

Button
Caption: Calculate x_0
GeoGebra Script:
1 SetValue[x_0, m (b - c) / (c - a)]

FIGURE 1.14 Button dialogue.

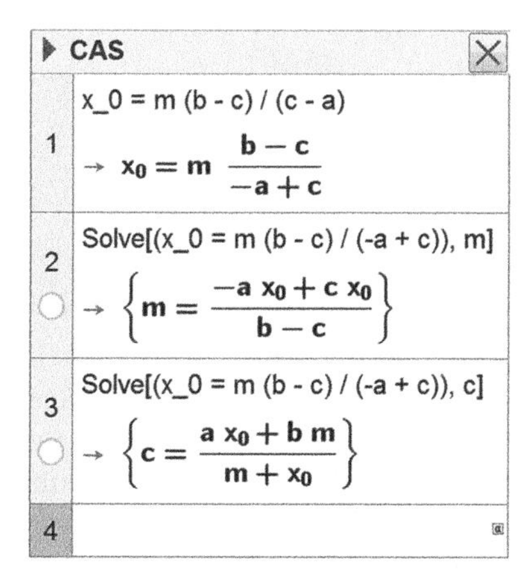

FIGURE 1.15 CAS window used to help you solve for variables in a formula.

text size, length for the field, and also lock this field. Enter an arbitrary value for x_0 so that the formula will disappear.

You can then create a command button by clicking on the Button tool and then somewhere in the graphics field. Fill in the dialogue box as in Figure 1.14.

The command `SetValue[x_0, m (b-c)/(c-a)]` calculates the value of the formula and updates x_0 to this value. It then creates four more buttons, one for each parameter. The formula that you should use may be found in different ways. You could use manual algebraic manipulations on paper, or you could open up a new window in GeoGebra with **Ctrl-N**, next open up the CAS windows with **Ctrl-Shift-K,** and last, write `x_0=m (b-c)/(c-a)` in line 1. In line 2 you write `Solve[$1, m]`. In this command `$1` is a reference for the equation in line 1. Figure 1.15 shows this process.

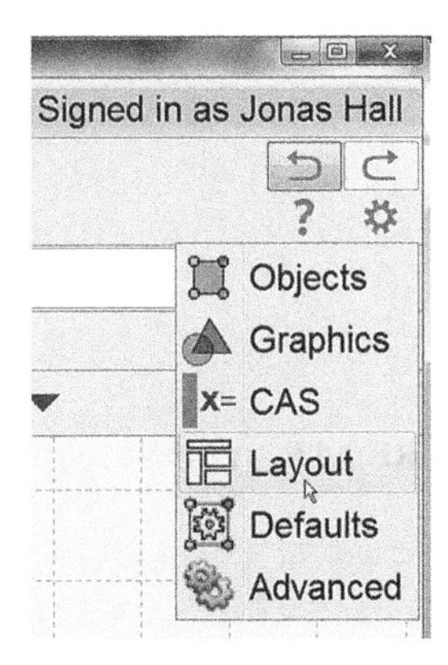

FIGURE 1.16 Where to find the layout settings.

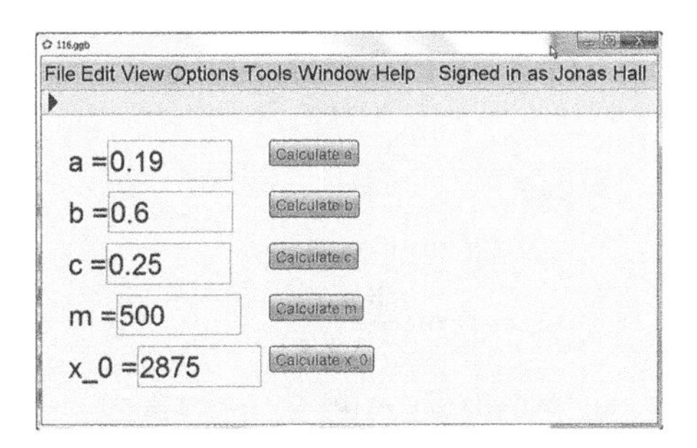

FIGURE 1.17 A "Solver" tool in GeoGebra.

When you have entered your formulas using the command buttons, you could—if you like—disable the visibility of the tool field and the input field by clicking on the property gear wheel in the top right corner and selecting **Layout** as shown in Figure 1.16.

In Figure 1.17 you can see the finished "Solver." Similar solvers can be created for other problems, and if you upload your solver to GeoGebraTube, it may be run as an app on a phone or a tablet.

1.4 SYDNEY HARBOR BRIDGE

Figure 1.18 and Figure 1.19 show the famous Sydney Harbor Bridge. If the bridge heads are 89 meters high, how long is the bridge between the bridge heads?

Proposed Solution

This is a common situation. You have photograph where you know one distance and you want to find another distance. As long as these distances are at the same distance from and close to perpendicular to the viewer, a linear proportional model can be

FIGURE 1.18 Sydney Harbor Bridge. By Yun Huang Yong via Flickr, CC BY-SA 2.0. https://www.flickr.com/photos/goosmurf/3001997390/sizes/o/.

FIGURE 1.19 One of the bridge heads stands prominent in the mist behind a ferry. Photo: Hpeterswald/Wikimedia Commons. https://commons.wikimedia.org/wiki/File:Sydney_Ferry_Queenscliff_1.jpg.

TABLE 1.3 Two by Two Table for Organizing Proportional Data

	Unknown	Known
Reality (m)	x	89
Photo (px)	1403	175

TABLE 1.4 Proportionality Is Invariant under Rotation and Mirroring Operations

	Reality (m)	Photo (px)
Known	89	175
Unknown	x	1403

used. By measuring the distances in the photo, you will get the remaining information you need. Assuming proportionality yields

$$\frac{\text{Unknown distance (real)}}{\text{Unknown distance (image)}} = \frac{\text{Known distance (real)}}{\text{Known distance (image)}}$$

In this simple model you can ignore that perception issues might cause the proportionality to break. Proportionality can also be visualized through a simple 2×2 table or box. If the unknown distance is x and you measure the known distances in the photo through some computer program to 175 pixels and to 1,403 pixels, you can then set up your data in Table 1.3.

One interesting property of Table 1.3 is that it can be rotated and mirrored and still be valid. You may, for example, turn it 90 degrees counterclockwise to get Table 1.4.

It is useful to show this with the table written on a bit of cardboard that can be physically flipped and rotated. Any of these configurations may be set up as an equation. In the example above you will get the equation:

$\frac{89}{x} = \frac{175}{1403}$, which may be written without denominators as the crosswise product

$$175x = 89.1403, \text{ and so } x = 715.5 \text{ meters.}$$

However, there are several ways to measure the distances in the photo. One way is to measure directly the distance in the photo using a ruler. The photo can be imported into GeoGebra with the Image tool, and then you can place points on the photo to measure the distances. If you want more precision, there is a small free program called ***Digitize It*** that gives much better precision for measuring distances using an image or photo.

In Figure 1.20 is an example of how several points' positions are captured by the software and put in the table to the left. Note the small magnifying glass window floating between the table and the image. You can download this program for free at http://download.cnet.com/Digitize-It/3000-20415_4-10489008.html.

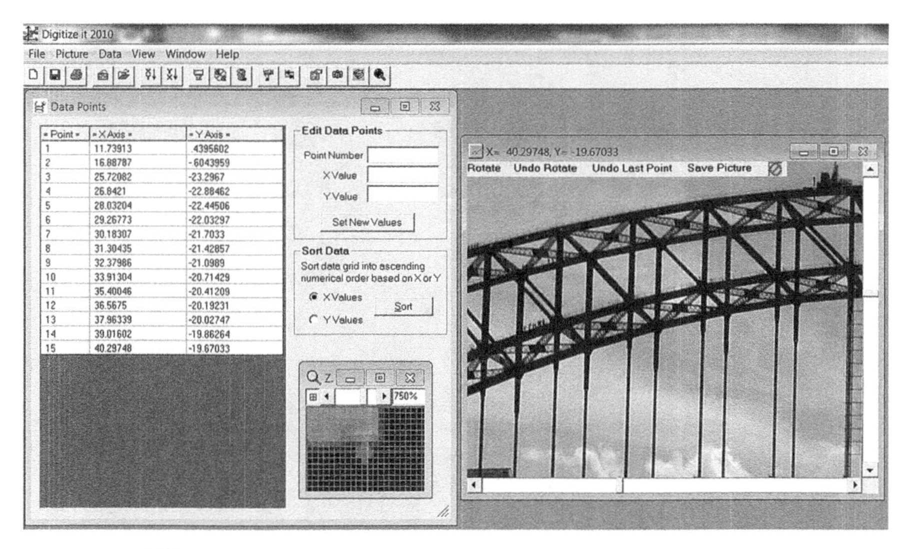

FIGURE 1.20 *Digitize It*, a program for digitalization of photos.

Many students today can use cameras in smart phones, tablets, and the like, to capture interesting situations or the objects themselves. It is best to zoom in and take a photo from far away to minimize nonlinear effects. One student could hold a meter stick and another jump as high as possible. How high was the jump? How tall is the student who holds the stick? Proportionality will give the answer.

1.5 PERSPECTIVE

In Figure 1.21 you can see the same girl photographed five different times. The perspective makes her look bigger when she is closer to you and smaller when she is far away.

If an object is far away enough, so that you experience it under a viewing angle that is less than about 30°, the product of the subject's size and distance will be approximately constant. Another way to express the same relation is to say that the size is inversely proportional to the distance.

If the distance to the girl at her closest instance is 3 meters, how far away was she at the most distant instance?

Proposed Solution

Start a new window in GeoGebra with **Ctrl-N.** Import your image by using the Image tool and next click on, for example, the origin. Then right-click on the photo and select **Properties**, the **Position** tab, and lock the image's both lower corners in the coordinate system as shown in Figure 1.22.

FIGURE 1.21 Perspective run. Photo: Nevit Dilmen (original work) CC-BY-SA-3.0 via Wikimedia Commons. http://commons.wikiwithia.org/wiki/File:08913-Perspective_Run.jpg.

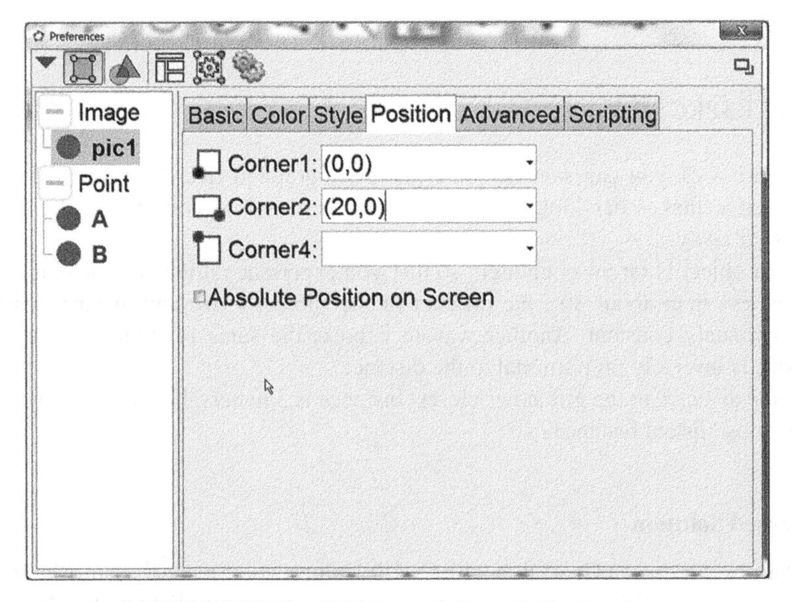

FIGURE 1.22 Lock the image in the coordinate system.

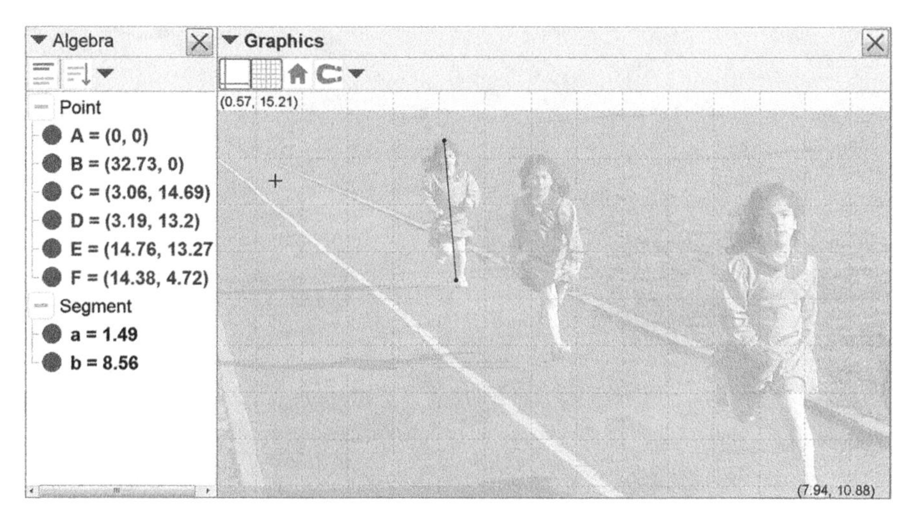

FIGURE 1.23 Measuring the girl's height in GeoGebra.

In the **Basics** tab you select **Background image**, and under the **Color** tab you adjust the opacity to 25% in order to more easily be able to see the coordinate system.

Measure the girl's height by using the Segment tool. Click on the girl's head and heel both when she is at her closest and when she is at her farthest. The lengths of the segments will be shown in the Algebra window in Figure 1.23.

If you would like to fine-tune the position of the points, first press **Escape** to activate the Select and Move tool; then right-click on the word **Point** in the Algebra window and select **Properties**, the **Algebra** tab, and set the step length **0.001** for all points at the same time. Next you select a point and use the arrow keys to move it.

You may also zoom in on the image as you place the points on the girl farthest away. If you want to stick with the proportions 1:1 between the x- and y-axes, it is best to zoom by right-clicking on an empty part of the graphic area and then selecting **Zoom > 400%.** That done, drag the graphic area to pan in order to find the girl again.

If the product of the distance d and the size s is constant, then $d_1 \cdot s_1 = d_2 \cdot s_2$. In this case $d_1 = 3$ m, $s_1 = 8.56$, and $s_2 = 1.49$; the result is $d_2 = 17.2$ m. Let the students practice this by photographing each other in groups of three and checking the distances with long tape measures.

1.6 LAKE ERIE'S AREA

Try to estimate the size or the area of Lake Erie from Figure 1.24 or a similar map from Google Earth in several different ways. Also try to estimate the error in the method you choose and show your work clearly.

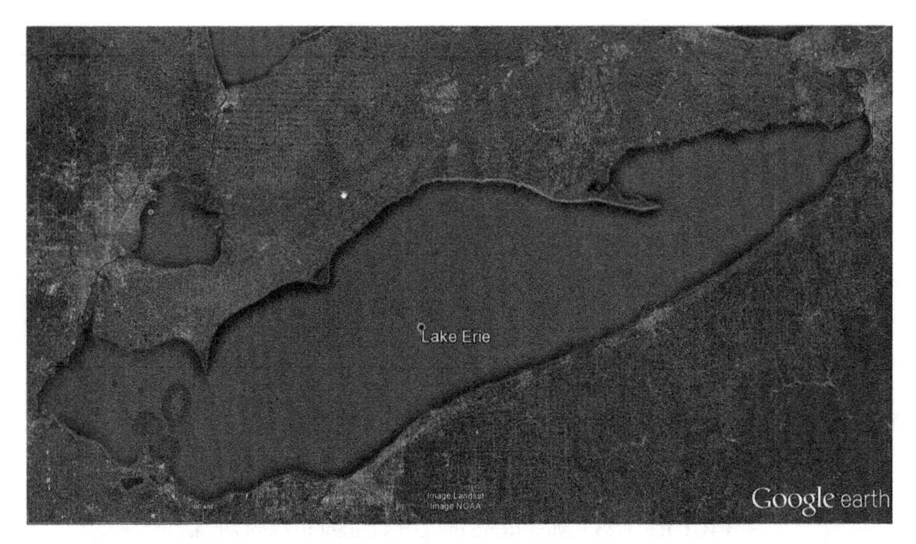

FIGURE 1.24 Lake Erie, Google Earth. Image Data from Landsat and NOAA.

Proposed Solution 1

The classical solution to cut out the image and estimate the weight only works if you have a very accurate scale available. Enlarge the image so that it almost fills an 8 × 11 inch sheet of paper. Cut out the lake. The surface will be something like 20 × 10 cm. Ordinary printer paper weights about 80 grams/m^2, so a square 10 cm across has a weight of 0.8 grams. Lake Erie should then have a weight of about 1.6 grams. With a weighing scale that can handle a resolution of 0.1 grams or maybe even 0.01 grams, the accuracy of the paper weight will be crucial. It may therefore be necessary to calibrate the weight of a sheet of paper in this particular batch and by a particular brand.

Assuming that an A4 sheet paper weighs in at 4.83 grams and the cutout of Lake Erie weights 0.94 grams, this will yield a surface weight of 4.83 grams/(21.0 cm × 29.7 cm) = 0.007744 grams/cm^2, from which the cutout's area is found to be 0.94 grams/0.007744 grams/cm^2 = 121.4 cm^2. You can wait until later to do any rounding.

Once you have an estimate of how large Lake Erie is, you can transform this to square kilometers. Assuming the 50 km tick mark to be 3.55 cm in the image, the distance scale can be estimated to 1 cm in the image = 50/3.55 = 14.08 km in reality, which give an area scale where 1 cm^2 on the image is 33.71^2 = 198.4 km^2 in reality.

With an area of 121.4 cm^2 you can get a value for the area of Lake Erie: $A = 121.4\,\text{cm}^2 \cdot 198.4\,\text{km}^2/\text{cm}^2 \approx 24{,}000\,\text{km}^2$. How accurate is this estimate? Which of your measurements has limited the accuracy most? How can you improve the accuracy? How well does your cutout conform to the actual shape of the lake?

Note also how units are handled together with the scales. This is an important skill for students to learn.

Proposed Solution 2

You can take the image or a screen dump of it and import it into GeoGebra in order to measure the area there. Import an image by selecting the Image tool and clicking somewhere down to the left in the graphics area. You then get a pop-up dialogue where you can select an image. You may also drop an image in the graphics area.

Two points are provided to control the image, and you can set the coordinates of B to be (10, 0). Then right-click on the image and select **Object Properties**; check the **Background image** on the **Basic** tab and the image will be placed behind the axis and the grid.

Now you can start placing points along the coast line. It might be a good idea to select **Options>Labeling>No new objects** from the menu before you start. When you have placed your points, you should select the Create list tool and then drag a rectangle around all points to create a list with these points. Last, create a polygon from all the points using the command **Polygon[list1]**. The area of the polygon is approximately 16.08 area units as shown in Figure 1.25. You could also have used the Polygon tool directly.

Now deal with the scale: zoom in on the scale as shown in Figure 1.26 and measure the 80 km tick to 2.04 length units by placing segment over the mark. One length unit = 80/2.04 = 39.216 km and one area unit is $39.216^2 = 1537.9\,\text{km}^2$. The area of the lake is $A = 16.08\,\text{a.u.} \cdot 1537.9\,\text{km}^2/\text{a.u.} \approx 24\,730\,\text{km}^2$.

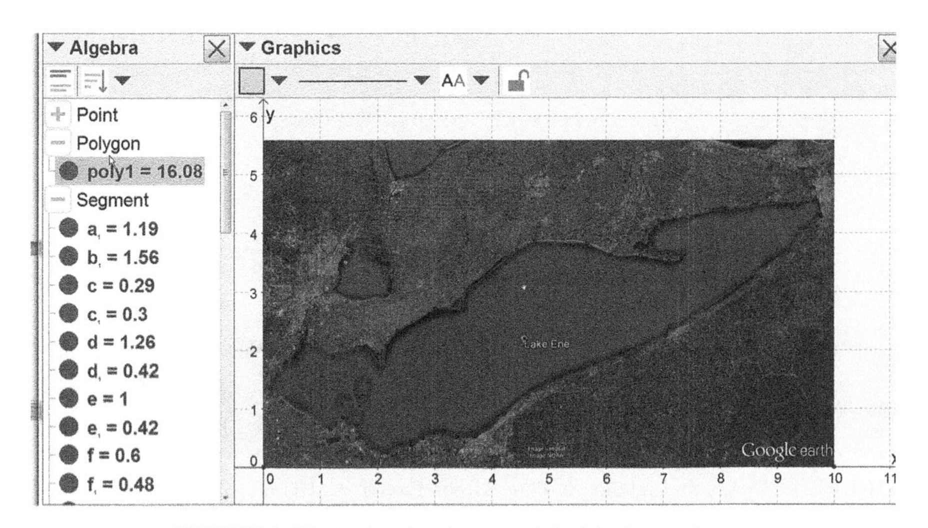

FIGURE 1.25 Estimating the area of the lake by a polygon.

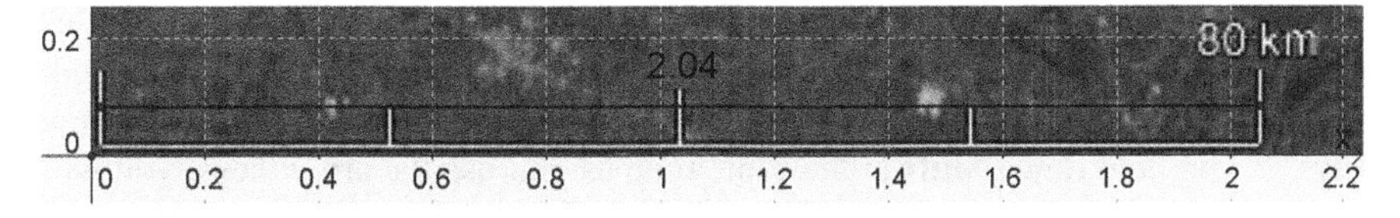

FIGURE 1.26 A segment used to measure a distance.

WolframAlpha computational knowledge engine

size lake erie

Examples Random

Input interpretation:

Lake Erie | surface area

Result: Show non-metric

25 800 km² (square kilometers)

Unit conversions:

2.58×10^{10} m² (square meters)

9970 mi² (square miles)

2.78×10^{11} ft² (square feet)

Comparisons as area:

≈ (0.0018 ≈ 1/542) × total area of arable land on Earth (≈ 1.4×10^{13} m²)

≈ 0.37 × Lake Victoria area (68 870 km²)

≈ 1.2 × total area of Wales (≈ 8023 mi²)

FIGURE 1.27 Wolfram Alpha can provide you with geographical data.

FIGURE 1.28 Lorain Lighthouse. Photo: Rona Proudfoot (CC BY-SA 2.0). https://www.flickr.com/photos/ronnie44052/2420422668/in/photostream/.

The error can be estimated by fine-tuning a few of the point's positions, noting the effect of this on scale and area. To fine-tune a point's position, open the properties dialogue box and then click on the **Algebra** tab and change the value of step length to 0.001. Thereafter you may fine-tune the position of the point by using the arrow keys. If you hold down **Shift** at the same time as you use the arrow keys, you will increase the accuracy 10 times and if you hold down **Ctrl**, you will lose accuracy by a factor 10, jumping faster.

Proposed Solution 3

Wikipedia, referring to the US Environmental Protection Agency, claims that Lake Erie has a surface area of 25,667 km^2. NOAA claims it to be 25,655 km^2, and Wolfram Alpha, using several sources, says in Figure 1.27 that the area is 25,800 km^2. Most likely these sources define the lake area differently, so let's say we are satisfied with 25,700 km^2. The Lorain lighthouse on the south coast of Lake Erie is depicted in Figure 1.28.

1.7 ZEBRA CROSSING

In Figure 1.29 you see people traversing a huge zebra crossing. You can assume that the white lines and the distances between them are the same everywhere and that the ground is flat. The camera that took the photo is located some distance behind the closest line and placed some distance above the ground.

Try to determine where the camera is placed, both in terms of height and in terms of depth, in relation to the first visible white line. Also try to determine how tall the single man with necktie and briefcase is, if the man with blue scarf and briefcase is 170 cm tall.

FIGURE 1.29 Zebra crossing. Photo: Flikr, user MrHayata (CC BY-SA 2.0). https://www.flickr.com/photos/mrhayata/6789615983/sizes/o/.

Proposed Solution 1

In Figure 1.30 you see a figure of what it might look like from the side the crossing when the photo was taken. You can see sections consisting of white lines and black ground that are represented by the segments labeled with d. To the left in Figure 1.30, or at the bottom of Figure 1.29, there is a small segment called d_0 that is not a whole section.

The points between these sections are projected on to the image plane to create an image. We now drastically simplify the situation and assume that the image plane is perpendicular to the ground plane in order to make calculations less difficult. But, in making this assumption, we introduce an error that we hope will not become too large. It is, however, better to have a simple model than no model at all. In general, always begin with a model sufficiently simple in order to get started!

In Figure 1.31 we concentrate on the incoming ray for just one of the section boundaries. Proportionality in triangles KCB and POB then gives

$$\frac{\mathrm{h}}{\mathrm{a}+\mathrm{d}_0+n\cdot\mathrm{d}}=\frac{p_n}{\mathrm{d}_0+n\cdot\mathrm{d}}$$

Here n is the serial number for the white line in the zebra crossing and p_n is the horizontal distance from the camera to the photo's bottom edge.

The proportionality can be rewritten in the following way by solving for p:

$$p_n=\frac{\mathrm{h}\cdot(\mathrm{d}_0+n\cdot\mathrm{d})}{\mathrm{a}+\mathrm{d}_0+n\cdot\mathrm{d}}$$

where n is an independent variable and the numbers h, a, d, and d_0 are unknown numbers that need to be measured.

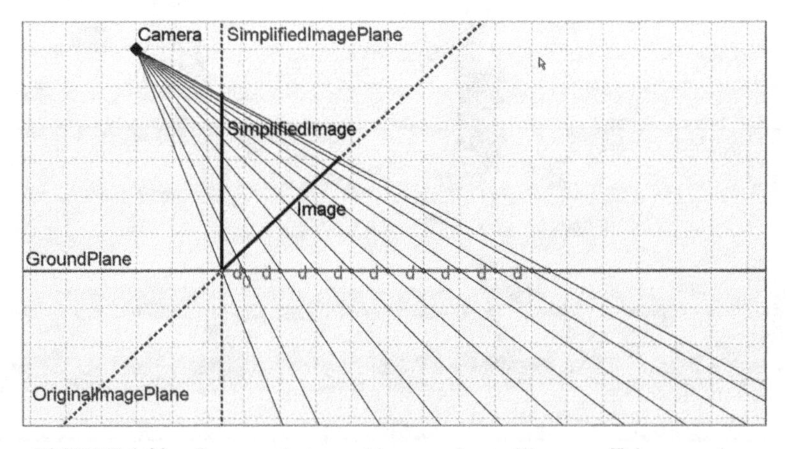

FIGURE 1.30 Camera photographing people strolling over Zebra crossing.

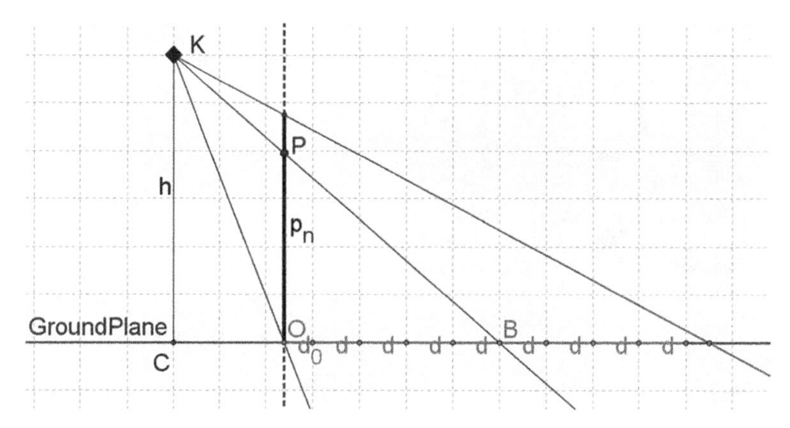

FIGURE 1.31 Triangles KCB and POB are uniform.

Start a new window in GeoGebra with **Ctrl-N,** and import your image by selecting the Image tool ; then click on, for example, the origin.

On the **Basic** tab of the image's properties dialogue, you should select **Background image** and on the **Color** tab you should set the opacity to 25% so that you can see the coordinate system.

Now zoom in at the image and start measuring the values. Measure the *y*-coordinate for the upper part of the white lines to the left of the image and enter the values into the spreadsheet that you open by pressing **Ctrl-Shift-S**.

The values can be seen in the spreadsheet in Figure 1.32. The first white lines upper edge corresponds to n = 1, and so forth. The *x*-coordinate is simply the number of the white line in this model.

To visualize these points in a graph, open up Graphics Area 2 by pressing **Ctrl-Shift-2**. Click once in the Graphics 2 area to make it active. Select the values (but not the headings) in the spreadsheet; right click on the selection and choose **Create... > List with points**. The points will appear in Graphics Area 2.

With this method, the image planes' *x*-coordinate will not be exactly 0, since the bottom edge of the image not is aligned with a white lines upper edge. So assume that the image plane is at where the *x*-coordinate = w. Then w is used in roughly the same way as d_0. Create the slider w to go from –5 to 5 with an increment of 0.001.

Now recreate the light rays. Light ray *n* comes from the ground at the point (*n*, 0) and goes through the vertical image plane in the point (w, p). Create these rays by typing

```
Ray[(A2, 0), (w, B2)]
```

in the spreadsheet cell C1 and copy this cell downward by dragging the fill handle. If you now adjust the value of w, you can make all rays go through the same point. That must be the position for the camera, as is clearly visible in Figure 1.32.

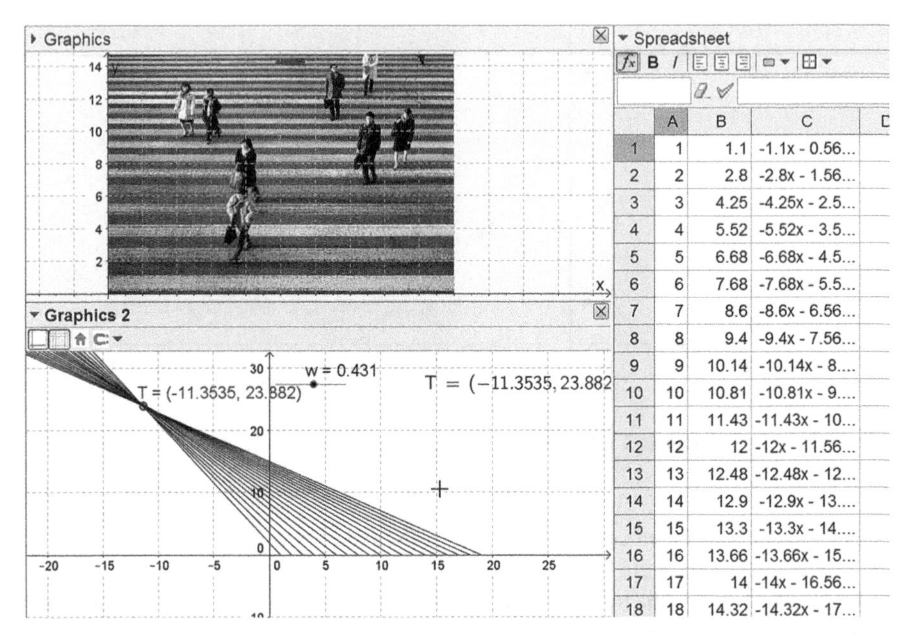

FIGURE 1.32 Distances from the bottom of the image to the upper edge of the white lines as shown in the spreadsheet.

You have now found the relative position of the camera, but you have not yet used any real units; therefore you do not know the scale for your figure. The only true measurement you have is the height of the man with the blue scarf who is 170 centimeters tall.

By zooming into the image and creating points at the man's feet and head, you can get the points' coordinates. In Figure 1.33 you can find the p-values, or y-coordinates, and so create the rays L2 and L3 by typing

`L2=Ray[K,(w,y(P2))]` and `L3=Ray[K,(w,y(P3))]`

After that, create a line perpendicular to the x-axis going through the intersection of L2 and the x-axis with the Perpendicular Line tool or with the command

```
PerpendicularLine[Intersect[L2, xAxis], xAxis]
```

If you do not see a specific construction in Graphics 2, it may have happen that it has been created in Graphics 1 instead. If so, just find the object in the Algebra window, right-click on it, and select **Properties**; click on the **Advanced** tab in the dialogue shown in Figure 1.34, and change the graphics area to where you want the object to be shown.

The intersection between the blue line and the ray L3 creates a point that corresponds to the head and the segment between head and foot has a length of 6.1391 length units. This corresponds to 1.70 m, which means that the scale will be that one

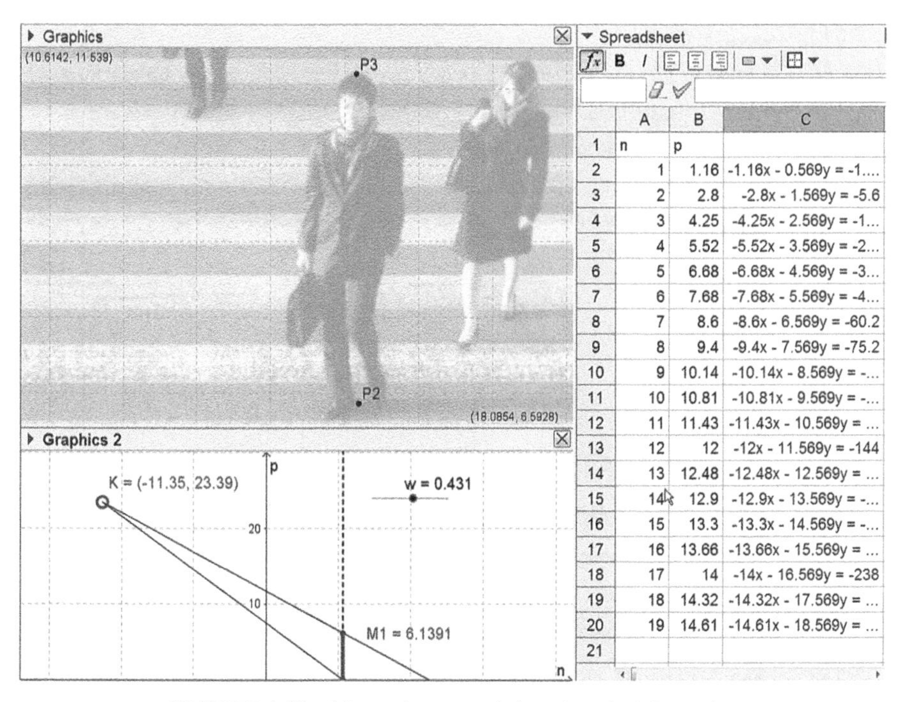

	A	B	C
1	n	p	
2	1	1.16	-1.16x - 0.569y = -1....
3	2	2.8	-2.8x - 1.569y = -5.6
4	3	4.25	-4.25x - 2.569y = -1...
5	4	5.52	-5.52x - 3.569y = -2...
6	5	6.68	-6.68x - 4.569y = -3...
7	6	7.68	-7.68x - 5.569y = -4...
8	7	8.6	-8.6x - 6.569y = -60.2
9	8	9.4	-9.4x - 7.569y = -75.2
10	9	10.14	-10.14x - 8.569y = -...
11	10	10.81	-10.81x - 9.569y = -...
12	11	11.43	-11.43x - 10.569y = ...
13	12	12	-12x - 11.569y = -144
14	13	12.48	-12.48x - 12.569y = ...
15	14	12.9	-12.9x - 13.569y = -...
16	15	13.3	-13.3x - 14.569y = -...
17	16	13.66	-13.66x - 15.569y = ...
18	17	14	-14x - 16.569y = -238
19	18	14.32	-14.32x - 17.569y = ...
20	19	14.61	-14.61x - 18.569y = ...
21			

FIGURE 1.33 Measuring a man's length to find the scale.

unit is equal to 0.2769 meter. If you use this to calculate the position of the camera, you will find that the camera is 3.1 meters behind the lower image edge and about 6.6 meters above the ground.

If you recalculate the construction for the man with the necktie, you find that his height is 6.2499 length units, which also corresponds to 1.73 m. With the accuracy you have now and the uncertainties in the measurements, due to the two men walking far apart when the photograph was taken, you would probably be hard pressed to justify saying anything other than that the two men are roughly the same height, with only a hint of the second man being taller.

Proposed Solution 2

You probably noticed that you never used the proportionality that you initially set out to do in the previous solution. You could have done the geometrical construction at the initial image plane, but that construction would have been a little more complicated than the one you did, since you used the coordinates for the point in a simpler way. This happens sometimes. Problem solving led you into different and unexpected paths where you probably did not plan to go.

Another path you might have followed, using proportionality, and shown in Figure 1.35, is to find the values for your parameters through regression analysis.

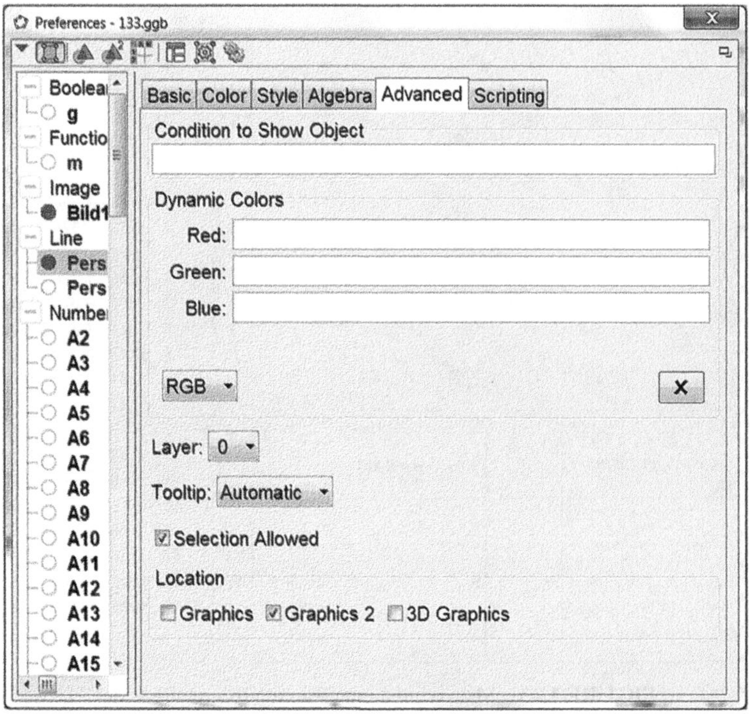

Preferences - 133.ggb
Boolean
g
Function
m
Image
Bild1
Line
Pers
Pers
Number
A2
A3
A4
A5
A6
A7
A8
A9
A10
A11
A12
A13
A14
A15
Basic
Color
Style
Algebra
Advanced
Scripting
Condition to Show Object
Dynamic Colors
Red:
Green:
Blue:
RGB
Layer: 0
Tooltip: Automatic
Selection Allowed
Location
Graphics
Graphics 2
3D Graphics

FIGURE 1.34 Window to change the graphics area you want the object to be visible in.

Graphics 2
f(x) = 23.82 · −90.82 + x 192.52 / 2218.69 − 90.82 + 192.52 x
18
16
14
12
10
8
6
4
2
0
0 2 4 6 8 10 12 14 16 18 20 22 24 26 28 30 32 34 36 38

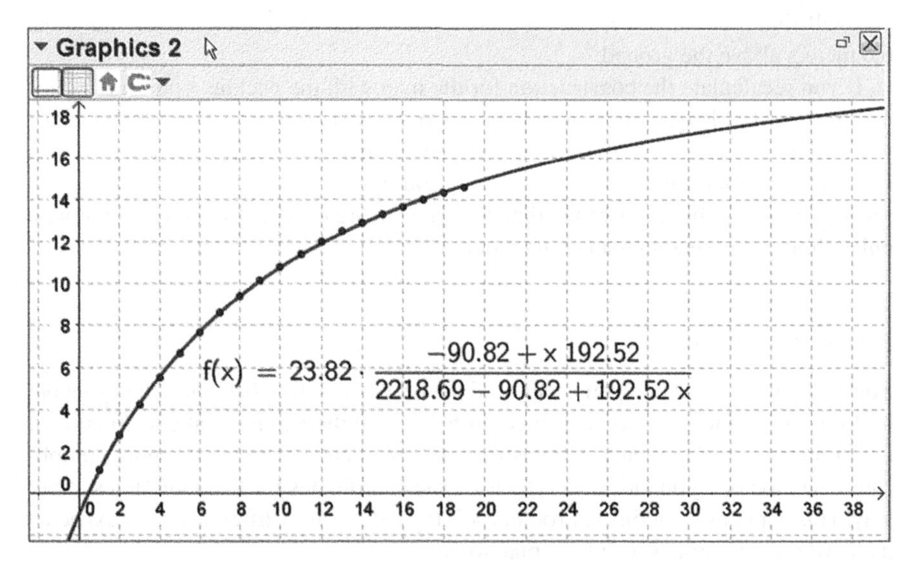

FIGURE 1.35 Regression used to find answers.

Through the definition of the numbers h, a, d, and d_0 together with a model function

```
m(x) = h (d_0 + x d)/(a + d_0 + d x)
```

you can use the command

```
Fit[list1,m]
```

to create a function that is fitted to the data points. The values for h, a, d, and d_0 should be reasonably accurate, so that the function m(x) at least has the correct form compared with the values for the data points. Using this method you get h=23.82, a=2218.7, d=192.5, and $d_0 = -90.82$. For images where the cameras direction is more or less parallel with the ground plane, you can use this function to determine the height scale for different objects at different distances from the camera.

The model function is of the type $(a \cdot x+b)/(c \cdot x+d)$, which has a horizontal asymptote at $y=a/c$. This corresponds to $y=h \cdot d/d=h$, or in other words the graph flattens out at $y=23.8$. This corresponds to the situation where the camera is positioned parallel with the ground plane, so the camera can see all the infinitely many white lines below the horizon.

You can identify other relations as well,: namely the graph intercepts the x-axis when $x=-d_0/d=0.47$, which you can compare with w=0.43 length units. The vertical asymptote here is $-(a+d_0)/d=-11.05$ length units, corresponding to the camera's position in the x-direction, as compared with the −11.35 length units you got earlier. To derive the scale, you have to do the same as you did in the earlier proposed solution with similar results.

These two different methods gave similar results, and the discrepancies are a measure of the size of the errors you make in your models. This quite complicated problem and the two suggested attempts to solve it is probably beyond most high school courses of mathematics in terms of complexity but is included in this introductory chapter as an example of a task well suited for longer coursework in modeling.

1.8 THE SECURITY CASE

A metal workshop has received an order for a security case. This case will have a square bottom and is required to have a volume of exactly 900 liters $=0.9\,m^3$. The material for the sides and the top cost 20 USD per square meter and the material for the bottom, which needs to be thicker, is 40 USD per square meter. What dimensions should the box have in order for the material costs to be as low as possible?

When you have a first solution to this problem, you should attempt to improve on your model, using Figure 1.36 as inspiration.

FIGURE 1.36 Type of security locker case. Photo: HenryWBee (own work) CC-BY-SA-3.0 via Wikimedia Commons. https://commons.wikimedia.org/wiki/File:HK_CWB_Victoria_Park_Sport_Court_Locker_Cages.JPG.

Proposed Solution 1

A routine algebraic solution to this typical optimization problem could look something like this:

Assume the sides of the square bottom to be s meters and the height to be h meters. Then $s^2 \cdot h = 0.9$ (cubic meters), leading to $h = 0.9/s^2$. The cost for the material can then be divided into the following parts:

Top:	$20 \cdot s^2$
Sides:	$4 \cdot 20 \cdot s \cdot h = 80s \cdot (0.9/s^2) = 72/s$
Bottom:	$40 \cdot s^2$

The total cost sums to $C(s) = 60s^2 + 72/s$.

It is obvious that there exists a minimum because the function is increasing both as it approaches 0+ ($72/s$ dominates) and for large values of x ($60s^2$ dominates). The minimum is found by setting $C'(s) = 0$, leading to the equation $120s - 72/s^2 = 0$, which can be reduced to $s^3 = 0.6$. So $s \approx 0.843$ meter, giving a height of 1.265 meter and a total cost of 128 USD.

Proposed Solution 2

GeoGebra is an excellent tool for modeling situations like these. You can then solve the problem without the use of derivatives. Begin by noting that $h = 0.9/s^2$, and let $x = s$ and $y = h$. You might want to draw a graph, but you should instead make a geometric model of the locker case. Because the case has a square bottom, you can make a reasonable model using two dimensions just by constructing a scale figure of one side.

Start by creating a point on the x-axis. You can do this using the point tool. Just click on •ᴬ and then on the x-axis. Next press **Escape** to avoid producing more points by mistake. You could rename the point by right-clicking on it or by simply typing "P" **immediately** after creating it, before pressing **Escape**. Alternatively you could create the point by typing

```
P = Point[xAxis]
```

in the Input bar. The P will locate itself at the origin, but it can be dragged anywhere on the x-axis. Your work will be easier if you create s and h, so type in

`s = x(P)` and `h = 0.9/s^2`

where the function x(P) gives the x-coordinate of the point P. Then create another three points by typing

`O = (0,0)` and `Q = (0,h)` and `R = (s,h)` and finally
`Polygon[O, P, R, Q]`

You can now drag P and see how the case changes its appearance, as indicated in Figure 1.37. Calculate the total material cost by typing

```
c = 600s^2 + 720/s
```

then drag the result from the Algebra window to the graphics view. You can now simply drag P to find the minimum cost. By selecting P, you can control it using the

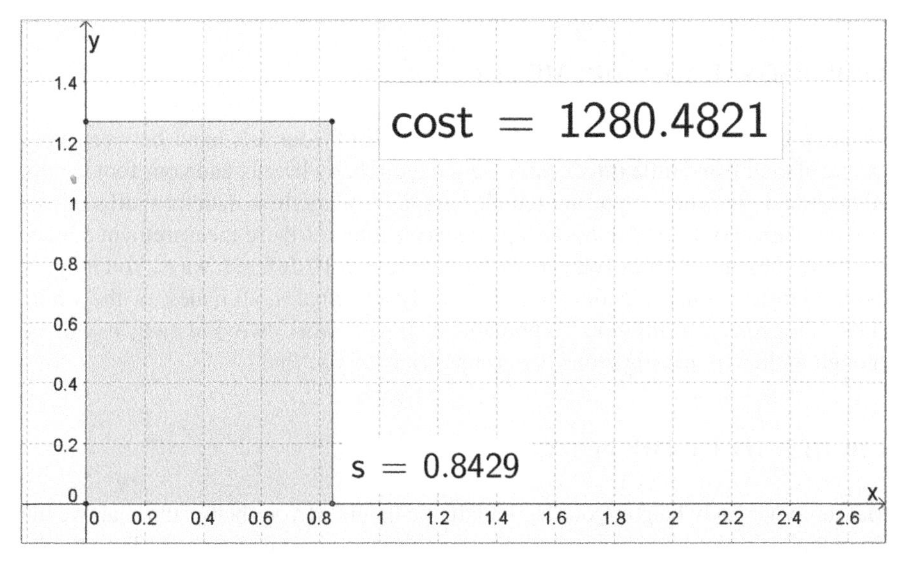

FIGURE 1.37 Dynamic model of the locker case.

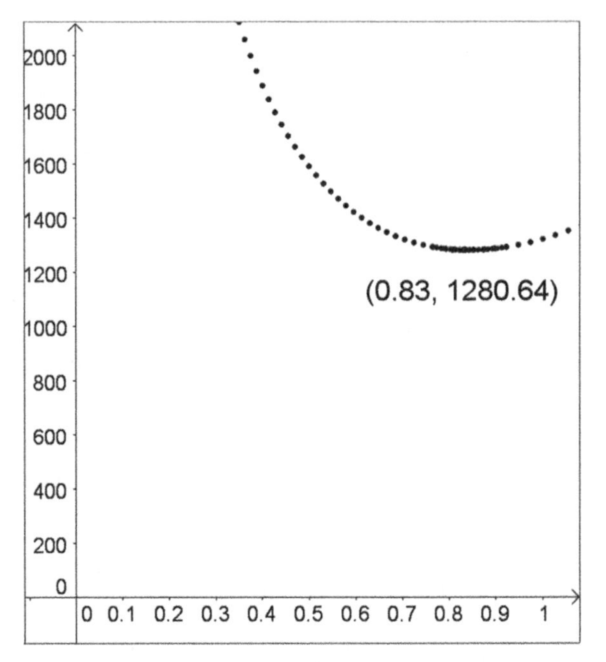

FIGURE 1.38 Plot of cost c against side s.

arrow keys. The step size is set to 1 for points by default, but this can be changed in the properties of the point. On the **Algebra** tab you set the **Increment** to be 0.001.

You have now built a dynamic simulation of the locker case. In Figure 1.38, Graphics view 2 is open where a point (s, c), having its **Trace On,** has been created. Moving P will now result in a trace, clearly showing the minimum.

1.9 PERSONAL MEASUREMENTS

Measure the length of your right foot, the width of your left hand between your extended thumb and little finger, your weight, your body length, and your foot length. Also write down whether you are female or male. Collect these data from all students in your class. Try to find relationships between some of these measurement values. With five measured values you can use two of them in 10 different ways. You may for these 10 different investigations select to study all females, all males, or the whole class, altogether 30 different combinations. If you work two and two, it will be enough to do two investigations per group. What do you find?

1.10 HEIGHT OF THE BODY

Measure your body height, your leg length, the height of your belly button above the floor, the length of your arm from your armpit to your middle fingertip. Then do the same measurement with 10 of your friends. What relationships can be found? Is it true that the length relates to the belly button according to the golden section, 1.618 ...? What about arm length and the length of the forearm? Also control if there is a relationship between length and shoe size.

1.11 LAMP POLE

Photograph lamp poles or fence poles with equal distances on a straight road. Carefully note from where you are taking the photo and the distance between the poles. Then measure the number of pixels from the edge to of the photo to every pole. Calculate the actual distance to every pole and make a diagram of the relationship between your two measurements. Can you find a point in the photo that is exactly 143 meters from your camera?

1.12 THE SKYSCRAPER

At http://www.flickr.com/photos/aaronescobar/2178776113/ there is a photo of a skyscraper in Miami where you cannot see the ground level. Is there a way for you to measure the number of floors in the building directly from the photo?

1.13 THE FENCE

At http://www.flickr.com/photos/lesphotosdejerome/3899807799/ there is a photo of a woman at a fence. Is it possible for you to cut out the woman, make her twice as large, and place her at the right position in the photo according to her new size?

1.14 THE CORRIDOR

At http://www.flickr.com/photos/ell-r-brown/5128688426/ there is a photo of a corridor with stone slabs and wall panels. Make an assumption regarding the distance to the first stone slab and the closest wall panel and then determine the length of the corridor and the length of the left wall.

1.15 BIRD FEEDERS

An ordinary cylindrical bird feeder with dual eating holes and perches at the bottom, two-thirds down and one-third down takes 8 hours to half empty. How full was it two hours after it was filled? When will it be completely empty?

If you were to construct a bird feeder with the same number of holes and perches that had the property that it was half emptied in exactly half the time it is completely empty, how would you construct it, meaning where would you place the holes and perches?

1.16 GOLF

An average golf player is out practicing how to play toward green. He is particularly interested in one specific part of the amazing golf course of the club in which he is a member. The green is small and circular with only about a 5 meter radius and is situated on the top of a high-rise hotel about 30 meters higher than the surrounding flat fairway.

Any golfer who wants to go for the green has to shoot the ball about 90 meters in such a way that it stays on the green. The way the shot should be executed depends on the wind. If there was no wind, a good shot toward green should be steep and short, with a good backspin to make the ball stop. Nevertheless, there is almost always a steady headwind at this part of the golf course when the balls and the players reach the fairway close to the hotel. So the player wants to have a low pinch that will stop dead in the wind above the platform and just drop down on the green. This will require a really strong backspin to make the ball "balloon."

Golf shots are difficult to model, but it can be done. One freeware program is called TrajectoWare Drive 1.0 and can be downloaded (2.5 MB) by registering at http://www.trajectoware.com/.

Another possibility is to download the CDF player (690 MB) from http://demonstrations.wolfram.com/download-cdf-player.html and then investigate the Wolfram Demonstration Projects.

Trajectory of a golf ball at http://demonstrations.wolfram.com/TrajectoryOfAGolfShot/

Flight of a golf ball at http://demonstrations.wolfram.com/FlightOfAGolfBall/

Study either or several of these simulation models of golf shots. Find reasonable parameters that will enable the golfer to succeed with the shot. Take measurements of the trajectories and plot the points in GeoGebra.

2

LINEAR MODELS

Central to empirical mathematical modeling is the concept of functions, both in the form of linear and nonlinear functions. Knowledge about different mathematical representations such as graphs, tables, algebraic and symbolical expressions, situations, and words needs to be established and practiced by students.

This chapter is mainly about linear relationships. Linear models are often the first choice of model because they are easily handled. However, linear models do present some limitations that the students need to be made aware of.

Yet, more than the limitations to the linear models, you, the teacher, need to concern yourself with the limitations the data may impose on you. Do you have access to all available data? Can you look beyond data? Are some of your data irrelevant or not suited to the choice of a linear model? These and other questions need to be constantly on your mind while working with real data.

A very specific thing to learn is that summary statistics is not everything. Anscombe's Quartet is a set of completely different data sets that have identical statistics and linear regressions. The lesson to be learned, and one that using GeoGebra will help you learn, is that you always need to plot your data and inspect it before analyzing it. You can view Anscombe's Quartet at http://tube.geogebra.org/student/m28709.

Mathematical Modeling: Applications with GeoGebra™, First Edition. Jonas Hall and Thomas Lingefjärd.

Companion website: www.wiley.com/go/Hall/MathematicalModeling

2.1 ARE WOMEN FASTER THAN MEN?

Will women outrun men in the future? How fast do you think women will run in 100 years? In 200 years?

In Table 2.1 you can see the results from the Olympic 200 meter gold medalists. Use these data and construct mathematical models for predictions and comparisons with respect to women's and men's records for the 200 meter run in future Olympic Games and world championships.

Then complete the table with results after 1988. How do these new data change your predictions?

Proposed Solution

Feed the table entries into the spreadsheet, which you can find from the menu **View>Spreadsheet** or by pressing **Ctrl-Shift-S.** Enter the years and the names of the male runners in the A and B column. In cell D2 type the formula

```
=(A2-1900, C2)
```

This will generate a point in the graphic area that represents the record for men in 1988. Notice that the x-axis shows the number of years after 1900 and that the point

TABLE 2.1 Olympic Winners in the 200 Meter Sprint

Year	Men	Time (s)	Women	Time (s)
1988	J. DeLoach, USA	19.75	F. Griffith-Joyner, USA	21.34
1984	C. Lewis, USA	19.80	V. Brisco-Hooks, USA	21.81
1980	P. Mennea, Italy	20.19	B. Wöckel, East Germany	22.03
1976	D. Quarrie, Jamaica	20.23	B. Eckert, East Germany	22.37
1972	V. Borzov, Soweet	20.00	R. Stecher, East Germany	22.40
1968	T. Smith, USA	19.83	I. Szewińska, Poland	22.5
1964	H. Carr, USA	20.3	E. McGuire, USA	23.0
1960	L. Berruti, Italy	20.5	W. Rudolph, USA	24.0
1956	B. Morrow, USA	20.6	B. Cuthbert, Australia	23.4
1952	A. Stanfield, USA	20.7	M. Jackson, Australia	23.7
1948	M. Patton, USA	21.1	F. Blankers-Koen, Holland	24.4
1936	J. Owens, USA	20.7		
1932	E. Tolan, USA	21.1		
1928	P. Williams, USA	21.8		
1924	J. Scholtz, USA	21.6		
1920	A. Woodring, USA	22.0		
1912	R. Craig, USA	21.7		
1908	R. Kerr, Canada	22.6		
1904	A. Hahn, USA	21.6		
1900	W. Tewksbury, USA	22.2		

will have the x-coordinate 88. Always be careful when defining the variable representing time.

Select cell D2 again and drag the fill handle down to generate the rest of the points for the men's results as shown in Figure 2.1.

If the points for some reason do not show, right-click on the word **Point** in the Algebra window and select **Properties**. Make sure that the **Show object** option is activated.

The set of points seem to show a weak decreasing trend. You can therefore try to do a linear regression on this set of points. This can be done in either of two different ways. Either select the points in the spreadsheet, right-click on them and select **Create...>List** and then enter the command **FitLine[list1]** in the input field, or alternatively, click on the Best Fit Line tool and then drag a rectangle around all the points in the Graphics window.

Repeat the same procedure for the women's data. With the Intersect tool you can click on the graphs in order to find the intersection between them. This is shown in Figure 2.2. It looks as if the women will start beating the men about 82 years after

Spreadsheet

D2 =(A2 - 1900, B2)

	A	B	C	D
2	1988	19.75	21.34	(88, 19.75)
3	1984	19.8	21.81	(84, 19.8)
4	1980	20.19	22.03	(80, 20.19)
5	1976	20.23	22.37	(76, 20.23)
6	1972	20	22.4	(72, 20)

FIGURE 2.1 Information stored in the GeoGebra spreadsheet.

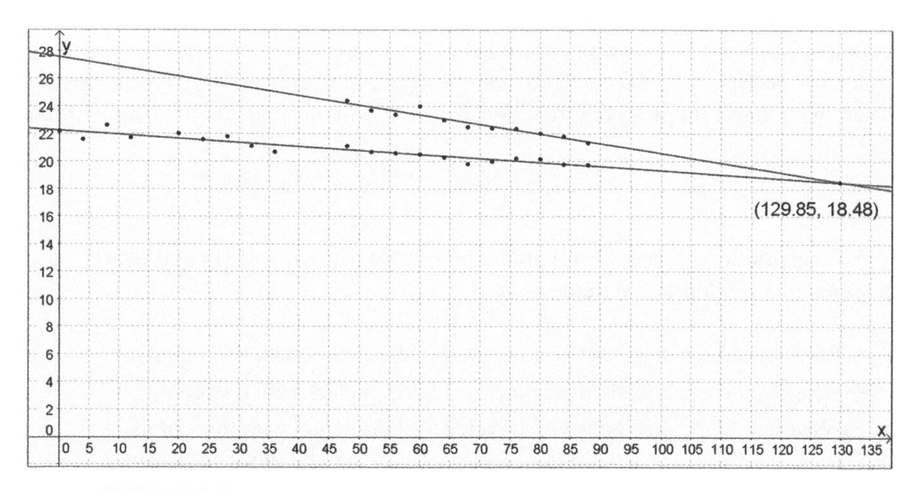

FIGURE 2.2 Women's record times as seen to decrease faster than men's.

1948, around 2030, when the record times for the Olympic 200 meters will have gone down to about 18.48 s for both men and women.

In about 100 years the women will run 200 meters in 17.20 s, and in 200 years they will run 200 meters in extraordinary 10.20 s. In 300 years the record will be 3.20 s and soon after that you will get negative values. While this simple model may lack some credibility, it may be useful to present to your class in small intervals. For example, what is a reasonable limit for human running speed?

Using the Internet, it is quite easy to find results from all the Olympic Games after 1992 and add these data on new rows in the spreadsheet. The English Wikipedia has an article at http://en.wikipedia.org/wiki/200_metres in which your students can investigate this situation even further. Instead of creating new lists, you can easily modify the old ones by double-clicking on them, and then adding your new points straight into the lists. When doing so, you also dynamically update the regression. You now get quite different results than before. The women's record from 1988 seem to hold still, but Usain Bolt ran 200 meters in 19.30 s year 2008 and on 19.32 s year 2012. These results move the point of intersection further into the future, and therefore it is quite doubtful that women will ever outrun men in the future.

This mathematical modeling exercise is an example of how important it is not to predict farfetched conclusions for simple trends. New data can easily discredit weakly supported theories.

2.2 TAXI COMPANIES

Many taxi companies today have a price model. In the price model the total price is written $P(x, y) = \mathrm{a} + \mathrm{b} \cdot x + \mathrm{c} \cdot y$, where a is the start-up, also known as pickup or flag fall cost, b is the price per kilometer (USD/km), c is the hourly cost, x is the length of the run, and y is the total time of the run. In addition, the parameters a, b, and, c typically vary with the time of the day and sometimes even with the number of customers. In some countries only distance, and not time, is used. A typical setup for a company in Paris is shown in Figure 2.3.

It may be easier on the students and their graphing capabilities if you start by assuming a much simpler price function $P(x)$ that only depends on the length of the taxi run:

$$P(x) = \mathrm{a} + \mathrm{b} \cdot x.$$

So imagine a company, Taxi1 that has a pickup fee of 5 USD and charges the passengers 3 USD per kilometer.

- What would the cost be if you traveled 3 kilometers with this company?
- What would the cost be if you traveled 6 kilometers with this company?
- What would the cost be if you traveled 12 kilometers with this company?
- If you never go any further than 20 kilometers, what will be the maximum price you need to pay?

PARIS TAXI FARES

Fare as at 03/02/2015

FARE CURRENTLY IN FORCE	
Pick-up	€ 2.60
Minimum journey	€ 7.00
Supplement per person over and above 3 passengers	€ 3.00
Supplement for second and further items of baggage	€ 1.00

	FARE PER KILOMETER	FARE PER HOUR
TARIFF A	€ 1.05 per km	€ 32.05 per hour
TARIFF B	€ 1.27 per km	€ 38.00 per hour
TARIFF C	€ 1.56 per km	€ 35.70 per hour

		PARIS CITY AREA	SUBURBAN ZONE *	OUTSIDE THE SUBURBAN ZONE
	Midnight to 7.00 am	B	C	
	7.00 am to 10.00 am			
Monday to Saturday	10.00 am to 5.00 pm	A	B	C
	5.00 pm to 7.00 pm	B		
	7.00 pm to Midnight		C	
Sunday	12.00 am to 7.00 am	C	C	C
	7.00 am to 12.00 am	B		
Public holidays	Midnight to 12.00 am	B	C	C

FIGURE 2.3 Taxis G7's Paris fares 2015.

- After several trips with this company, you learn that a 5 kilometer trip will cost 20 USD and a 13 kilometer trip will cost 44 USD. Enter these values as points in the coordinate system. How large is the difference between these two points with respect to the point's x- and y-values? Is there any connection between these differences?
- A line through these points may be a way to show the general relationship. In the Algebra window you can read the symbolic representation for this line. What does the intersection between the line and the y-axis represent? What does the intersection between the line and the x-axis represent? What units should these intersection points be expressed in?
- A competing taxi company, Taxi2, is introduced and is advertising a 0 USD run-up fee. But the price is 5 USD per kilometer. When is it better to travel with Taxi2 or Taxi1?
- After some time, it turns out that Taxi1 dominates the long taxi rides while Taxi dominates the short taxi rides. Taxi1 is thinking about winning back customers. How should they do that?

Proposed Solution

The full model may be viewed as a plane in three dimensions, and in using the GeoGebra 3D window, you can do everything proposed in this problem in 3D. But again, let's start with the simplified model.

The input command field works as a simple calculator. By entering

`5+3 3` (the space bar works as a multiplication sign)

you can see (from the result in the Algebra window) that it costs 14 USD to travel 3 kilometers. In the same way the

`5+3 6`

gives the price 23 USD for a travel of 6 kilometers, and

`5+3 12`

equals 41 USD for 12 kilometers.

You can also create a simple table in the GeoGebra spreadsheet, which is opened by **Ctrl-Shift-S** or from the menu **View>Spreadsheet**. Enter 1 and 2 in the cells A1 and A2. Select the two cells and copy down by dragging the fill handle—the small square in the lower right corner. In cell B1 you type `=5+3 A1` and press **Enter**. The cost for one kilometer is thus calculated. Select cell B2 and copy this downward in the same way. This will create a simple table with values where you can find the price for travel of different lengths as shown in Figure 2.4.

If you want to represent this in a coordinate system, you can do this by entering as the cell points within parentheses and a separating comma sign just as you would do in the Input bar. You type

`(5,20)`

in the Input bar and press **Enter** and thereafter

`(13,44)`

and **Enter** again. Your two points are created: A and B.

In order to construct the line between them, you either click on the Line tool and then the two points, or you type the command

`Line[A,B]`

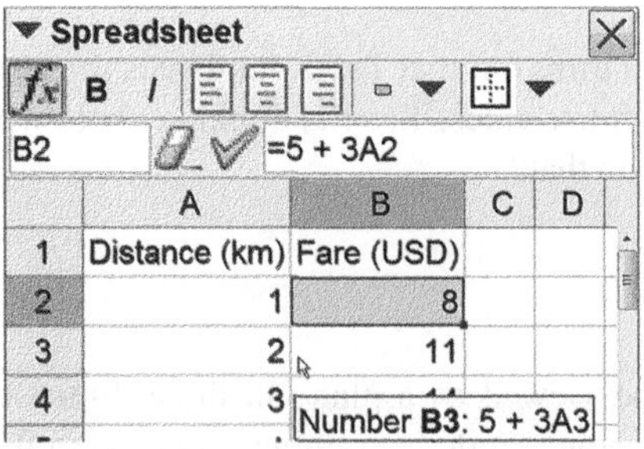

FIGURE 2.4 Using the spreadsheet in GeoGebra.

in the input command field and then press **Enter**. You can also type the function expression

```
5+3x
```

in the input command field and then press **Enter**. When you create the graph for the taxi company Taxi2, you enter

```
5x
```

and press **Enter**.

The intersection point between the two lines is created by clicking first on the Intersect tool and then on the two lines. The coordinates for the point now appear in the Algebra window. If you travel 5 kilometers, it will cost 20 USD using either company. Taxi2 is cheaper for short travels and Taxi1 is cheaper for long travels. Of course, you can also make a table for Taxi2 in the spreadsheet by entering `4*A1` in cell C1 and dragging this formula downward.

In Figure 2.5 the points and lines are connected by a thicker line. Most visual properties can be changed using the Style Bar, which is accessed by the arrow to left of the word **Graphics**, shown in Figure 2.6.

You could also have changed the names on the lines to indicate which line belongs to which company by right-clicking and selecting **Rename**. Whenever you need to click on an object, you can click on the object either in the Graphics window or in the Algebra window.

You can also solve the problem algebraically by solving the equation $5+3x=5x$. It is well worth mentioning to your students that the left-hand and right-hand parts of the equation represent two lines whose intersection provides the solution to the equation. It may be useful to mention that when algebraic solutions are hard to find,

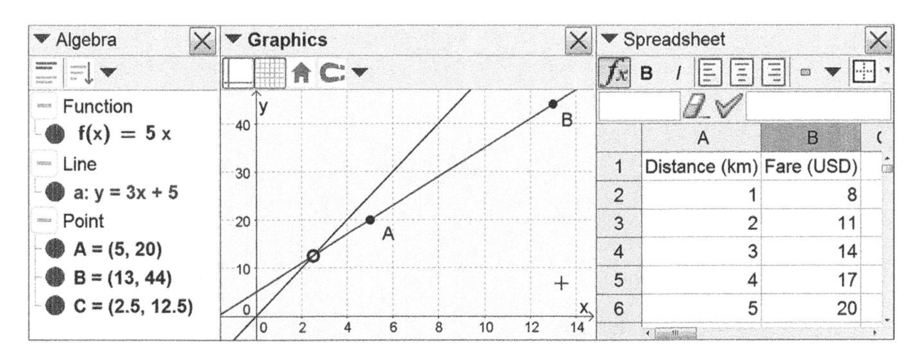

FIGURE 2.5 Comparing two taxi companies.

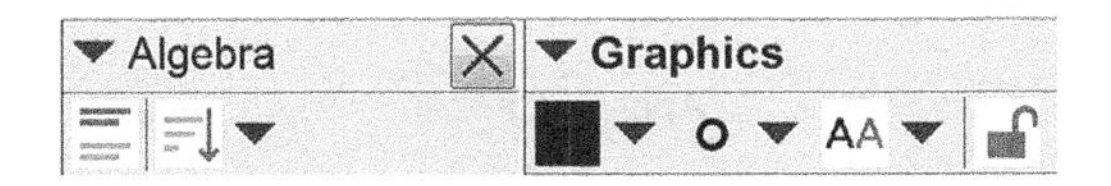

FIGURE 2.6 Style Bars for the Graphics and Algebra windows.

such as for the equation $x^8 - x^3 = 25x$, you can always find a graphical solution. (Try it!)

In the real world, as in the work of engineers for instance, both numerical and graphical solutions are much more common than algebraically exact solutions.

For the analysis of the next question, how the pickup and the kilometer costs affect the position of the intersection point (= its x-coordinate), you will use a separate graphics area.

GeoGebra has two graphics areas, but normally you just see the first of them when you start the program. From the menu, you may select **View>Graphics 2** or press **Ctrl-Shift-2** to enable this area. The first time you show it, the Graphics 2 window is probably not docked to the main window, but floating around and without a grid. Dock it using the docking control in the upper right of the window, and drag it to where you want it. Then activate the grid by pressing in the Style Bar.

The two 2D graphics areas and the 3D window are really just different windows onto the same universe. It is the objects that are set to be visible or not visible in each window. Every object has visibility controls to reflect this in the **Properties dialogue**, at the bottom of the **Advanced** tab.

The next step is to create sliders for the run-up cost and for the kilometer cost. The sliders are created by clicking the slider tool and thereafter click in the graphics area where you want the slider to appear. GeoGebra will then open up a dialogue where you can enter a suitable name, decide minima and maxima values, and so on. In Figure 2.7 the sliders have been placed in the bottom part of Graphics 2. They are called u and k and go from 0 to 60 with an increment of 1.

A slider is a geometrical representation of a variable number. The current value of a number may be seen in the Algebra window. When you change the slider, the number in the Algebra window is changed too.

Our first line for Taxi1 was defined as the line through two points. Double click it and redefine it as

```
u + k x
```

shown in Figure 2.8, and click on **OK**.

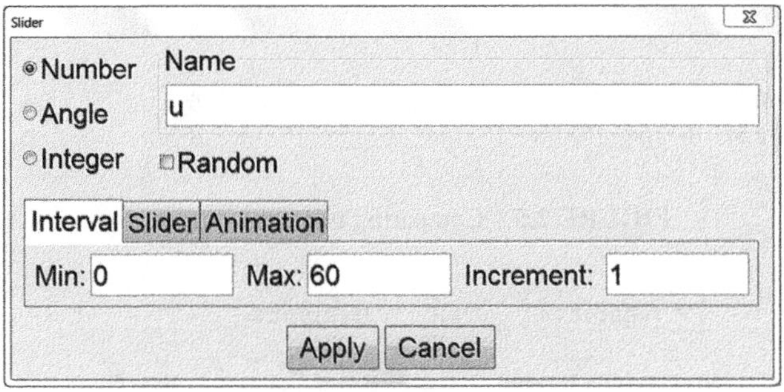

FIGURE 2.7 How to define a slider.

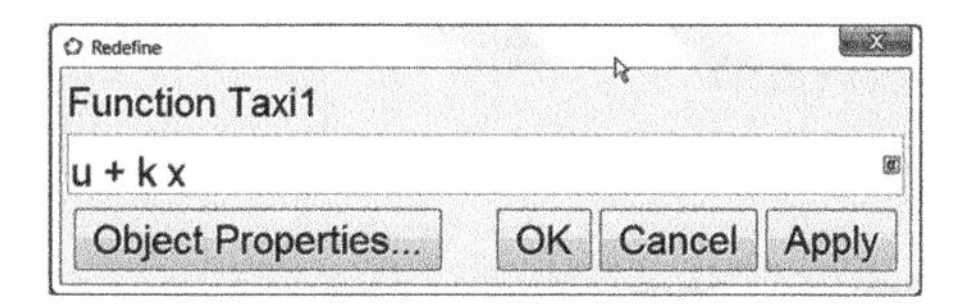

FIGURE 2.8 Objects may be redefined by a double-click on them.

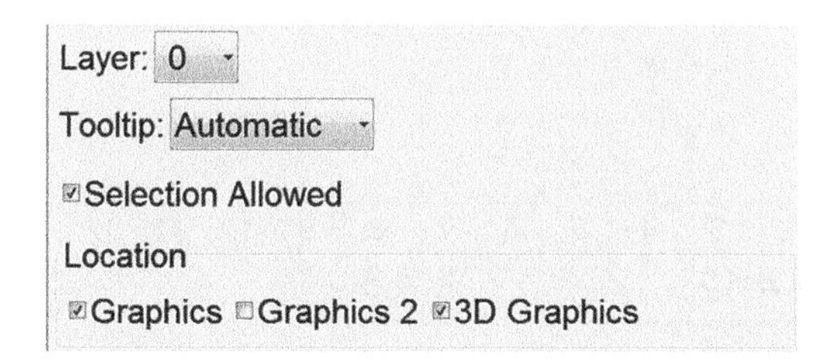

FIGURE 2.9 Any object may be visible in both graphic areas at the same time, in one of them at a time, or be completely hidden.

Play with the sliders and see how the intersection point C is moving. You will now make a construction that shows the intersection point's x-coordinate as a function of u. In Graphics Area 2 you want the x-axis to represent u and the y-axis to represent x(C), the x-coordinate for a point.

First, click once in Graphics area 2 to make it active, and then enter

```
(u,x(D))
```

in the input field (and press **Enter**). This creates a point, D, that moves at the same time as you move the slider u. Should Graphics area 1 have been active the point would have been created there and not be visible in Graphics 2. This is a common mistake but easy enough to repair as shown in Figure 2.9. Just right-click on the point in the Algebra window and select **Properties**, on the **Advanced** tab and change the check boxes at the bottom.

You can understand the relationship between the objects better if you right-click on the point and select **Trace on**. This will make the point leave a trace as in Figure 2.10 when you change the slider u. The trace can later be erased with **Ctrl-F**.

The relationship is linear: through the origin and from the coordinates for D you can see that the relationship can be written as $x(\text{D})=0.5u$. This is only true for k=3 though. If you change k to 4, then the relationship seems to be $x[\text{D}]=u$ instead.

The point D will move on a line when you vary u. This is a way to present a function to your students so that they can see that a function is the *locus* for a set of points. It is, however, not practical to manually vary u for every new value for k. Instead, show the *locus* for the point D when u varies by clicking on the Locus tool, then clicking on the point E, and last clicking on the slider u. The locus is shown as a line that is changed when you change the value for k. If you like, you could trace the locus while you are varying k. Right-click the locus itself, and select **Trace on**. The result may be seen in Figure 2.11.

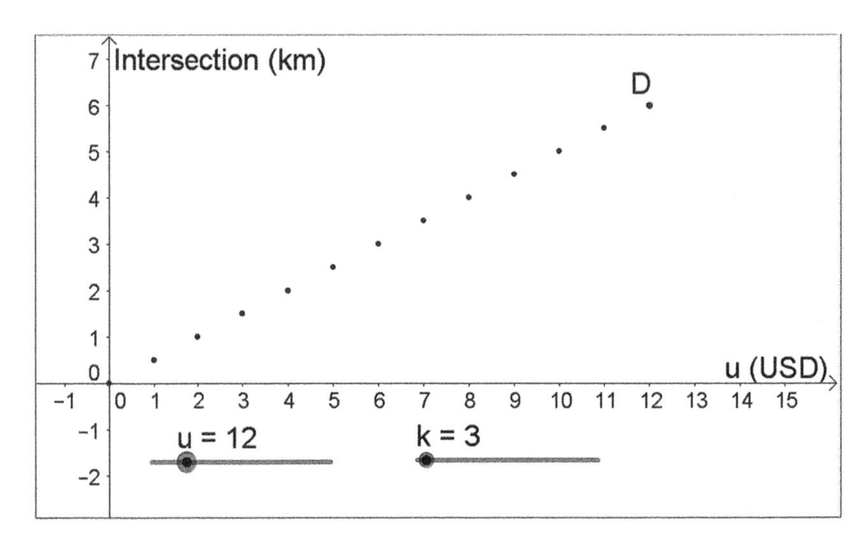

FIGURE 2.10 A point that leaves a trace indicates the functional relation.

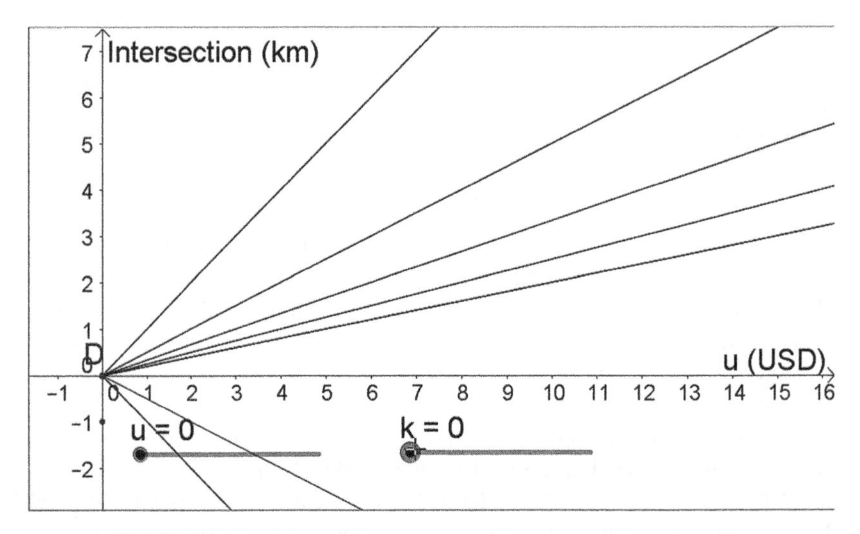

FIGURE 2.11 Linear relations vary with respect to the value of k.

The relationship between the intersection's x-coordinate and the parameter k is slightly more complicated than the linear relationship you just studied. So create a new point E, depending on the k-value by clicking in Graphics 2 for activating that area, and then type

```
(k,x(C))
```

in the input field. Create a locus for the new point by clicking the Locus tool, the point, and the slider k. Erase the existing traces by **Ctrl-F**, disable the trace for the

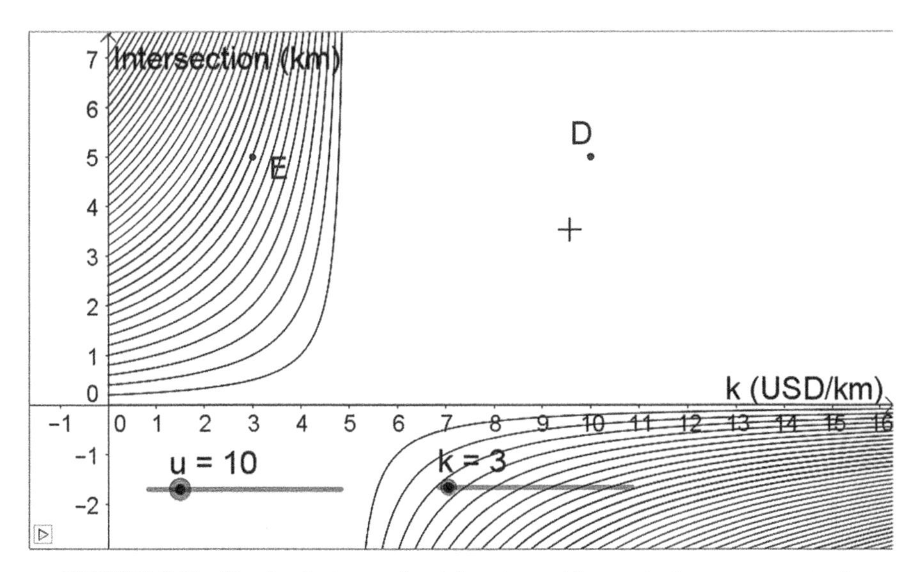

FIGURE 2.12 Varying k gives rational functions with a vertical asymptote at k = 5.

locus through D, and activate the trace for the locus through E. Now vary u again, this time by right-clicking u and selecting **Animation on**. When you want to stop the animation, you can use the play/pause-button that has been created at the bottom left of Graphics 2.

The rational function shown in Figure 2.12, which for all values for u has a vertical asymptote at k = 5, could possibly be inversely proportional to 5 – k. For all values of k, it is also proportional against u. A qualified guess therefore would be that it is proportional to u/(5 – k), something you can confirm when you solve the equation algebraically. From $u + kx = 5x$ you get $x = u/(5 - k)$. Being able to guess a symbolic representation from a graphical representation is a useful skill, but this requires experience with different types of functions.

In this simplified situation we have only considered how the intersection varies, depending on the parameters u and k. What parameter values should actually be selected depend on an analysis of the profitability as well as the policy of the taxi company. Always consider the relative benefits and limitations of your model in relation to the real world.

2.3 CRIME DEVELOPMENT

How many crimes will take places in Sweden in 1998, 2013, and 2020?

The increase of crimes in Sweden from 1950 to 1990 is described in Table 2.2. Observe that the numbers show crimes reported to the police, but there may be unreported crimes as well. Regardless of the unreported crimes, figures like these are used for statistical analyses and thus for planning relevant social infrastructure.

TABLE 2.2 Number of Reported Crimes in Sweden, 1950 to 1990

Year	Crimes
1950	195,261
1951	232,252
1952	225,169
1953	208,632
1954	218,391
1955	243,786
1956	255,106
1957	277,945
1958	301,135
1959	300,291
1960	297,874
1961	308,952
1962	322,316
1963	339,626
1964	367,649
1965	448,619
1966	473,397
1967	505,237
1968	567,063
1969	611,265
1970	656,042
1971	713,822
1972	691,129
1973	655,383
1974	675,276
1975	755,405
1976	799,228
1977	844,360
1978	803,275
1979	816,108
1980	928,277
1981	935,825
1982	983,758
1983	959,127
1984	983,175
1985	1,018,349
1986	1,095,357
1987	1,093,417
1988	1,086,211
1989	1,144,800
1990	1,218,820

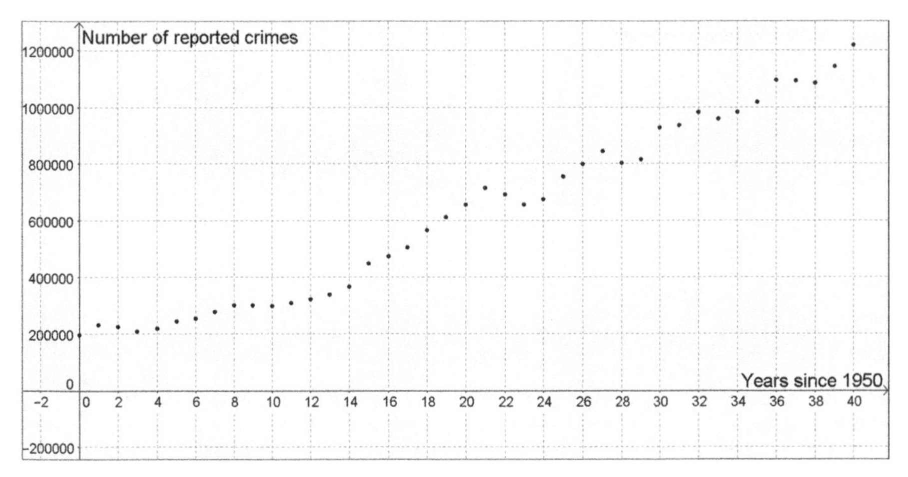

FIGURE 2.13 Crime development in Sweden, 1950 to 1990.

Use the data set in the table and recreate a mathematical model that can predict the number of reported crimes for 1998. Check the result from your model against some reliable source. Then discuss the reliability of your model and see if you can strengthen your model enough for it to be useful for predicting the crime development for the year 2012.

How certain is your prediction for 2020?

Proposed Solution

Start by just having a look at the data. In Figure 2.13 the number of years since 1950 are on the x-axis and the number of crimes are on the y-axis. Notice that there are periods when the crimes levels are low, despite the overall trend showing crime levels to be increasing.

The increase is probably connected to the change of the population during the same time. The population of Sweden increased from about 3 million people in 1950 to about 5 million in 1990. In order to balance the data set for this population growth, crimes rates (per one hundred thousand residents) are often used instead of actual numbers.

Here you can ignore this and make a simpler analysis, keeping with the actual number of crimes. But if you look at how the population has grown during the same time, it seems to have grown exponentially up to the 1970s and then almost linear after that. It is not totally unrealistic, looking at the data, to believe that this growth also is a valid explanation for the crime increases.

But if the number of crimes has followed two different mechanisms, then you should only use the latest data points. The earlier ones must have followed mechanisms that are no longer valid. So restrict the investigation to a linear analysis of the data from 1970 onward.

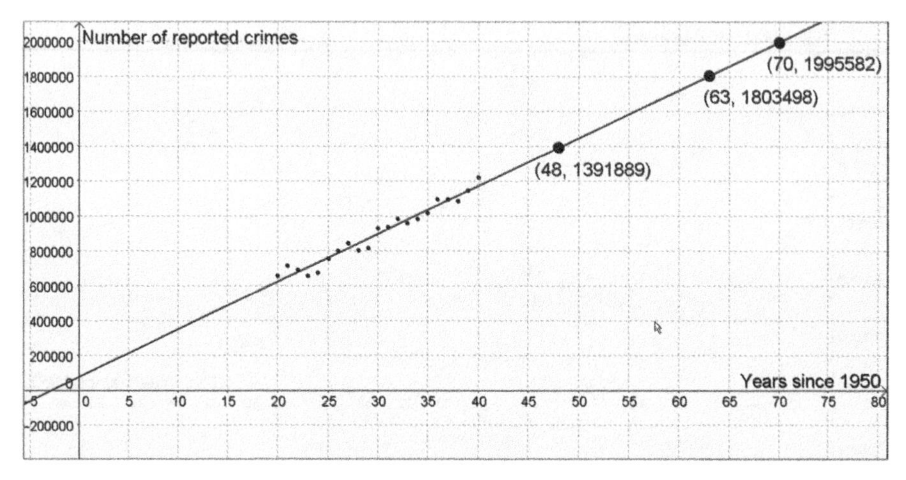

FIGURE 2.14 Linear regression of the data from 1970 and forward.

Select these data in the spreadsheet and create a separate list of these data points. Afterward do a linear regression, typing in the command

```
FitPoly[list2,1]
```

to get the result shown in Figure 2.14. Use **FitPoly** rather than **FitLine** in order to get a function, rather than just a geometrical line, as you want to be able to make function calls.

In the diagram, the points that predict the assumptions for the years 1998, 2013, and 2020 were plotted by entering

```
(48,f(48))
```

and so on, in the command field.

By Googling on some of the numbers in the table, you can find updated information. In 1998, 1,181,056 crimes were registered and 2013, 1,401,982 crimes were registered in total. You can also see that the increase in crime development has stalled and is no longer following the projection. This is a typical example illustrating that extrapolation is generally not recommended when dealing with complex situations. Always be careful when extrapolating from data without an underlying theory.

How can you evaluate the validity of the model? There are several methods. Most convenient is perhaps to select a point among the 20% first and a point among the 20% last points and draw line through these using the Line tool and compare it to the regression line. Do this both ways, both upward and downward. These lines will give an *estimate* of the error limits of the model. Using Figure 2.15, you can see that the error margins for the 1998 estimate are about ±100,000 crimes, or about 7%. The reality is, however, well beyond these limits.

It is important that you select your points for the error lines to get as extreme lines as you can. You may also experiment by moving a point in the data set so that the

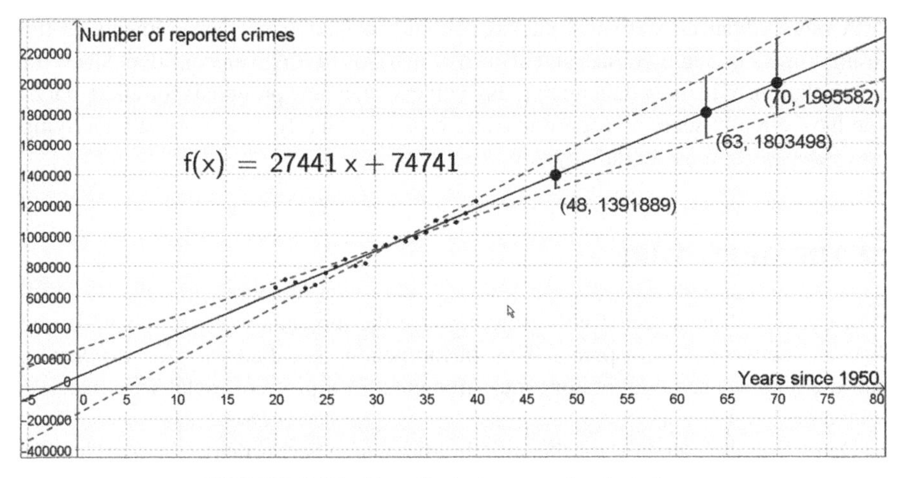

FIGURE 2.15 Error limits by manual calculations.

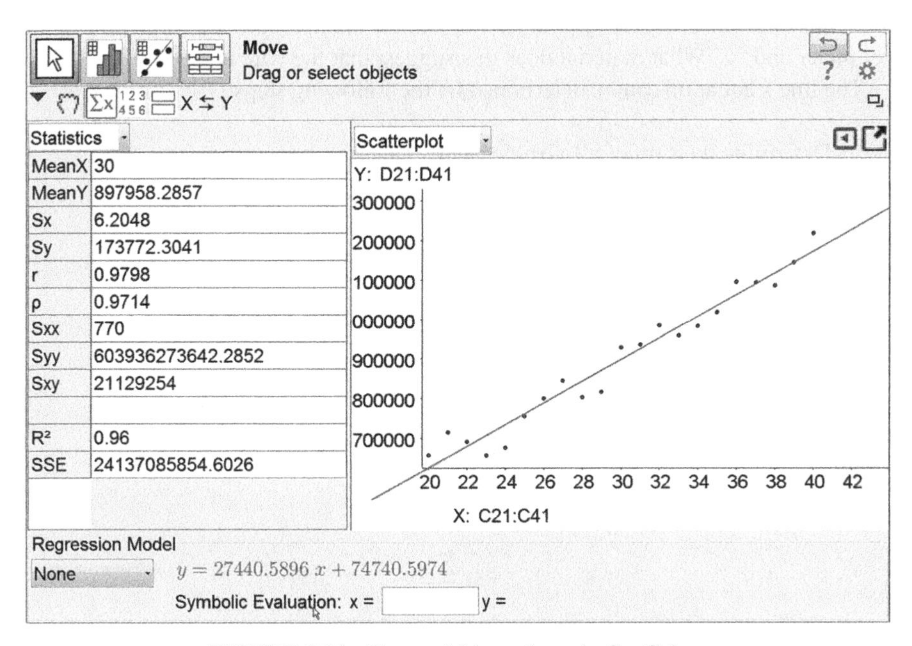

FIGURE 2.16 Two variable analyses in GeoGebra.

regression line is altered. The correlation coefficient, *r*, or SSE, the *sum of squared errors*, are quantities normally used as measurements for how well a regression fits a set of data. These tools can be found in GeoGebra in the tool for **Two Variable Analysis** dialogue when you have chosen to show statistical data and the spreadsheet is active. You find the **Two Variable Analysis** tool in the spreadsheet tool set. Select the relevant data in the spreadsheet, click on ⁖, and click **Analyze**. Select a regression model and click on **Σx** to show the statistical information shown in Figure 2.16.

A more advanced statistical package should include error measurements on the straight line's k- and m-values and also error margins (with given confident level) for extrapolation. With new data up to today you can do a new prognosis for 2020, but as you have already seen, a prognosis solely based on merely data, not the underlying processes, will probably fail anyway.

2.4 THE METAL WIRE

When you heat an object it will grow because the molecules' vibrations make the distance between them spread. This phenomenon is called *thermal expansion.* Table 2.3 shows some laboratory measurements for a piece of metal wire that was heated to over 100 °C.

Determine the equation for the straight line, which according to the *least square fit method* is the best line to fit these values. Furthermore use this *regression line* to decide what length the wire will have at the temperature 123.3 °C. Calculate the *thermal expansion coefficient;* in other words, how much longer the wire will extend per meter and °C. What material does this suggest that the wire is made of?

The line's equation can be determined by the following steps:

1. Determine the sum of all x-values and y values

$$\Sigma x$$

and

$$\Sigma y$$

2. Determine the mean values for your data values

$$\overline{x} = \frac{\Sigma x}{n}$$

and

$$\overline{y} = \frac{\Sigma y}{n}$$

TABLE 2.3 Thermal Expansion of a Wire

Temperature (°C)	Length (mm)
110.0	1000.05
120.0	1000.17
130.0	1000.30
140.0	1000.41
150.0	1000.52

3. In a table then calculate the following values for each data point (x, y):
   The differences from the means

$$(x-\bar{x})^2$$

   and

$$(x-\bar{x})\cdot(y-\bar{y})$$

4. Then calculate the sums

$$\Sigma(x-\bar{x})^2$$

   and

$$\Sigma(x-\bar{x})\cdot(y-\bar{y})$$

   for all data points
5. The line's *slope* a is the quotient between these sums:

$$\mathrm{a} = \frac{\Sigma(x-\bar{x})\cdot(y-\bar{y})}{\Sigma(x-\bar{x})^2}$$

6. The line's equation can now be written as

$$(y-\bar{y}) = \mathrm{a}\cdot(x-\bar{x})$$

   which you can simplify to the well-known form $y = \mathrm{a}x + \mathrm{b}$.

This line will minimize the sum of the squares of all errors, where *error* means the distance from a data point vertically up or down to the line.

Proposed Solution

Enter the data into GeoGebra's Spreadsheet, shown by selecting **View > Spreadsheet** in the menu or by the keyboard command **Ctrl-Shift-S.** Start by summing the data in cells A7 and B7 where you can write the formulas

```
=Sum[A2:A6]
=Sum[B2:B6]
```

respectively. In row 8, create the mean values using the formulas

```
=A7/5
=B7/5
```

Then move on to the rest of the calculations step by step.

In cell C2, create the differences between the x-values and their mean value with the formula

```
=A2-A$8
```

The dollar sign makes the 8 a *static*, or *absolute* reference to the row holding the mean so that it is not changed later when you copy this down by dragging the fill handle at the bottom right of the cell. Using the dollar sign like this invokes *absolute references*, as opposed to the usual, *relative references*.

In cell D2, create the squares of the values by typing

```
=C2^2
```

You can get a nice looking square by pressing **Alt-2** instead of **^2**.

In cell E2, create the differences between the y-values and their mean values:

```
=B2-B$8
```

In cell F2, create the product of the differences:

```
=C2*E2
```

Note that it is possible to use the space key as multiplication sign.

Now select cells C2 to F2 and grab the small the small fill handle in the lower right of the selected cells. Then drag the selected area down to row 6. In cell D7 and F7 you sum row 2-6 as before:

```
=Sum[D2:D6]
=Sum[F2:F6]
```

This will result in Figure 2.17. In the command field you can now calculate

```
a = F7/D7
```

You can view the data points graphically by selecting cells A2 to B6, right-click on them and select **Create...>List with points**. The line will be drawn if you type its equation in the command field:

```
y - B8 = a (x - A8)
```

	A	B	C	D	E	F	G
1	Temperature (°C)	Length (mm)					
2	110	1000.05	-20	400	-0.24	4.8	
3	120	1000.17	-10	100	-0.12	1.2	
4	130	1000.3	0	0	0.01	0	
5	140	1000.41	10	100	0.12	1.2	
6	150	1000.52	20	400	0.23	4.6	
7	**650**	**5001.45**		**1000**		**11.8**	**Sums**
8	**130**	**1000.29**					**Averages**

FIGURE 2.17 Calculations in the spreadsheet.

Observe the space after a to avoid having GeoGebra think a is a function. But as seen in Figure 2.18, the line appears very flat. You need to rescale the graphics to see the line. You do that by right-clicking in the back ground of the Graphics window and select **Graphics…** to get the dialogue shown in Figure 2.19.

Here you can choose the intersection point of the axes. On the x-axis tab, enter an intersection point at $y=999$, and on the y-axis tab, set up an intersection at $x=100$. You now get the view shown in Figure 2.20.

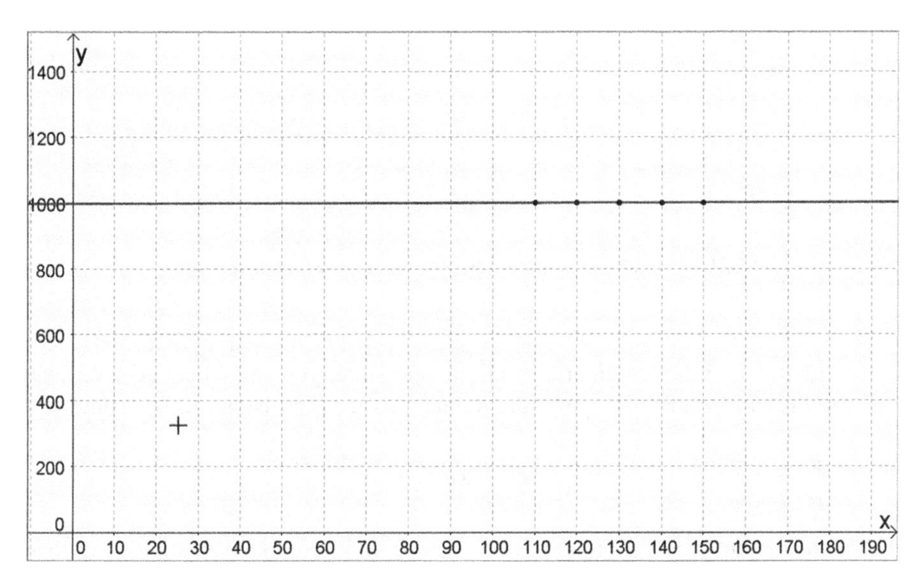

FIGURE 2.18 Measured values in a diagram.

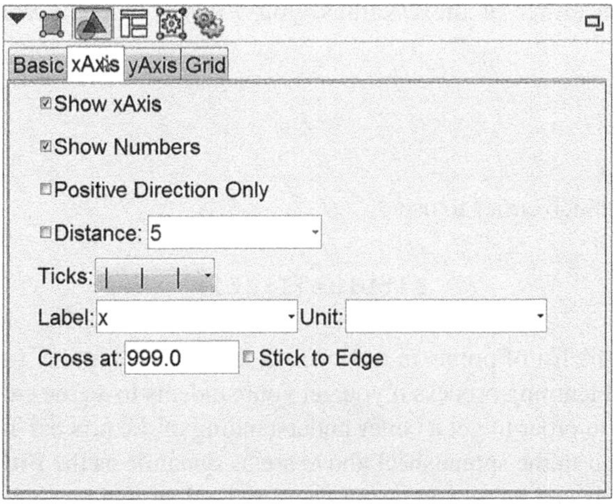

Basic xAxis yAxis Grid
Show xAxis
Show Numbers
Positive Direction Only
Distance: 5
Ticks:
Label: x Unit:
Cross at: 999.0 Stick to Edge

FIGURE 2.19 Make sure that the x-axis will intersect the y-axis at $y=999$.

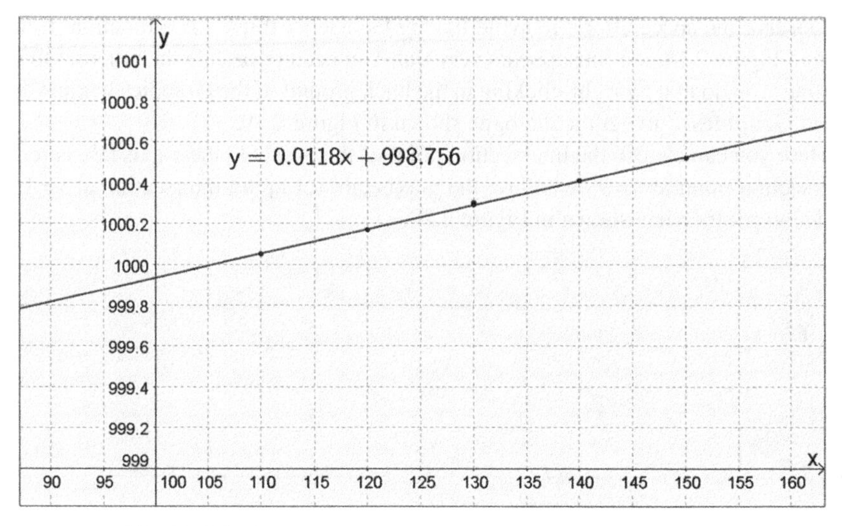

FIGURE 2.20 Data points and the calculated regression line.

In order to find the length at a certain temperature, you should create a vertical line for this temperature and see where it intersects the regression line by typing

```
x = 123.3
```

Create a point with the Intersect tool and from its coordinates you can see that the length will be 1000.21 mm.

The *thermal expansion coefficient* is related to the slope and may be written as 0.0118 mm/(m · °C) or $11.8 \cdot 10^{-6}$ °C^{-1}. If you Google *thermal expansion coefficient* with either of these values, you will see that the wire is probably made of iron.

Comments

It is, of course, much easier to use

```
FitLine[list1]
```

once you have the list of points in order to get the regression line. It is, however, in line with a good learning process if you get your students to do the calculations more or less by hand in order to get a better understanding of the process The calculations you implemented in the spreadsheet above are as dynamic as the **FitLine** command is: the data points can be moved or the measured values can be changed in order to see the effect this will have on the line.

2.5 OPTIONS TRADING

In the São Paulo Stock Exchange, shown in Figure 2.21, as well as in every stock exchange, you trade with economical constructs like bonds, shares, stock, and so on. One way of trading with these is to buy or sell *options* to buy or sell bonds.

In finance, a *bond option* is an option to buy or sell a bond at a certain price on or before the option's expiry date.

A *call option*, often simply labeled a "call," is a financial contract between two parties, the buyer and the seller of this type of option. The buyer of the call option has the right, but not the obligation to buy an agreed quantity of a particular commodity or financial instrument (the underlying) from the seller of the option at a certain time (the expiration date) for a certain price (the *strike price*). The seller (or "writer") is obligated to sell the commodity or financial instrument to the buyer if the buyer so decides. The buyer pays a fee (called a *premium*) for this right.

A "put," or *put option* is a stock market device that gives the owner of the put the right, but not the obligation, to sell an asset (the underlying), at a specified price (the strike), by a predetermined date (the expiry or maturity) to a given party (the buyer of the put). Put options are most commonly used in the stock market to protect against the decline of the price of a stock below a specified price. If the price of the stock declines below the specified price of the put option, the owner/buyer of the put has the right, but not the obligation, to sell the asset at the specified price, while the seller of the put has the obligation to purchase the asset at the strike price if the owner uses the right to do so (the owner/buyer is said to exercise the put or put option). In this way the buyer of the put will receive at least the strike price specified, even if the asset is currently worthless. A *sell option* gives the owner the right to sell shares to a given price. Anyone who buys sell options wishes that the share price should go down so that he or she can sell shares relatively high.

FIGURE 2.21 Trading at the São Paolo stock exchange. Photo: Rafael Matsunaga/Flikr CC BY 2.0.

The price the share may be sold or bought for is often called the *strike price*. This is the price the shares real value must exceed or fall short of before it is wise to *redeem* the option. The price the option is sold for is called the *premium*.

The commercial trade with shares can be compared to a game where you bet money on the possibility that the stock market will go up or down.

For example, a share is right now sold for 35 pounds. For the premium 200 pounds you buy a call option for the possibility to buy 100 shares for the price 37 pounds/share. If the price for the share goes up to 40 pounds/share before the period is over you can call the option and buy 100 shares for 3,700 pounds. Since you principally can sell them directly afterward for 4,000 pounds, you have gained a profit of 300 pounds – 200 pounds = 100 pounds.

Investigate how the profit is connected to the actual price of the share when you call the option for both selling and buying.

A common way for commercial trade to deal with options is the combined trade of call and sell options of the same underlying share. This is called making a *bull spread* or a *bear spread*. Bull and bear are synonyms for if the stock market is raising (bull) or sinking (bear). The Bull and the Bear adorns the entrance of Frankfurt's stock exchange in Figure 2.22.

For example, a trader combines a call option for a share with the strike price 30 pounds for a premium of 300 pounds/100 shares with a sell option with the strike price 35 pounds for a premium of 100 pounds/100 shares. If the price for

FIGURE 2.22 Bull and Bear in front of Frankfurt's stock exchange. Photo: Eva K. (Eva K.). [FAL or CC BY-SA 2.5 (http://creativecommons.org/licenses/by-sa/2.5)], via Wikimedia Commons.

the share at the end of the period is 33 pounds, the trader's counterpart will not redeem the share for 35 pounds, so the trader will redeem the share himself for 30 pounds. That will yield a profit according to (sell of the 35 pounds-option) – 300 pounds (for the buying of the 30 pounds-call option) + 300 pounds (3 pounds profit per share) = 100 pounds.

Investigate how the profit is connected to the factual price of the share when you call the option for both selling and buying. Is it possible to avoid being totally divested?

Proposed Solution

Start by investigating what will happen with the profit if you buy options. Call the *redeem* for a share R, the *strike price* K, the *premium* K_p. the profit V_{buy}, and the final price for the share for P. All units are *pounds* per share. You can now translate the first example using this mathematical notation.

In this example $K = 37$, $K_p = 2$, $P = 40$, and $V_{buy} = 40 - 37 - 2 = 1$ pounds per share. In general, it seems as if $V_{buy} = P - K - K_p$, but you may remember that this is only valid if you actually redeem the option. There is no point in doing this if the current final price P is less than the strike price K. If, in this example, the share does not rise above 36 pounds per share, you won't want to buy it for 37 pounds per share, so you don't. In this case your profit $V_{buy} = -2$ pounds per share, which is the premium you paid when buying the option. So $V_{buy} = -K_p$ when $P < K$. Combine these findings in a piecewise defined function:

$$\begin{cases} V_{buy} = P - K - K_p & P < K \\ V_{buy} = -K_p & P \geq K \end{cases}$$

Piecewise defined functions like these are excellent for discussing inequalities and domains.

You now have enough information to graph this solution in GeoGebra. Start by creating sliders for K and K_p (to create a slider or other object named K_p, enter K_p for the name—the underscore signals a one-letter index). Let them be whole numbers from 0 to 100 and then adjust their values so that $K = 37$ and $K_p = 2$. In GeoGebra you then create the piecewise defined function by typing

```
Vbuy(P) = If[P<K, P-K-Kp, -Kp]
```

The function $V_{buy}(P)$ now has different definitions depending on the value of P. In Figure 2.23 a text label for the function has been created by dragging the expression from the Algebra window to the graphics area.

You can do the same analysis for a call option that you sell. Start by defining the sliders S for the strike price and S_p for the premium.

An option sold yields a premium to the seller. If the share price P does not increase above the strike price S, there is no point for the buyer to redeem the option and the profit for the seller will be $V_{sell} = S_p$.

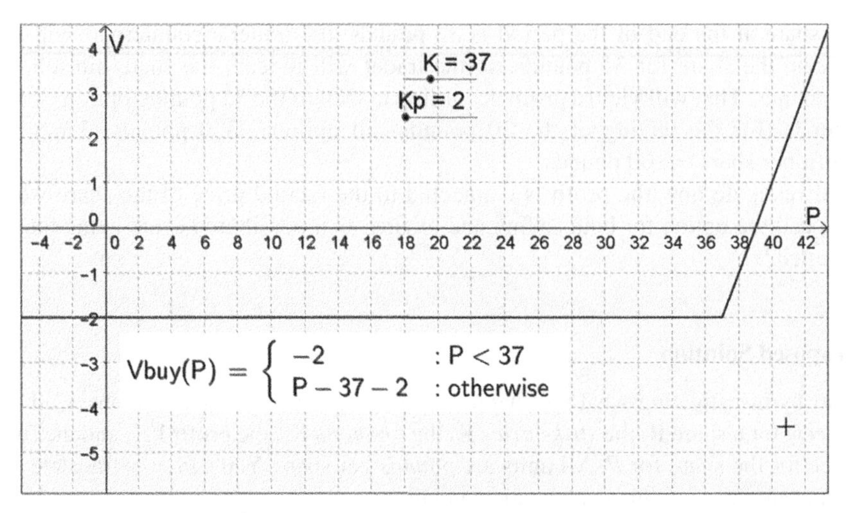

FIGURE 2.23 Profit of buying a call option as a piecewise defined function.

However, if the price P rises above the strike price S, then the owner of the option will redeem the shares and can buy these shares at a lower price S than the actual price P. The profit then becomes $V_{\text{sell}} = \mathrm{S} - P + \mathrm{S_p}$, so

$$\begin{cases} V_{sell} = \mathrm{S_p} & P < \mathrm{S} \\ V_{sell} = \mathrm{S} - P + \mathrm{S_p} & P \geq \mathrm{S} \end{cases}$$

In GeoGebra, this is entered as

```
Vsell(P) = If[P<S, Sp, S-P+Sp]
```

Both functions are shown in Figure 2.24.

When you buy a call option, you can make a profit if the price is rising, and when you sell a call option, you can make a profit if the price is falling. With put options, the opposite is true. The graphs will then look mirror imaged. Except for this remark, we will now ignore put options.

You can combine a sale of a call option with an acquisition of another call option to a *spread*. The total profit will be the sum of the profits for the sale and the acquisition, which you can create in GeoGebra by typing

```
V(P) = Vbuy(P) + Vsell(P)
```

This function is shown as a function with two "knees" in Figure 2.25.

The function is piecewise defined as

$$\begin{cases} V(P) = \mathrm{S_p} - \mathrm{K_p} & P < \mathrm{K} \\ V(P) = P - \mathrm{K} + \mathrm{S_p} - \mathrm{K_p} & \mathrm{K} \leq P < \mathrm{S} \quad (\text{for } \mathrm{S} > \mathrm{K}) \\ V(P) = \mathrm{S} - \mathrm{K} + \mathrm{S_p} - \mathrm{K_p} & P \geq \mathrm{S} \end{cases}$$

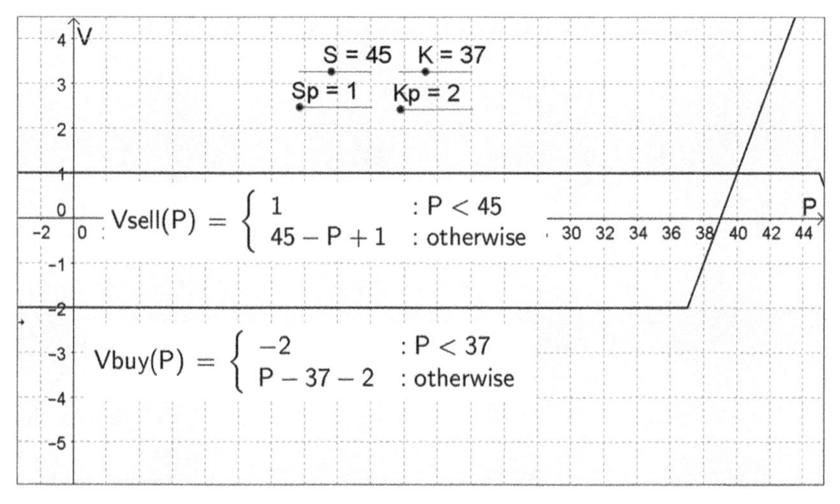

FIGURE 2.24 Profit of selling a call option is also a piecewise defined function.

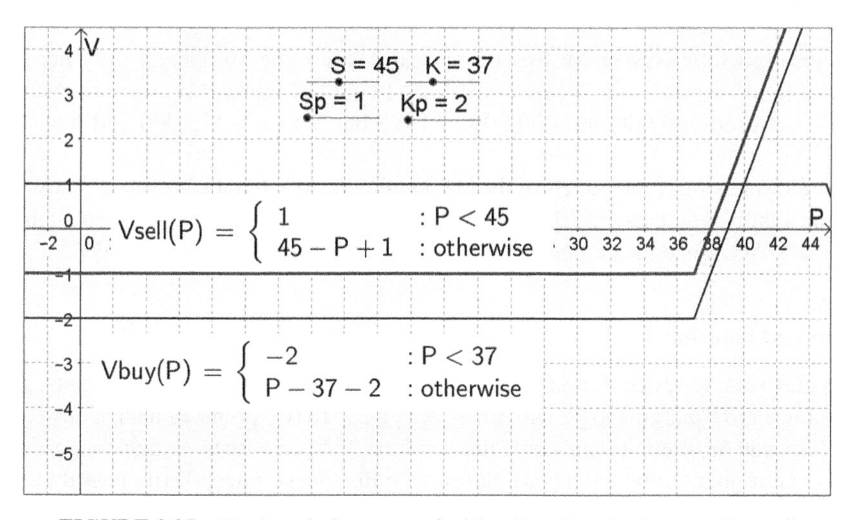

FIGURE 2.25 Total profit for a *spread* with call options having two "knees."

It is theoretically possible to construct a spread so that the profit never goes below zero. That requires $S - K + S_p - K_p \geq 0$ or $K_p - S_p \leq S - K$. An example is shown in Figure 2.26.

We leave to the reader the task of working out the corresponding details for the case of put option spreads.

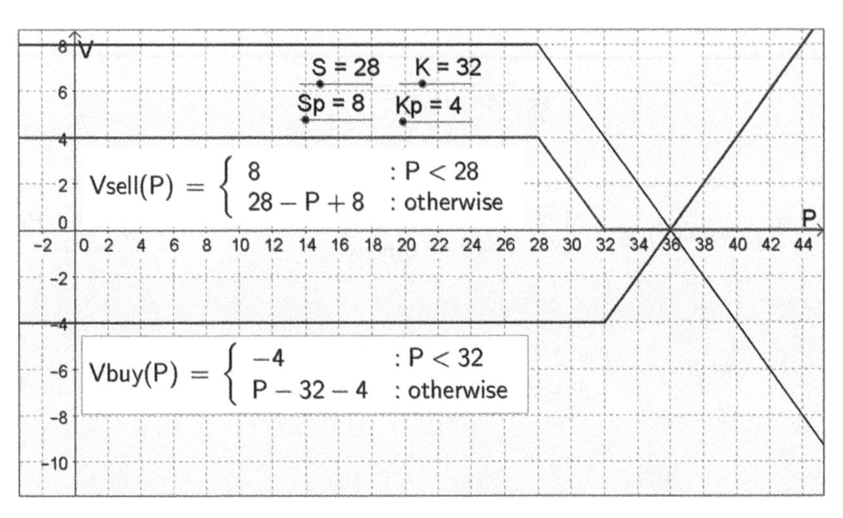

FIGURE 2.26 A spread where the profit never is negative.

2.6 FLYING FOXES

According to the Australian Environmental Agency the number of grey-headed flying foxes, a particular species of bat as shown in Figure 2.27, were estimated to lie between 360,000 and 400,000 in 1998 but between 425,000 and 674,000 in 2005.

Build some different models of the bat population and attempt to make an estimate of the number of bats in 2020. Which is the most probable lowest and highest estimate? What results are possible if even improbable numbers are included?

Proposed Solution 1

You may want to reflect for a moment that in this problem only two data points, and with huge error margins to go with them, are given. To two points of data, almost any function can be fitted. It seems unrealistic to fit a function with more than two parameters and it ought to be an increasing function that doesn't necessarily pass through the origin. This implies linear and exponential functions. With more data, you might use more complex models, such as logistic functions, but with only two points of data, this would be a meaningless exercise.

A theoretical maximum is to fit four parameters to two data points, for example, a cubic function, is completely determined by its two points of inflection.

Start by investigating a linear model. In Figure 2.28 the known facts about the bat population has been visualized by feeding points into the input field one by one in typing

```
(-2,360)
```

FIGURE 2.27 Grey-headed flying fox. Photo: Mike Lehmann (own work), licensed under CC BY-SA 3.0 via Commons. https://commons.wikimedia.org/wiki/File:GreyHeadedFlyingFoxWingspan.jpg#/media/File:GreyHeadedFlyingFoxWingspan.jpg.

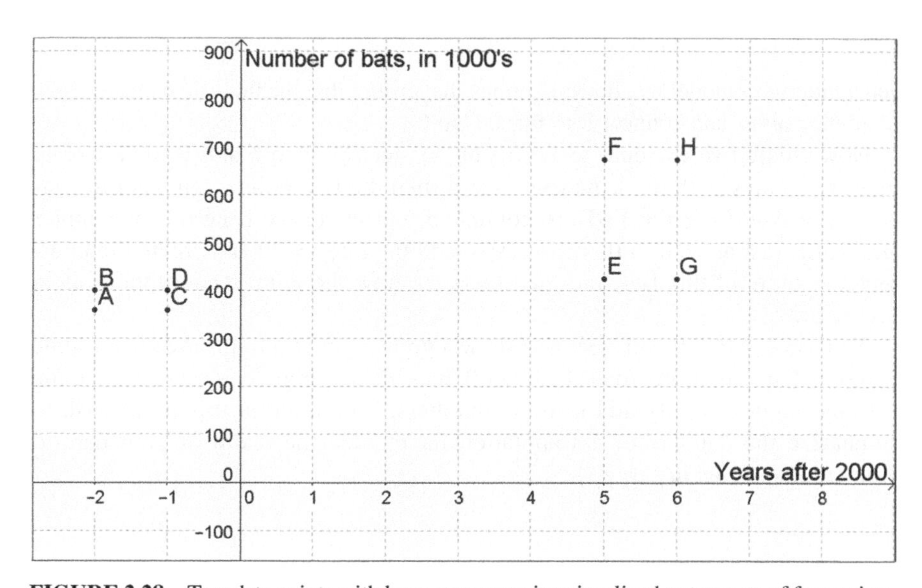

FIGURE 2.28 Two data points with large error margins visualized as two sets of four points.

and so on. Instead of creating points with error bars, four points are given for each data point, signifying the possible extreme values. As you can see, the years are interpreted as intervals from the beginning to the end of the year, and the x-axis shows the number of years since the year 2000.

If you work with the maximum values, you can create one model, and if instead you use the minimum values, you can create another model. So create a maximum

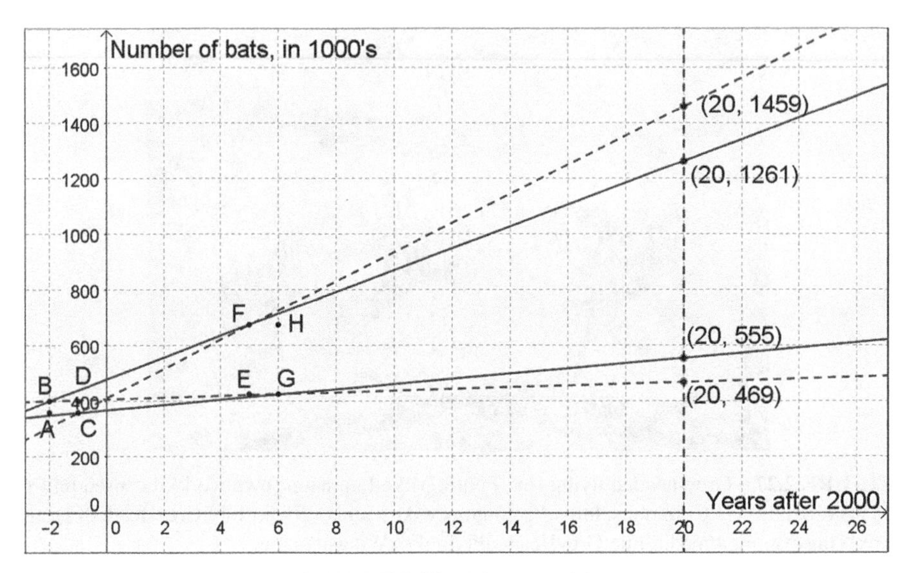

FIGURE 2.29 Linear models.

and a minimum model by choosing points that ensure that the lines lie on the outside of all the points, and connect B to F and C to G.

Now create two extreme scenarios by connecting C to F and B to G. Create these four lines with the Line tool and then the two points you wish to connect. You *could* use the **FitLine**-command, but using the Line tool is simpler and works just as well in this case. Also add the line $x = 20$ and create the intersection points of this line and the others to show the point's coordinates as in Figure 2.29.

You can see that the projections show a great span of possibilities from some 550,000 bats to more than 1,200,000 bats. The range is so great that the extreme models hardly add to these numbers. You might at this point wish to emphasize the dangers of extrapolation far outside the realm of your data to your students—again.

Proposed Solution 2

If instead you try to create exponential functions in the same way as in Solution 1, except that you use the exponential fitting command **FitExp[B, F], and so on,** you get some truly fascinating results, shown in Figure 2.30. The cautious prognosis lands on just less than 600,000 bats, whereas the top range goes up to a bit more than 2 million bats. Three, if you allow for the extreme models.

Notice that every model tried so far has been increasing. This is difficult to avoid considering the meager number of data points, but it is telling that these simple models do not even consider a decrease as a reasonable alternative.

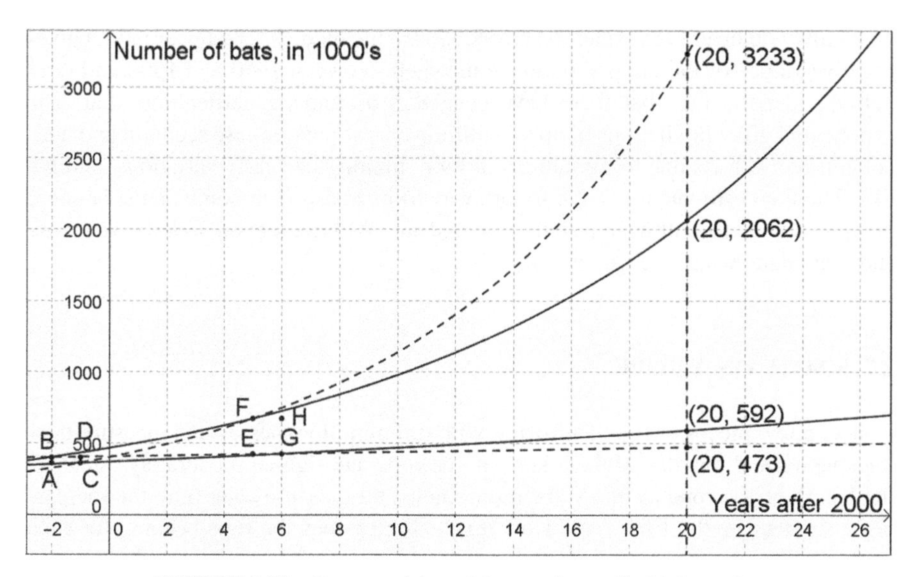

FIGURE 2.30 Exponential models can give really high results.

Proposed Solution 3

Since you only have two points and are going to fit models with two parameters, you can do so algebraically. To find the line $y = a \cdot x + b$ through the points (−2, 400) and (5, 674), first determine the slope, $a = \Delta y / \Delta x \approx 39.14$ (thousands of bats per year). Then find the y-intercept, b, by using the values for a, x and y (from a data point) in the equation for the line, $y = a \cdot x + b$. You get the equation $400 = 39.14 \cdot (-2) + b$, giving $b = 478.3$ (bats year 2000). So the equation is $y = 39.14x + 478.3$.

To find the exponential function $y = C \cdot a^x$ passing through two given points, you set up a (nonlinear) system of equations where each equation represents a data point. You solve it by dividing the equations with each other in order to eliminate C:

$$\begin{cases} 400 = C \cdot a^{-2} \\ 674 = C \cdot a^{5} \end{cases} \Rightarrow \frac{400}{674} = a^{-7}, \quad \text{giving} \quad a \approx 1.07739$$

After this, the first equation gives $400 = C \cdot 0.8615$; that is, $C \approx 464.3$. The equation in this case becomes $y = 464.3 \cdot 1.07739^x$.

You can split all the errors in this process in three different categories. First, you have the errors *within* each model. These can roughly be estimated through the upper and lower estimates *within* each model: 555,000 to 1 260,000 bats using the linear model and 592,000 to 2,000,000 bats using the exponential model.

Then you have the errors *between* the models. By comparing the models, you can see that the errors arguably must be larger than the errors that only the linear model suggest. Also note that the predictions overlapping the interval between 600,000 and 1,200,000 bats are slightly more credible.

Finally you need to consider the errors *outside* the models. The linear model gives negative values of the bat population somewhere between 1960 and 1990 and does not consider the fact that there have been bats around far earlier than that. The exponential gives predictions of up to 3 million bats as a worst case scenario and does not consider effects due to limitations in food, hunting or yearly variations, and the like. The final estimate for 2020, so far away from so few data points, must be more of a guess than a result from a statistical analysis. Without a good underlying theory and more data, what else can be said?

2.7 KNOTS ON A ROPE

Collect a few pieces of rope and string with different thicknesses and measure their thicknesses and lengths. Make a knot in one rope and tighten moderately. Measure the length of the rope again. Make more knots and measure each time the increasingly shorter length of the rope. How much shorter does the rope become for each knot? Collect data values and sketch a diagram in GeoGebra. Fit a straight line to your data. What is the meaning of the slope?

Redo this for ropes of different thickness. In what way is the thickness of the rope affecting the increasingly shortening of the rope for every knot? Make one diagram where you graph the shortening of the rope against the thickness of the rope. How do you think that curve will look (before you measure)? Can you fit a graph to your data points?

How much shorter will one rope with the diameter 8 mm be if you tie 8 knots in it?

2.8 THE CANDLE

Measure the length of a burning candle and plot the length versus time. Do this for candles of different thicknesses and makes. Make assumptions and try to predict how long a time it will take for an 18 cm long candle with a diameter of 2.4 cm to burn down.

2.9 HOOKE'S LAW

Borrow a spiral spring, a long ruler, and a dynamometer that can measure force. The more you pull the spring with the dynamometer, the longer the spring becomes. Measure the force and length for some different lengths of the spring. Plot your measurements in a diagram and fit a straight line to them. Determine from the straight line what the *spring constant* is for your spring, that is, what force it would take to get the spring one meter longer (if it could take it).

Redo the experiment with a rubber band instead. What happens now?

TABLE 2.4 Points versus Rank

Point	Rank
30	2
29	1
29	3
29	6
28	4
26	8
25	5
23	7
20	10
18	9
14	11
8	12

2.10 RANKING

Two teachers studied 12 students during seminars and held group discussions regarding these students' diligence, ambition, and knowledge. Then the two teachers ranked the students according to their expectations for their upcoming examination. After the examination, the rankings were compared. Determine the two teacher's abilities to predict a student's ability by measuring the correlation between the two data sets given in Table 2.4. It was possible to get a maximum of 30 points for the entire examination. The two teachers gave the rank number 1 to the student they predicted to be the best and the rank number 12 to the student they predicted to be the worst.

2.11 DOLBEAR'S LAW

Dolbear's law identifies the relationship between air temperature and the rate at which crickets chirp. This relation was initially formulated by Amos Dolbear and first published in 1897 in an article called "The Cricket as a Thermometer" in *The American Naturalist* (31: 970–971). The formula is expressed as

$$T_F = 50 + \frac{N_{60} - 40}{4}$$

where the formula estimates the temperature T_F in degrees Fahrenheit from the number of chirps per minute N_{60}. For more representations of this formula, see the Wikipedia article at https://en.wikipedia.org/wiki/Dolbear%27s_law.

Table 2.5 shows some measured data. Do they confirm Dolbear's law?

TABLE 2.5 Cricket Chirps at Different Temperatures

Temperature (°F)	Chirps per 13 s
69	28
74	34
60	19
77	39
80	45
66	23
71	30
57	18
63	22

TABLE 2.6 Heights and Weights for Male Office Workers

Name	Height (cm)	Weight (kg)
Anders	187	90
Lars	183	85
George	190	85
Bengt	189	85
Jonas	190	95
Steve	191	93
Lennart	176	74
Thomas	182	81
Bertie	181	83
Ingmar	178	80

2.12 MAN AT OFFICE

In Table 2.6 these are listed the heights and weights of 10 men from the same office. Determine a linear relation between the weight and the length.

Make an interpretation of what the slope means in this context.

If you were to change the axes, so that weight and height exchange places, what is the value of the slope now, and what are the units, and interpretation?

2.13 A STACK OF PAPER

Take a stack of copying paper and measure the height and weight of the stack as accurately as you can for 6 to 10 different heights of the stack. Plot these data and fit a line from the origin to them. What is the value and unit of the slope? How can you check this? Can you calculate the density of the paper?

2.14 MILK PRODUCTION IN COWS

At https://datamarket.com/data/list/?q=provider:tsdl you will find a list of several hundred different data sets concerning all sorts of things. One of these is data sets concerns the weight of the milk produced per cow from 1962 to 1975. If you average the data to show a running average of 12 months to get rid of the yearly variations you get the diagram at https://datamarket.com/data/set/22ox/monthly-milk-production-pounds-per-cow-jan-62-dec-75#!ds=22ox&display=line&f=rolling:12. These data can be exported in a variety of formats. Do that, and then copy the data into GeoGebra, produce data points and fit a line to it. What are the units of the slope and what does it mean?

As an extension for nonlinear modeling; fit the original non-averaged data to a suitable function. By browsing this and other similar websites for data sets, you will find many examples of both linear and nonlinear data.

3

NONLINEAR EMPIRICAL MODELS I

We now cast away the limitations of linear models and consider polynomial, exponential, trigonometric, logarithmic, and other functions used to model data. While the variety to choose from now is large, a word of caution is needed. With so much to choose from, it is easy to find some function that will fit any data, but without a theoretical model underlying your empirical findings, you still need to be very careful when it comes to extrapolation. The fact that a run of random choices by a cat (e.g., https://en.wikipedia.org/wiki/Mr._Nuts) seemingly can predict the outcome of previous years' elections does not mean that it will be able to predict future elections any better than a random choice.

A large branch of modeling concerns itself with population dynamics. Using exponential and logistic models, you can start to get realistic results from model predictions provided that you are careful as to what data you use, as is pointed out in Problem 3.7 *Modeling the Population of Ireland*.

Students who have embraced the study of nonlinear functions can often tackle more complicated and comprehensive problems, and given enough time, they can stay the task longer. The written reports they produce may open the way toward deeper discussion. Typically, you want to give some sort of general instructions to students working on these tasks. One example may be:

> *Write a short report that describes your hypothesis, your analysis, and your conclusions in a satisfactory way. Be distinct and clear in your written language, number figures, and tables, and separate your hypothesis from your conclusions. Use correct*

Mathematical Modeling: Applications with GeoGebra™, First Edition. Jonas Hall and Thomas Lingefjärd.

Companion website: www.wiley.com/go/Hall/MathematicalModeling

nomenclature and typesetting for the mathematical portions of the text. Also make some sort of appropriate estimate of the size of the errors in your results.

By giving the students time to go into deeper modeling problems, you will train them in important skills, such as perseverance, writing, and typesetting skills, advanced use of technology, working with real world problems, and communication. Many of these lie at the center of the lifelong skills students need to master in today's world.

3.1 GALAXY ROTATION

Table 3.1 contains data describing points on a theoretical rotation curve for a spiral galaxy, such as M88 in Figure 3.1. You have been given the distances, in kiloparsecs, from the rotating galaxy's center and the circular velocity, in kilometers per second, relative to the galactic center.

It is a curious fact that if all stars were to rotate freely, only under the influence of their own gravitation force, we might not see any spiral arms at all in the galaxies. The fact that they exist indicates that there is more mass in the galaxies than what we are able to see, and formed of so-called dark matter.

Fit a function to the data in the table and calculate at what distance from the center of the galaxy the circular velocity is the least.

Proposed Solution

Open the spreadsheet with the keyboard shortcut **Ctrl-Shift-S** and enter the values there. Then select these cells, right-click, and select **Create... >List of points** from the pop-up menu. After zooming to find the points, you will see that they seem to follow some quite complicated relationship. Fortunately, you are only tasked with finding the coordinates of a minimum. You can do this by fitting a polynomial to the

TABLE 3.1 Galaxy Rotation Values

Distance (kiloparsec)	Circular velocity (km/s)
1	244.0
2	221.0
3	208.0
4	208.0
5	211.5
6	216.0
7	219.0
8	221.0
9	221.5
10	220.0

FIGURE 3.1 Messier 88 Galaxy. Photo: Jschulman555 (own work) [CC BY 3.0 (http://creativecommons.org/licenses/by/3.0)], via Wikimedia Commons.

points closest to the minimum, here colored red. Often a quadratic, second-degree polynomial is sufficient but here a cubic, a third-degree polynomial is better suited to the task, since the points lie asymmetrically round the minimum.

Create a new list by typing

```
list2 ={B,C,D,E,F}
```

and fit a cubic function to these points by typing

```
FitPoly[list2,3]
```

Then use one of the commands

Extremum[f]	(US English)
TurningPoint[f]	(UK and AU English)

to find local minima and maxima. The Style Bar has been used to show these points as circles, and the coordinates for the minimum are displayed in Figure 3.2. The speed is at a minimum at around 3.6 kpc (kiloparsecs) from the center of the galaxy. It is worth noting that this is a rather shallow minimum and that the error margins in the x-direction are quite large.

The students need to develop strategies for selecting appropriate functions for different data sets. Here it was possible to simplify the problem, allowing the selection of a polynomial to fit a limited range of the data. Read more about galaxy rotation and dark matter at http://en.wikipedia.org/wiki/Galaxy_rotation_curve.

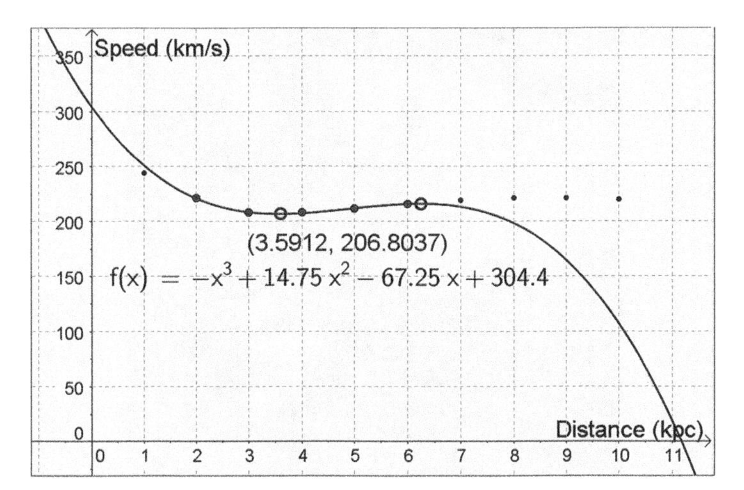

FIGURE 3.2 A minimum is easiest found by fitting a polynomial to the points closest to the minimum.

3.2 OLYMPIC POLE VAULTING

Figure 3.3 shows Théo Mancheron in the pole vaults completions of the 2013 Athletics Games in France. Pole vaulting is a version of high jumping where the jumper uses a long pole to gain height. This is a sport that has been known since ancient times and has been included in the Olympic Games since 1896 for men and since 2000 for women. Looking at Table 3.2, you can see that Olympic records change over time. Make a model based on these values and calculate the winning height for the pole vault in the Olympic Games in 2020. Construct different models and argue why some models might be more reasonable than others.

Steve Hooker from Australia won the gold medal in the pole vault at Olympic Games in Beijing OS 2008 with 5.90 meters, but when the gold medal was won, Steve Hooker asked for the rib at the Olympic record height of 5.96—and cleared it in his third jump. If you add this height to the list, do you get a more realistic model? Try!

Compare your models with the result for summer Olympics 2012, easy enough to find on the Internet. What model gives the best results? How will your models be affected if you add the results from 2008 and 2012 in your calculations? Are there any clear conclusions to be made from this?

Proposed Solution 1

Enter the table values into the spreadsheet of GeoGebra which you find through from the menu **View>Spreadsheet**. You might need to adjust your coordinate system by dragging the axes to see the points. In Figure 3.4 the y-axis is set to intersect the x-axis at $x = 1900$.

FIGURE 3.3 Théo Mancheron in the pole vaults completions in the Athletics Games France 2013. Photo: © Marie-Lan Nguyen/Wikimedia Commons, via Wikimedia Commons. https://upload.wikimedia.org/wikipedia/commons/8/83/Men_decathlon_PV_French_Athletics_Championships_2013_t141910c.jpg.

TABLE 3.2 Olympic Records in Pole Vaulting for Men, 1900 to 1996

Year	Height (m)
1900	3.30
1912	3.95
1924	3.95
1936	4.35
1948	4.30
1960	4.70
1972	5.64
1984	5.75
1996	5.92

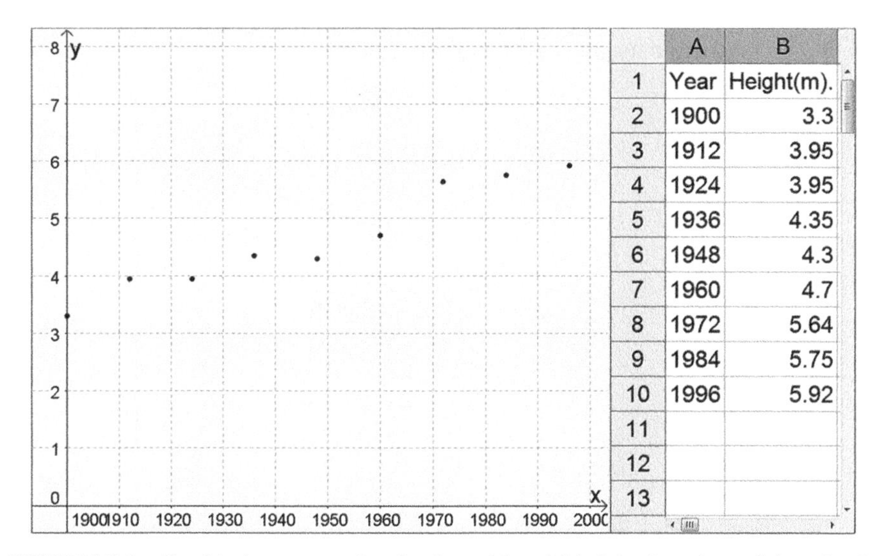

	A	B
1	Year	Height(m).
2	1900	3.3
3	1912	3.95
4	1924	3.95
5	1936	4.35
6	1948	4.3
7	1960	4.7
8	1972	5.64
9	1984	5.75
10	1996	5.92
11		
12		
13		

FIGURE 3.4 Graphical representation for the gold medal heights in pole vault jumping in the Olympic Games.

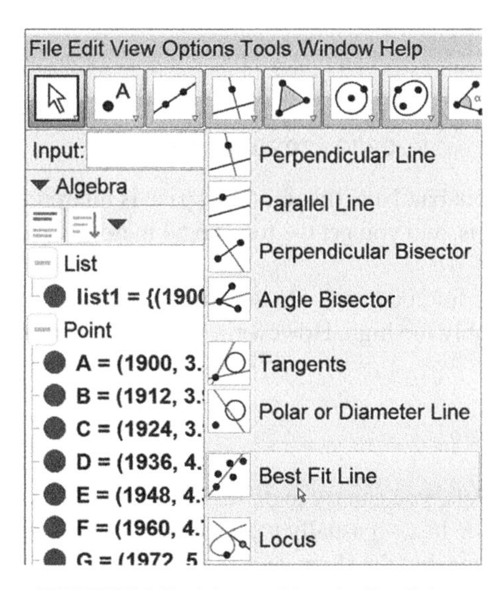

FIGURE 3.5 Line tool box in GeoGebra.

Now ask GeoGebra for a Best Fit Line in the Line toolbox, shown in Figure 3.5. Click the tool and then select the points by right-dragging to get a selection rectangle. This is an alternative way to create a line of best fit to using the **FitLine[]** command. The result is shown in Figure 3.6.

The function $f(x) = 0.0272x - 48.4048$ seems to fit the set of points rather well. Nevertheless, it is not known if the model will predict good results for future Olympic Games.

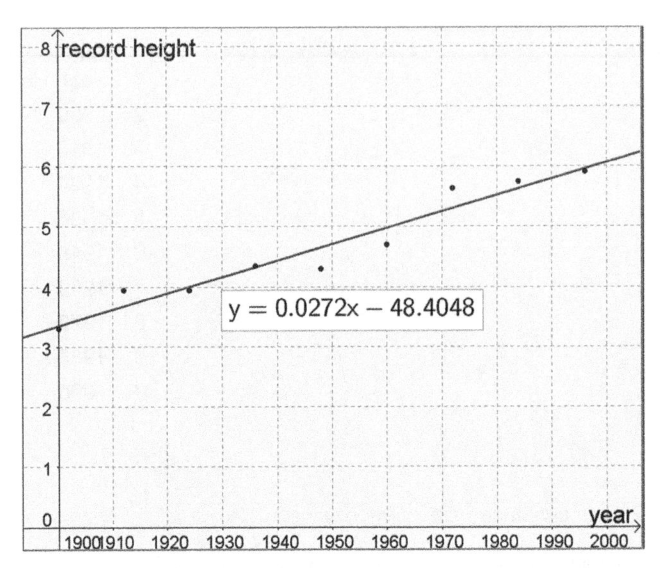

FIGURE 3.6 Simple linear model describing how the height of pole vaults has increased in 1912 to 1996.

Test the mathematical model *Height*=0.0272 *Year* – 48.4048 and see what height it will give for the Summer Olympics 2020. You do this by typing

```
0.0272 2020 - 48.4048
```

into the GeoGebra Input bar. Note that the first space is interpreted as multiplication, just as the asterisk (*) is, and you get the result 6.54 meters, which seems just a touch unrealistic.

If you add the value for 2008 and redo the process, you will get 6.48 meters, which is better but still probably too high. However, you will not get a better result with this simple linear model.

Proposed Solution 2

Going past linear models, you can try to do this with a nonlinear model. Logarithmic models of the type $y = \mathrm{a} \cdot \ln x + \mathrm{b}$ usually works well with athletic records. They grow strongly in the beginning, but the slope decreases quickly. A logarithmic model never stops growing entirely.

To do this in GeoGebra, you first must convert the coordinates into a list. The simplest way is to select the values in both columns (but not the headings) in the spreadsheet and then right-click and select **Create...>List with points**. The list will be named **list1** or **list2**.

There is nevertheless a problem with the regression model $y = \mathrm{a} \cdot \ln x + \mathrm{b}$. There is a pre-defined command for this type of regression: **FitLog[list1]**. But if you type this into the Input bar, you will get a function, shown in Figure 3.7, that looks just like a straight line.

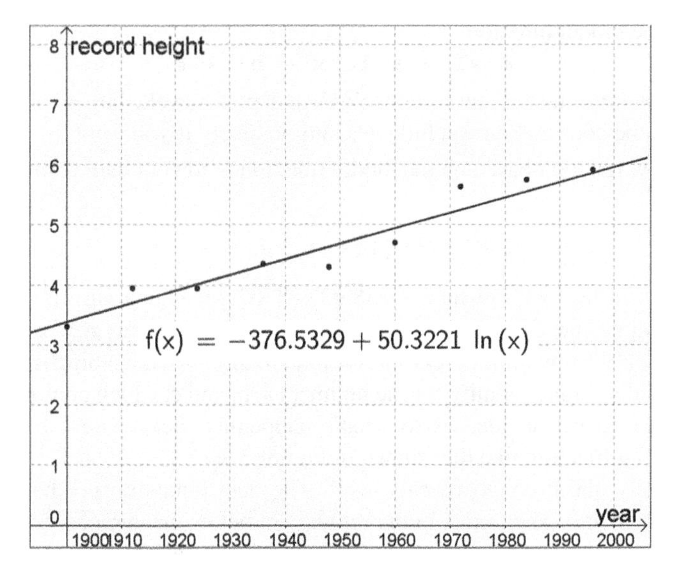

FIGURE 3.7 A first attempt to fit a logarithmic function to our data.

The difference between this model and a linear model is almost nothing. This model gives the prognosis 6.46 meters for year 2020, a result you can get by typing

```
f(2020)
```

into the input field.

Why do you get an almost straight line? The whole idea behind the logarithmic function is to get a visual representation that curves strongly. But wait … it curves most close to the origin … aha—the x-values are too large. By using the correct values for years, you have forced the points far away from the origin, out to the regions where logarithmic functions are almost linear.

Maybe it is possible to subtract 1900 from all year values. That should give a more curved function, surely. But why subtract exactly 1900? The choice of this value seems arbitrary.

Maybe it would be best if GeoGebra could optimize this value too. You can do this if you use the model $y = \mathrm{a} \cdot \ln(x - \mathrm{b}) + \mathrm{c}$. By exchanging x for $(x - \mathrm{b})$, you will shift the curve b steps to the right. There is no built-in model to do this, so you have to do it in another way.

Start by setting three numbers for a, b, and c. After that you define the model function. Then you can do the regression. This is the procedure used to do a *general fit,* and you will use it frequently throughout this book:

- Create the numbers—enter the following values as guesses.

`a = 2`	If a is in use already, please use another free letter
`b = 1890`	must be less than 1900, to avoid log(0)
`c = -3`	

- Enter the model function

```
m(x) = a ln(x - b) + c
```

  It is not important if curve doesn't fit the points well, but its shape should roughly be correct. You can hide $m(x)$ immediately if you want to.
- Fit a new function based on the model function m to your data points in list1 by typing

```
Fit[list1, m]
```

The function $h(x)=41.59 \ln(x - 338.92) - 302.44$, which surprisingly also is almost a straight line. The record height for year 2020 with this new model is also 6.46 meters, only a few millimeters shorter than before. What is going on here?

Use the sliders' functionality for the numbers a, b, and c, which control the model function $m(x)$, to get an idea as to what's happening. Set a=1.84, b=1870, and c=−3.2 to get a function like that shown in Figure 3.8.

Now, finally, the problem reveals itself. The data points do not have the right shape for this model. The points in the middle are below the curve, while the points in the beginning and in the end are above the curve. The effect is that the points are forcing the function to an almost straight line.

Looking more closely on the data, you may notice something that perhaps should have been seen before: between 1960 and 1972 something happened in the sport that made the results much higher than previous results would have made us believe. During that period pole vaulting went from using stiff poles to more flexible fiberglass poles. In 1963 alone, the world record was raised by an amazing 25 cm. In the class it is useful to allow the students to find this out for themselves, using subtle prompts.

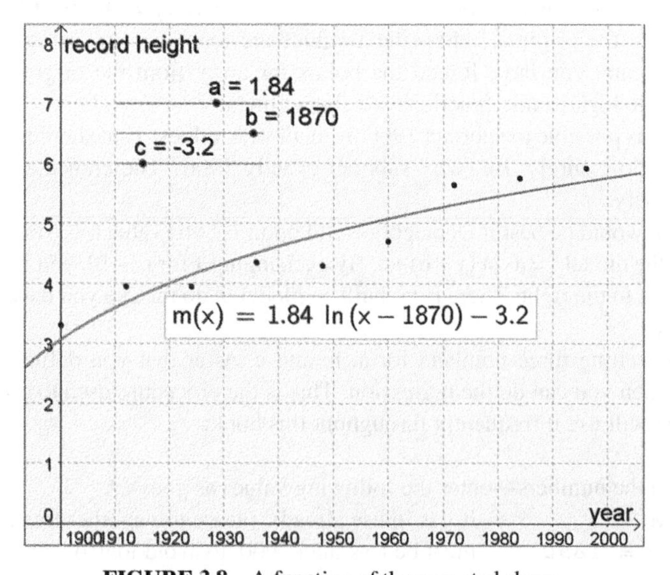

FIGURE 3.8 A function of the expected shape.

FIGURE 3.9 Pole vaulting with stiff poles a long time ago. Photo: USA Library of Congress det.4a15081.

The data wouldn't let us analyze it in a simple way, since the data essentially come from two different sport activities, pole vaulting with a stiff pole and pole vaulting with an elastic pole. This tells us that it is extremely important to really understand the characteristics of the data when doing mathematical modeling. It is also very important to allow students take part of this discussion.

In order to create a reasonable mathematical model, you can only use data from 1972 onward. You can do this by creating a new list with points—you only select the interesting values from 1972 to 2008—and thereafter type the regression command again. The results, in using the logarithmic regression, now come to $h(x)=0.4893 \ln(x-1936.24)+3.884$, and hence 6.05 meters in 2020.

RenaudLawellenie managed to clear 5.97 m in 2012. If you add that value to the list, then the special prognosis will go down to 6.03 m for the 2020 Olympic Games. Maybe we just have to wait and see. When making a prognosis for a particular year, our models do get better with time! In Figure 3.9 you can see a glimpse of pole vaulting the way it used to be.

3.3 KEPLER'S THIRD LAW

In Table 3.3 are listed the planets' average distances to the Sun and the time it takes for each planet to orbit around the Sun, which is equal to the length of a year at that planet. Find a relation between each planet's average distance to the Sun and the time

TABLE 3.3 Planetary Average Distances from the Sun and Orbital Times in Our Solar System

Planet	Average distance (AU)	Orbital time (year)
Mercury	0.387	0.241
Venus	0.723	0.615
Earth	1.000	1.000
Mars	1.523	1.881
Jupiter	5.203	11.861
Saturn	9.541	29.457
Uranus	19.190	84.008
Neptune	30.086	164.784
Pluto in the Kuiper Belt	39.507	248.350

it takes for a year. You may also wish to look at Titus–Bode's law, which connect the planet's "number" to its average distance to the sun.

Here the average distances are measured in *astronomical units* (AU), where 1 AU = 149,600,000 km is equal to the Earth's average distance to the Sun. For example, 5 AU means that a planet is placed five times as far out from the Sun as the Earth is.

Try to fit a straight line to the data points, a straight line that describes the relation between the average distance to the Sun and the time for a rotation around the Sun. Since the points do not lie on a straight line, you are to find the best fit for the line $y = ax + b$.

You might even be able to find another function that fits your data points better.

Proposed Solution Using a Linear Function

In Figure 3.10 the data table has been entered into GeoGebra's Spreadsheet, which you can view either from the menu alternative **View > Spreadsheet** or with the keyboard shortcut **Ctrl-Shift-S**. If you experience problem with the formulas, it may be due to invisible spaces that are sometimes introduced around values and make GeoGebra interpret the numbers as text instead of numbers. Copy and enter the text into a word processor and be sure to show all signs (in Microsoft Word you use the ¶ button). Remove all spaces or tabs if there are any.

Then enter

```
=(B2,C2)
```

into cell D1. The parenthesis and the comma sign show that you are creating a point. This definition is possible to copy down in the usual way by dragging the fill handle. You then select the points and proceed as follows:

- Create a list of these points, by right-clicking on them and selecting **Create... List**. The list will be called **list1**, but you can rename it if you like.

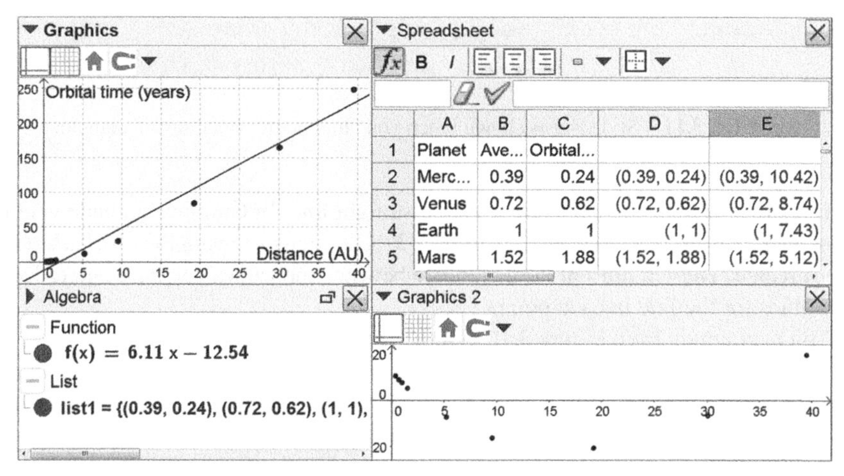

FIGURE 3.10 Planet's orbital time against their average distance to the Sun represented in GeoGebra.

- Change the color and the size of the points by right-clicking the points and select Properties. You may also use the Style Bar. Furthermore you may format the axes by right clicking in an empty area in the Graphics window and selecting **Graphics...**

To find the best straight line to the data points, you can use either of the following methods:

Manual method: Enter

```
y = a x + b
```

and accept to create sliders for a and b. Vary the sliders until you are satisfied. Observe that the space between a and x works as a multiplication sign.

Automatically version 1: Enter

```
FitLine[list1]
```

A straight line is drawn, and in the Algebra window the equation is shown. The name of the line is **a**. You can show the equation in another form by right-clicking it and selecting another alternative under **Equation**.

Automatically version 2: Enter

```
FitPoly[list1, 1]
```

This command will give you a first-degree polynomial, which mathematically is the same thing as a straight line. It might require more explanations for the students but may also allow more possibilities to work with the function f. You will get the same result but the equation for the line will be given as $f(x) = \ldots$

FIGURE 3.11 Style Bar is shown when you click on the small upper triangle.

It is obvious that the points are not on a straight line. In Graphics 2, made visible from the menu alternative **View > Graphics 2**, new points showing the *residuals* have been created. These represent the difference between the values you have entered and the values for the line in each point.

You may notice from Figure 3.10 that the program windows are rearranged. Each window may be docked to the main window or released to become free-floating if you click on the small symbols up to the right in the windows name list. All docked windows might be rearranged if you grab and drag the title bar.

Click in Graphics 2 and make sure the axes and grid are shown. You can use the buttons in the Style Bar, shown in Figure 3.11, for this.

With Graphics 2 selected, all new objects will be created there. In cell E2, type

```
=(B2,C2-f(B2))
```

and copy it down. This will create a point with the same x-coordinate as before, but with the difference between the data point's y-values and the function values as y-coordinate. After you have zoomed the axes by dragging them, one at a time, you will get a nice looking residual diagram. The deviations look systematic, which normally indicates that there is some basic fault with the theory, rather than a random measurement error.

Proposed Solution Using a Power Function

It seems as if a bent curve through the origin would fit the data points better. The students could, of course, try a quadratic function, but they will then discover that while it will indeed be better, the function will not fit the shape of the data well. In cases like this, when a function goes through the origin, it might be better to try a power function such as $y = C \cdot x^a$. In GeoGebra the only difference from the previous description is that the regression command is **FitPow[list1]**.

Drag the function expression from the algebra window and drop it in the Graphics window to get a text label with the function expression neatly next to the graph. The resulting function, shown in Figure 3.12, indicates that the relation might be written

$$\textit{Orbital Time} = \textit{Distance}^{1.5}$$

or, if you want to avoid exponents that are not integers:

$$\textit{Orbital Time}^2 = \textit{Distance}^3$$

This beautiful relationship is indeed what Kepler showed in his third law in 1619.

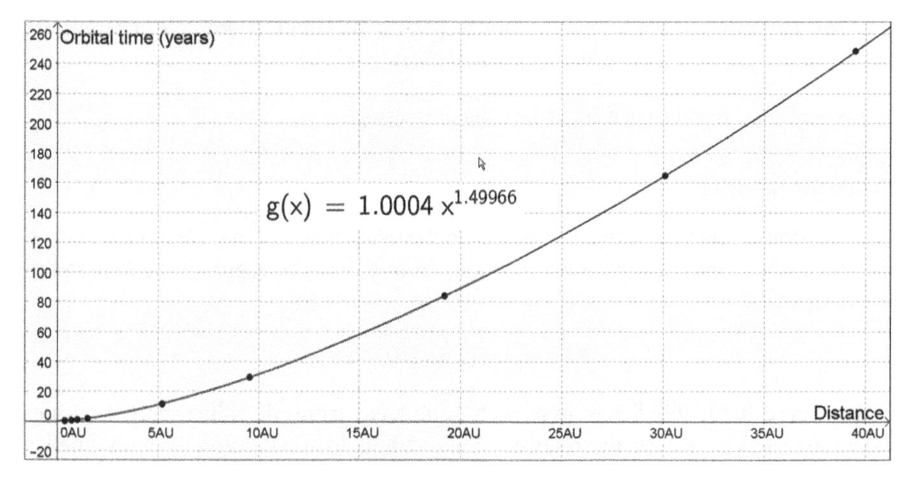

FIGURE 3.12 A power function fits the data well.

TABLE 3.4 Density of Water at Different Temperatures

Temperature (°C)	Volume (cm³)
0.00	999.8700
0.50	999.8400
1.00	999.8142
1.50	999.7925
2.00	999.7750
2.50	999.7614
3.00	999.7519
3.50	999.7464
3.80	999.7449
3.90	999.7447
4.00	999.7447
4.10	999.7448
4.20	999.7451
4.50	999.7469
5.00	999.7528
5.50	999.7625
6.00	999.7759

3.4 DENSITY

Table 3.4 shows the volume of one kilogram of water at different temperatures, measured in degrees Celsius.

Try to find two different functions that fit these data points. Also carry out some form of suitable error analysis. Determine the best data fit, that is, which one of your functions fits the data points best, as you see it.

Also determine at what temperature water has its smallest volume, meaning its largest density. Comment on the following statement:

"Water has its maximal density at the point of freezing"

Proposed Solution

Fit a quadratic function to the data points in the usual way, by entering data values in the spreadsheet and creating a list of points from them. The regression command to use in this case is

```
FitPoly[list1, 2]
```

In Figure 3.13, 999.5 has first been subtracted from all values before fitting. Generally, it is easier for the regression algorithms to work correctly with smaller values than with very large values that might cause lack of precision. In write-ups and reports as well as in everyday work, it is important to separate the variable, here y, from the values given in the table, here the volume values V.

Depending on your background and experience, it may be difficult to imagine a better function, but in principle, a quartic—a fourth-degree polynomial function—should theoretically give a better fit than a quadratic—a second-degree polynomial function. For comparing models with each other, the GeoGebra tool *Two Variable Regression Analysis* can be especially useful.

Select your data points in the spreadsheet and click on the button to arrive at the Two Variable Regression Analysis, as shown in Figure 3.14. Click on **Analyze**. You will then arrive at something like Figure 3.15a and Figure 3.15b. By clicking on the buttons for Show The Other Diagram and Show Statistics, and selecting a polynomial regression model, you will see a lot of information displayed.

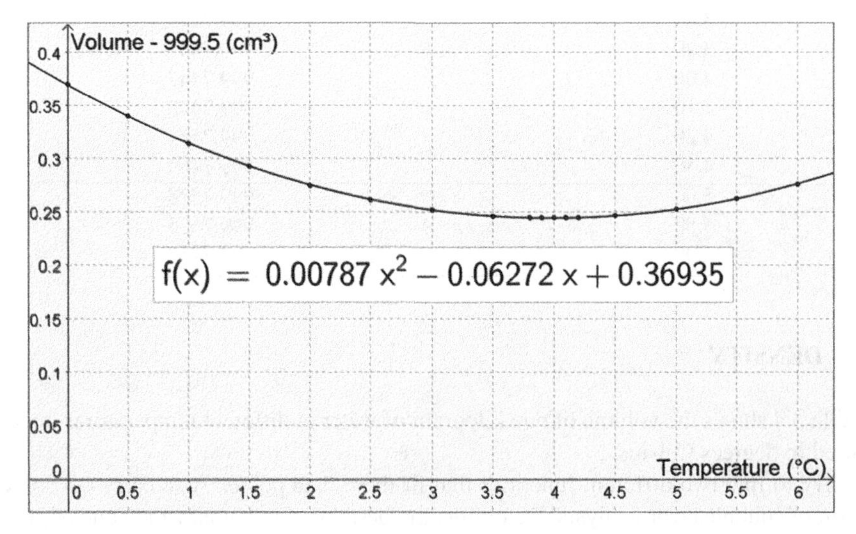

FIGURE 3.13 A quadratic function fits the data almost perfectly.

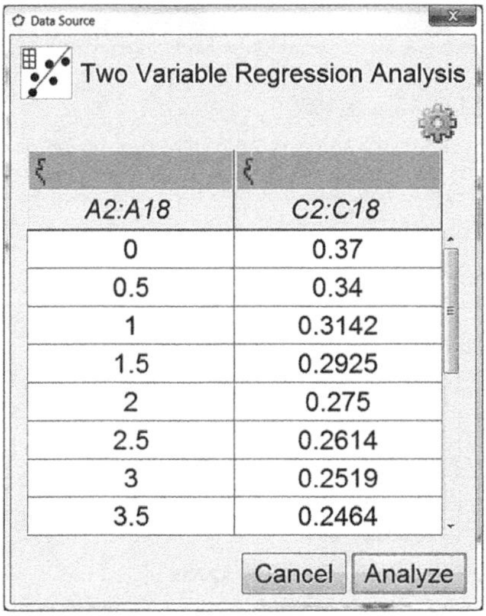

Data Source

Two Variable Regression Analysis

A2:A18	C2:C18
0	0.37
0.5	0.34
1	0.3142
1.5	0.2925
2	0.275
2.5	0.2614
3	0.2519
3.5	0.2464

Cancel Analyze

FIGURE 3.14 Two Variable Regression Analysis dialogue.

Look, for instance, at the value for SSE at the bottom of the left column with statistical data. SSE is an abbreviation for the *sum of squared errors*. This value is often considered to be the best measurement for how well a function has been fitted to a set of data points. The lesser the SSE value, the better fit you have. The fourth-degree polynomial, the quartic, is better in this sense than the second-degree polynomial, the quadratic.

Then study the residual diagrams. A *residual* is something that remains. In mathematics a residual diagram is a diagram that shows the difference between the data points and the function you have fitted. In a sense, it shows the details you weren't able to fit. It is useful to look at when the fits are so good that you really do not see any difference in the original plots. Notice that the second-degree polynomial leave a trace of systematic errors, while the fourth-degree polynomial show no systematic errors at all, just a small amount of randomized noise.

The conclusion is that even if the second-degree polynomial gives a perfectly adequate fit, the fourth-degree polynomial is obviously better. Executing the command

```
FitPoly[list1, 4]
```

in the main window gives you the quartic as $g(x)$. With the command

```
Extremum[g]          (US English)
TurningPoint[g]      (UK English)
```

the two different function's minimum points are found to be 3.983 °C for the quadratic and 3.966 °C for the quartic. While you're at it, also create a regression for a cubic function. Its minimum value is 3.967 °C. All together 3.966 °C seems to be a stable value with small errors.

(a)

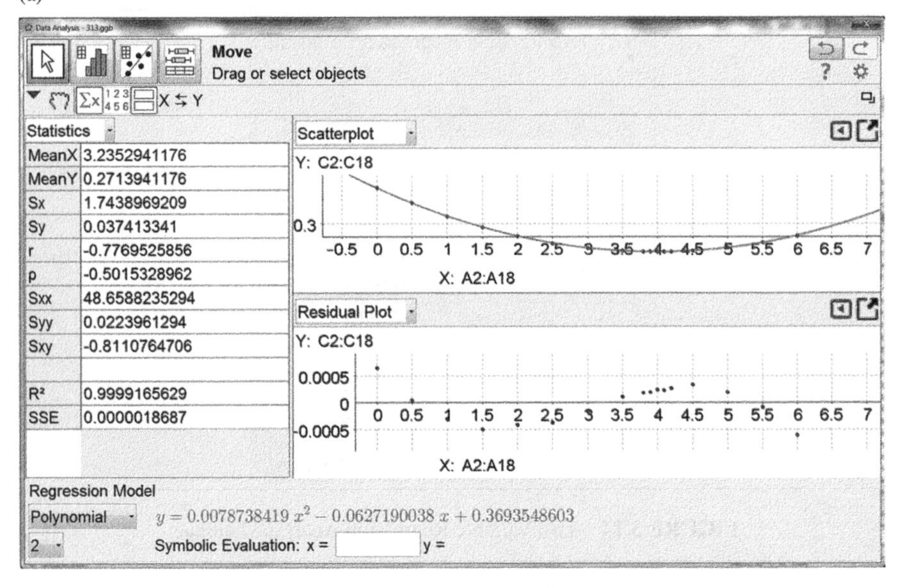

(b)

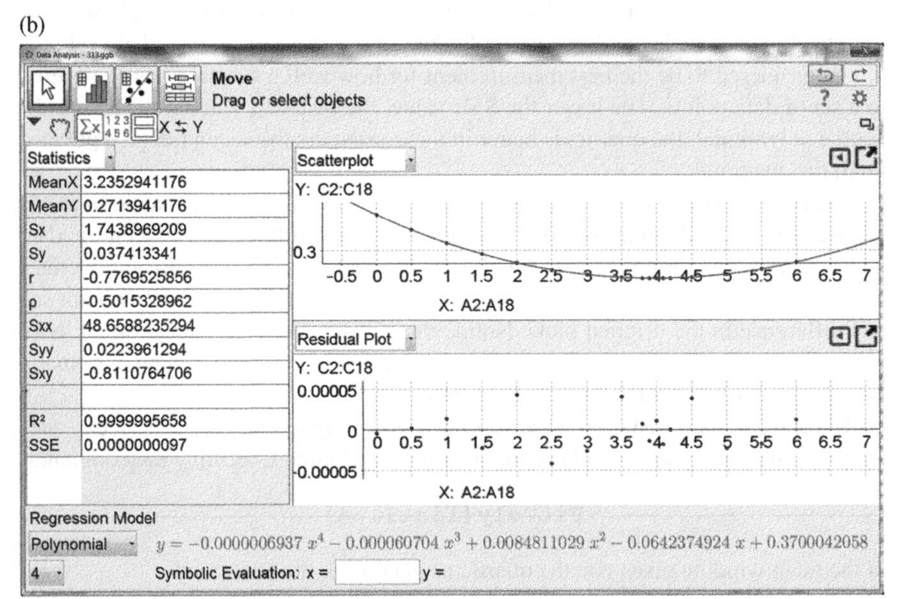

FIGURE 3.15 Two Variable Regression Analysis with (a) second- and (b) fourth-degree polynomial. Pay attention to the difference between the residual diagrams.

How sensitive is this value to measurement errors? You can test this question by changing one of the data points. If you change the last measurement value from 999.7759 to 999.7758, you will find that the minima value is changed by about 0.0007 °C. Even if you have a good model, it may seem that the model still is sensitive for errors in the measurement process. Unfortunately, the sizes of these measurement errors are unknown, but there are methods for estimating a model's sensitivity.

One such method is called *jackknife resampling,* meaning that you systematically take away one data point at the time. Every time you do this, you will get a new value for the minima point (or whatever you are currently trying to determine or measure). Jackknife resampling is a useful general method for estimating errors for parameters that you have determined from data.

Removing each data point in turn yields a list of values for the minimum from which you can determine an error estimate by calculating the standard deviation. This procedure can take some time and be tedious work, but it is quite possible to do this. The result in this case is (3.9662±0.0007)°C with 95% confidence. This error is smaller than the current precision that the temperatures are measured with, so you should simply say that the temperature you are looking for is 3.97 °C.

The claim that water has its maximal density when freezing is obviously wrong. If it were correct, all ice would sink to the bottom of the ocean and the oceans would freeze. It is easy to verify this by freezing water in a plastic bottle in your own freezer.

3.5 YEAST

Table 3.5 shows the population data for a yeast culture. The number of yeast cells were estimated every hour for 18 hours.

Use GeoGebra to illustrate the relationship between time and population. You could try the logistic function

$$P(t) = \frac{\mathrm{C}}{1 - \mathrm{a} \cdot e^{-\mathrm{b} \cdot t}}, \quad \mathrm{b} > 0$$

Use the model to determine the *saturation level* of the yeast culture, that is, the maximum number of cells the population can sustain.

Proposed Solution

For this problem you can use the built-in tool Two Variable Analysis, described earlier in the proposed solution to Problem 3.4, *Density*. Enter the data in the spreadsheet,

TABLE 3.5 Number of Yeast Cells Does Not Grow without Bounds

Time (h)	Population (1000/s)	Time (h)	Population (1000/s)
0	10	10	513
1	18	11	560
2	29	12	595
3	47	13	629
4	71	14	641
5	119	15	651
6	175	16	656
7	257	17	660
8	351	18	662
9	441		

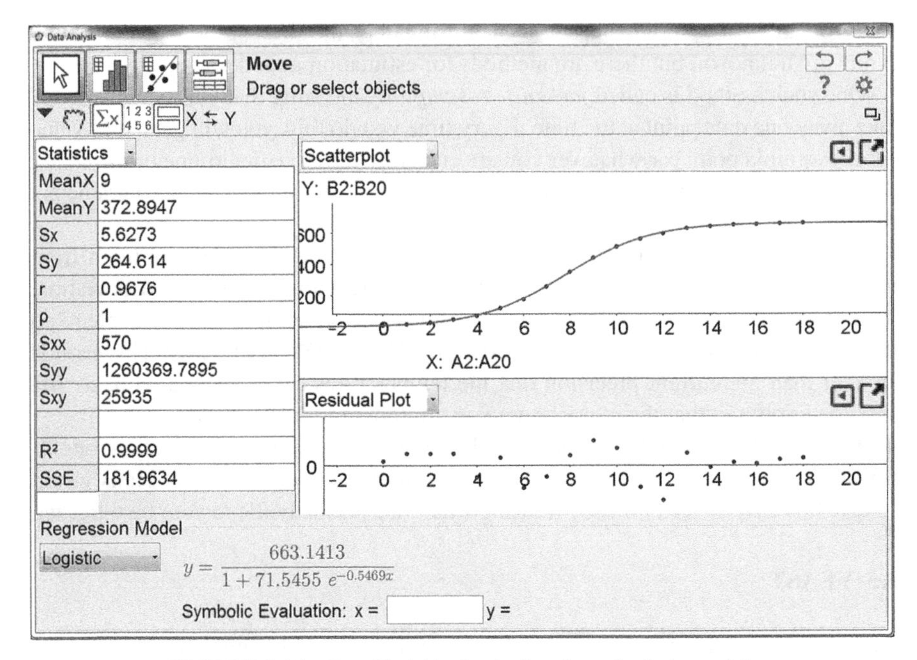

FIGURE 3.16 Two Variable Analysis using a logistic model.

then select it and click on the button and then on **Analyze**. Click Show The Other Diagram and next Show Statistics. Selecting the logistic fit suggested in the bottom dropdown menu will yield Figure 3.16. In the given model, C represents the sought maximum value, which is found to be 663.14 ≈ 663 thousand cells.

3.6 COOLING I

A small amount of water cools rapidly after being poured from a kettle into a cup as in Figure 3.17. The time t, in seconds, and the temperature T, in °C, are given in Table 3.6. The ambient temperature was, at the time of the experiment, measured to be 22.3 °C.

Find a model that fits these data well. Use this model to determine the time it will take to cool the water all the way down to 26 °C.

Proposed Solution 1

Enter the data in the GeoGebra spreadsheet, activated by **View>Spreadsheet** from the menu or with the keyboard command **Ctrl-Shift-S**. Select the data, right-click, and select **Create…>List of Points**. Then study the distribution of these points.

It should appear that the data are decreasing exponentially, but not down to zero. The values come closer and closer to the ambient temperature. The standard models that graphing calculators and GeoGebra offer include an exponential fit $f(x)=C\cdot a^x$

FIGURE 3.17 Hot water (cropped). Photo: {N}Duran /Flikr CC BY 2.0 http://www.flickr.com/photos/nduran/5385001700/.

TABLE 3.6 Cooling Temperatures

t (s)	T (°C)	t (s)	T (°C)	t (s)	T (°C)
0	69.58	360	43.96	720	35.00
30	66.11	390	42.92	750	34.53
60	61.41	420	41.95	780	34.04
90	58.07	450	41.05	810	33.59
120	55.60	480	40.18	840	33.20
150	53.58	510	39.40	870	32.76
180	51.66	540	38.70	900	32.37
210	50.05	570	38.00	930	32.00
240	48.52	600	37.32	960	31.64
270	47.24	630	36.67	990	31.30
300	46.00	660	36.08		
330	44.96	690	35.50		

that decreases down to zero. You therefore have to polish the data before you can use this model.

So now create two new columns in the spreadsheet. In cell C2, type

```
=A2
```

just to duplicate the time and make the subsequent selection of two columns easier. In cell D2, type

```
=B2-22.3
```

to calculate the temperature difference above the ambient temperature. Copy these formulas down and create new points and a new list in the same way as before.

If you initially had typed `T_0 = 22.3` into the input field and then `=B2-T_0` in cell D2, you would now be able to see how the model would change if you change the ambient temperature. You can change the value of a number either by selecting it and using the arrow keys or by clicking on the white circle to the left of the number's name in the Algebra window, thereby making the number's slider visible.

You can now fit an exponential function to the new data by entering

```
FitGrowth[list2]
```

in the input field as shown in Figure 3.18. **FitExp** would have given you the base e.

The fit may not be as good as you might have initially expected. There may be several explanations for this. An exponential decay is really only valid for a singular, point-like object in a surrounding that keeps a constant temperature, regardless of the object's temperature. In this case some of the water's heat will warm up the cup and thus act as a buffer toward the rest of the environment. Perhaps you can get a better fit if you ignore some of the data at the start of the measurement where the fit works least well?

This is not quite the same as throwing away data. What you are doing here is limiting the interval in which your model is valid. If the cup gobbles up more heat initially than later, then it is only reasonable to create a model from the point where the

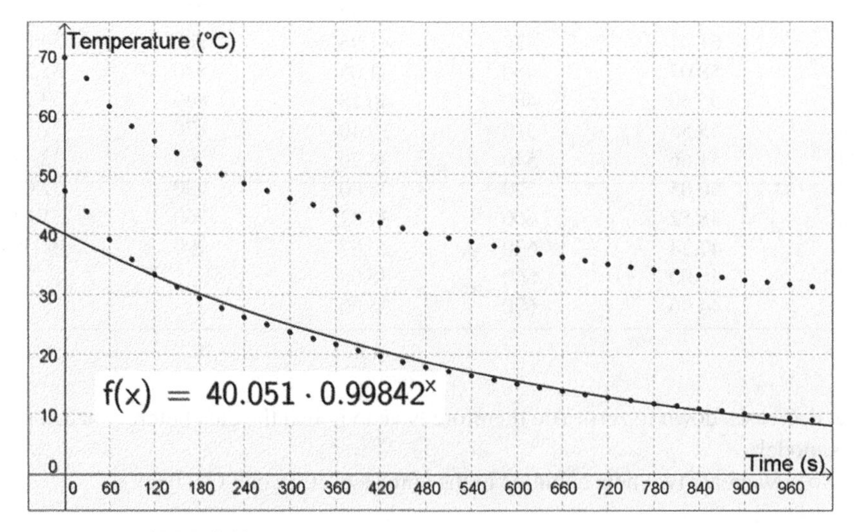

FIGURE 3.18 Exponential fit to data not decreasing toward zero.

cup is already warm. So let's (admittedly somewhat arbitrarily) decide to ignore the first six data points. To avoid creating new points unnecessarily, simply take the points you have and create a new list from them, using the List tool {1,2}.

Select list2 in the Algebra window and press **F3**. The definition of list2 will then be copied to the input field. Press **Home** to jump to the start of the line and change list2 to list3 as in Figure 3.19. Erase the first three points and press **Enter** to create the new list.

Then press the down arrow twice to fetch the fitting command in the list of previous command, change list2 to list3 once again, and press Enter: `FitGrowth[list3]`.

In Figure 3.20 you see the latest model. This one seems acceptable. To answer the question, type

```
y=26-22.3
```

in the Input bar to create a horizontal line. Use the Intersect tool and click on both the line and the graph. A point is created at the intersection, but you can't yet see it until you drag the background to pan right. Then right-click the point and in the properties, show its value as a label. The x-coordinate is 1,582 seconds or a little more than 26 minutes. Because of the shallow angle of intersection, this value has probably a hefty error.

FIGURE 3.19 First six points are selected for removal.

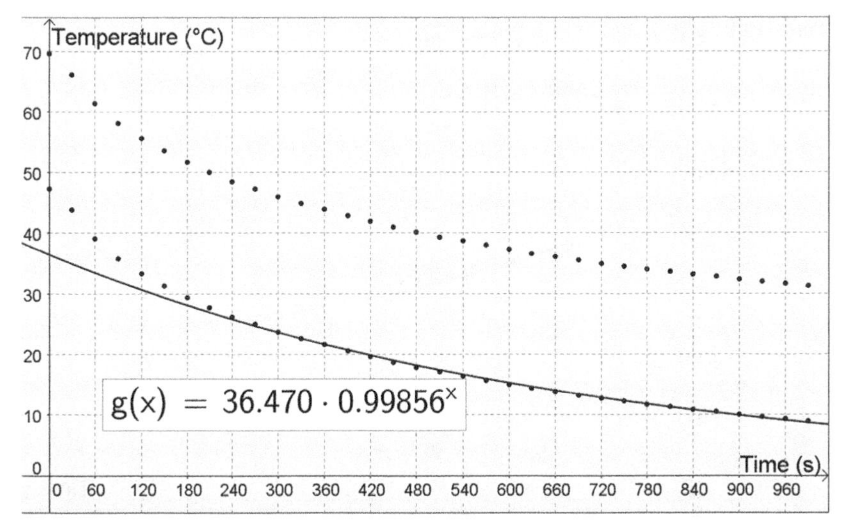

FIGURE 3.20 New graph fits the data much better because the first six data points are excluded.

Proposed Solution 2

You really don't have to fiddle with the data to make use of GeoGebra's standard exponential model. Indeed, GeoGebra can fit *any* function to a list of data, in this case it is a model $m(x)=C \cdot r^x + t_0$, where t_0 is the ambient temperature. After entering the values in the spreadsheet, you create this model function by typing

```
m(x)=c r^x + 22.3
```

in the input field. Notice the blank space between c and a. The blank space works as a multiplication sign. When asked if you want to create sliders for c and a, click **Yes**.

Using the sliders, you can adjust the model function to lie roughly along the data points. The fitting algorithm will use these current values of the sliders as reasonable starting values.

To adjust any parameters, you need to go into their properties, and on the slider tab, you need to change the maximum value to at least 50. While you're in the properties dialogue, you might as well change the step size to 0.0001 to be able to fine-tune the parameters later.

When $m(x)$ look reasonably like the data, for instance, when c=43 and a=0.998, you can hide the model function $m(x)$ by clicking on the blue button to the left of m in the Algebra window. Then create a fit for the data in list1 of the same form as $m(x)$ by entering

```
Fit[list1,m]
```

Observe in Figure 3.21 that just as in the previous solution, the fit isn't as good as what was hoped for. Therefore create a new list without the six first points and make a new fit for a much better result. Enter the horizontal line

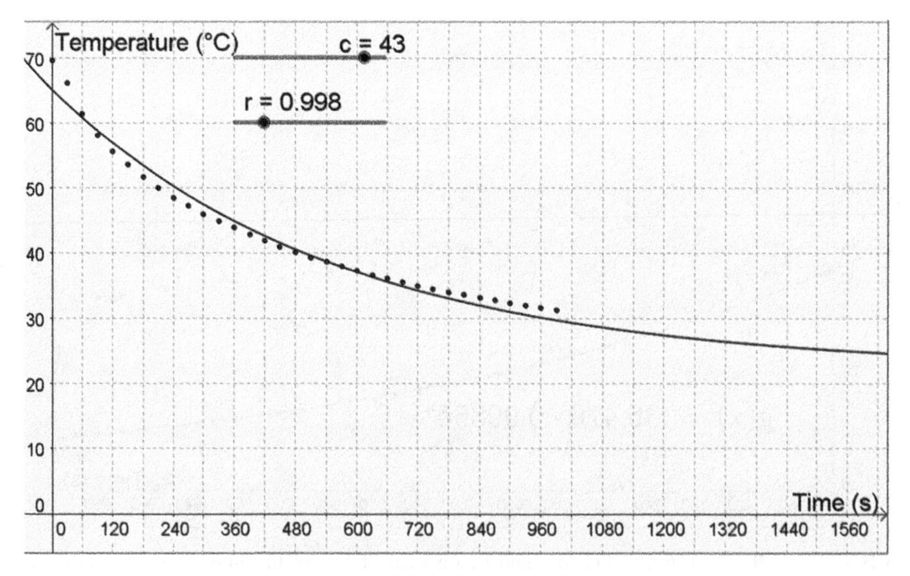

FIGURE 3.21 General fit to the data, using the model $m(x)$.

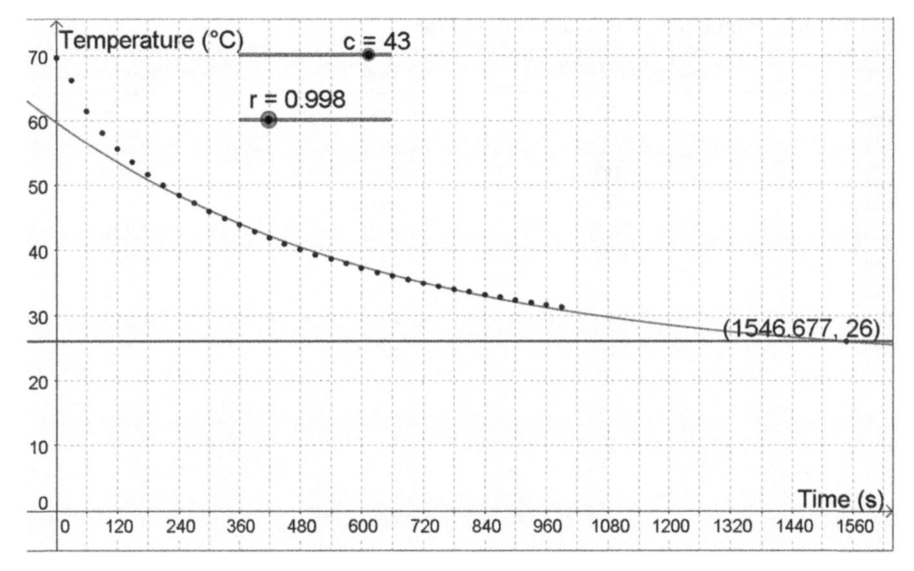

FIGURE 3.22 Intersection will give us the answer to our question.

```
y = 26
```

and use the Intersect tool to create the intersection of the line and the graph. The result can be seen in Figure 3.22.

This time is 1,547 seconds, or just below 26 minutes. That you get slightly different results is due to the fact that the two different fitting commands use slightly different algorithms. In a sense, the difference in results is a small indication of the size of the error.

3.7 MODELING THE POPULATION OF IRELAND

Table 3.7 gives the population of Ireland from 1672 to 1966. It is an interesting set of data because it raises a lot of questions. Some questions are related to the emigration years, here illustrated by Figure 3.23. Try to model this population over the last 350 years in a sensible way; also formulate your own questions and try to answer them.

Start by plotting the points in a coordinate system. Try to fit a function to the data. Can you find any? Try to answer the questions your diagram suggests.

Proposed Solution

Many Irish emigrated to the United States from approximately 1845 onward. This is shows up clearly in Figure 3.24. At first, the triggering events were a series of years with crop failures, but once the Irish colony in America had become stable, it kept attracting people from Ireland for the rest of the nineteenth century.

TABLE 3.7 Population of Ireland, 1672 to 1966

Year	Population (1000 : s)	Year	Population (1000 : s)
1672	1,100	1881	5,175
1770	3,000	1891	4,705
1791	4,100	1901	4,459
1821	6,802	1911	4,390
1831	7,767	1926	4,229
1841	8,175	1936	4,248
1851	6,552	1951	4,332
1861	5,799	1961	4,243
1871	5,412	1966	4,369

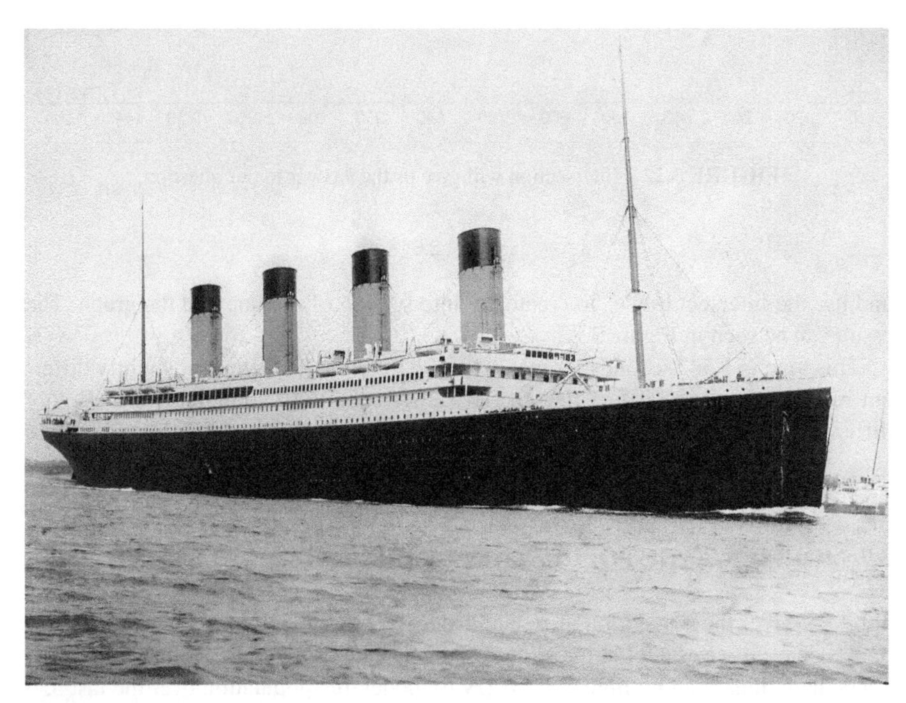

FIGURE 3.23 120 Irish passengers were on the Titanic in 1912. Photo: F. G. O. Stuart (1843–1923) [Public domain], via Wikimedia Commons. https://sv.wikipedia.org/wiki/RMS_Titanic#/media/File:RMS_Titanic_3.jpg.

To be able to make realistic predictions about the future population of Ireland, it is no use to consider data older than the First World War. Also, because our data don't go beyond 1966, you would be well advised to find more recent data.

But you can still make some rudimentary observations from the data you have been presented with. By fitting different exponential functions for the separate

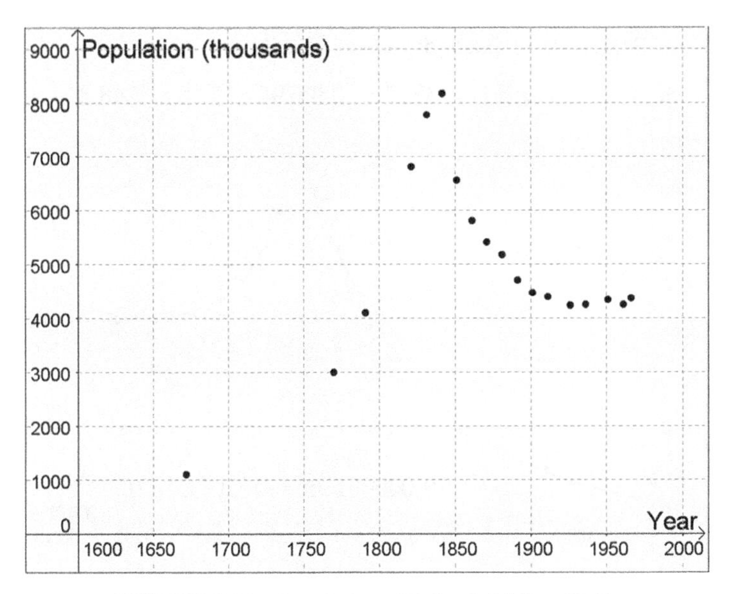

FIGURE 3.24 Population of Ireland, 1672 to 1966.

halves of the data, you will see that the population in Ireland was increasing by 1.2% annually before the emigration began. It does not seem to be a lot, but it does represent almost a ninefold increase in 200 years. This is a clear example of the fact that exponential growth is not viable in the long run. After the emigration gathered momentum the population was reduced by 4.2% per year for over 50 years, which near enough halved the population.

In Figure 3.25 $f(x)$ is fit with one of GeoGebra's standard alternatives for regression functions, **FitGrowth[<List with points>]**. In contrast, $h(x)$ does not flatten out toward zero, and here a so-called general fit has been done. A detailed description of how to do this can be found in Section 3.6, proposed Solution 2. In this case a model function with four parameters of the form

$$m(x) = \mathrm{a} \cdot \mathrm{b}^{(x-\mathrm{c})} + \mathrm{d}$$

was used. The parameter c makes the fitting algorithm work better; otherwise, the exponents would become so large that you would lose precision.

From these function expressions you can find the annual rate of change in the periods from the parameter b.

The population in the Republic of Ireland was 4.6 million in 2011, but the total population of the Island of Ireland the same year was 6.4 million. So, clearly, using 50-year-old data to model the current situation is not going to yield especially revelatory results.

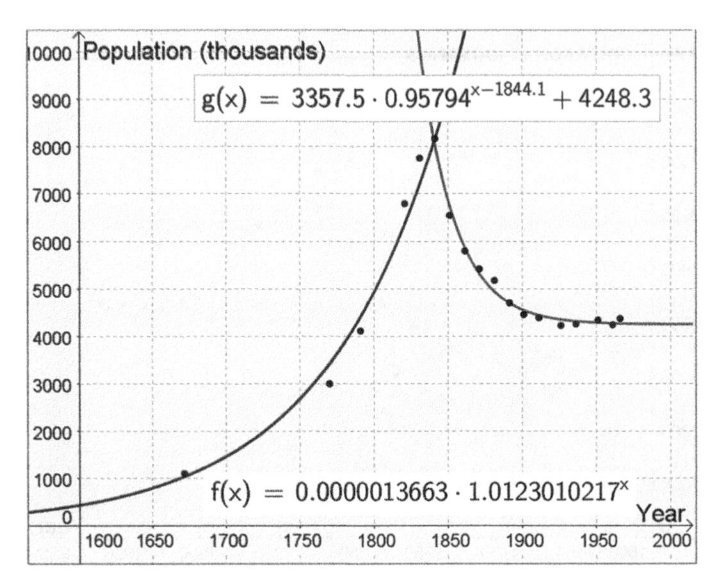

FIGURE 3.25 Two different exponential functions fit to different parts of the data.

FIGURE 3.26 Money, making more money. Photo: Jericho [CC BY 3.0 (http://creativecommons.org/licenses/by/3.0)], via Wikimedia Commons. https://upload.wikimedia.org/wikipedia/commons/f/f9/Money_Cash.jpg.

3.8 THE RULE OF 72

Figure 3.26 illustrates the concept of money increasing its value, seemingly by itself. As a rule of thumb, you can calculate the doubling time for a capital deposited in a bank with p percent interest by $T_2 = 72/p$ years. Make an exact calculation of the doubling speed and fit the function $f(x) = a/x$ to the exact solution. What is the size of the error for different values of p? Can the value 72 be improved upon?

Proposed Solution

The basic exponential relationship, often written as $y = C \cdot a^x$, can now be written as $FV = PV \cdot r^t$, where FV is the *future value*, PV is the *present value*, and $r = (1 + p/100)$ is the rate of change with p being the interest in percent, and t is the time, typically measured in years.

In mathematics we often study structures, since talking about x and y can often be too abstract for students. They need to learn to spot similarities and differences between the abstract structures, where "we don't know what we're talking about," and the specific applications, where "we know what we are talking about." This is one example of such an application.

For a doubling at the time $t = T_2$, $FV(T_2)/PV = 2$, which gives us

$$r^{T_2} = 2$$

or

$$T_2 = \frac{\lg 2}{\lg r} = \frac{\lg 2}{\lg\left(1 + \frac{p}{100}\right)}$$

We let GeoGebra draw this function by typing the command

```
T_2(p) = lg(2)/lg(1 + p/100)
```

and add the function

```
f(p) = 72/p
```

at the same time.

There is hardly any difference between the functions. In Figure 3.27 the lower graph belongs to the true process and the upper graph belongs to $f(p) = 72/p$. By calculating and graphing the difference, you can get a better idea of how close these values are.

Hide the two functions you have so far, and type in

```
D(p) = T_2(p)-f(p)
```

Now re-zoom until you get something like Figure 3.28. For large values of p you get an error not exceeding a quarter of a year until the interest rate exceeds 30%. However, for small values of p, smaller than some 3 percent, you get errors in excess of 6 months, rapidly increasing as p grows less.

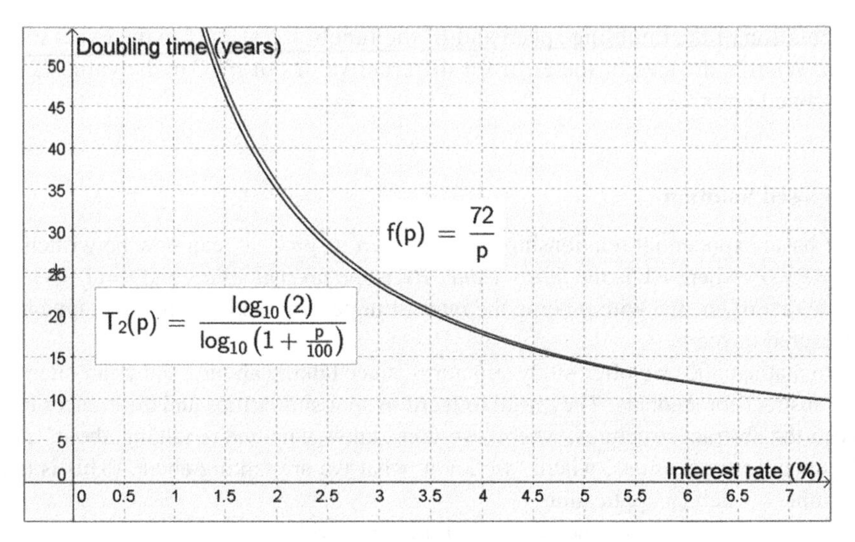

FIGURE 3.27 Both functions are very close in a large interval.

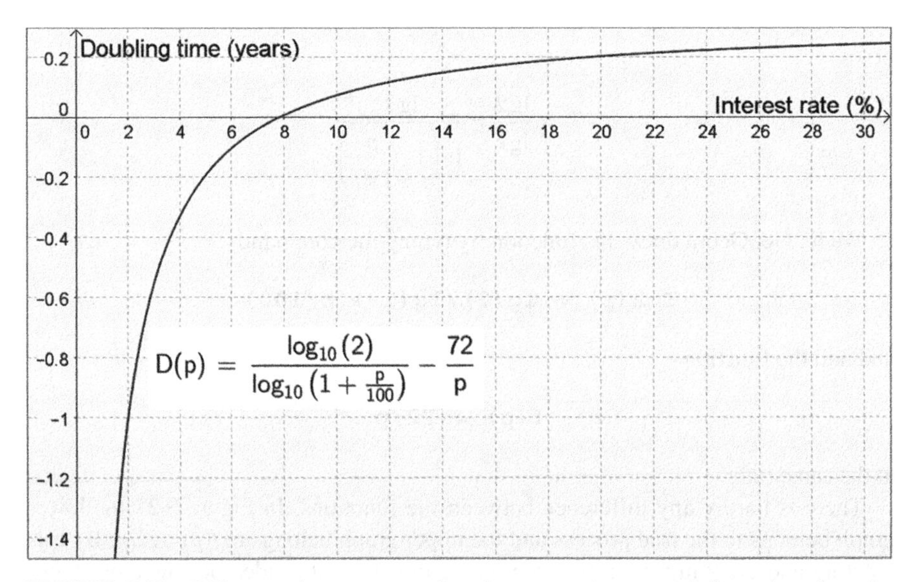

FIGURE 3.28 Difference between two functions can give more information than the individual functions if they are close.

Seeing this, you may decide to try to optimize a new function, better adapted to low interest rates. Hide $D(p)$ and show $T_2(p)$ again. Then place some points on the function for interest rates between 1% and 4% and fit a function to these points by typing

```
Fit[{A, B, C, E, F, G, H}, {1/x}]
```

Note that GeoGebra skips D while naming the points. This is because you already have a function by that name.

In Figure 3.29 you see that for low interest rates you could simply use the function $g(p) = 70/p$ using the constant 70 instead of 72. How large will the errors be now, and for what interest rate will the two functions be equally good?

Once more, hide what you see and show $D(p)$ again, while perhaps wishing you'd drawn it in Graphics Area 2 from the start. Anyway, type in **D2(p) = T_2(p) – 70/p** showing both of these difference functions in the same graphics area as in Figure 3.30.

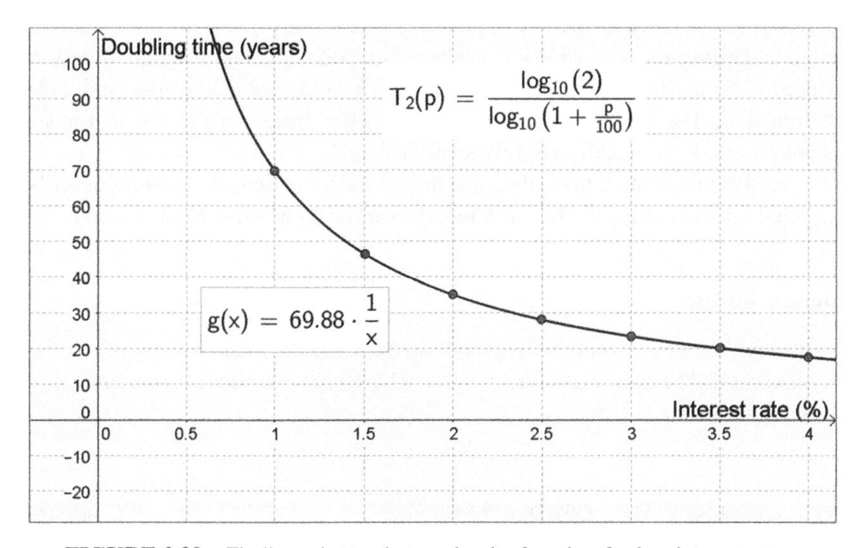

FIGURE 3.29 Finding a better, but as simple, function for low interest rates.

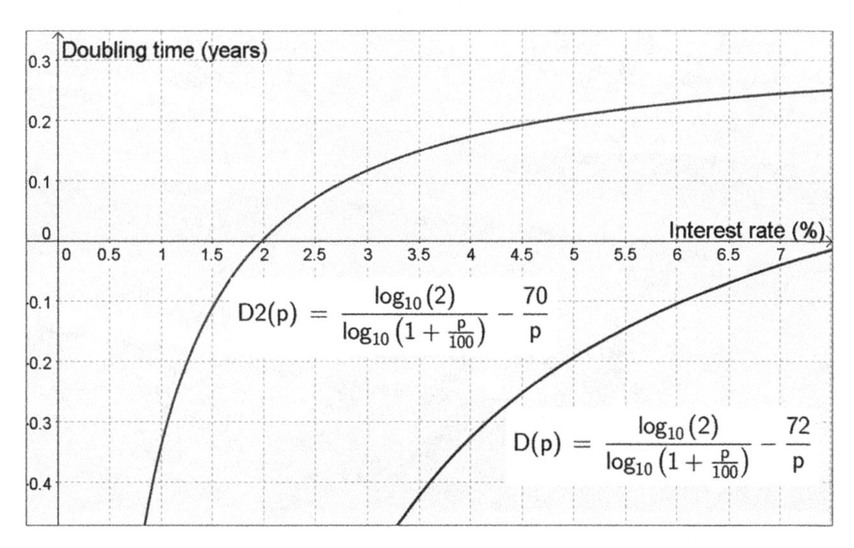

FIGURE 3.30 70/p works best in the interval 1 to 5%.

For really small rates of interest, below 1%, there are still large errors, but for interests in the interval 1% up to 5% you now have much smaller errors than before. For rates larger than 5%, 72 still works better than 70 as a parameter. What happens if you try to redo these calculations to find a function that works best for a rate of 0.1%? Is this even relevant?

3.9 THE FISH FARM I

Fish in a fish farm, such as the one depicted in Figure 3.31, grow toward a maximum length a according to $l(t) = a(1 - k^t)$, where k is a constant. The weight of the fish is measured to be approximately 15 g for 10 cm fish, 500 g for 30 cm fish, and 2,000 g for 50 cm fish. The number of fish remaining after time t are $N = N_0 \cdot r^t$ for some constant r. r and k are usually slightly smaller than 1.

Convert this information into a dynamic model and try to find relationships describing the optimal time to extract the fish as a function of the parameters N_0, a, k, and r.

Proposed Solution

The optimum time to extract the fish ought to be when the total biomass of the fish is at its maximum. The total mass is $N(t) \cdot m(t)$, where $m(t)$ is some function giving mass as a function of time.

FIGURE 3.31 Fish farming in Turkey. Photo: Vera Kratochvil via Wikimedia Commons. https://upload.wikimedia.org/wikipedia/commons/2/20/Fish_Farm_Site.jpg.

You have $l(t)$, so the piece of the puzzle missing is a function giving you the mass as a function of the length, $m(l)$. So far you only have a few measurements for this, but the mass ought to be reasonably proportional to the volume of the fish, which in turn should be reasonably proportional to the cube of the length. Therefore assume $m(l) = C \cdot l^{d}$ and try to find this function.

Put the data into the GeoGebra spreadsheet and create points from them by right-clicking the selection of data and then selecting **Create...>List of points** from the pop-up menu. Find a suitable function by fitting a power function to these data points by typing

```
FitPow[list1]
```

The fit is reasonably good, and shown in Figure 3.32.

Now show Graphics Area 2 by pressing **Ctrl-Shift-2**. With this area active, create the other pieces of this puzzle:

```
a = 100
k = 0.9
l(t) = a (1 - k^t)
N_0 = 500
r = 0.9
N(t) = N_0 r^t
```

Do take some time to play around with the values of the parameters and to see how the different functions behave. This is both part of the fun as well as part of the learning process. Are the functions suitable for modeling length and number? Can you make changes to the model to make it more realistic? What are the limitations of the model?

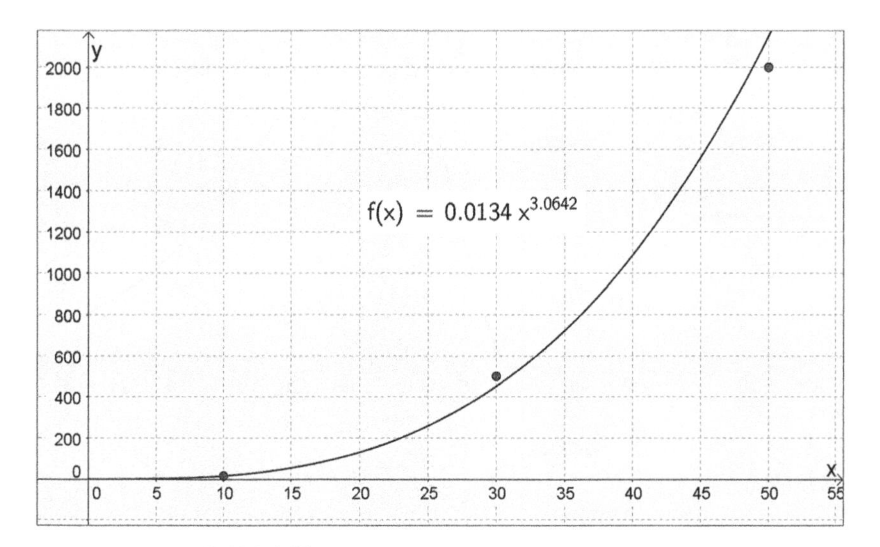

FIGURE 3.32 Nearly cubic function fits the data well.

Then continue with

```
m(t) = f(l(t))
TotalMass(t) = N(t) m(t)
```

You can now hide all functions except TotalMass, by clicking on the blue buttons to the left of the functions names in the Algebra window.

Zoom into the axes until you see a clear maxima, and then find its coordinates by typing either of the commands

```
Extremum[TotalMass,0,100]       (US English)
TurningPoint[TotalMass,0,100]   (UK English)
```

Read the coordinates from the newly created point in the Algebra window or right-click on the maximum point and select Object Properties. Then set the label to show the value, as in Figure 3.33.

Now play with the parameters again to see their effect on the optimal time for a catch. The algorithm finding the maximum is not entirely stable and does not always give a point. You need to change the value of, say N_0, until you get a point. You can still easily see that a (maximum length of each fish) and N_0 (initial number of fish) do not affect this time at all, whereas r and k do.

Analyzing how extreme values depend on the parameters of a model forces the students to really understand the inner workings of the model. Why do some parameters affect the maximum and not others? Have your students answer this question.

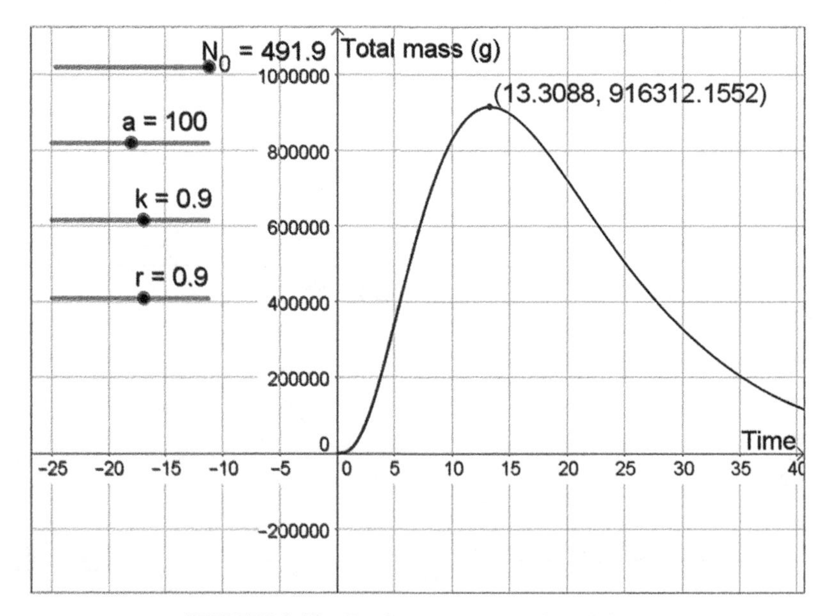

FIGURE 3.33 Total mass as a function of time.

Ideally, you should now have had yet another graphics area. Failing that, you can hide (don't delete anything—and remember **Ctrl-Z**) everything in Graphics Area 1. Adjust the axes to get a domain of 0 to 1 on the x-axes and a range of 0 to 20 on the y-axes. You are going to draw functions showing how the optimal time changes with respect to the parameters r and k, and do so in the following way:

First make sure Graphics Area 1 is active by clicking somewhere in it. Then check the name of the point in Graphics Area 2 showing the maximum. In this case it is called D. With Graphics Area 1 still active, type

```
(r,x(D))
```

to create a point in Graphics Area 1. By varying r, the point moves. If you want to, you can right-click the point and select **Trace On**. Then you can right-click r or its slider and select **Animation On** for a while to see the shape of the relationship, shown in Figure 3.34.

Both the graph and the theory indicates that there is a vertical asymptote at r = 1. What might be a suitable function to model this relation with? Probably some rational function where the denominator has a zero at r = 1. As r gets smaller, so does the time, so the numerator's degree should be smaller than the denominator's degree. This leads to a function of the type $T(\text{r}) = \text{c}/(1 - \text{r})$ for some constant c. Type in

```
T(r) = c/(1-r)
```

And accept to create a new slider for c. The agreement is not as good as one might have hoped, so create the number n = 2 and change T to

```
T(r) = c/(1-r)^n
```

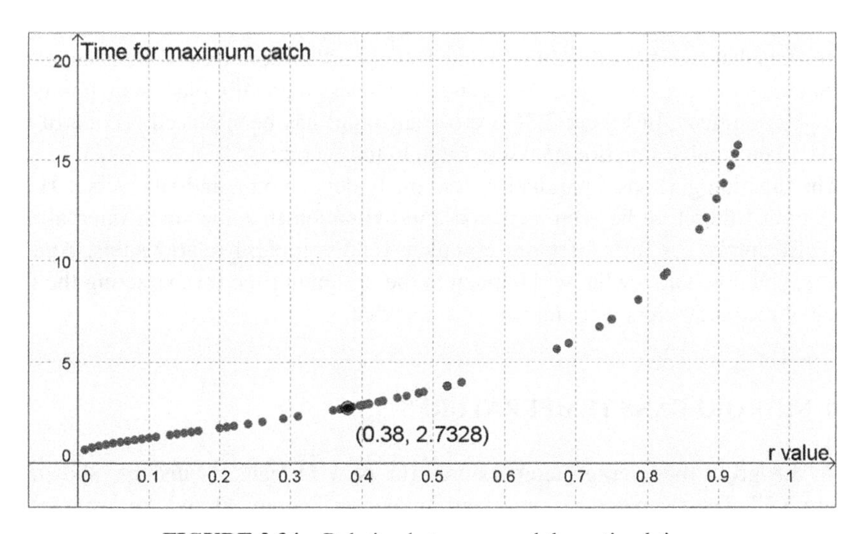

FIGURE 3.34 Relation between r and the optimal time.

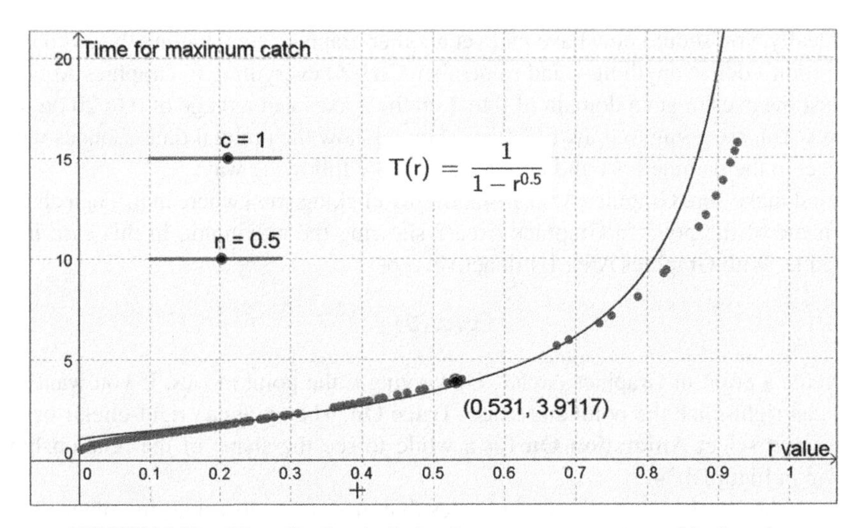

FIGURE 3.35 Manually fitted relation between a parameter and its function.

After this, play around a little. You get a relatively adequate fit for c = 1.88 and n = 0.88. You can also try

```
T(r) = c/(1-r^n)
```

and get a reasonable fit, shown in Figure 3.35, for c = 1.00 and n = 0.5, but only up to r = 0.7.

Hide *T*(r) and the point E and press **Ctrl-F** to erase the tracks from the point. Then repeat this procedure for the parameter k. Reset r = 0.9, create a point **(k, x(D))** in Graphics Area 1, activate its trace, and animate k.

This time the shape resembles a third-degree polynomial. These contain four parameters that are very difficult to manually fit. Furthermore GeoGebra is unable to fit functions to traces of points. You can nevertheless manually place new points on top of these traces. In Figure 3.36 a crosshair point has been placed on one of the traces. Then a cubic function g(x) was fitted to these points.

The function g(x) has 4 parameters that likely depend on r, and vice versa. These results must therefore be seen as a first approximation in some small interval, but they also imply the form of more complete (and complex) relationships. And of course, you now know what will happen to the optimum time for extracting the fish if one of the parameters were to change somewhat.

3.10 NEW ORLEANS TEMPERATURES

Table 3.8 gives the average temperatures for New Orleans, Louisiana, shown in Figure 3.37. The temperatures are given in degrees Fahrenheit and are averages of measurements between 1951 and 1980.

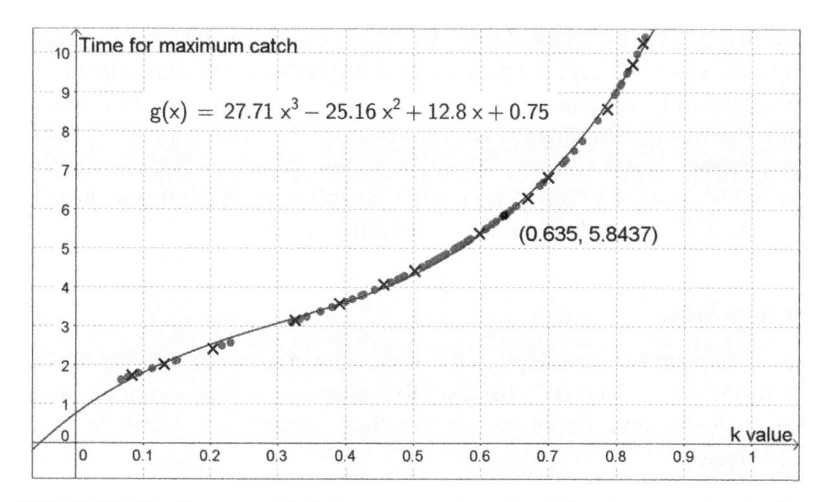

FIGURE 3.36 Parameter k's influence over the optimal time for extracting the fish.

TABLE 3.8 New Orleans Temperature Averages, 1951 to 1980

Month	Temperature (°F) (1951–1980)
January	52
February	55
March	61
April	69
May	75
June	80
July	82
August	82
September	79
October	69
November	60
December	55

FIGURE 3.37 New Orleans town center. Photo: Michael Maples, US Army Corp of Engineers Digital Visual Library/Wikimedia Commons Public Domain. https://en.wikipedia.org/wiki/Crescent_City_Connection#/media/File:USACE_New_Orleans_skyline.jpg.

Show these data in a diagram. Construct a mathematical model describing the relationship between month and temperature, and graph this in the same diagram. Also convert temperatures to centigrade. Discuss how well your model fits to your data.

Using this model, determine what fraction of the year New Orleans experiences an average temperature exceeding 20°C. Given time, you may want your students to find precipitation data for New Orleans and construct a complete climate diagram for extra credit. Different students may work on different cities.

Proposed Solution

A climate diagram for a particular city or position consists of a continuous curve describing the temperature and a bar chart for precipitation. In this example, a similar diagram, using the temperature data for both the continuous graph and the bar chart, has been created.

As you can see from Figure 3.38, The GeoGebra spreadsheet has been shown with **Ctrl-Shift-S**, and the month and the corresponding temperature in Fahrenheit has been placed in columns B and C. In column A, the number of the month has been added.

You can then convert temperatures from Fahrenheit to Celcius in column D by typing

```
=5(C1-32)/9
```

in cell D1, and copy this down by dragging the fill handle. In the last column, column E, create points by typing

```
=(A1,D1)
```

into cell E1, and copy this down.

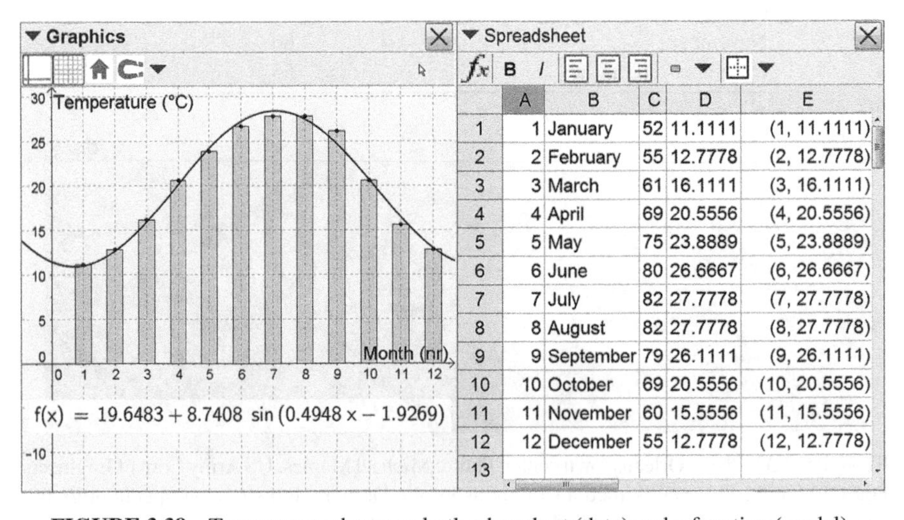

FIGURE 3.38 Temperature shown as both a bar chart (data) and a function (model).

TABLE 3.9 Rocket elevations

Time (s)	Height (m)
0	0
0.5	6.23
1.0	9.56
1.5	11.05
2.0	9.26
2.5	8.61

Then create three lists from columns A, D, and E by selecting the entries in one column at a time, right-clicking, and selecting **Create...>List** (not list of points). The bar chart can be created by typing

```
BarChart[list1,list2,0.5]
```

The concluding 0.5 indicates the width of the bars, and may be omitted.

The trigonometric model was found by typing

```
FitSin[list3]
```

While it is possible to work on this some more to get a better model, we leave that as an exercise to the reader.

3.11 THE RECORD MILE

Find information for world records for running 1 mile = 1,609 meters. You might, for instance, have a look at http://en.wikipedia.org/wiki/Mile_run_world_record_progression.

Try to predict what year the time for the world record drop below 3.40? In what way is the model for men and women different?

3.12 THE ROCKET

A homemade PET bottle rocket is launched from the ground. The launch is filmed with a mobile camera and subsequent analysis allows the rocket's height above the ground to be measured every half second and entered into Table 3.9.

Determine a suitable model that describes the rockets height above the ground. Calculate when the rocket will hit the ground.

3.13 STOPPING DISTANCES

According to https://www.gov.uk/government/uploads/system/uploads/attachment_data/file/312249/the-highway-code-typical-stopping-distances.pdf the stopping distance at 40 mph or 64 km/h is about 36 meters and 60 mph or 96 km/h about

73 meters. You may assume that the stopping distance at 0 mph or 0 km/h is 0 meters. You also know that the relationship between stopping distance and speed is quadratic. Determine a model for calculation of stopping distance for any speed and create a table for different speed from 30 km/h till 110 km/h.

3.14 A BOTTLE WITH HOLES

Fill a large PET bottle with water. Make a small hole in the cork and a small hole close to the bottom. Arrange the bottle close to a ruler and video film when the water is drained out from the bottle. Measure the height of the water from the video. Fit a function to your measurements. Which function works best?

3.15 THE PENDULUM

Investigate how a pendulum's oscillation time T varies with some different parameters:

- Vary the mass of the pendulum's weight from about 20 g to 200 g, but keep the length and the angle constant.
- Vary the length of the rope from about 10 cm to about 100 cm, but keep the weight and the angle constant.
- Vary the angle from which the weight is dropped from about 10° to about 45°, but keep the weight and the length constant.

Sort out the parameters that do not seem to be important for the oscillation time. Adjust a suitable model to the rest of your data. Find the theoretical relationship for the oscillation time and try to determine the value for the Earth's gravitational acceleration from your measurements by comparing theory with your function.

3.16 RADIO RANGE

The range for radio broadcasting at sea normally is presented as $d = 2.6\sqrt{h}$, where d is the range in nautical miles and h is the height for the antenna in meters above the sea level. The formula is approximate but fairly accurate. Construct a model of the real, geometrical situation where you can determine the error and the relative error for some different values for h. Finally determine a better value for the constant 2.6.

3.17 RUNNING 400 METERS

A promising athlete student is practicing short-range running. Her favorite distances are 100 and 200 meters, but now and then she also practice 400 meters.

By a friend at the training club, she is asked to join the 4×800 meters theme for a competition next Sunday. The student who is studying both athletics and mathematics decides to make a mathematical model of her possibilities for the 800 meter race.

Her personal best records are as follow:

100 meters	10.91 seconds
200 meters	21.35 seconds
400 meters	47.85 seconds

Make a mathematical model and predict the results for the student if she runs the 800 meter race. The student knows that an 800 meter race is quite different than a short-range race. In order to better understand her own capacity, she makes a couple of training runs over the distances of 300 meters and 500 meters. She records the following average results:

300 meters	34.15 seconds
500 meters	63.90 seconds

Now make a new model and predict the results for the student when she is running the 800 meter race. Now you have two different results for the 800 meter race. Which do you believe most in? You might also use the two models to predict the student's time for a 1,500 meter race. You can compare that time with common results for 1,500 meters, which is an international distance.

3.18 BLUE WHALE

The largest animal on Earth is the blue whale. During the last hundred years the number of blue whales has decreased heavily depending due to human whale hunters. In 1900, there were about 239,000 blue whales in the large oceans, and one hundred years later, there were just 2,300 left. It is fair to assume that the number of blue whales is decreasing exponentially over time.

Determine what year the number of blue whales for the first time will be fewer than 200 if the decrease continues at the same pace.

3.19 USED CARS

Decide on a car model and browse the Internet and the classifieds in the local newspapers for ads on this model. Plot the price versus age for similar mileages and price versus mileage for similar ages of this particular model. Try to find a function that describes these relationships. Note that the prices in the ads are for the asked price, not the actual selling price.

As an extension, for a model where there are many ads, try to find a function of two variables, *Price*(*age*, *mileage*) that describe how the price varies with both age and mileage simultaneously, and plot this function in the 3D-window.

3.20 TEXTS

During 1998, 44 million text messages were sent in Sweden. In a sense, this year marked a starting point and the following years the numbers of text messages grew.

During 2011, more than 18 times the number of text messages than in 2001 were sent, and for the first time surpassed the number of phone calls.

During 2012, 16,514 million text messages were sent. Assume that the yearly percentage increase in the number of text messages has been about the same during the whole period of time.

Construct a GeoGebra model that can be used to find the yearly rate of change. What does this model predict for the number of texts last year in Sweden? What are the actual figures? Explain.

4

NONLINEAR EMPIRICAL MODELS II

In this chapter are collected some problems in which we try to develop techniques that go further than just fitting a function to a set of data points. We introduce a more realistic, reliable, and visible technique to evaluate models: the *sum of squared errors (SSE)*, which measures the deviations of predicted values from actual empirical values. SSE is a measure of the discrepancy between the data and an estimation model. A small SSE indicates a good fit of the model to the data. It is often used as an optimality criterion in parameter selection and model selection. Closely connected to the SSE is the concept of residuals. We will discuss this in relation to intuitive and informal ways to estimating errors.

Plotting data in lin-log or log-log diagrams is an old but still useful method, using nonlinear scales, that is convenient to use when there is a large range of empirical data. This is a common choice when modeling earthquake strength, sound loudness, light intensity, or pH, for example.

In this chapter we will also work more with functions, for instance, function compositions, f(g(x)), where we apply one function to the results of another function. We will also fit implicit functions and conics to data, and we will explore how the number of parameters affects the fit.

Mathematical Modeling: Applications with GeoGebra™, First Edition. Jonas Hall and Thomas Lingefjärd.

Companion website: www.wiley.com/go/Hall/MathematicalModeling

4.1 COOLING II

A copper sphere is heated to 100 °C by letting it sit in boiling water. At time $t=0$ the sphere is placed in a tank where water of the constant temperature 30 °C is flowing past it. The surface temperature $T(t)$ of the sphere is measured with a small but accurate thermometer, soldered onto the sphere. The temperature for the first 12 whole minutes is recorded in Table 4.1.

Use the data and construct at least two different models describing the cooling of the copper sphere. Make some suitable error estimation to try to evaluate the models. Discuss similarities and differences of your models. What is the basis for the differences?

Find the time when the surface temperature reaches 31 °C.

Proposed Solution 1

Newton's law of cooling shows that an exponential decay of temperature occurs according to $y=\mathrm{c}\cdot e^{-\mathrm{b}\cdot x}+\mathrm{d}$. Cooling models are not built in as standard models in GeoGebra but can be achieved with a general fit. Start by creating the parameters and the model function, and type into the input bar:

```
b = 0.1
c = 30
d = 70
m(x) = c e^(-b x) + d
```

Don't forget to type **Alt-e** to get the base for the natural logarithms. GeoGebra will sometimes interpret an ordinary e as an ordinary number. The model function

TABLE 4.1 Cooling a Cupper Sphere

Time (min)	Temperature (°C)
0	100.0
1	88.1
2	78.2
3	70.0
4	63.2
5	57.1
10	40.9
15	34.4
20	31.3
25	30.6
30	30.3
60	30.1

can now be hidden by clicking the blue button to the left of its name in the Algebra window. The numbers for b, c, and d are reasonable guesses, but you can start with any set of values that is reasonable enough.

You must also feed the data into GeoGebra. **Ctrl-Shift-S** will show the spreadsheet. Enter the data, select it, right-click the selection, and select **Create...>List of Points**. **Ctrl-Drag** the axes to zoom and drag the background to pan. By right-clicking on an empty part of the background and selecting **Graphics...** and the tabs for the axes, you can set the axes labels.

Fit a new function of the same type as the model function $m(x)$ to the data in list1 by typing

```
Fit[list1,m]
```

This function seems to fit the data well. Drag it from the Algebra window to the graphics area end; then set the resulting text label's background color to white in its properties dialogue.

To find the point corresponding to the 31 °C, you can create a horizontal line by typing

```
y = 31
```

in the input field. Find the point of intersection by clicking on the Intersect tool and then clicking on the line and the graph. Alternatively, you can type the command

```
Intersect[f,e]
```

if e is the name of the line and *f* the name of the function. The coordinates can either be found in the Algebra window or by right-clicking the point, selecting **Properties** and activating a label showing the points value. Colors, line styles, and the like, can be set in the Style Bar or in the properties dialogue for each object. The result is shown in Figure 4.1.

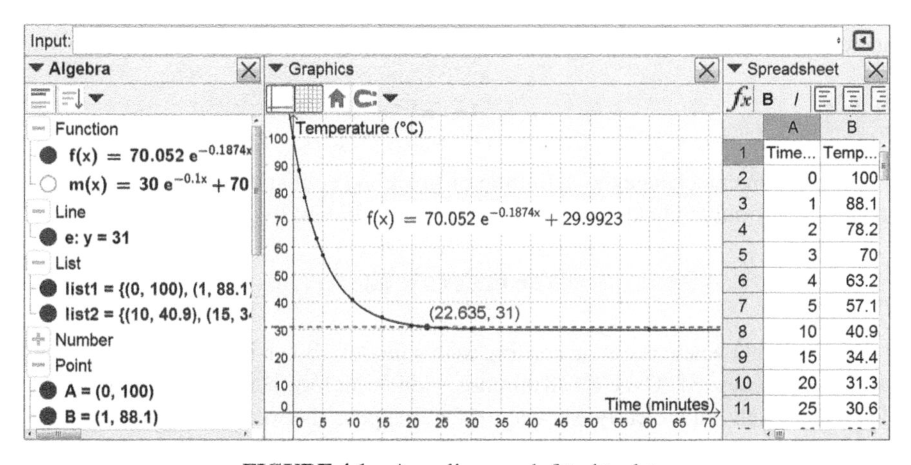

FIGURE 4.1 A cooling graph fitted to data.

The surface temperature for the copper sphere is 31 °C at t=22.635 minutes, that is, after 22 minutes and 38 seconds.

Proposed Solution 2

You could also try an alternative model. If the water really holds 30 °C, it ought to be possible to lock this parameter down and use the model function

```
m(x) = c e^(-b x) + 30
```

However, this function will not yield a significantly different result, since the earlier model already identified this value to 29.9923 °C. This new model gives the time t=22.6694 minutes, or 22 minutes and 40 seconds, only a two-second difference from the previous attempt.

Proposed Solution 3

If you for an instant conveniently forget everything about the theory of cooling, you might conceivably imagine that the first part of the graph, up until t=some 20 odd minutes, reminds you of a fourth- or sixth-degree polynomial, while the rest of the graph consists of a horizontal line. The polynomial appears to be in basic form but translated. Thus assume that the polynomial can be written $y = \mathrm{k} \cdot (x - \mathrm{a})^6 + 30$ and attempt to use all data up to t=20 minutes for this new model. Select these particular data again and create a new list of points, list2. Enter a starting value for the new parameter by typing

```
k = 1
```

and edit your model function by entering

```
m(x) = k (x-a)^6 + 30
```

Finally you can change the definition of function f so that it applies to list2 rather than list1 by double-clicking it and changing its definition to

```
Fit[list2, m]
```

Now you get a new result: according to this model, shown in Figure 4.2, the temperature reaches 31 °C at t=18 minutes, which unfortunately does not fit the data very well and clearly shows that it happens between t=20 minutes and t=25 minutes.

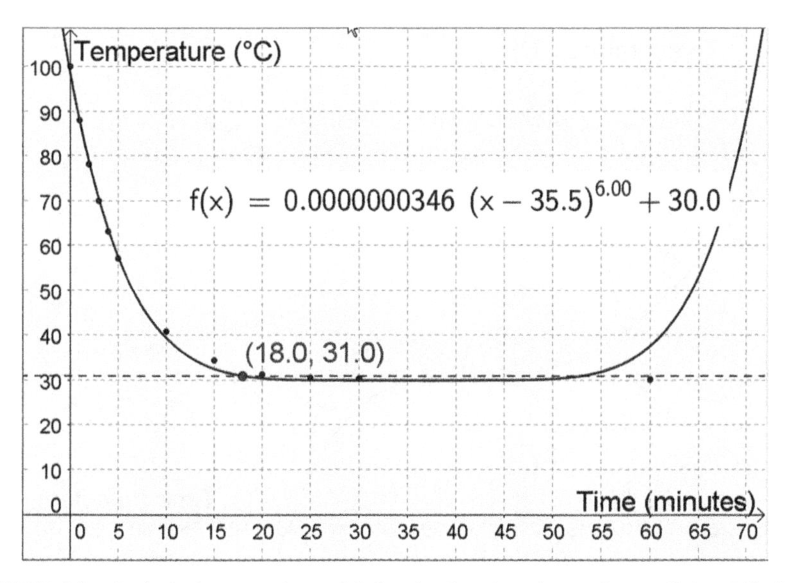

FIGURE 4.2 A sixth-degree polynomial fits the first few data points well but fails in the region where we are most interested in exact results.

Proposed Solution 4

If you only wish to find the time when the temperature is 31 °C, then perhaps a better alternative is to only take into consideration the closest data points and fit a simple polynomial to these points. There are 5 data points between $t=10$ minutes and $t=30$ minutes. Create a new list with these points, for example, by typing

```
list3 = {G,H,I,J,K}
```

then change function f to use this list and change the model function to a fourth-degree polynomial (for 5 data points, this is the maximum degree polynomial that can be fitted). Define the model function to be

```
m(x) = a x^4 + b x^3 + c x^2 + d x + k
```

(since k is already defined since before). The result is shown in Figure 4.3.

The time now becomes 21.075 minutes, or 21 minutes and 4 and one-half seconds.

Comments

Of these four different solutions, numbers 1, 2, and 4 seem reasonable enough. Unfortunately, these results differ by more than a minute. The first two solutions have a better theoretical background but consider many points far away and may possibly

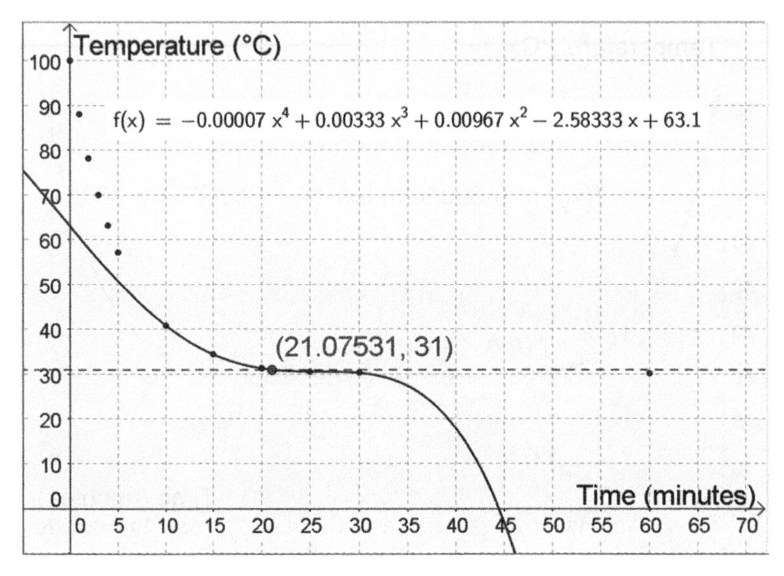

FIGURE 4.3 A fourth-degree polynomial with 5 parameters passes exactly through 5 data points.

be less exact at the target temperature. Solution 4 may yet have slightly better precision round the target temperature, but it is unable to predict anything outside the interval 10 minutes $\leq t \leq$ 20 minutes. This model clearly shows obvious mismatches immediately outside this interval, and it is not unreasonable to think that this has spread at least part way inside this interval as well.

The difference between 22 minutes 38 seconds and 21 minutes 4 seconds is quite large, around 7%. You can make a crude final result by taking the average of these two results and give half the distance as some sort of error estimate: $t = 21$ minutes and 51 s $\pm$ 47 s. However, you know nothing of the confidence of this error estimate, so you might as well simply give the final answer as approximately 22 minutes.

In a last attempt, go back to the exponential model in the first solution, but let it work with the short list of 5 data points in the interval 10 minutes $\leq t \leq$ 20 minutes. This yields $t = 22.3437$ minutes, or 22 minutes and 21 seconds. Perhaps this answer is preferable? It is based on ignoring most of the data to build a limited model in a small interval, which perhaps isn't the best of practices.

4.2 BODY SURFACE AREA

In some medical situations when a doctor is treating a patient, the *body surface area*, BSA, is used instead of the total body weight to determine accurate dosages. This is obviously difficult to measure, so doctors much prefer to calculate dosages from the patient's weight and length.

Relationships between different bodily sizes and bodily functions can often be described as power functions, that is, function of the type $y = C \cdot x^a$ for some constants,

C and a. In this case you want to find the body surface area as a function of two variables, the height H and the weight W, so you are looking for functions of the type $BSA = C \cdot H^a \cdot W^b$ for some constants C, a, and b.

But you can know more about this function from looking at the units involved. Performing a *unit analysis* like this can frequently disclose details about the shape of a function. Looking at the units in this case, you can see that it is quite impossible to get the unit square meter from the units meter and kilograms unless b=0 and a=2. But this does not correspond to a reasonable solution. Likely density is somehow involved, but because density is more or less constant for most populations, it is part of the constant C. It will nevertheless be important for the units and because of this, also for the choice of a and b. Momentarily therefore, write the function as being dependent on the density *D:*

$$BSA = C_1 \cdot H^a \cdot W^b \cdot D^c.$$

Looking at the units for mass and length, you can see that the following equation must be true:

$$[m]^2 = [m]^a \cdot [kg]^b \cdot [kg/m^3]^c$$

For the units of lengths to work out, you must have

$$2 = a - 3c$$

Looking at units of weight, the corresponding equation is

$$0 = b + c \quad \text{or} \quad c = -b$$

Combining these and solving for a gives

$$a = 2 - 3b$$

In other words you cannot select the exponents a and b freely, so you must write the model, for example, as

$$BSA = C \cdot H^{2-3b} \cdot W^b, \quad \text{where C and b are two parameters to be determined}$$

Table 4.2 lists the data for 16 patients giving their total body surface area, BSA, measured together with information on these patients' lengths and body mass indexes, BMI. Use this information to find an empirical formula that you can use to determine other, similar, patient body surface areas.

Proposed Solution

If you look up *body surface area* on Wikipedia, you will see an abundance of different formulas. Say that, from the top of the stack, you pick up Du Bois's formula stating

$$BSA = 0.007184 \cdot W^{0.425} \cdot H^{0.725}$$

In all formulas presented, H is measured in cm, W in kg, and BSA in m^2.

TABLE 4.2 Physical Exam Data for 16 Different Patients

Weight (kg)	BMI (kg/m^2)	BSA (m^2)
61	23	1.65
65	25.2	1.69
72	23	1.88
59.5	22.7	1.63
61.4	23.7	1.63
63.7	21.6	1.75
66.1	25.3	1.69
68.48	26.67	1.71
72.7	23.1	1.89
75.7	24.7	1.91
77.4	23.1	1.99
80.03	26.89	1.92
80.6	31.9	1.82
81.3	26.1	1.96
89.9	29.8	2.04
91.1	29.2	2.08

Start by entering the data into GeoGebras spreadsheet, made visible by the menu alternative **View > Spreadsheet**, or the keyboard command **Ctrl-Shift-S.** You have not been given any data about the height of the patients but know that weight, length, and BMI are connected by $BMI = W/L^2$, where L is the height in meters, allowing you to recreate the patient's heights in the spreadsheet. So in cell D2, type

```
=100·sqrt(A2/B2)
```

to calculate the height in centimeters. This formula is copied down by dragging the fill handle in the bottom right of the cell as has been done in Figure 4.4.

In this case you will not be able to fit some function to the data because you have a function of two variables and need to fit a surface to a cloud of three-dimensional points. A more powerful mathematics environment, such as *Mathematia,* would be able to do this, but using GeoGebra, you have to fall back to more basic methods.

Start by inspecting Du Bois's formula and create two sliders, C and b. Then set C to go from 0 to 1 in steps of 0.000001 and b to go from 0 to 1 as well, but in steps of 0.001. You also need to set GeoGebra to display more decimals, so from the menu you should select **Options > Rounding > 10 decimal places**.

In cell E2 using the model, create a theoretical value for *BSA* by typing

```
=C A2^b D2^(2 - 3b)
```

Copy this formula down!

As you select the sliders, using the arrow keys to change their values, you see how the calculated values gradually change, moving closer to or farther from the measured values in column C in the spreadsheet. A good estimate of the overall

Spreadsheet

D2 =100sqrt(A2 / B2)

	A	B	C	D
1	Weight (kg)	BMI (kg/m^2)	BSA (m^2)	Length (cm)
2	61	23	1.65	162.8549634811
3	65	25.2	1.69	160.6040186099
4	72	23	1.88	176.9303473859
5	59.5	22.7	1.63	161.8995174313

FIGURE 4.4 Calculating the length of the patients in the spreadsheet.

Graphics

b = 0.4281043

C = 0.00740963

Spreadsheet

F18 =Sum[F2:F17]

	A	B	C	D	E	F
2	61	23	1.65	162.8549634811	1.6483952121	0.0000025753
3	65	25.2	1.69	160.6040186099	1.6770419736	0.0001679104
4	72	23	1.88	176.9303473859	1.8778101464	0.0000047955
5	59.5	22.7	1.63	161.8995174313	1.6240650088	0.0000352241
6	61.4	23.7	1.63	160.9570532651	1.639203906	0.0000847119
7	63.7	21.6	1.75	171.7286835119	1.7442332549	0.0000332553
8	66.1	25.3	1.69	161.6368838274	1.6969010457	0.0000476244
9	68.48	26.67	1.71	160.2397903181	1.7121230253	0.0000045072
10	72.7	23.1	1.89	177.4031044595	1.8892087059	0.0000006261
11	75.7	24.7	1.91	175.0650544208	1.9040329802	0.0000356053
12	77.4	23.1	1.99	183.0477902257	1.9845537576	0.0000296616
13	80.03	26.89	1.92	172.5166464607	1.9295602051	0.0000913975
14	80.6	31.9	1.82	158.9542628565	1.8252773012	0.0000278499
15	81.3	26.1	1.96	176.4919977998	1.9745430706	0.0002115009
16	89.9	29.8	2.04	173.6887596677	2.0379107947	0.0000043648
17	91.1	29.2	2.08	176.6313396229	2.074302906	0.0000324569
18					Sum:	0.0008140673

FIGURE 4.5 Calculating the sum of squared errors, the SSE.

fit is the sum of squared errors, SSE, which you can calculate separately. In cell F2 type

```
=(E2-C2)^2
```

and copy this down. E2 – C2 is the actual error but can be either positive or negative. By squaring their values, you create positive estimates of error. Finally sum these in cell F18 by typing

```
=Sum[F2:F17]
```

Yet again, take a look at Du Bois's formula and set C=0.007184 and b=0.425 as the starting values. Then adjust C until the value for SSE in cell F18 is as small as it gets. Next adjust b in the same way, then C again, then b again, …. This kind of iterative procedure, or looping, is not uncommon in mathematics. After a while of doing this, you are done, having reached C=0.0074096 and b=0.4281. You can now write the formula as $BSA = 0.00741 \cdot W^{0.428} \cdot H^{0.716}$. In Figure 4.5 you see the minimum value of SSE.

Try to calculate your own total body surface area using this, and other, formulas.

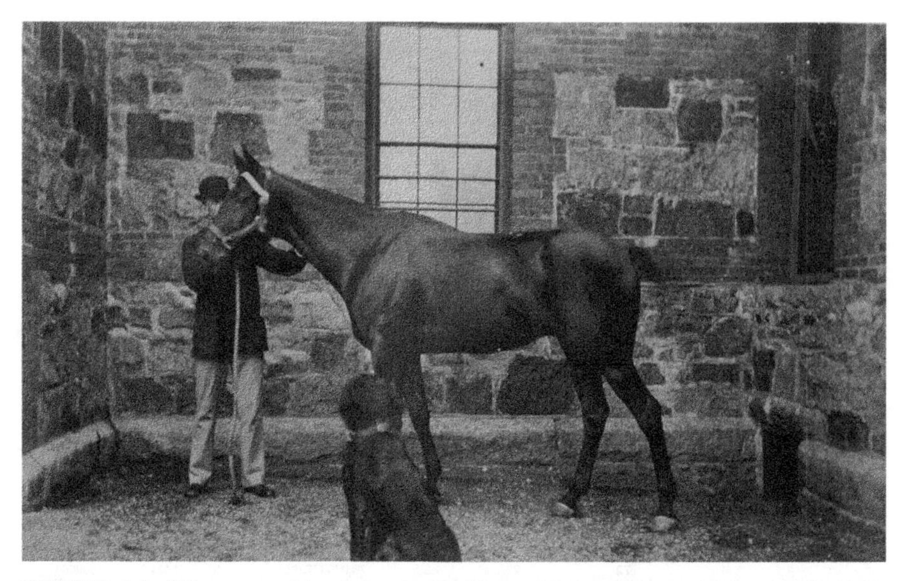

FIGURE 4.6 Three warm-blooded animals. Photo: Marian Hooper Adams (1843–1885) (Massachusetts Historical Society) [Public domain], via Wikimedia Commons. https://upload.wikimedia.org/wikipedia/commons/e/e4/Brooks_Adams_with_horse_and_dog%2C_photograph_by_Marian_Hooper_Adams%2C_ca._1883.jpg.

TABLE 4.3 Heat Production of Some Animals

Animals	Body weight (kg)	Heat production (kJ/day)
Rat	0.5	130
Hen	2.2	410
Marmot	2.8	320
Dog	18	2,000
Man	70	6,000
Horse	500	30,000

4.3 WARM-BLOODED ANIMALS

Figure 4.6 shows three warm-blooded animals. How much heat do you think is produced by an animal with a weight of 100 kilograms?

Construct a model that fit the data in Table 4.3 and determine from your model how much heat an animal with a weight of 100 kg produces.

Proposed Solution—Linear Relation

If you assume that the heat production is proportional to the mass of the animal and that it consequently is a directly linear relationship, you will get the graph shown in Figure 4.7.

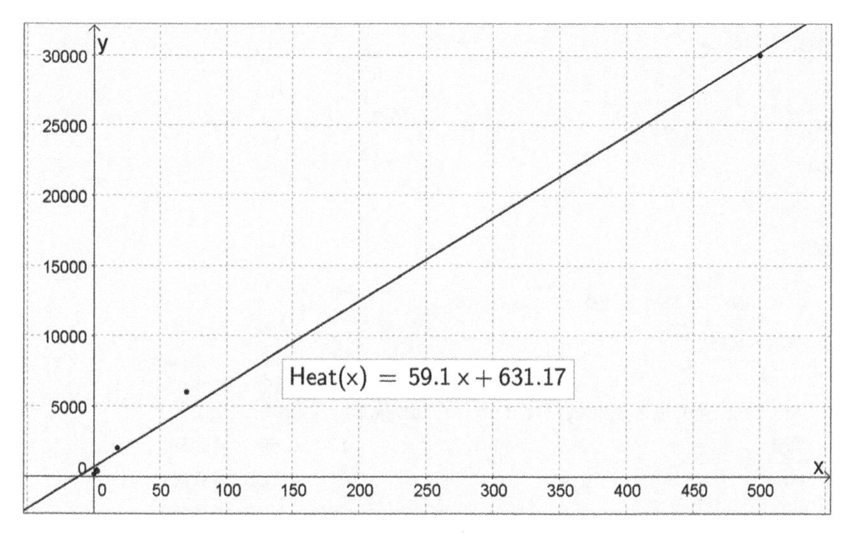

FIGURE 4.7 A linear fit seems to work well enough.

Using a linear fit, you get a heat production of 6.5 MJ per day for an animal with a weight of 100 kg. It may appear at first as if the data points lie rather close to the straight line, but a closer look reveals that the data points are above the line in the middle and under the line at the ends. You can see that this is a sign of some sort of systematic error. You want to find a function that goes from the origin and continues to grow, though at a slower and slower pace. Functions with this behavior might more reasonably be root functions or other power functions, $y = \mathrm{C} \cdot x^{\mathrm{a}}$. This is possible to discover on closer inspection of the residual diagram as well.

Proposed Solution—Power Relation

Since there is a huge difference between the smallest and the largest values both for body weight and for heat production, it may be hard to see many of the points in an ordinary coordinate system. One standard solution is to first calculate the logarithms of the values before you graph them. In GeoGebra, these values can then be graphed in Graphics Area 2 as shown in Figure 4.8. You can see that a linear relationship between these logarithmic values is likely, which also indicates that it is actually a power function that you are looking for. To use logarithmic values for analysis might seem obsolete, but it is still handy to use sometimes.

In Figure 4.8 the different windows from the **View** menu have been opened and then dragged to their current locations. To plot the points, you select the data values (columns B and C for the left diagram and columns D and E for the right), right-click on the selection, and then select **Create...>List of Points** from the pop-up menu.

Starting with the power function $y = \mathrm{C} \cdot x^{\mathrm{a}}$ and taking the logarithm of each side of the relationship, you get $\log y = \log \mathrm{C} + \mathrm{a} \cdot \log x$. This is a linear relation between the variables $\log y$ and $\log x$, intersecting the y-axis at $\log \mathrm{C}$ and with slope a. When you

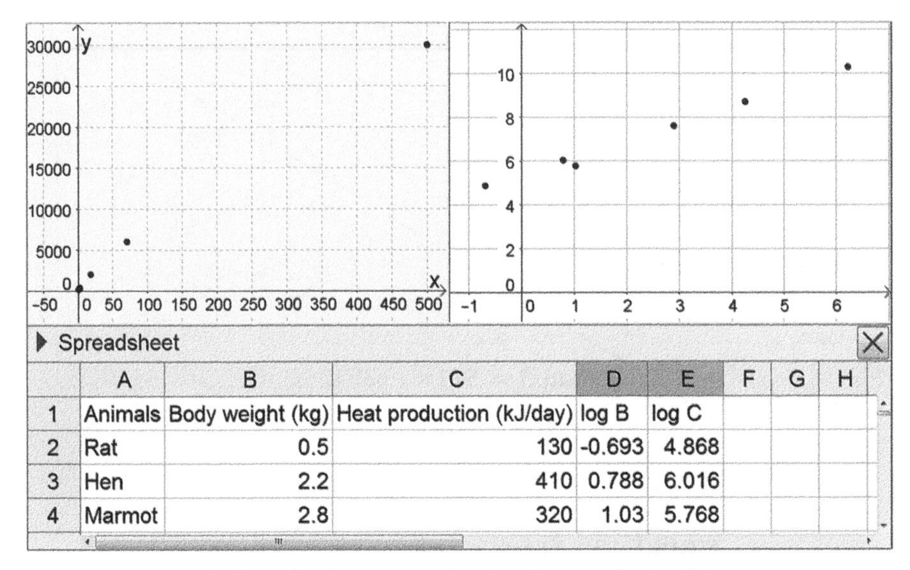

	A	B	C	D	E	F	G	H
1	Animals	Body weight (kg)	Heat production (kJ/day)	log B	log C			
2	Rat	0.5	130	-0.693	4.868			
3	Hen	2.2	410	0.788	6.016			
4	Marmot	2.8	320	1.03	5.768			

FIGURE 4.8 Constructing log-log diagrams in GeoGebra.

plot the points (log x, log y) in Graphics Area 2 of Figure 4.8, this will also show the points better than in Figure 4.7.

To fit functions to your data, make sure the left diagram is selected and then type

```
FitPow[list1]
```

Then select the right diagram and type

```
FitPoly[list2, 1]
```

Using the **FitPoly** command to create a line makes sure that it is returned as a function and not a geometric line. This is useful if you later want to evaluate the function for some value of x.

From either of the two fits shown in Figure 4.9, you should be able to calculate the heat production of a 100 kilo heavy animal to 7.9 MJ. In the initial left-hand diagram, you will see that g(100) = 7 900 kJ. In the right-hand diagram, it is more difficult to find the same result. First you must take the logarithm of 100 kg and log 100 = 2, so you have $f(2) = 3.9$. Since $\log y = 3.9$, you finally arrive at $y = 10^{3.9} = 7{,}900$ kJ. The exponent, the value for the parameter $a \approx 0.8$ can be read directly in the two diagrams.

Spontaneously one might think that the heat production should be proportional to the muscle mass, which should be proportional to the total mass. This should give a linear relationship. The data points indicate that it is more complicated than this and that other effects are at play here. For instance, it is likely that a large animal does not move as much as a small animal.

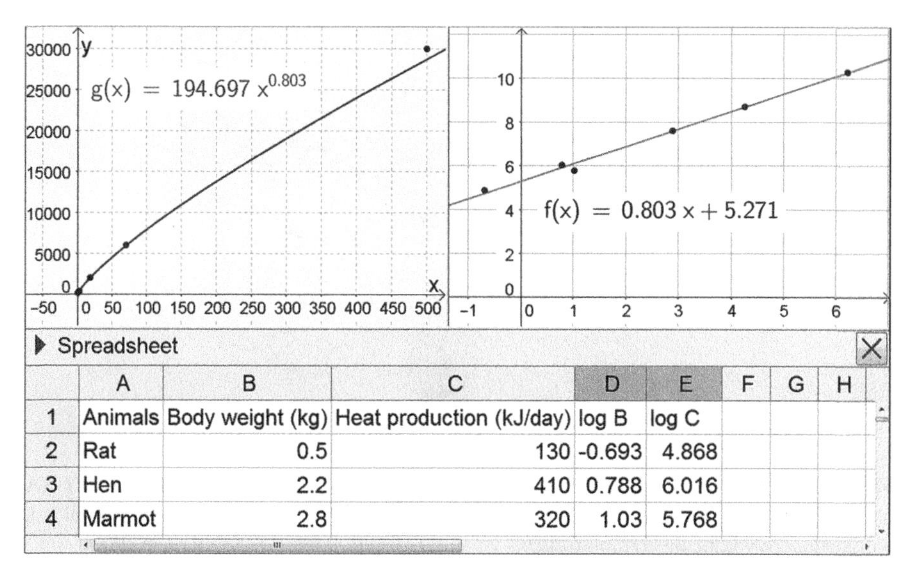

	A	B	C	D	E	F	G	H
1	Animals	Body weight (kg)	Heat production (kJ/day)	log B	log C			
2	Rat	0.5	130	-0.693	4.868			
3	Hen	2.2	410	0.788	6.016			
4	Marmot	2.8	320	1.03	5.768			

FIGURE 4.9 Initial and logarithmic values with a power function and corresponding linear function fitted to the data.

4.4 CONTROL OF INSECT PESTS

The larvae to the gypsy moth, *Lymantria dispar*, seen in Figure 4.10, creates a lot of damage to forests and fruit plantations. At mating, the male moths are attracted to the female moths through a sexual stimulant odorant. The same odorant can be used to trap the male moths into insect traps, to keep the population of moths under control. During a project with insect traps the number of trapped male moths were recorded as presented in Table 4.4.

Graph these data in a coordinate system and try to fit a function to the data points. Make an estimate of how many male moths you would trap if you used 150 μg of odorant.

Proposed Solution

You can see that the data points are not linear. Start by entering them into the spreadsheet in GeoGebra and create a list of points as in Figure 4.11.

A reasonable function should pass through the origin with a steep slope, which will then decrease. The function should not have a maximum value. This is typical for power functions. With the same methods that were used for Problem 4.3 *Warm-Blooded Animals*, you will get the results shown in Figure 4.12. Putting a point on the graph is done by typing

```
(150, f(150))
```

in the input bar. For 150 μg you should be able to trap approximately 23 male moths.

FIGURE 4.10 *Lymantria dispar*, the gypsy moth. Photo: Vitaman (own work) [CC BY-SA 3.0 (http://creativecommons.org/licenses/by-sa/3.0)], via Wikimedia Commons. https://upload.wikimedia.org/wikipedia/commons/d/d9/Lymantria_dispar_female_user_vitaman.JPG.

TABLE 4.4 Trapped Male Moths as a Function of the Odorant Amount

Odorant amount in trap (μg)	Number of trapped male moths
0.1	3
1	7
10	11
100	20

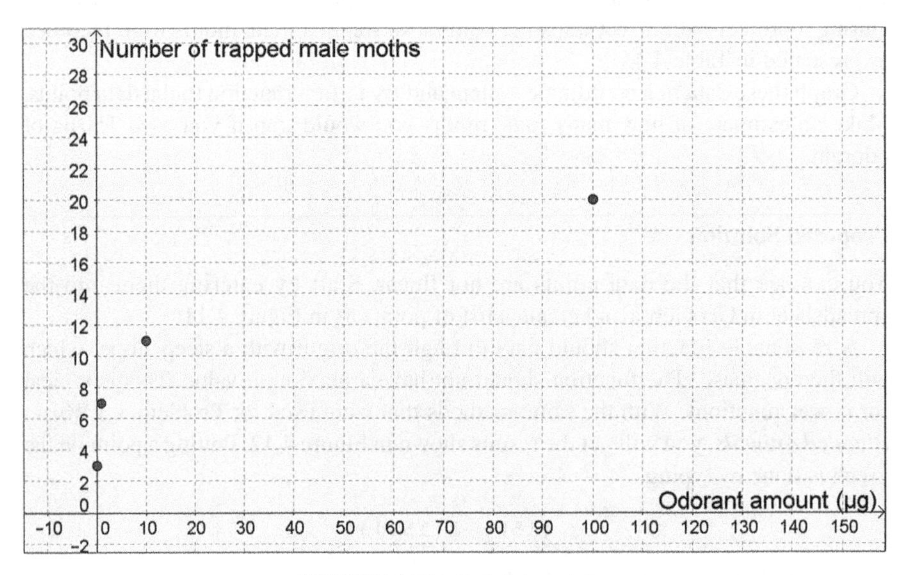

FIGURE 4.11 Just data points.

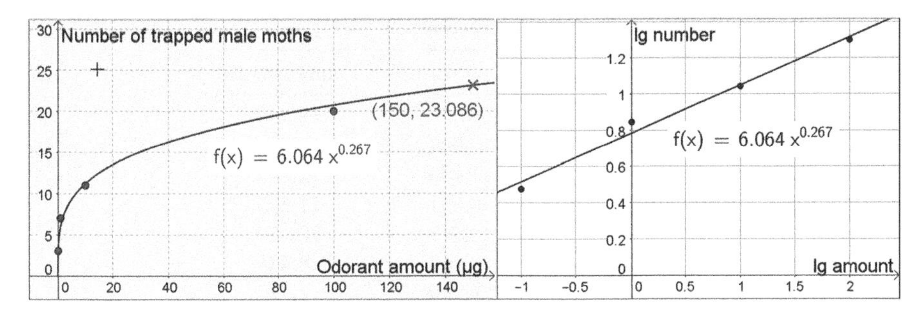

FIGURE 4.12 Finding a power function to fit data.

4.5 SELLING MAGAZINES FOR CHRISTMAS

In Sweden, right after school starts in September, the Christmas magazine companies distribute thick envelopes filled with information to hopeful eleven-year-olds detailing just how much they might earn by selling books, magazines, and other things to their parents, friends and neighbors for Christmas. In Sweden, there is a strong tradition of young children earning a bit of money for themselves in this way, and while the red tape nowadays is handled by their parents, those that participate are set up with their own web shops and get a bit of real experience in the door-to-door selling business.

One of these companies used the following business model in 2013: the amount of money you sell for is converted to points. Points can be used for premiums to obtain stuff like kick bikes, surf pads, and quadcopters but can also be converted to money.

If you disregard premiums, the sales are converted to points as shown in Table 4.5, and these points are converted to money, as in to Table 4.6. Because of this roundabout system, it is not exactly clear just how much money the children will earn given that the magazines sell for a certain amount. What will the direct relationship between your sales and your earnings look like?

If you instead focus on premiums, one of the top premiums in 2013 was an iPhone 5, shown in Figure 4.13, reportedly worth SEK 5,995, which one could get with the required 4,010 points. The amount of money earned from the sales then drops because a large portion of the earned points gets used up. How much exactly will the points drop, and what does this mean for the "price" of the iPhone?

Proposed Solution 1

Start by entering the tables in the GeoGebras spreadsheet, shown by clicking on **Ctrl-Shift-S**. It is a good idea to work with the tables from left to right, so place both tables in column A, the first table starting on row 1 and the second on row 51.

Now show Graphics Area 2 with **Ctrl-Shift-2** and set the axes as seen in the lower left of Figure 4.14. Click in Graphics Area 1 to make it active and then type `=(A2,C2)` in cell D2; copy this down to create points representing the lower end of the interval.

TABLE 4.5 Points from Sales

Sales from (SEK)	to (SEK)	Points awarded
12,999	—	13,800
11,994	12,998	13,040
10,882	11,993	12,110
10,418	10,881	11,230
9,964	10,417	10,020
9,520	9,963	8,435
9,086	9,519	7,720
8,663	9,085	7,210
8,520	8,662	6,315
7,847	8,519	5,930
7,454	7,846	5,545
7,071	7,453	4,465
6,698	7,070	4,010
6,335	6,697	3,720
5,982	6,334	3,545
5,639	5,981	3,220
5,306	5,638	3,015
4,987	5,305	2,820
4,673	4,986	2,625
4,373	4,672	2,310
4,068	4,372	2,115
3,797	4,067	1,905
3,521	3,796	1,700
3,266	3,520	1,540
3,007	3,265	1,415
2,769	3,006	1,310
2,534	2,768	1,200
2,311	2,533	1,005
2,101	2,310	905
1,893	2,100	730
1,704	1,892	660
1,522	1,703	565
1,349	1,521	485
1,197	1,348	420
1,045	1,196	385
905	1,044	305
776	904	255
653	775	230
537	652	211
437	536	202
313	436	193
245	312	180
0	244	0

TABLE 4.6 Earnings from Points

Points from	to	Earnings (SEK)
13,800	—	10,000
12,252	13,799	8,574
11,322	12,251	7,923
10,461	11,321	7,320
9,663	10,460	6,762
8,924	9,662	6,244
8,240	8,923	5,766
7,607	8,239	5,322
7,021	7,606	4,912
6,478	7,020	4,532
5,975	6,477	4,180
5,510	5,974	3,855
5,079	5,509	3,553
4,680	5,078	3,274
4,311	4,679	3,015
3,969	4,310	2,776
3,652	3,968	2,554
3,359	3,651	2,349
3,087	3,358	2,158
2,836	3,086	1,983
2,603	2,835	1,820
2,387	2,602	1,668
2,187	2,386	1,528
2,002	2,186	1,399
1,831	2,001	1,279
1,673	1,830	1,169
1,526	1,672	1,066
1,390	1,525	971
1,264	1,389	882
1,148	1,263	801
1,040	1,147	726
940	1,039	656
848	939	591
762	847	531
683	761	476
610	682	425
542	609	377
479	541	333
421	478	292
360	420	254
317	359	219
283	316	187
228	282	157
188	227	129
151	187	103
117	150	79
86	116	58
57	85	37
30	56	19
1	29	10

FIGURE 4.13 A popular phone. Photo: Brett Jordan/Flickr CC BY 2.0. https://www.flickr.com/photos/x1brett/8074153124/sizes/o/.

Graphics: Points vs. Sales (SEK)

Graphics 2: Earnings (SEK) vs. Points

Spreadsheet

	A	B	C	D
1	Sales...	to (SEK)	points	
2	13000	-	13800	(13000, 13800)
3	12000	13000	13000	(12000, 13000)
4	10900	12000	12100	(10900, 12100)
5	10400	10900	11200	(10400, 11200)
6	9960	10400	10000	(9960, 10000)
7	9520	9960	8440	(9520, 8440)
8	9090	9520	7720	(9090, 7720)
9	8660	9080	7210	(8660, 7210)
10	8520	8660	6320	(8520, 6320)
11	7850	8520	5930	(7850, 5930)
12	7450	7850	5540	(7450, 5540)
13	7070	7450	4460	(7070, 4460)
14	6700	7070	4010	(6700, 4010)

FIGURE 4.14 Both tables represented graphically.

Next click in Graphics Area 2 to make it active and type `= (A51,C51)` in cell D51, copying this down as well to create points for the second table. After rearranging the window's positions, you will get Figure 4.14.

You should immediately notice a couple of differences. While the conversion from points to earned money seems to be linear, the conversion from sales to points is

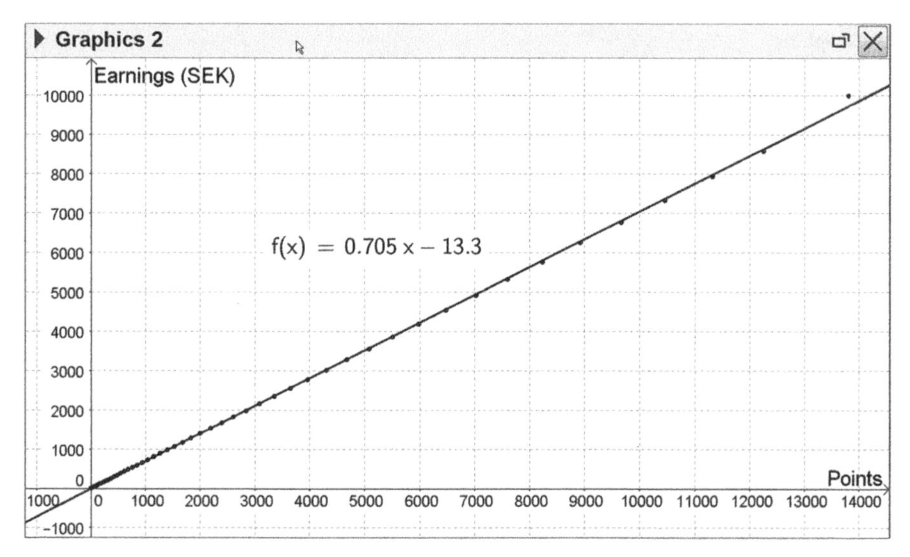

FIGURE 4.15 Earnings are a linear function of the points received.

anything but linear. It starts out relatively linear, or quadratic, but has unexpected jumps in the middle, only to finish somewhat like a negative quadratic.

Drag a selection rectangle round all the points in Graphis 2 with the right mouse button to select them, and then create a list from them by clicking on the Create List tool {1,2}. You can then fit a linear function by typing

```
FitPoly[list1,1]
```

Use **FitPoly** rather than **FitLine,** since **FitPoly** returns a function whereas **FitLine** returns a line. You can see in Figure 4.15 that you have received some 70% of the points as money.

So what to do with the relationship between sales and points? As a first approximation, attempt to fit a quadratic function to the first part of the relationship, before the jump at SEK 7,454.

Only select the points to the left of this jump, and click on the Create List tool {1,2} again. Then type

```
FitPoly[list2,2]
```

in the input field. This results in the quadratic shown in Figure 4.16. Since the quadratic term is very small, you should set GeoGebra to display at least three significant figures, in the **Options > Rounding** menu.

Combine both of these relationships. You can now calculate the earnings y as a function of the sales x by $y=f(g(x))$. You could do this algebraically by hand, or let GeoGebra do it with **Ctrl-Shift-K** by showing the CAS window, typing **`f(g(x)),`**

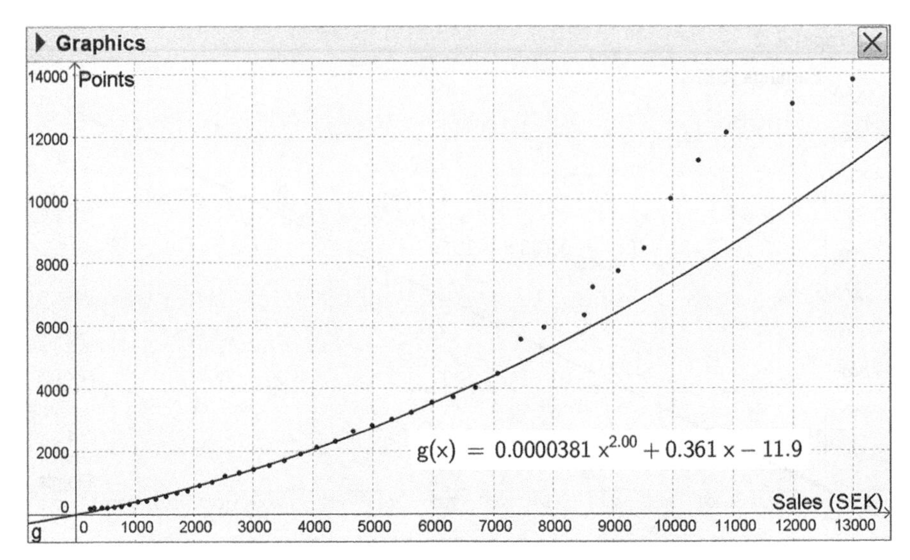

FIGURE 4.16 A quadratic function is fit to the first part of the table.

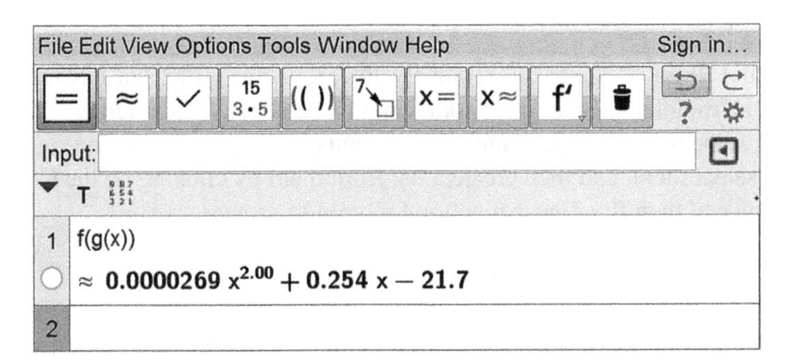

FIGURE 4.17 Using CAS to calculate $f(g(x))$.

and clicking the button for a numeric solution ≈ as shown in Figure 4.17. The earnings, y, can be calculated as

```
y = 0.000 0026 9 x² + 0.254 x - 21.7
```

In an empty part of the spreadsheet you can type some values, that is, 500, 1500, 2500, …, 7500. To the right of the topmost of these entries, type

`=f(g(X9)` (where X9 is the cell containing the value 500)

and copy this down. Some additional calculations give Table 4.7.

This method can produce quite sizable errors. You haven't taken into consideration that the tables aren't continuous functions and you have based the model on only one of the endpoints of the sales intervals, so you are stuck with systematic errors as well.

TABLE 4.7 Calculating the Size of the Errors of the Model

Sales (SEK)	Calculated earnings (SEK)	Actual earnings (SEK)	Error (SEK)	Relative error (%)
500	112	129	–17	–13
1,500	420	333	87	26
2,500	782	656	126	19
3,500	1,200	1,066	134	13
4,500	1,670	1,528	142	9
5,500	2,190	1,982	208	10
6,500	2,770	2,554	216	8
7,500	3,400	3,855	–455	–12

Proposed Solution 2

Obviously, the tables need to be represented in a better way. Mathematically the tables are piecewise-defined functions. With x in a given interval, the function has a value, given by the table. In GeoGebra piecewise defined functions can be constructed in two ways. You can directly input a function definition and its conditions like this:

```
f(x) = x2, -1 <= x < 1
```

Alternatively, for more complicated situations like the current one, you can use the **If** command.

Start by clicking in Graphics 1 to activate it. Next fill cells B2 and B52 with an arbitrarily large value, say 99,000, in order to replace the dashes or empty cells. In cell E2, you then type

```
=If[A2 <= x <= B2, C2, 0]
```

This means, that if x is in the right interval, the function is evaluated to whatever is in cell C2; otherwise, it is set to 0. The cell now contains a piecewise-defined function dealing only with row 2. Use $\leq$ both at the left and right ends of the interval, since the tables are constructed in this way and only use integer numbers.

Copy this formula down to cell E44. In Graphics 1 all these functions are shown simultaneously, but there is no greater use for them as individual functions. Therefore type

```
Sum[E2:E44]
```

in the input field, creating the function $h(x)$ shown in Figure 4.18. This function is a true representation of the table, and you can calculate values in the ordinary way, for example, $h(80000) = 5930$. You can now hide the functions in cells E2 to E43 to avoid duplication by selecting them, right-clicking the selection and selecting **Show object**.

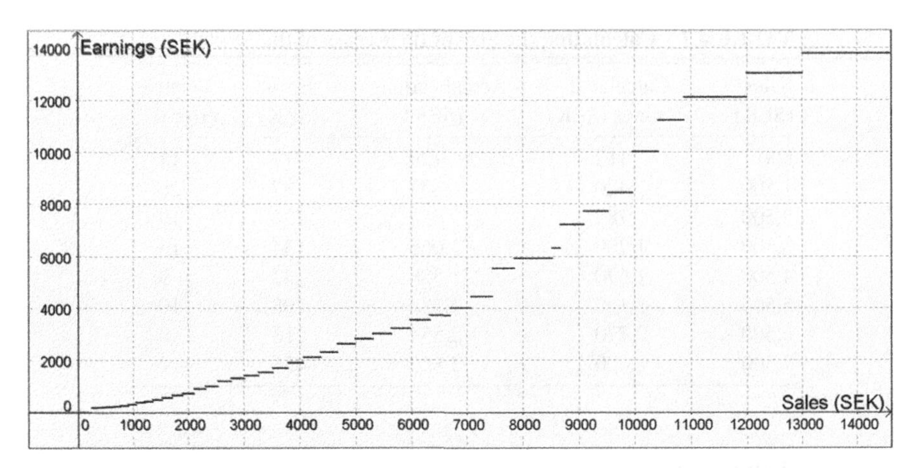

FIGURE 4.18 Table represented as a piecewise defined function.

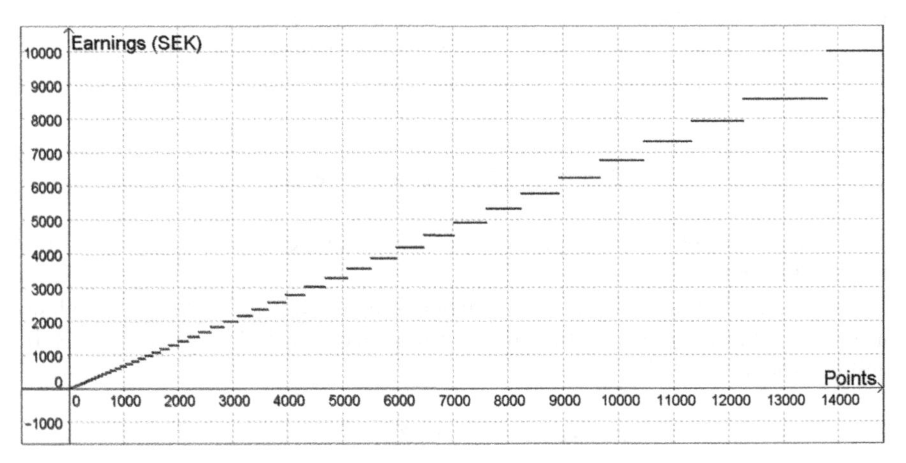

FIGURE 4.19 Converting points to earnings, as a piecewise defined function.

Now do the same for the other table. Click in Graphics 2 to make it active so that new objects are created there. In cell E52 you type

```
=If[A52 <= x <= B52, C52, 0]
```

which you copy down to E101. Then create the function $p(x)$, shown in Figure 4.19, by typing

```
Sum[E52:E101]
```

in the input field.

Now take the time to save your work! Read that sentence again and do it!

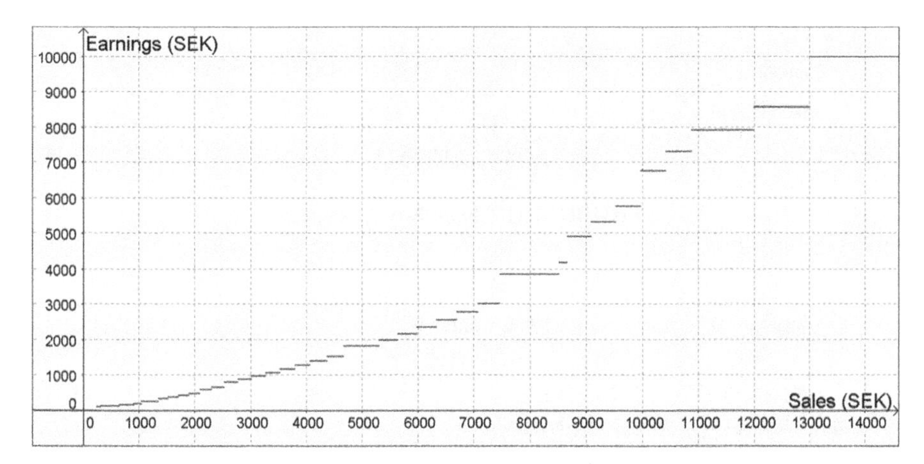

FIGURE 4.20 Earnings as a piecewise defined function of sales.

What you are about to do demands a lot of internal resources from GeoGebra. Remember that you are not dealing with two functions but some 45 + 50 functions and that when you try to combine these, GeoGebra will have to try to sort out more than 2,000 combinations of functions.

It is useful to save what you have done so far in one file first, and then again in a new file by a different name. Should the file become completely corrupt, you can then go back to one of these files.

Click the blue circles in the Algebra window next to the functions h and p so that they become white and the functions are hidden. Also close Graphics 2 and change the y-axis label to "Earnings (SEK)" and adjust the drawing window. Save again!

Now type this into the input field and press **Enter**:

```
q(x) = p(h(x))
```

GeoGebra may very well take some time do calculate this new function and the fan in your computer may start. GeoGebra may even crash, but if successful, you will be met by the graph in Figure 4.20.

This graph shows how much you earn if you convert your points to money without taking any premiums first. Notice that the intervals have become smaller as you have combined the basic functions.

As GeoGebra will now be bogged down with this monster function it will take a long time to complete even the simplest task. You should therefore attempt to save (which may, or may not work) and close this file and then open the previous file again.

It is actually more practical to create a calculator, or app, that calculates the earnings for a given sale. You might do this with a slider, but a slider can be difficult to set exactly, and it can result in lots of values being calculated as it slides toward a specific number.

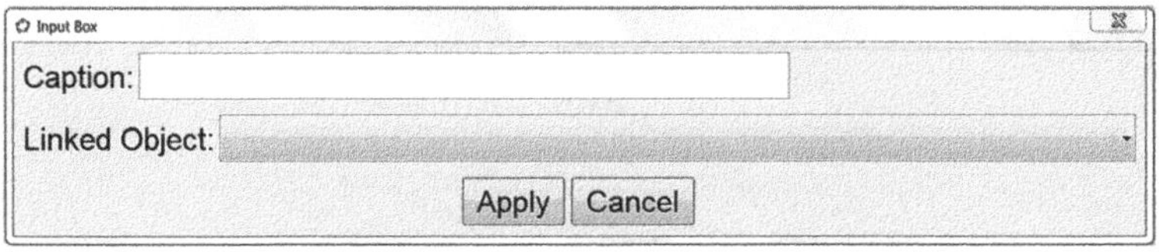

FIGURE 4.21 Input Box dialogue.

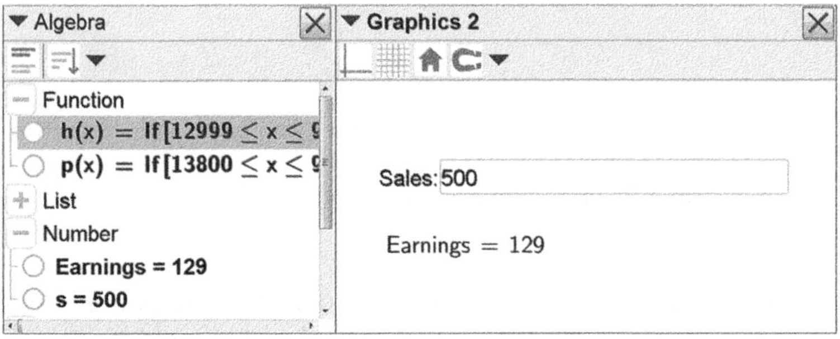

FIGURE 4.22 An "app" to calculate our earnings.

A better way is to use an input box. Type

```
s = 1
```

in the input field and then click on the Input Box tool and click in either graphics area where you want the input box to appear. You will get the dialogue shown in Figure 4.21.

In **Caption** you type "Sales," and then you select s in the list of objects shown under **Linked Object** and click **Apply**. You will get an input box where you can control the value of s. Now type `Earnings = p(h(s))` in the input field. Look for the number **Earnings** in the Algebra window and drag it out to the graphics area as shown in Figure 4.22. You can now calculate the earnings for any given sale.

By closing the Algebra window and the Style Bar, and by using the layout options to remove the input field and the toolbar, you can create a minimal, app-like look, shown in Figure 4.23. By uploading it to GeoGebraTube, you have actually created an online app to calculate your earnings from your sales. This app, shown in Figure 4.24, can be viewed online at http://tube.geogebra.org/m/1752501.

You can now attempt to answer the question of how the choice of an iPhone 5 affects the earnings. You must sell goods for at least 6,698 SEK to get the 4,010 points required for an iPhone 5. After this you have no points left and your earnings are 0 SEK. If you hadn't chosen the iPhone, those 4,010 points could have been converted to 2,776 SEK in earnings. You might therefore say that the price you pay for the iPhone is 2,776 SEK. Since an iPhone sells for more than this—even if

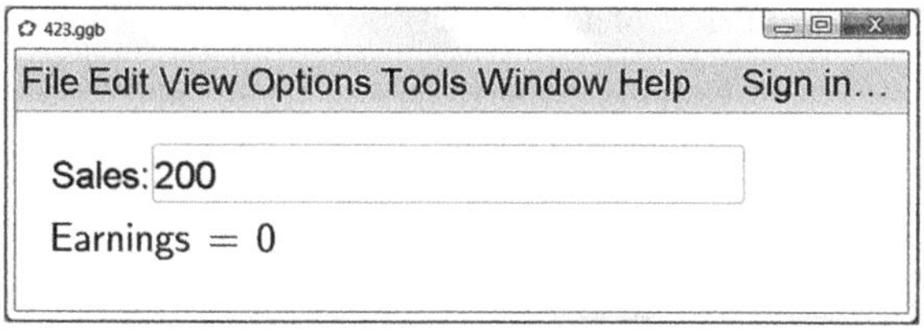

FIGURE 4.23 Creating a minimalistic look.

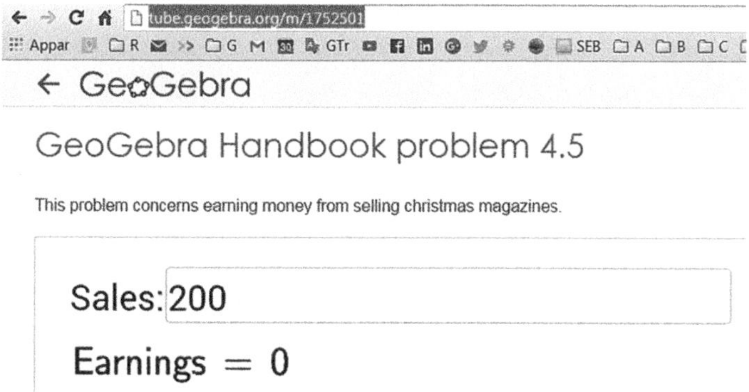

FIGURE 4.24 Using the app online.

not for 5,998 SEK—you have also received an extra bonus by selecting this particular premium.

How do you calculate this with the functions h and p? $h(6698)=4{,}010$ p and $p(4010)=2{,}776$ SEK. Also $p(4010-4010)=0$ SEK. So the "price" P of the phone can be written

$$P = p\big(h(s)\big) - p\big(h(s) - 4010\big)$$

Here s represents the sales in SEK, as before. By typing this into the input field, you get the number P. Drag this number out from the Algebra window to the graphics area. Then try out some different values for $s>6{,}998$ SEK to see the effect on the "price" and organize the results in Table 4.8.

It would appear that the "price" for individual items can be considered neither constant nor consequent. Probably, the two-table business model has been tweaked in various ways over the years without considering the overall effects.

It is also doubtful if the primarily eleven- and twelve-year-olds who take part in these schemes can really make sales in the 10,000 SEK range in the two months they have at their disposal. To make a 7,000 SEK sale, you must sell 770 SEK each week on average. Most of these children will probably not get an iPhone—at least not by selling Christmas magazines.

TABLE 4.8 "Price" of an iPhone

Sale (SEK)	Earnings (SEK) (if no iPhone claimed)	"Price" of iPhone (SEK) (if iPhone claimed)
7,000	2,776	2,776
8,000	3,855	2,576
9,000	4,912	2,754
10,000	6,762	2,582
11,000	7,923	2,601
12,000	8,574	2,330
13,000	10,000	3,238

4.6 TUMOR

A disc-shaped tumor has been discovered with X-ray diagnostics as shown in Figure 4.25, and it is to be surgically removed. In order for the surgeon to know how much tissue to remove, x- and y-coordinates in centimeters are determined for six points at the edge of the tumor. These values are presented in Table 4.9.

Graph the data points in a coordinate system. Adjust a circle to the data points and determine the center and size of the tumor.

Proposed Solution 1

A good start is to enter the data points into the coordinate system of GeoGebra. In Figure 4.26 they seem to lie on an imaginary circle – but *how* should you adjust a circle to the data points? And how should you decide which circle is the best one?

One way to compare the quality of adjustments is to use the concept of SSE, squared sum of errors. As an estimate of the error, you can take the radial difference from every point respectively in or out from the circle. The SSE on these distances should work as a comparative measure between different circles.

A circle can be defined in several ways:

- Any unique circle goes through three different points. By selecting these three points differently, you will get several different circles to compare.
- A circle may be defined with the x- and y-coordinate for its center and its radius r. By creating three sliders x_0, y_0, and r and varying the values for these sliders, you can manually try to find a minimum for the error.
- The data points can be combined into pairs and connected to the point that seems to be most opposite; these three points will form the vertexes of a triangle. A center point (though there are many to choose from) to that triangle could be a candidate for a center point of the circle.

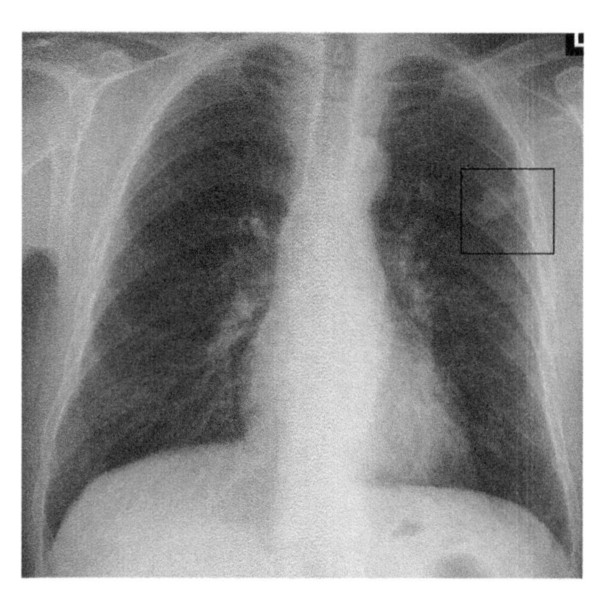

FIGURE 4.25 Possible cancer in left lung. Photo: Lange123 at the German language Wikipedia [GFDL (http://www.gnu.org/copyleft/fdl.html) or CC-BY-SA-3.0 (http://creativecommons.org/licenses/by-sa/3.0/)], via Wikimedia Commons. https://upload.wikimedia.org/wikipedia/commons/b/bf/Thorax_pa_peripheres_Bronchialcarcinom_li_OF_markiert.jpg.

TABLE 4.9 Coordinates of Tumor Edge

x	y
4.56	1.36
–0.28	1.43
–2.96	–2.92
–0.64	–7.21
4.73	–7.18
7.10	–2.88

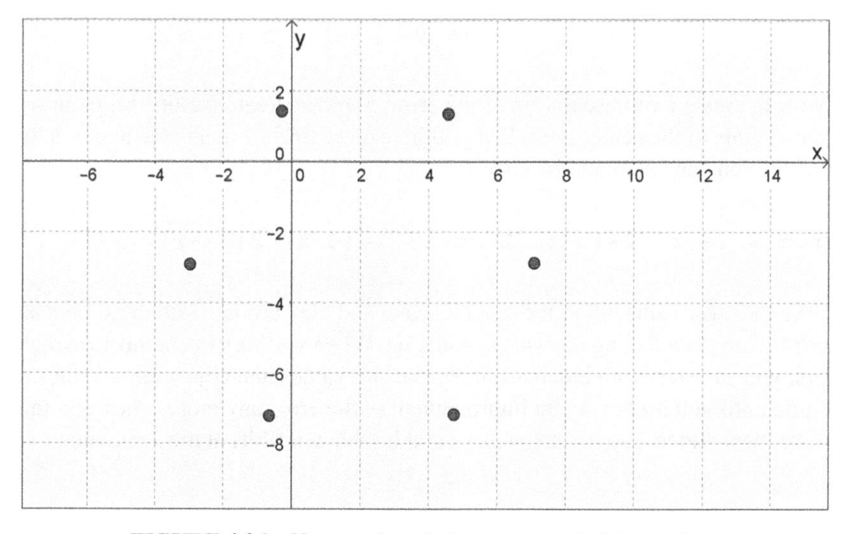

FIGURE 4.26 How we fit a circle to geometrical data points.

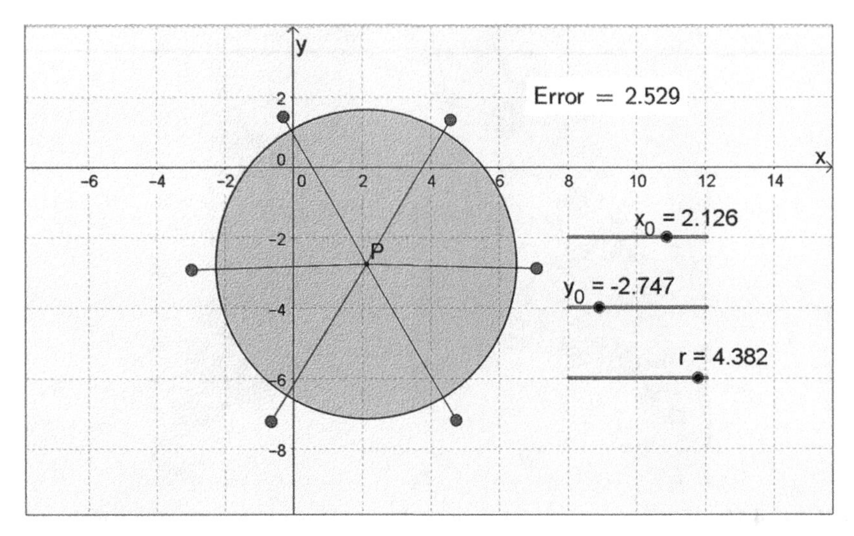

FIGURE 4.27 Circle in the process of being adjusted to the data points.

The method using sliders seems to be the most interesting. Therefore create the sliders x_0, y_0, and r (use underscores to indicate indexes in GeoGebra object names) and also construct a circle with the help of the sliders by typing

```
c1 = Circle[(x_0,y_0), r]
```

Alternately, you could use the Circle with Center and Radius tool (second in the circle tool box menu) after manually creating the sliders. Set the increment for the sliders to 0.001. Also define the center point of the circle by typing

```
P = (x_0, y_0)
```

Now, to create a representation of the error, you can create the line segment from the data points to the center point P. If you give these line segments the names a, b, c, d, e, and *f* you can calculate the error as

```
Error = (a-r)^2+(b-r)^2+(c-r)^2+(d-r)^2+(e-r)^2+(f-r)^2
```

Drag the error value out to the graphics area and play around with the sliders as in Figure 4.27 to get a feeling for what is going on. When you vary the numbers using the sliders, you can start with any reasonable starting values and adjust these values one at a time until you do not get an improvement in the error any more. Then you move on to the next slider. Use the arrow keys and hold down **Shift** at the same time to get

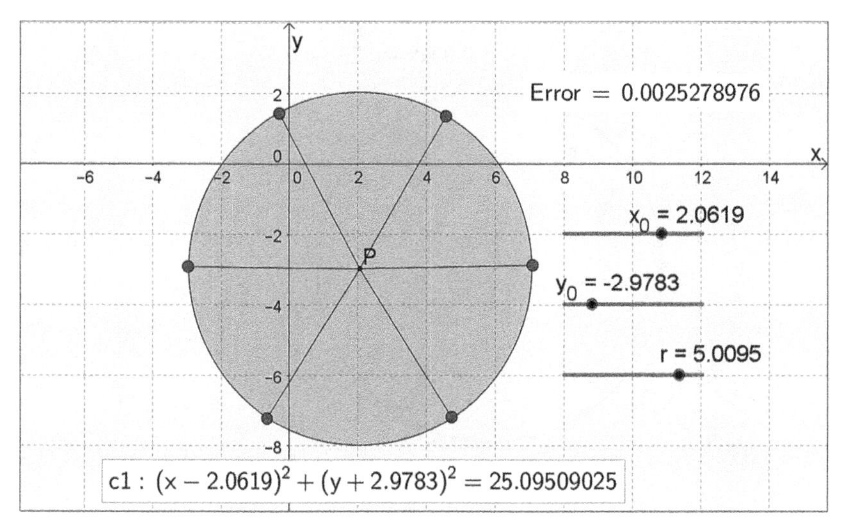

FIGURE 4.28 Final, best circle.

even better precision. You could also set GeoGebra to show 10 decimal places in order to track the error accurately.

In Figure 4.28 the center of the circle lies at (2.06, –2.98) and the radius is 5.01 cm.

Proposed Solution 2

There is a possibility in GeoGebra to fit implicit polynomial curves of a given degree to a set of points. A circle is a polynomial curve of degree 2, but so are ellipses, parabolas, and hyperbolas. If you accept that the solution might be an ellipse rather than a circle you can use this possibility by typing

```
FitImplicit[{A,B,C,D,E,F},2]
```

This creates the rather circular ellipse shown in Figure 4.29. But what are its properties, specifically its center and size?

The ellipse created is a curve, or path. GeoGebra doesn't know any of its properties. But you can work around this. By placing five new points on the path, you can create a new ellipse on top of the first one, with the difference that GeoGebra now knows it is an ellipse and can give you information about it.

So select the Conic through 5 Points tool and click on the path of the ellipse to place 5 new points, G to K, evenly spread out, on top of it. This creates a conic exactly overlapping the path. The equations of the path and the conic seem to differ in Figure 4.30, but this is only because of a multiplicative constant.

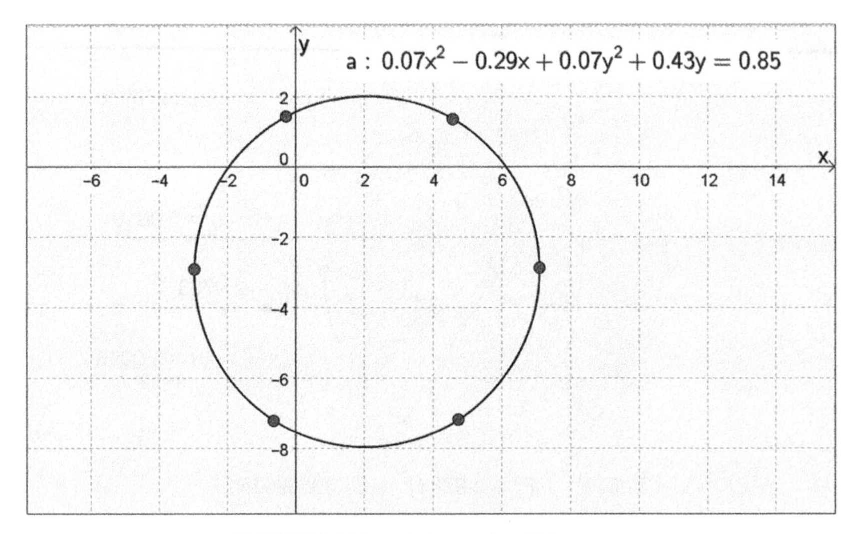

FIGURE 4.29 Fitting an implicit curve.

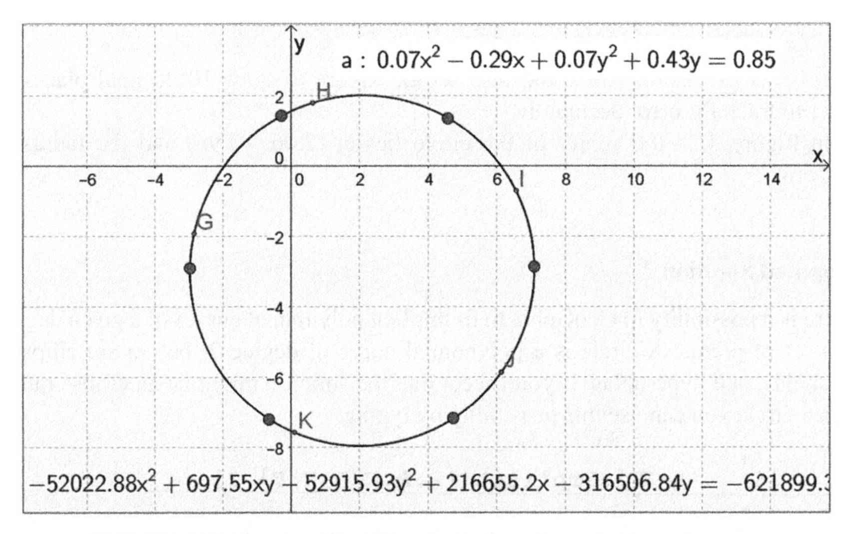

FIGURE 4.30 Letting GeoGebra re-discover the conic from the curve.

With conic c you can now access some of its properties. Draw its axes and find their intersection as a center point, and then find the length of the axis by typing the following commands:

```
Axes[c]
Intersect[b,d]
SemiMajorAxisLength[c]
SemiMinorAxisLength[c]
```

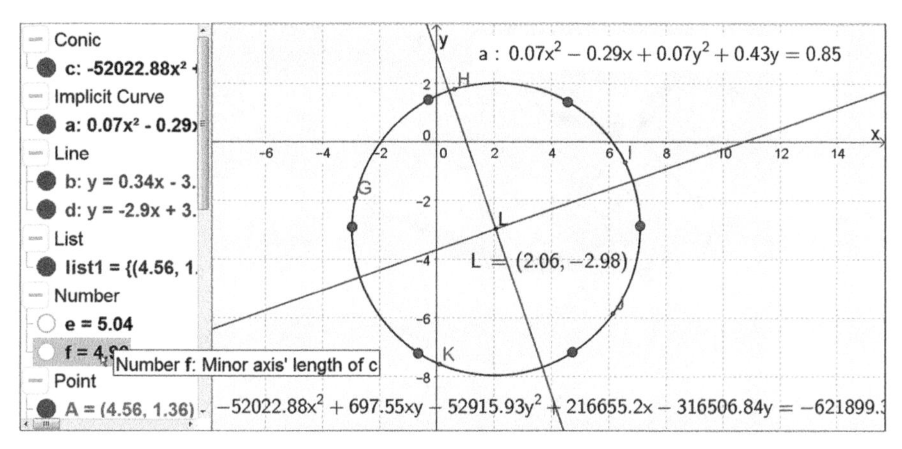

FIGURE 4.31 Analyzing the conic.

From Figure 4.31 you can see that the center of the ellipse is at (–2.06, –2.98) and the lengths of the semi-axes are 5.04 cm and 4.98 cm respectively, corresponding very well to the previous solution.

4.7 FREE FALL

Figure 4.32 hints at a functional relationship between the time in free fall, here indicated by the number of jumpers in the air behind you, and the vertical distance fallen so far. To investigate this relation, let's move into the lab. Table 4.10 describes the motion of a freely falling body in the laboratory. The entire fall takes place within one second.

Fit an exponential function of the form $d(t) = A + Be^t$ to these data. Also make some sort of analysis with regard to the suitability of the fit.

Proposed Solution 1

Enter the data into GeoGebra's spreadsheet, right-click the selection of these data, and choose **Create...>List of Points**. Adjust the Graphics window so you can see the data points clearly.

The function suggested is not one of the standard models implemented in GeoGebra, so you have to do a general fit of the kind described in Appendix B.5. In the input field you type

```
a = 1
b = 1
```

`m(x) = a + b e^x` This model function you can hide immediately.

```
Fit[list1, m]
```

FIGURE 4.32 Sky diving. Photo: Vernon Pugh (http://www.history.navy.mil/planes/v-22.html) (link broken on Wikimedia) [Public domain], via Wikimedia Commons. https://upload.wikimedia.org/wikipedia/commons/1/14/Aircraft.osprey.678pix.jpg.

TABLE 4.10 Distance Fallen as a Function of Time

Time (s)	Distance fallen (cm)
0.16	12.1
0.24	29.8
0.25	32.7
0.30	42.8
0.32	55.8
0.36	63.5
0.50	124.6
0.57	150.2
0.61	182.2
0.68	220.4
0.72	254.0
0.83	334.6
0.88	375.5
0.89	399.1

Note that you should press **Alt-e** to get the particular *e* representing the number 2.718…. If you enter a normal e, GeoGebra may sometimes interpret it as a not yet defined parameter, wanting to create a slider. This typically happens when e is already used once. Also don't forget that a space can work as a multiplication sign.

Figure 4.33 shows a reasonable fit, at least at a first glance. To do some error analysis and find the sum of squared errors, SSE, you may have wanted to use the

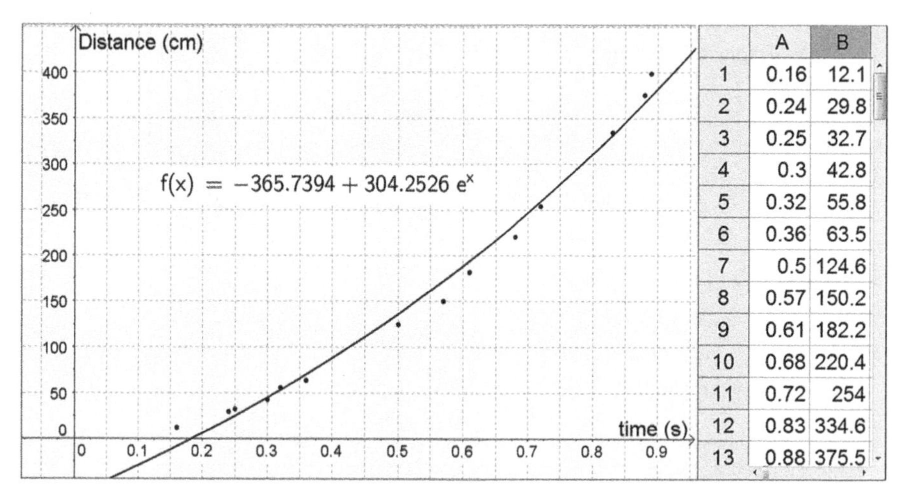

	A	B
1	0.16	12.1
2	0.24	29.8
3	0.25	32.7
4	0.3	42.8
5	0.32	55.8
6	0.36	63.5
7	0.5	124.6
8	0.57	150.2
9	0.61	182.2
10	0.68	220.4
11	0.72	254
12	0.83	334.6
13	0.88	375.5

FIGURE 4.33 Exponential function fitted to the data.

C1 =B1 - f(A1)

	A	B	C	D	E	F
1	0.16	12.1	20.7957	432.4623	(0.16, 20.7957)	
2	0.24	29.8	8.7586	76.7131	(0.24, 8.7586)	
3	0.25	32.7	7.7714	60.3945	(0.25, 7.7714)	
4	0.3	42.8	-2.1586	4.6595	(0.3, -2.1586)	
5	0.32	55.8	2.5448	6.4758	(0.32, 2.5448)	
6	0.36	63.5	-6.8547	46.9874	(0.36, -6.8547)	
7	0.5	124.6	-11.2883	127.425	(0.5, -11.2883)	
8	0.57	150.2	-22.0604	486.6604	(0.57, -22.0604)	
9	0.61	182.2	-12.0166	144.398	(0.61, -12.0166)	
10	0.68	220.4	-14.418	207.8776	(0.68, -14.418)	
11	0.72	254	-5.3272	28.3788	(0.72, -5.3272)	
12	0.83	334.6	2.5913	6.7147	(0.83, 2.5913)	

FIGURE 4.34 Calculating errors in the spreadsheet.

Two Variable Regression Analysis tool, but unfortunately this only works with the standard models.

However, the two most important functions from the analysis tool can easily be recreated in the spreadsheet. In Figure 4.34 two new two columns of data are added. In cell C1 you can type

```
=B1-f(A1)
```

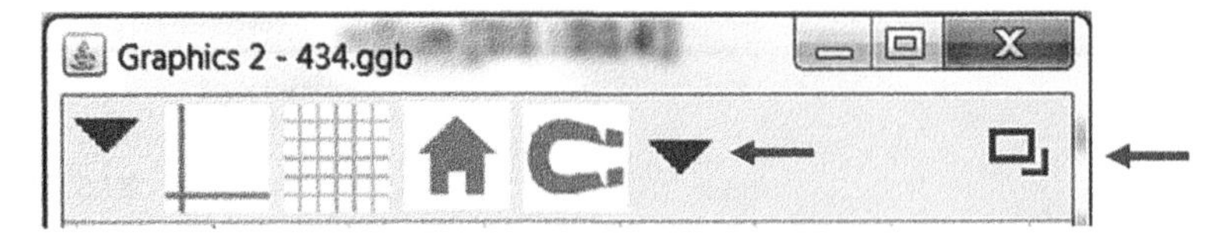

FIGURE 4.35 Docking button allows you to dock or undock a window from the rest.

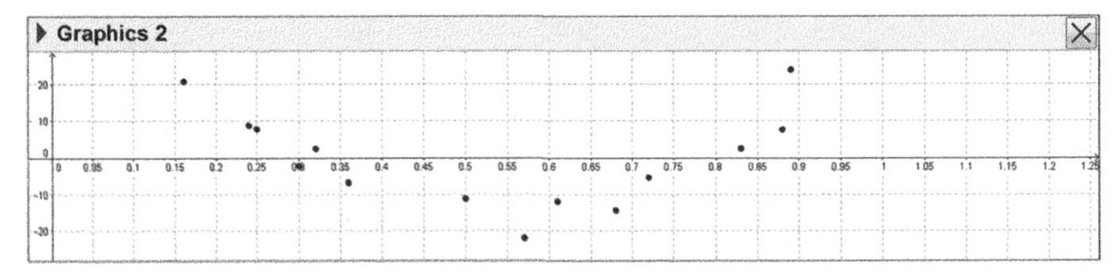

FIGURE 4.36 Residual diagram created manually.

and in cell D1

```
=C1^2.
```

Then copy these formulas, dragging the fill handle in the lower left of the selection of the two cells. In cell D15, create the sum of the squares of the errors, the *SSE*, by typing

```
=Sum[D1:D14]
```

This sum is approximately 2,262, which can later be compared with the sums of other models.

The other important tool that can be used from the Two Variable Analysis tool is the possibility to view a residual diagram. You can do so by showing Graphics Area 2, either from the **View** menu or by pressing **Ctrl-Shift-2**.

First, a note about GeoGebra windows: if this is the first time you use this window, it may be shown floating without axes or grid. But you can dock it to the main window by clicking the docking button in the top right of the window, shown in Figure 4.35. To the top left you will find Style Bar buttons for toggling axes and grid, zooming will show all objects and changing the point will capture the style.

Once Graphics Area 2 is docked, and set up the way you want it to look, you can create the points from the spreadsheet. Make sure that Graphics 2 is active; then type `=(A1,C1)` in cell E1 and drag this formula down. This is a quick way of creating points from the spreadsheet. The spreadsheet can have all kinds of objects in its cells—unlike most other spreadsheet programs that only can handle numbers.

You can immediately see from Figure 4.36 that there may be systematic errors, since the differences are positive at the endpoints of the interval and negative in the middle. Save this file for later.

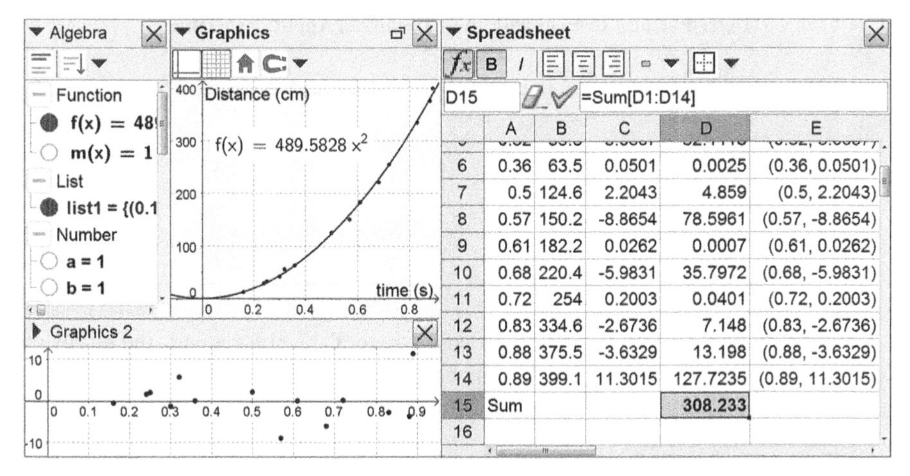

FIGURE 4.37 Same construction with another model.

Proposed Solution 2

Continuing from where you stopped in the previous proposed solution, you can now change the model function to a quadratic expression with only a quadratic term, by typing

```
m(x) = a x^2
```

The entire construction, including all calculations in the spreadsheet will immediately update as you press **Enter**.

Note from Figure 4.37 that the sum of squared errors, *SSE* = 308, is much less than the exponential solution. Besides, the residual diagram shows smaller and unsystematic errors.

Of course, this is due to the fact that for short falling distances—where air resistance is negligible—underlying theory states that the falling distance $s = gt^2/2$. You can even use the data to measure $g = 2 \cdot 489.6 \approx 980\,\text{cm/s}^2$.

In this case the theory gave a better result than a pure guess. Unfortunately, this is not true every single time, so you might find yourself choosing between a bad fit supporting the theory and a good fit from an ad hoc guess. If the measurements have been done carefully, this may later lead to new and better theories.

4.8 CONCENTRATION

On the company you currently work at, a chemistry engineer has just finished an experiment and now requires your help.

The engineer has added a valuable—but secret—agent called *X*-37 to a complex chemical solution. Using electronic probes, the engineer has measured the concentration $C(t)$ of the dissolved *X*-37 for different times, and the results are given in Table 4.11.

TABLE 4.11 Concentration of Dissolved Agent over Time

t (min)	$C(t)$ (mol/kg)
3	4.1
9	4.3
12	3.9
18	3.4
24	3.1
30	2.7

In order to optimize a production process where X-37 plays a part, the engineer wishes to know at what time t that the concentration is greatest. From previous experiments it is known that a suitable model for these data is $C(t) = \mathrm{c} + \mathrm{a} \cdot e^{-0.47 \cdot t} + \mathrm{b} \cdot e^{-0.06 \cdot t}$, for some values a, b, and c.

You mission, should you choose to accept it, is to find the time of greatest concentration. You must first find a good model for your particular set of data. You must also supply reasoning and a relevant error estimate supporting your conclusion.

Proposed Solution 1

Having entered the data in GeoGebra's spreadsheet and created a list of points from them you can perform a *general fit* by typing the following commands into the input field:

```
a = 10
b = 10
c = 10
m(t) = c + a e^(-0.47t) + b e^(-0.06t)
Fit[list1, m]
```

The result is shown in Figure 4.38. This is a type of function rarely encountered in standard textbooks, but nevertheless useful to know about since it can model pulses. The term whose exponent has the least absolute value will dominate for large x, while the other one will dominate for small x. If you graph the terms separately, you can see how they interact with each other.

Since the function is not a polynomial you have to add two parameters to the command for finding the maximum value and type

```
Max = Extremum[f,0,10]        (US English), or
Max = TurningPoint[f,0,10]    (UK and AU English)
```

to let GeoGebra search for the first maximum found in the interval [0, 10]. The concentration seems to reach a maximum value after 5 minutes and 38 seconds.

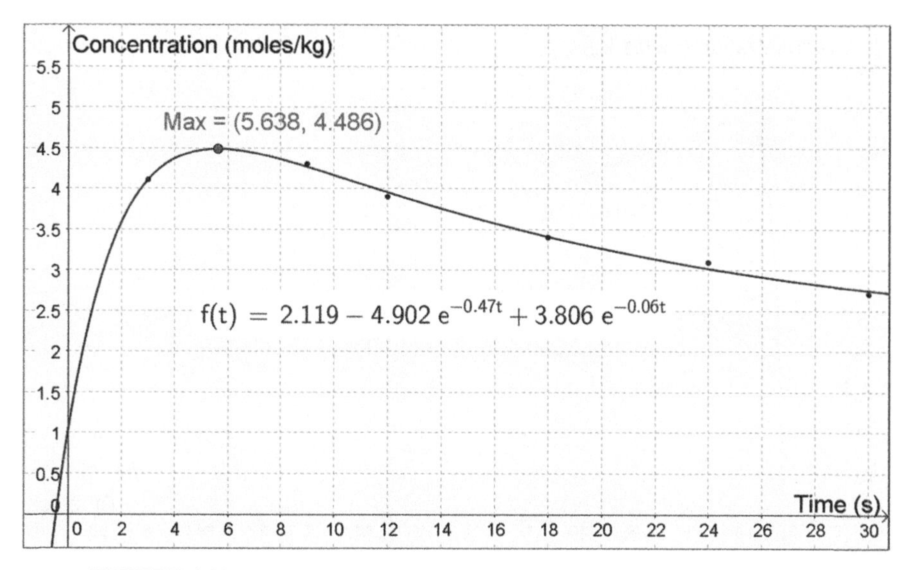

FIGURE 4.38 Sum of two exponential functions can model an initial pulse.

TABLE 4.12 ***Jackknife Resampling*** **Results**

Removed point	Time for maximum concentration (min)
A	Undefined
B	5.552
C	5.707
D	5.643
E	5.633
F	5.614

Since there are only six measurements, you can get an error estimate using a method called *jackknife resampling*. By removing one measurement at a time and noting the time for the maximum, you get a list of values from which you may calculate a standard deviation and confidence intervals. The practical bit of this is performed by editing **list1** where the data points are listed. Removing the first entry, you get the new list **{B, C, D, E, F}**, and then you simply re-edit this list to get the values in Table 4.12.

You may notice that the differences are greatest when removing points close to the maximum. Since there is only one measurement during the initial increasing phase, the model collapses when you remove this point.

Try to do the best of the situation and calculate the standard deviation from the five values you have to get $\sigma = 0.0499$ minutes. With 95% confidence and using the average of the mean with $5 - 1 = 4$ degrees of freedom, the time for the maximum concentration is $t = (5.64 \pm 0.05)$ minutes.

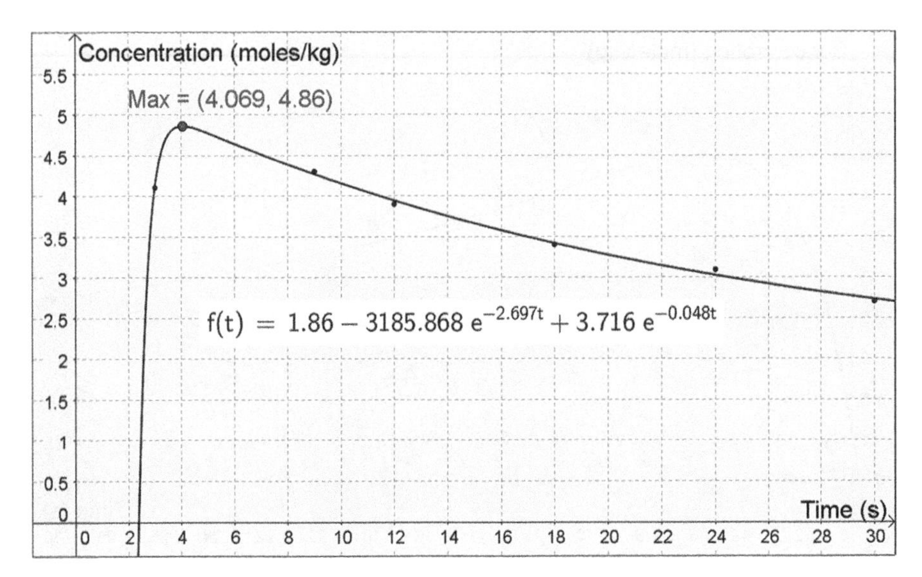

FIGURE 4.39 Danger of having too many parameters.

Proposed Solution 2

If there is no theory supporting the model, then it has probably been derived empirically from other sets of measured data. It might then be interesting to try to determine the exponents in the model as well. Perhaps this would lead to a better model? Try to fit a function $C(t)=c+a\cdot e^{-g\cdot t}+b\cdot e^{-h\cdot t}$ and see what you get. In this case the result seems to be sensitive to the starting values of the parameters, so you should set g=–0.47 and h=–0.06. The shape of the graph as shown in Figure 4.39 is now different and unfortunately much less realistic.

The graph comes up very sharply and does not give any realistic values for the concentration for times less than some two minutes. By fitting a model having too many parameters, you can indeed make the model cling tightly to the measured points, only to then do what it pleases outside the data set (e.g., or even between the points as is the case of high-order polynomials). You would probably get better results if you could find the concentration at $t=0$ minutes and preferably several more measurements on the rising flank and round the maximum. Unfortunately, this is not the case, and you must simply dismiss this model for the benefit of the previous result.

Proposed Solution 3

Since the *X*-37 agent apparently is destroyed by the solution the concentration will eventually decline toward 0. If you assume that the initial concentration is C_0, you should be able to use a model $C(t)=c+a\cdot e^{-g\cdot t}+b\cdot e^{-h\cdot t}$, where c=0, g>h>0, and $a+b=C_0$. Substitute $b=C_0-a$ and type

```
m(t) = a e^(-gt) + (2-a)e^(-ht)
```

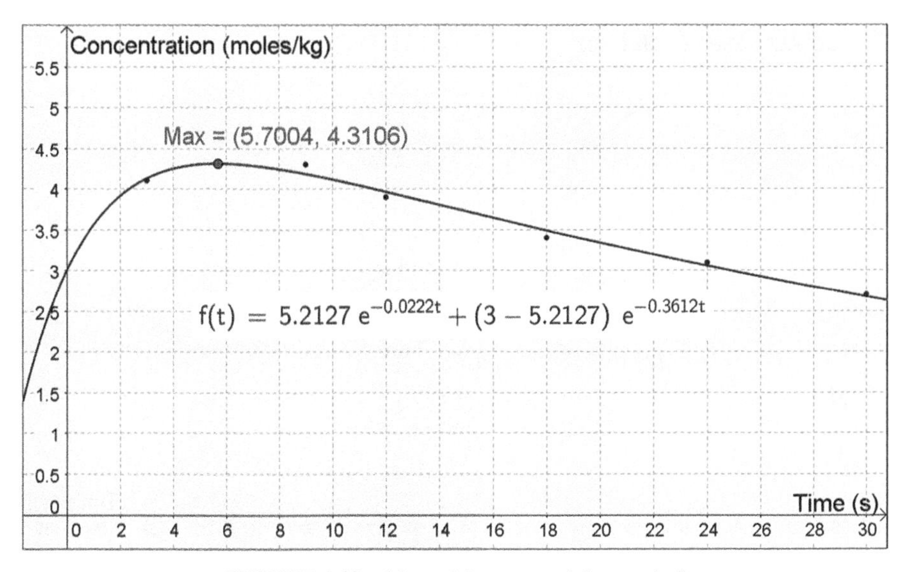

FIGURE 4.40 More elaborate model—maybe?

TABLE 4.13 Varying the Initial Concentration to Get a Range of Possible Values

Initial concentration (mol/kg)	Time for maximum concentration (min)
0.0	5.3633
0.5	5.4099
1.0	5.4624
1.5	5.5216
2.0	5.5872
2.5	5.6546
3.0	5.7004

Assume that $C_0 = 2$ mol/kg. The resulting graph is shown in Figure 4.40.

You can now investigate what will happen if you change C_0. Unfortunately, you cannot make C_0 a slider that you can change dynamically, since the fitting algorithm then will change it. Instead, change it manually in steps of 0.5 moles per kilogram to get the results shown in Table 4.13.

Now what to use for a result? Here it may be helpful to seek out and speak with someone who knows a bit of chemistry in order to better explain these disparate results. Still it may be reasonable to assume that the concentration is 0 when $t=0$, when you add the agent but while it still haven't had time to spread in the solution. What role do the exponents play in this model? Can this experiment update them or are they fixed by theory and is your work in vain?

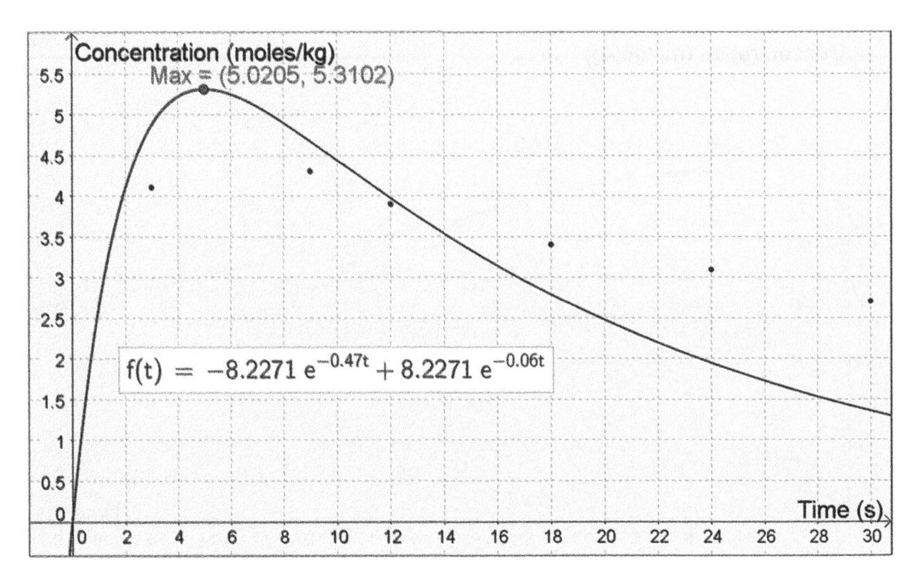

FIGURE 4.41 Too few parameters fit the data poorly.

Proposed Solution 4

Finally try for a synthesis of the first and third solution. Return to fixed exponents, –0.47 and –0.06, but combine this with the assumption that the concentration is 0 at the start of the process and that the concentration decays down to 0 with time. In this case the model should be formulated $C(t)=\mathrm{a}\cdot e^{-0.06\cdot t} - \mathrm{a}\cdot e^{-0.47\cdot t}$. Only one parameter remains. The results for fitting this function to the data give $t=5.0205$ minutes for the maximum concentration, but the graph, seen in Figure 4.41, does not seem to fit the data particularly well. Yet another disappointment! It would seem that the given model fits the data best, after all.

4.9 AIR CURRENT

In Table 4.14 experimental data are calculated for an air current, measured above the tip of an airplane wing as shown in Figure 4.42.

Here x represents R/C, where R is the distance from the core of the current, C is the position of the edge of the wing, and y represents V_0/V_∞, with V_0 the tangential speed of the current and V_∞ the surrounding air's speed above the aircraft.

Based on the laws of fluid dynamics, you know that the relationship

$$y = \frac{A}{x}\left(1 - e^{-\lambda x^2}\right)$$

is a feasible model for data sets of this sort.

TABLE 4.14 Air Current Data

x	Y
0.73	0.0788
0.78	0.0788
0.81	0.064
0.86	0.0788
0.875	0.0681
0.89	0.0703
0.95	0.0703
1.02	0.0681
1.03	0.0681
1.055	0.079
1.135	0.0575
1.14	0.0681
1.1245	0.0575
1.32	0.049
1.385	0.049
1.43	0.0532
1.445	0.0532
1.535	0.0511
1.57	0.049
1.63	0.0532
1.755	0.0426

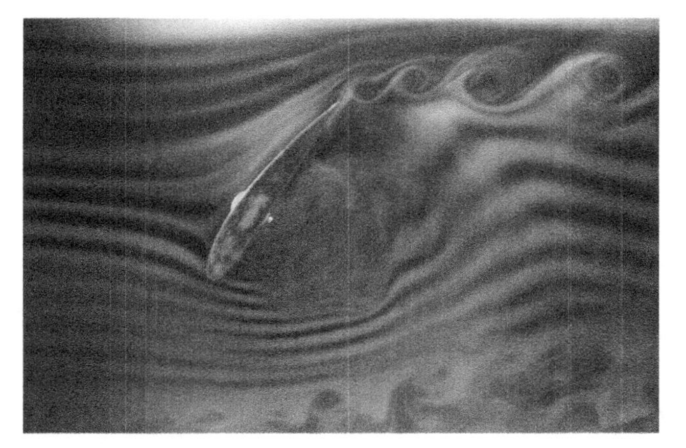

FIGURE 4.42 Wing profile in a wind tunnel. Photo: Georgepehli (own work) [CC BY-SA 3.0 (http://creativecommons.org/licenses/by-sa/3.0)], via Wikimedia Commons. https://upload.wikimedia.org/wikipedia/commons/8/8b/Fog_visualization.jpg.

Determine the parameters A and λ so that you get a good fit to the set of data points (x, y). Estimate the error in some reasonable way.

Proposed Solution

Enter the data in the spreadsheet and create a list of data points. Type in

```
a = 1
lambda = 1
m(x) = a/x (1-e^(-lambda x²))
Fit[list1, m]
```

and zoom into the axes to get the result shown in Figure 4.43 where the data points are shown as crosshairs.

Even before doing the fit, you may notice something odd about the data. Several of the y-values are identical, despite the seemingly large precision. This could be a sign of measuring equipment having a resolution less than advertised. Another thing you may notice is that the data are spread out and have measurement errors on the order of 0.1 in the y direction. You might just as well fit a straight line to the data as the complicated function you are given.

It is clearly very important in cases such as these, with a very scattered set of data, to have a sound theory to support the different functions that may be suitable.

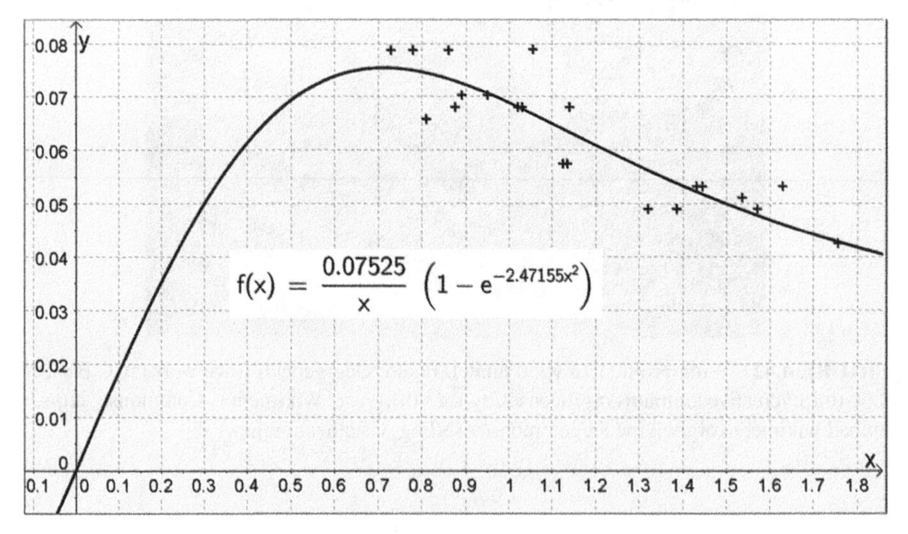

FIGURE 4.43 Scattered data requires a good theory to fit.

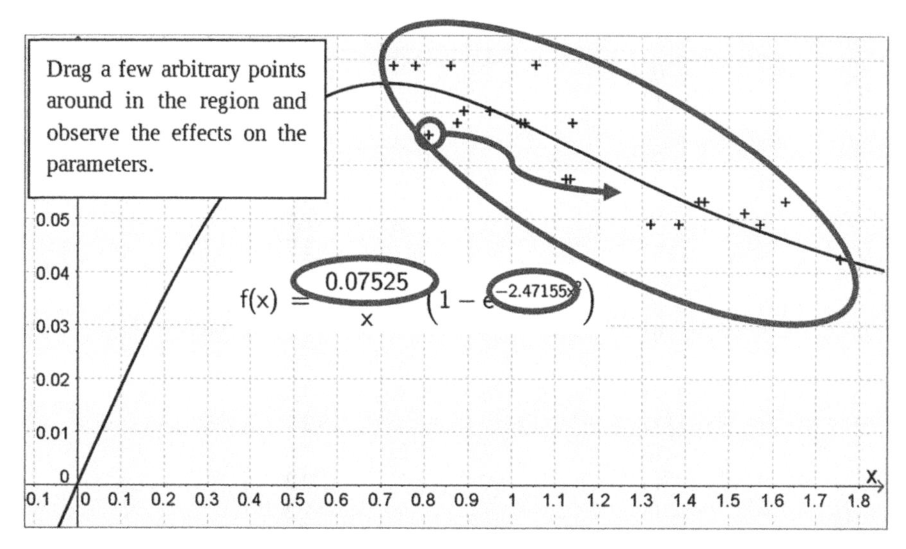

FIGURE 4.44 Intuitive way of estimating the error of the fitted parameters.

What can be said about error estimates? There are more than 22 data points and performing jackknife resampling manually would be very tedious. You could, however, do a simpler, more intuitive—though, of course, less accurate—variation of it, by taking some arbitrary data points (from different parts of the set) and simply dragging them round in the area where the other data points lie and then note the variation of the calculated parameters. This process is indicated in Figure 4.44.

After dragging a point, you may return it to its previous position by pressing **Ctrl-Z** immediately after you let it go. As you do this, you may note that you can get A to vary from approximately 0.072 to 0.079 and λ from approximately 2.35 to 2.75. In other words, you could set the values to $A=0.0755\pm0.0035$ and $\lambda=2.55\pm0.2$, which corresponds to relative errors of 5% and 8%. If you really tried to swing the points round in the area, most possibilities would be exhausted and these error estimates would probably be at 95 % confidence or so.

4.10 TIDES

In Table 4.15 the tidal data for Hall's Harbor in Nova Scotia are shown. This site is one of many with large variations of water level due to tides. In Figure 4.45 you see a jetty at low tide. In one session the water level was measured each hour for two full days from midnight July 1 to 23.00 July 2. The water level is measured in meters from a permanently mounted level indicator at the jetty.

Find a close-fitting function for this data set. Also try to make a reasonable error estimate.

TABLE 4.15 Tidal Data for Nova Scotia

Time (h)	Height (m)
0	2.4
1	1.2
2	–0.1
3	–1.5
4	–2.5
5	–3.0
6	–2.7
7	–1.6
8	0.2
9	2.1
10	3.4
11	3.6
12	2.9
13	1.6
14	0.2
15	–1.2
16	–2.4
17	–3.0
18	–3.1
19	–2.3
20	–0.7
21	1.3
22	2.9
23	3.9
24	3.1
25	2.0
26	0.6
27	–0.9
28	–2.2
29	–3.0
30	–3.2
31	–2.5
32	–0.9
33	1.1
34	2.9
35	3.9
36	3.6
37	2.5
38	1.0
39	–0.5
40	–2.0
41	–3.0
42	–3.4
43	–3.0
44	–1.7
45	0.2
46	2.2
47	3.5

FIGURE 4.45 Hall's Harbor, Nova Scotia, at low tide. Photo: RobNS (own work) [Public domain], via Wikimedia Commons. https://upload.wikimedia.org/wikipedia/commons/4/4c/Halls_Harbour_low_tide_view.JPG.

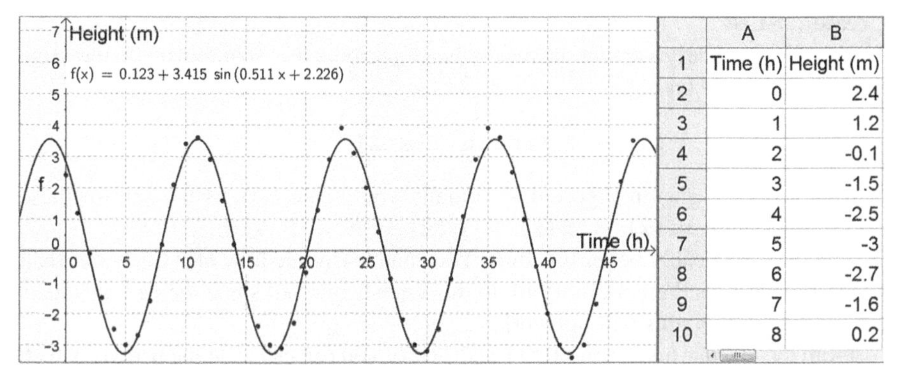

	A	B
1	Time (h)	Height (m)
2	0	2.4
3	1	1.2
4	2	-0.1
5	3	-1.5
6	4	-2.5
7	5	-3
8	6	-2.7
9	7	-1.6
10	8	0.2

FIGURE 4.46 A trigonometric fit.

Proposed Solution

Add the data to the spreadsheet and create a list of points. The points should show a clear periodicity. Also you could opt to make a trigonometric fit by typing

```
FitSin[list1]
```

The result, $f(x)$, in Figure 4.46, shows a relatively good resemblance to the data, but you can clearly improve on this.

You will now use a technique where you fit new functions to the residuals in order to build a complex model step by step. This is a useful way to increase the precision of a model, but ideally it should be supported by theory.

From data in columns A and B you can create new points in column C representing the residuals, which are the differences between data and function values. In cell C1 you type

```
=(A1,B1-f(A1))
```

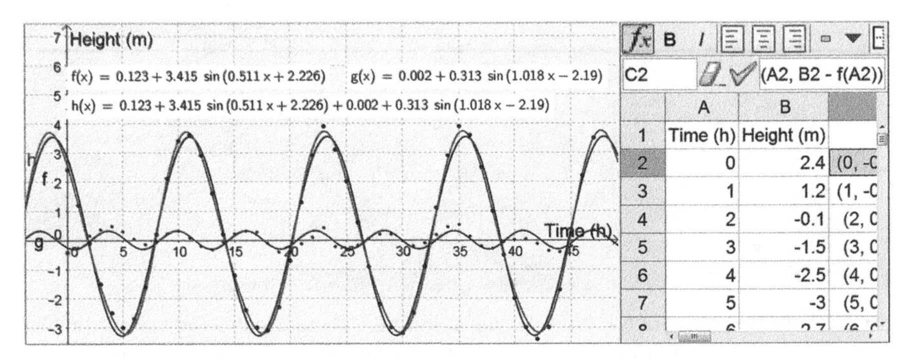

FIGURE 4.47 Successive regression analysis.

and copy this down. Then select all these points in column C, right-click, and select **Create...>List**.

You should clearly see that the residuals are periodic too, so now you do the same fit again, this time typing

```
FitSin[list2]
```

This gives the $g(x)$ in Figure 4.47. Finally you create $h(x)$, as the sum of these residuals by typing `h(x) = f(x) + g(x)`, or `h = f + g`.

The fit is noticeably better this time. You can clearly see how $h(x)$ fits better than $f(x)$ in most places. You could try to do this once again, but since the new residuals are mostly random, this will not work.

From the angular frequencies 0.511 and 1.0179 you can calculate the periods, which are $2\pi/0.511 = 12.3$ h and $2\pi/1.0179 = 6.17$ h. Orbiting round the Earth, relative to the Sun in 29.5 days, the Moon appears to "trail behind" each night, requiring approximately another 1/29.5 of a day $\approx$ 4/5 of an hour. This makes 24.8 h, close to the 24.6 h that the data suggest. The residuals have roughly the same period if multiplied by 4, suggesting they too are of lunar origin, possibly to do with irregularities due to the elliptic orbit.

At NOAA, http://www.tidesandcurrents.noaa.gov/map/you can search for tidal data. The Key West station http://www.tidesandcurrents.noaa.gov/stationhome.html?id=8724580 and most other stations have information on their *Harmonic Constituents* with links to these constituents at the bottom of the page. Here are the links to the Key West website and the site for the 37 identified Key West harmonic constituents:

Key West site: http://www.tidesandcurrents.noaa.gov/stationhome.html?id=8724580

Key West Harmonic Constituent's site: http://www.tidesandcurrents.noaa.gov/harcon.html?id=8724580

4.11 FITNESS

Let a student sit and rest for five minutes. Then measure the students' pulse, or heart rate. Next let the student perform some sort of physically demanding exercise until the pulse is high (though not exceeding 190 beats per minute). Then allow the student to rest again. Measure the pulse every other minute until it is down to the previously recorded resting heart rate.

One other student can do the recording and a third can document these measurements and film the entire process. Find a model for the relaxation of the pulse. What differences or a similarities to a pure exponential model do you find? What are the relative strengths or weaknesses of your model?

4.12 LIFE EXPECTANCY VERSUS AVERAGE INCOME

On Gapminder's website http://www.gapminder.org/world/you can find how the average life span correlates to the average income for different countries. In the data menu you can find the tables for life expectancy that visualize this. You can also download the data tables and fit a function in GeoGebra to model this relationship. Investigate the residuals for traces of systematic errors.

4.13 STOCKHOLM CENTER

Clicking this link, http://www.stockholmsfoto.se/bildarkiv/?imageID=121552 will take you to a photograph of the fountain at Sergels Torg, in the center of Stockholm, Sweden. It is shaped as a super ellipse whose equation may be written as

$$\left(\frac{x}{\mathrm{a}}\right)^{\mathrm{k}}+\left(\frac{y}{\mathrm{b}}\right)^{\mathrm{k}}=1$$

Determine the parameters a, b, and k, and the fountain's area and circumference.

4.14 WORKFORCE

A successful business have increased their number of employees according to Table 4.16.

Fit a suitable function to these data, and determine an estimate of the number of employees for 2016, 2020, and 2030.

TABLE 4.16 An Increasing Workforce

Year	Number of employees
2008	10
2009	12
2010	14
2011	17
2012	20
2013	24
2014	29
2015	34

TABLE 4.17 Population of Sweden

Year	Population (millions)
1750	1.78
1800	2.23
1850	3.48
1900	5.14
1950	7.04
200	8.87

4.15 POPULATION OF SWEDEN

The population of Sweden from 1750 onward is shown in Table 4.17.

Fit a suitable function to these data and make a prognosis for the years 2020, 2050, and 2100.

You can find similar data from any other country. Have each student work with a different country.

4.16 WHO KILLED THE LION?

A lion has been shot by a poacher. As the chief local investigator, the case has found itself to your desk. The three chief suspects for the crime—Darth Vader, The Joker, and Al Capone—all have alibis for the day in question, except for the following times.

- Darth Vader has no alibi between 8 am and 11 am.
- The Joker has no alibi between 11 am and 3 pm.
- Al Capone has no alibi between 3 pm and 9 pm.

The suspects could only have committed the crime in the time span when they did not have an alibi. You mission, should you choose to accept it, is to determine the exact time of the crime and thereby find out who committed the crime.

To determine the time of death, you measure the lion's body temperature at two different times. You do the first measurement at 9.00 pm the same evening the lion was found dead. The temperature was then 28.0 °C. Exactly three hours later, the lion's temperature has dropped to 25.6 °C. You may assume that the temperature decreases exponentially and that a living lion has a body temperature of 36.9 °C.

Are the assumptions reasonable? Are there assumptions that haven't been stated?

Who has committed the crime? Write a full report to the chief of police.

TABLE 4.18 AIDS Cases in the United States

Year	Number of AIDS cases
1982	295
1983	1,374
1984	4,293
1985	10,211
1986	21,278
1987	39,353
1988	66,290
1989	98,910
1990	135,614
1991	170,851

4.17 AIDS IN UNITED STATES

Table 4.18 gives the number of AIDS cases in the United States in the 10 years following its discovery in 1981. Try modeling these data and find a "prognosis" for 1995 and 2000 using this model.

Then plot the logarithms of the numbers against time. You should recognize the shape as an exponential rise. Now model the data again and find new estimates for 1995 and 2000. What is the symbolic function expression for the number AIDS cases? Look up the actual figures and discuss the results.

As an extension, look at the residuals. They exhibit an interesting systematic and periodic error. Could you use this to produce an even better model? Would this make your predictions better?

4.18 THERMAL COMFORT

On Autodesk's sustainability workshop page http://sustainabilityworkshop.autodesk.com/buildings/human-thermal-comfort you can find a complete app to determine your thermal comfort. All the variables and all mathematics are explained. Make a GeoGebra app that copies the behavior of the original app, complete with dynamic color to illustrate hot or cold.

Try to work backward from the given math and see if you can explain what the reasoning behind the math must be.

4.19 WATTS AND LUMEN

As old-fashioned energy-consuming lightbulbs are slowly phased away for the benefit of energy-saving alternatives such as halogen lightbulbs and LED lights, we are slowly getting used to converting from Watts (the energy a lamp is using per

TABLE 4.19 Watts to Lumen

Power consumption of an "old" lightbulb (Watts)	Light output (lumen)
15	120–135
25	220–235
40	410–470
60	700–805
75	920–1055
100	1330–1520
200	3010–3450

second) to lumens (a measure of the actual light output from the lamp). Table 4.19 gives both values for the "old" type of lightbulbs.

Try to find several different ways to model this relationship. Find the SSE for each model to determine which model is the best. Select a model and try using the minimum, average, and maximum values. Which of these has the best SSE?

Can you find a linear approximation that works in the range from 40 to 75 W (where most household lamps are) and that is simple to remember?

4.20 THE BEAUFORT SCALE

The Beaufort scale is a measure of wind speed as created at the beginning of the nineteenth century by Sir Francis Beaufort. Each step on the scale is represented by an integer, called the Beaufort number. The table found at the Wikipedia article https://en.wikipedia.org/wiki/Beaufort_scale#Modern_scale shows wind speeds, descriptions, and sea conditions for different Beaufort numbers.

The relationship between wind speed v m/s and the Beaufort number B is given by the empirical formula

$$v = 0.8365 \cdot B^{3/2}$$

The storm Hilde struck large parts of Sweden on November 16, 2013, and the highest wind speed was measured to 29 m/s, which is equal to 64.9 miles per hour. In contrast, the hurricane Katrina in 2005 had wind speeds up to 78 m/s (equal to 174.5 miles per hour).

When calculating B, the value is rounded to an integer. Calculate the Beaufort number B for the wind speed 29 m/s and 78 m/s.

For extreme wind forces, there are other scales. One of them is the TORRO scale, used for wind forces up to 130 m/s. The relation between wind speed v m/s and the number T according to the TORRO scale is given by the formula

$$v = 0.8365\sqrt{8} \cdot (T + 4)^{3/2}$$

Express the wind speeds 29 m/s and 78 m/s in both Beaufort and TORRO scales. What is the relation between the Beaufort and TORRO numbers?

4.21 THE VON BERTALANFFY GROWTH EQUATION

The *von Bertalanffy* growth equation is often used when measuring the length of a certain fish as a function of the fish's age, and one general version of this growth equation is given by

$$L(t) = L_{\infty} \cdot \left(1 - e^{-K \cdot (t - t_0)}\right)$$

The mathematical model, expresses the length, L, as a function of the age of the fish, t. The right-hand side of the model contains the age, t, and some parameters such as L_{∞} (maximum length), K (a specific curvature parameter), and t_0. Different growth curves can be created for each different set of parameters; therefore we can use the same basic model to describe the growth of different species simply by using a special set of parameters for each species. Compare this to Problem 3.9, *The Fish Farm I.*

What different species is this model useful for? Does it work for humans? Daffodils? Find growth data on the Internet or make experiments with plants. Decide which species it can be used for, apart from fish?

5

MODELING WITH CALCULUS

Why do we study calculus? This is perhaps best answered by asking back: What is calculus about? In short, one could say that calculus is the study of mathematically defined change. Two words in that definition requires further explanation: *mathematics* and *change*.

What is mathematics? One view is that mathematics is nothing more than the language of science. If science is a systematic study of nature, mathematics is a concise form of communication used to represent nature. We humans observe, dissect, and hypothesize nature and all its ongoing processes such that the end result of this orderly analysis is mathematics. In many cases we humans need a systematic way to study and control change. This is why we also need calculus.

The present is mainly dependent on the conditions existing within the short frame of time that it occupies. But when the present extends to the future, our studies becomes more complicated, and the need for calculus arises. To understand change, we need to explain the concept of time.

By definition, time is a passage of events, such that for time to pass, something must change with respect to itself. It can be a moving object that implies a changing distance covered from a reference point. This requires an event-defining time. A rising temperature will imply that the temperature is changing, thus occupying time. Changes are in fact the results of actions in a situation.

But even if calculus is the study of mathematically defined change, it is not necessarily the study of time alone. Other dimensions can be changing with respect

Mathematical Modeling: Applications with GeoGebra™, First Edition. Jonas Hall and Thomas Lingefjärd.

Companion website: www.wiley.com/go/Hall/MathematicalModeling

to each other. If one dimension is changing with respect to itself, we say it changes with respect to time. When factors change with respect to each other, we disregard the effect time has on the factors and proceed only to analyze interacting dimensions, assuming that our factors are constant with change with respect to each other, not with time. Calculus is thereby the branch of mathematics used to study any phenomena involving change, and change is a relative concept possibly involving any pair of dimensions, time, force, mass, length, temperature, and so forth.

In this chapter then, we will investigate change such as it is usually described in mathematics, as derivatives and antiderivatives, or as integrals. For some problems involving curve length, we will use Wolfram Alpha to calculate the results, but otherwise, we will continue using GeoGebra, especially since this gives us the possibility to investigate dynamic solutions.

5.1 THE FISH FARM II

You are the proud inheritor of a fish farm, such as the one shown in Figure 5.1, where you farm salmon. You have fixed expenses (rental for storage, wages for those who rinse and pack the fish, etc.) of approximately 2,200 USD per month. The variable

FIGURE 5.1 Loch Ainort fish farm. Photo: Richard Dorrell [CC BY-SA 2.0 (http://creativecommons.org/licenses/by-sa/2.0)], via Wikimedia Commons. https://upload.wikimedia.org/wikipedia/commons/5/59/Loch_Ainort_fish_farm_-_geograph.org.uk_-_1800327.jpg.

expenses (mostly fish food) are approximately 2.25 USD per salmon. The salmon are allowed to grow until they weigh approximately 3 kg and then harvested for sale.

Your great grandmother, who left you the fish farm told you this much:

> "It seems that nowadays no one will buy a salmon if it costs more than 25 USD. On the other hand, you won't ever sell any more than 3,000 salmon per month in this small village, even if you give them away."

What price should you set to maximize your profit, and how sensitive will this price be to changes in the different parameters of this problem?

Proposed Solution 1

There are some simple basic relations for all economical activities. Here are the most basic, linear relations:

Total revenues (income) = number of sold units · price per unit + fixed revenues
Total costs (expenses) = number of sold units · cost per unit + fixed costs

Fixed revenues could be subsidies, grants, and so on. Rewriting the above as functions of the price per unit, x, you get

$$Revenues(x) = Number(x) \cdot x + R_0$$

$$Costs(x) = Number(x) \cdot C + C_0$$

$$Profit(x) = Revenues(x) - Costs(x)$$

The so-called supply function, *Number*(x), tells you how many salmon are sold per month for a given price x USD. Normally this is a decreasing function, meaning the higher the price, the fewer salmon you will sell. In this case all you know initially about this function is that:

$$Number(0) = 3000$$

$$Number(25) = 0$$

There are many possible functions that satisfy these conditions:

$$NumberLinear(x) = 3000 \cdot (1 - x/25) \quad \text{(linear function)}$$

$$NumberCubic(x) = -(3000/25^3) \cdot (x - 25)^3 \quad \text{(cubic function)}$$

A lot of what has been done so far is really general information that the students could be given before actually starting to solve the problem. In GeoGebra, the supply functions can be placed in Graphics Area 2, leaving the economical functions for revenues, costs, and profit in Graphics Area 1. This makes sense for a couple of reasons. The scales and units on the y-axes are different, and both graphic areas will be less cluttered.

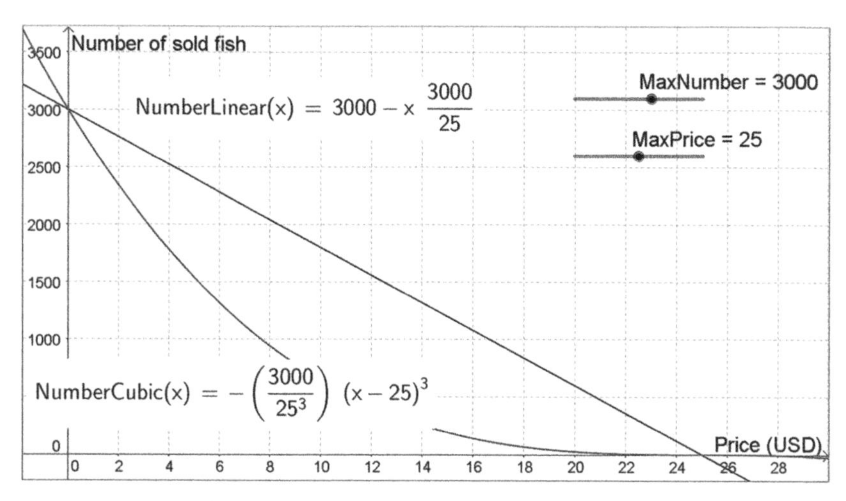

FIGURE 5.2 Two possible supply functions.

So make sure that both graphic areas are visible and that Graphic Area 2 is active. Then type in (remember that a space works like a multiplication sign)

```
NumberLinear(x) = MaxNumber - x (MaxNumber/MaxPrice)
```

Accept the suggestion to create sliders for MaxNumber and MaxPrice, and then type

```
NumberCubic(x) = -(MaxNumber/MaxPrice³) (x - MaxPrice)³
```

The cubes can be typed using the keyboard shortcut **Alt-3**.

In Figure 5.2 these functions are shown together. Both have their own advantages, and it is not à priori known which might be the more realistic of the two. The cubic function is on the form $y=\mathrm{k}\cdot(x-\mathrm{b})^3$, which guarantees that the derivative $y'=0$ when $x=\mathrm{b}$.

Now activate Graphics Area 1 and type

```
CostPerFish = 2.25
FixedCost = 2200
```

This creates two parameters that can be changed later. Doing it this way has the advantage of setting the correct value from the start. Make sure they are visible as sliders anyway by checking that the visibility indicators to the left of the names in the Algebra window are blue. The economic functions are entered like this:

```
Number(x) = NumberLinear(x)
```

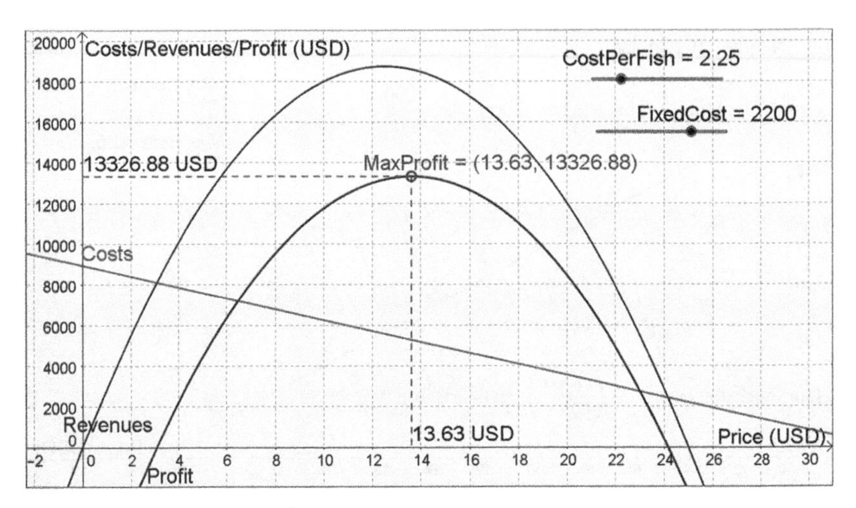

FIGURE 5.3 Finding the maximum profit from a linear supply function.

This way it easy to later change to the cubic function.

```
Revenues(x) = Number(x) x
Costs(x) = Number(x) CostPerFish + FixedCost
Profit(x) = Revenues(x) - Costs(x)
```

Note that GeoGebra show the resulting functions after substitutions have been made in the Algebra window. If you rather would like to see the definitions, as above, you need to go into the menu and select **Settings>Algebra>Definition.**

You now need to change the scale on the axes so that you can see the functions by zooming into (**Ctrl-Drag**) the axes and panning (**Drag**) the background. From the Style Bar you can add colors to the functions and find the maximum profit by typing

```
MaxProfit = Extremum[Profit]
```
(US English)

or

```
MaxProfit = TurningPoint[Profit]
```
(UK and AU English)

This yields Figure 5.3

Change the supply function by typing

```
Number(x)=NumberCubic(x)
```

so that you get completely different values for the maximum profit and the optimum price. In Figure 5.4 you can see that the profit is reduced to only a quarter of what it was. Clearly, there is insufficient information to draw any sort of conclusions about the profit. The optimum price varies at least from 8 to almost 14 USD. This is quite a wide range.

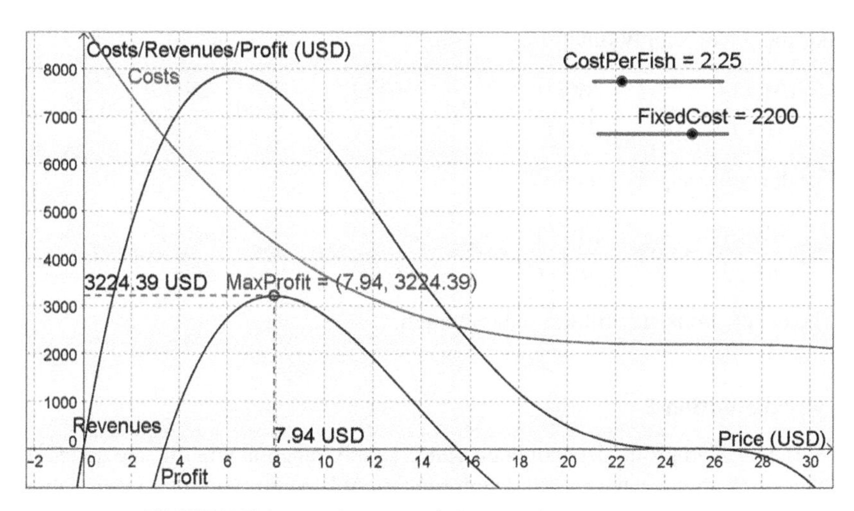

FIGURE 5.4 Maximum profit for a cubic supply function.

You are also supposed to investigate the sensitivity of the optimum price and maximum profit to the different costs and other parameters. If, for simplicity, you call the profit W (as your winnings, so not to confuse it with the price), the optimum price P, the fixed costs F, and the cost per unit C, you then need to investigate the size of the derivatives

$$\frac{dW}{dF}, \quad \frac{dW}{dC}, \quad \frac{dP}{dF}, \quad \text{and} \quad \frac{dP}{dC} \quad \text{for both supply functions}$$

You can do this without using any algebra by varying the costs a small amount and noting the change this produces in W and P. This is simply measuring the difference between the quotients for these pairs of variables. You can make the cost parameters CostPerFish and FixedCosts visible by clicking on the white circle next to their name in the Algebra window, and they will show up as sliders in the graphic area. Some simple observations can be made:

$$\frac{dW}{dF} = -1, \quad \frac{dW}{dC} \approx -1350, \quad \frac{dP}{dF} = 0, \quad \text{and} \quad \frac{dP}{dC} = 0.5 \quad \text{for the linear supply function}$$

$$\frac{dW}{dF} = -1, \quad \frac{dW}{dC} \approx -1040, \quad \frac{dP}{dF} = 0, \quad \text{and} \quad \frac{dP}{dC} = 0.75 \quad \text{for the cubic supply function}$$

The profit seems to be very sensitive to small changes in the costs per fish. A reasonable business strategy therefore would be to try to minimize these costs.

Studying the sensitivity to the *relative changes* will give slightly different results. For a small relative change, say, 1%, in both F and C, you get relative changes in W and P:

For the linear supply function:

$$\frac{dW/W}{dF/F} = -6.6, \quad \frac{dW/W}{dC/C} \approx -23.1, \quad \frac{dP/P}{dF/F} = 0, \quad \text{and} \quad \frac{dP/P}{dC/C} = 0.16$$

For the cubic supply function:

$$\frac{dW/W}{dF/F} = -6.8, \quad \frac{dW/W}{dC/C} \approx -56, \quad \frac{dP/P}{dF/F} = 0, \quad \text{and} \quad \frac{dP/P}{dC/C} = 0.21$$

The conclusions are still the same, though.

Proposed Solution 2

The differences in profit, depending on which supply function you use, are quite large, and thus imply that you have too little information. When you start selling your fish, you should carefully keep a list to document your sales. These numbers can give you the information you need as to what the supply function really looks like.

So after a while, you list your price and the number of sold fish at that price. Say you had a price of 15 USD for each salmon of 3 kg and managed to sell 1,000 fish. You now have yet another point on your unknown supply function. This point lies between the other functions. How can you find a reasonable function that pass through all three points? One simple method is to find the unique quadratic function that passes through all three points. You may find this quadratic function easy enough by typing the command

```
FitPoly[A,B,C,2]
```

where A, B, and C are the names of the points.

Using this method, you can find an optimal price of 12.80 USD, which yields a maximum profit of 11,000 USD as shown in Figure 5.5.

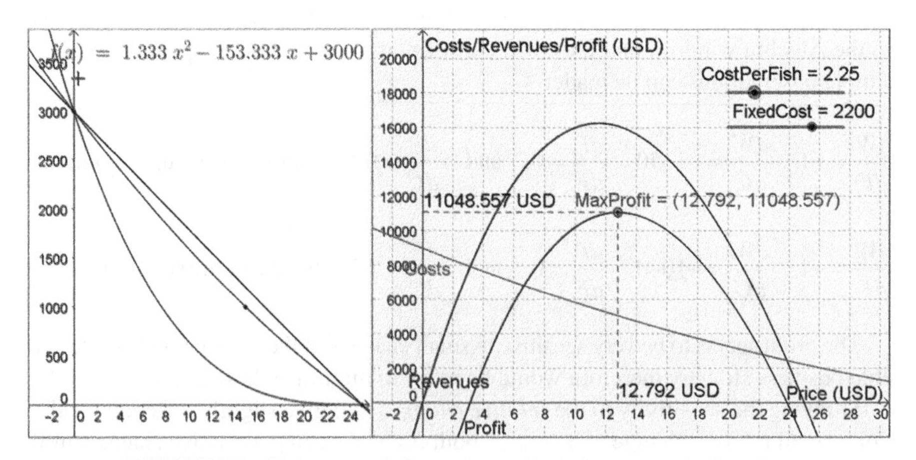

FIGURE 5.5 Using more information to find a better supply function.

Comments

There are obviously many faults with this model. Primarily, fish farms are seasonal businesses, both in demand and supply. The supply function will fluctuate drastically with time and also be strongly influenced by competition and advertising. In reality, analyses like these are not much used by small businesses. They are important tools of large companies and national and international companies where the economic models used are, of course, much more complicated than this simple model. Nevertheless, this simple model does gives some valuable insight into basic economic structures and prepares students for concepts such as derivatives and partial derivatives.

5.2 TITRATION

You allow 25.00 ml of an unknown, weak acid to be titrated with 0.200 M NaOH. You measure the pH for the solution in Table 5.1. You notice that the pH-values increase substantially when you add around 35 ml NaOH. This is the *point of equivalence*, where the sodium hydroxide completely neutralizes the acid. From this value it is possible to determine the concentration of the acid. If you also identify the pH-value at the half the added amount of NaOH, at about 18 ml, where the graph is very flat (or least steep), you will get the pKa-value for the acid, which you can use to identify the acid. This point is called the *half-titration point*.

Identify the *point of equivalence* with as high precision as possible, preferably in several different ways. Identify the acid and calculate the concentration.

Proposed Solution

Start by activating the spreadsheet with **Ctrl-Shift-S** or through the menu option **Show>Spreadsheet**. Enter the values and then create a point in cell C2 by typing

```
=(A2,B2)
```

Now copy this formula down by dragging the fill handle as indicated in Figure 5.6.

The points are represented graphically in the graphics area, as seen in Figure 5.7. In order to see the points better, the size of the points can be changed to 1 by selecting all the points in the spreadsheet, right-clicking, and then selecting **Object properties**, on the **Style** tab. You can also do this using the Style Bar.

It is then possible to see where approximately the *point of equivalence* is located, but what can you do if you want to be more accurate? The standard procedure is, of course, to find the maximum of the derivative.

The derivative for a function given in a table is obtained by using the central difference quotients between every pair of values. For example, in cell D2, you can create the x-coordinate for a point right in between the actual point and the next point by typing

```
=(A2+A3)/2
```

TABLE 5.1 Titration Results

NaOH (ml)	pH	NaOH (ml)	pH
0.00	1.93	31.00	5.62
0.50	2.92	31.50	5.68
1.00	3.23	32.00	5.75
1.50	3.41	32.50	5.83
2.00	3.54	33.00	5.92
2.50	3.64	33.50	6.03
3.00	3.73	34.00	6.18
3.50	3.80	34.50	6.40
4.50	3.92	34.60	6.46
5.50	4.03	34.70	6.53
6.50	4.11	34.80	6.62
7.50	4.19	34.90	6.72
8.50	4.26	35.00	6.85
9.50	4.33	35.10	7.04
10.50	4.39	35.20	7.39
11.50	4.44	35.30	9.06
12.50	4.50	35.40	9.86
13.50	4.55	35.50	10.16
14.50	4.60	35.60	10.36
15.50	4.65	35.70	10.46
16.50	4.70	35.80	10.56
17.50	4.75	35.90	10.66
18.50	4.80	36.00	10.66
19.50	4.85	36.50	10.86
20.50	4.90	37.00	11.06
21.50	4.95	37.50	11.16
22.50	5.01	38.00	11.26
23.50	5.06	38.50	11.36
24.50	5.12	39.00	11.36
25.50	5.18	40.00	11.46
26.50	5.24	41.00	11.56
27.50	5.31	42.00	11.66
28.50	5.38	44.00	11.76
29.50	5.47	46.00	11.76
30.00	5.51	48.00	11.86
30.50	5.56	50.00	11.86

In cell E2, you then calculate the derivative for this point by typing

```
=(B3-B2)/(A3-A2)
```

Next you select Graphics Area 2, using the keyboard shortcut **Ctrl-Shift-2** or by selecting **View>Graphics 2** from the menu. With Graphics 2 selected, you can now create points representing the derivative by typing

```
=(D2,E2)
```

▾ Spreadsheet

C2 =(A2, B2)

	A	B	C
1	Added NaOH (ml)	pH	
2	0	1.93	(0, 1.93)
3	0.5	2.92	(0.5, 2.92)
4	1	3.23	(1, 3.23)

FIGURE 5.6 Points are created when you type the coordinates within parentheses.

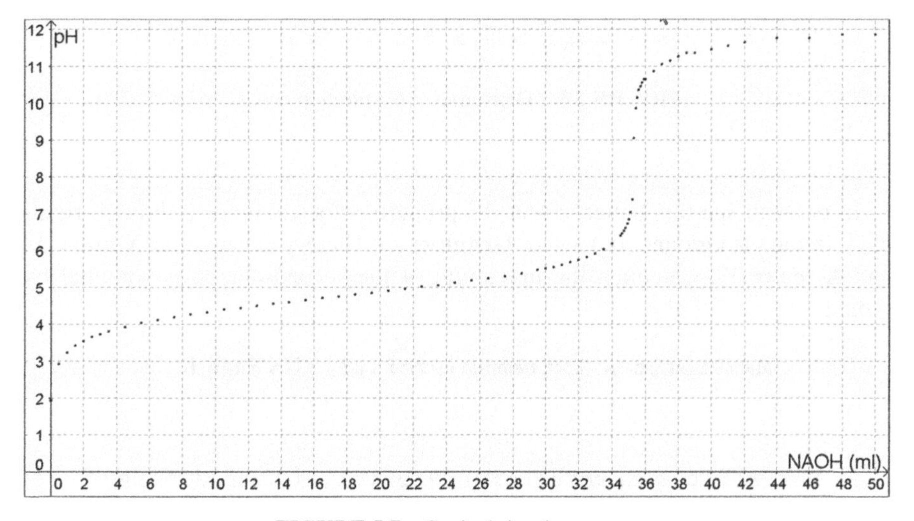

FIGURE 5.7 Typical titration curve.

into cell F2. Then select the cells D2, E2, and F2 and copy these down by dragging the fill handle. The result you get will be the same as shown in Figure 5.8.

You cannot just select the highest point. There may be larger values in between the two highest points. A better method is to fit a function to the points you have created around the maximum.

As a first attempt, you can try to fit a second-degree polynomial to points around the maximum. You can see in the table that there might be enough for you to do a regression using points F51, F52, and F53. Select the Graphics Area 2 and type the following command in the input field:

```
FitPoly[{F51, F52, F53}, 2]
```

where the number 2 means a second-degree polynomial.

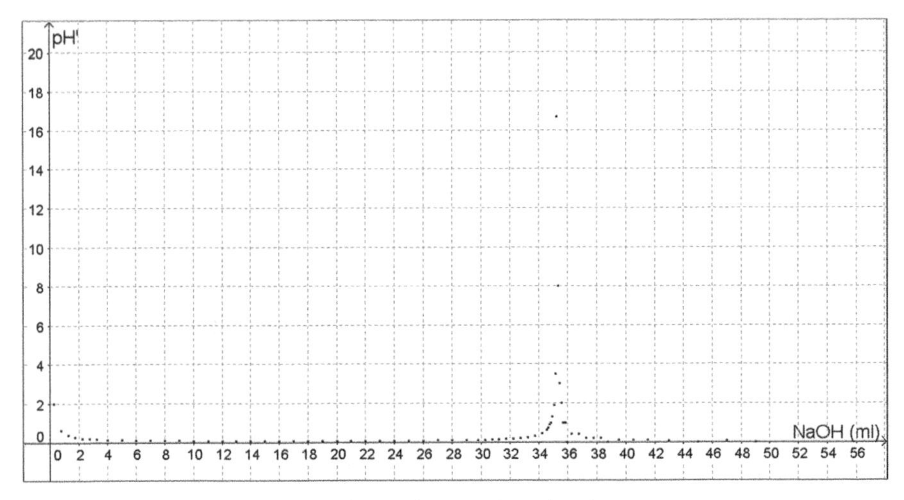

FIGURE 5.8 Derivative of the titration curve.

In order to see this better, select the properties for Graphics 2 through right-click in the background and select **Graphics....** There you can enter **X min = 33** and **X max = 37**. Calculate the maximum of the second-degree polynomial by typing

```
MaxProfit = Extremum[Profit]
```
(US English)

or

```
MaxProfit = TurningPoint[Profit]
```
(UK and AU English)

It is useful to label these properties as **Value**. In Figure 5.9 the color has been changed too. You can see that the *point of equivalence* is reached when you add 35.260 ml NaOH.

Now try another method for finding the extreme point. It is often prudent to work with several methods in order to get some insight into the magnitude of the errors.

You should see from the measured points that they appear more like a symmetric pulse than a second-degree polynomial. In Chapter 9 there are two examples of functions that resemble symmetric pulses. Use the first example as the model function and type

```
m(x) = a/(1+c (x - d)^2)
```

Notice the space between c and the parenthesis: otherwise, GeoGebra will interpret that c is a function. Select **Create sliders** in the box that appears. Select the three

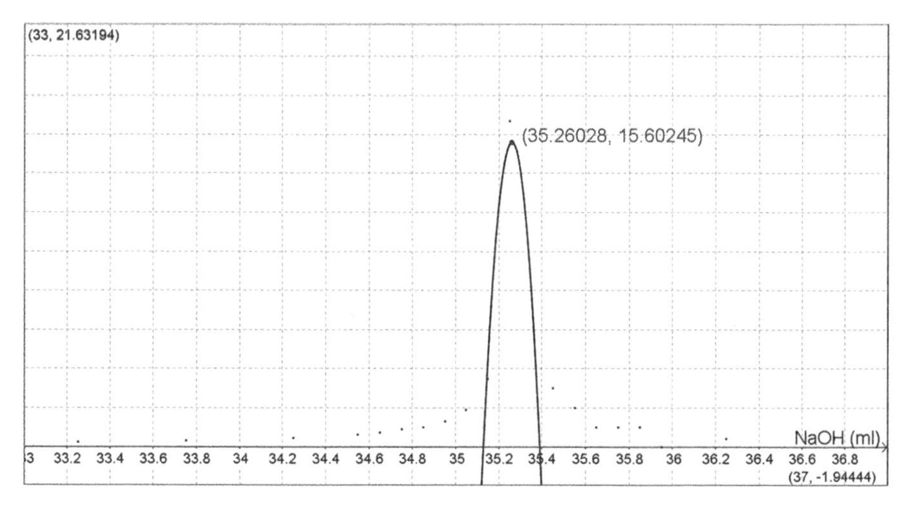

FIGURE 5.9 Second-Degree Polynomial Fit to Three Points for the Derivative.

numbers, right-click them, and select **Properties>Sliders** and enter **Max** = 50. Close the properties dialogue. Enter:

```
d = 36
a = 16
c = 50
```

in order to get the curve to look about the same as the data points. After that you might hide m(x) by clicking in the blue visibility indicator in the Algebra window to the left of m(x). The model function is just there to help GeoGebra understand which parameters to fit.

You can now use more points in the regression process. The points F45:F58 seem to be a reasonable choice. Start by creating a list. First type

```
list1 = F45:F58
```

Then create the regression function

```
Fit[list1, m]
```

It might be that the results are not at all what you expected. That in turn might depend on the values for the sliders, which are the start values for the algorithm that finds the regression function. If poorly placed, the algorithm may fail. If you adjust the values for the sliders, you may get a better result.

Since you no longer have a polynomial to work with, you can find the maximum point by typing

`Extremum[g, 35.2, 35.53]` (US English)

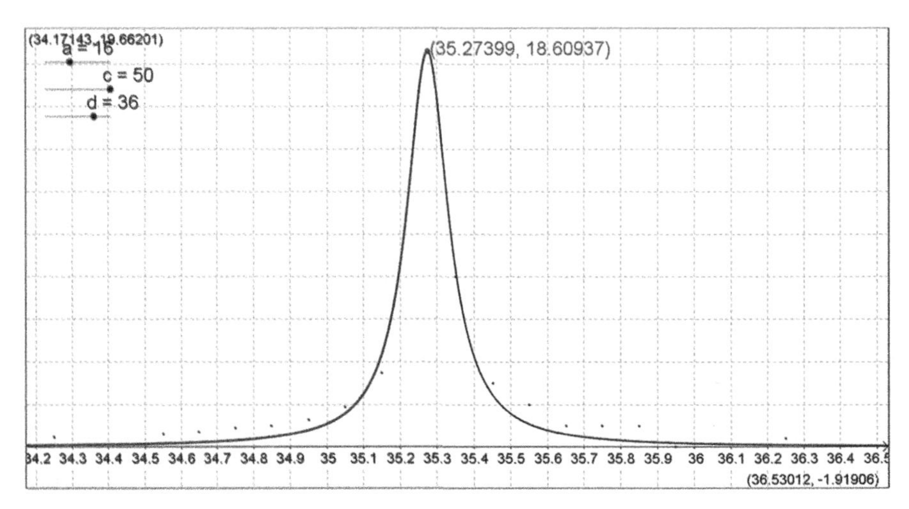

FIGURE 5.10 Pulse function of the type $1/(1+x^2)$ adjusted to the derivative of the titration curve.

or

```
TurningPoint[g, 35.2, 35.53]
```
(UK and AU English)

where you let GeoGebra know the interval the algorithm should search in. This time you get 35.274 ml NaOH; see Figure 5.10.

You might also try a normal distribution curve adjusted to the same set of points:

```
n(x)=c1e^(k1 (x-d1)^2)
```

Notice once again the space between k1 and the parenthesis. Accept to create sliders when prompted. Select these three sliders, right-click, and select **Properties>Sliders**, and enter **Min**=–50 and **Max**=50. Close the properties dialogue and enter the values d1=35, c1=16, and k1=–50. Hide n(x), select Graphics 2, and type:

```
Fit[list1, n]
```

This time you can find the maximum value by inspecting the function. The function is symmetric around d1, something you can see by dragging the function from the Algebra window to the graphics area. You can see in Figure 5.11 that the maximum value occurs for 35.268 ml NaOH.

The other part of the assignment was to find the function value for the *half-titration point* that gives the acid's pKa-value. Compare the values you have found so far in Table 5.2.

In order to find the function value for the *half-titration point,* you need to fit some reasonable function to the original data points. You can try this with a straight line and with a cubic function. Select the points C13–C33 in order to get points

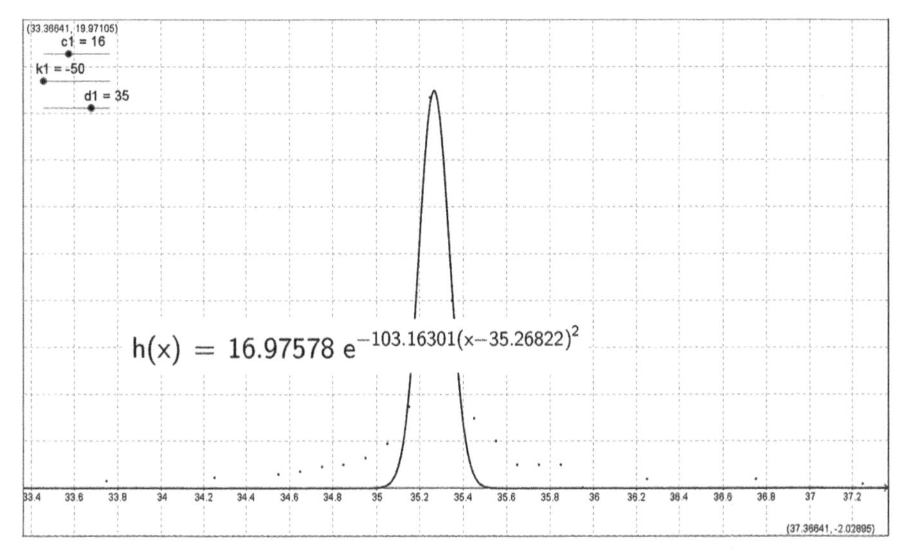

FIGURE 5.11 Normal distribution curve adjusted for our purposes.

TABLE 5.2 Titration Points by Different Models

Adjusted function	Titration point (ml NaOH)	Half-titration point (ml NaOH)
Quadratic	35.260	17.630
$1/(1+x^2)$	35.274	17.637
Normal distribution	35.268	17.634

symmetrically spread around the *half-titration point*. Create a list of these points by typing

```
list2 = C13:C33
```

Now find the straight line and the cubic function by first selecting Graphics 1 and then typing

```
FitPoly[list2,1]
```

and

```
FitPoly[list2,3]
```

This is shown in Figure 5.12. Function values can be obtained by function calls:

```
p(17.63)
```

You should arrive at the summary presented in Table 5.3. The results are near each other and you can use the average 35.267 ml NaOH for the chemical

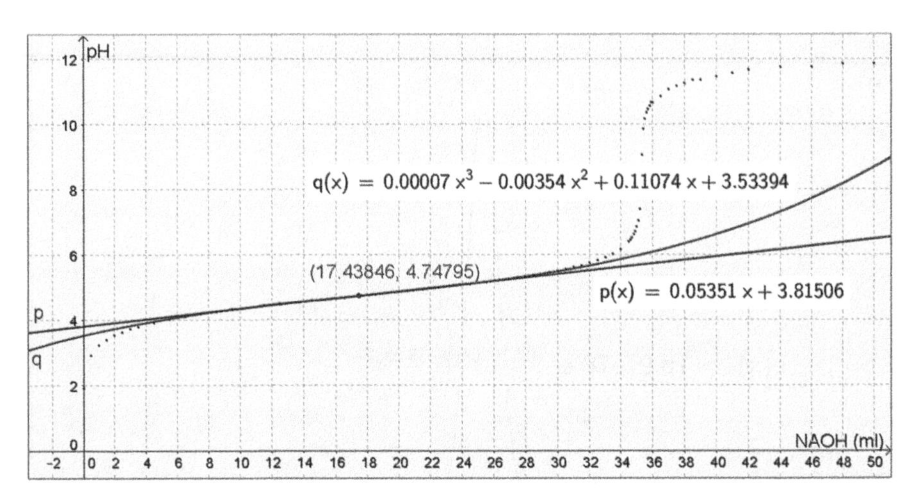

FIGURE 5.12 Linear and cubic regression functions from the points around the half-titration point.

TABLE 5.3 Results So Far

Adjusted function	Titration point	Half-titration point	pKa linear	pKa cubic
Second-degree function	35.260	17.630	4.75838	4.75735
$1/(1+x^2)$	35.274	17.637	4.75876	4.75769
Normal distribution	35.268	17.634	4.75860	4.75755

analysis. A look in a chemical table or a search on the Internet shows that the acid is most likely acetic acid, having pKa = 4.756. You also know that 25.00 ml of the acid is neutralized by 37.267 ml 0.200M NaOH, which gives an acid concentration of $32.267/25.00 \cdot 0.200 = 0.258$ M.

5.3 THE BOWL

A manufacturer of kitchen equipment produces different types of plastic bowls. The plastic bowls have all been generated from rotating a curve around a central axis in the middle. In Figure 5.13 you can see an example of such a curve, and in Figure 5.14 you can see a 3D rendering of its rotation around the y-axis.

The bowls are based on an exponential function, $f(x) = ae^{bx}$, which is rotated around the y-axis. Hence a rotation volume is generated:

$$V = \int B(y)\,dy, \quad \text{where } B(y) \text{ is the area of the circular sectional area at the height } y$$

For a specific plastic bowl, the function $f(x) = ae^{bx}$ goes through the points in Table 5.4. The plastic bowl should have an inner height of 10 centimeters.

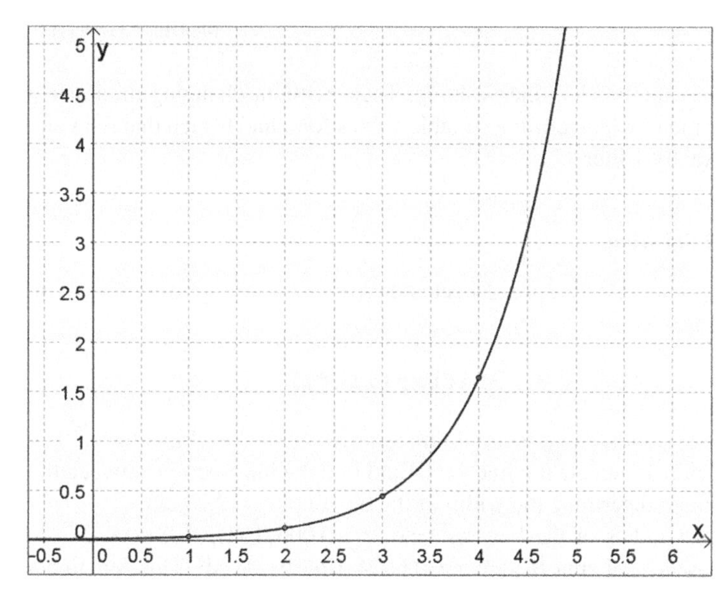

FIGURE 5.13 Exponential function is used to trace the inside of a bowl.

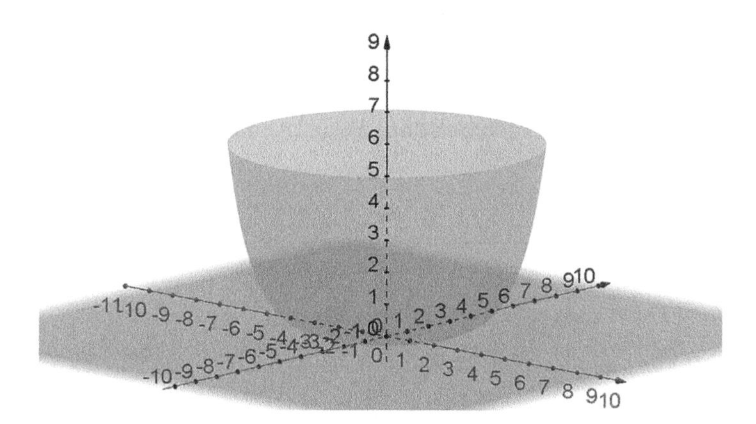

FIGURE 5.14 3D rendering of a bowl created from an exponential function.

TABLE 5.4 Coordinates for a Future Bowl

x (cm)	y (cm)
0.1	0.007
0.5	0.011
1.0	0.021
1.5	0.038
2.5	0.126
3.5	0.417
4.5	1.384
5.5	4.594
6.0	8.371

Calculate the bowl's inner volume. Then, without changing the inner width and height for the bowl, determine suitable values for a and b such that the volume of the bowl is exactly 1 liter.

Proposed Solution

Enter the points into the spreadsheet and create a list with points of them in the usual way. Typing

```
FitExp[list1]
```

will give the standard exponential function with a = 0.00621 and b = 1.20178.

In Figure 5.15, create the line y = 10 and find the intersection between the function and the line to determine the width of the bowl up to = 12.29 cm.

The inner radius of the bowl is $r = x = \ln(y/0.00621)/1.20178 = 0.8321 \cdot \ln(161.03y)$. The cross-sectional area $B(y) = \pi x^2 = 0.6924 \cdot [\ln(161.03y)]^2$. This function should be integrated from $y = 0.00621$ to 10, since $f(0) = a$.

You can use GeoGebra for this calculation, but because x and y are reserved names in GeoGebra, you need to type the following in the input field when Graphics Area 2 is active:

```
CrossSectionalArea(x) = 0.6924π ln(161.03x)^2
Volume = Integral[CrossSectionalArea, 0.00621, 10]
```

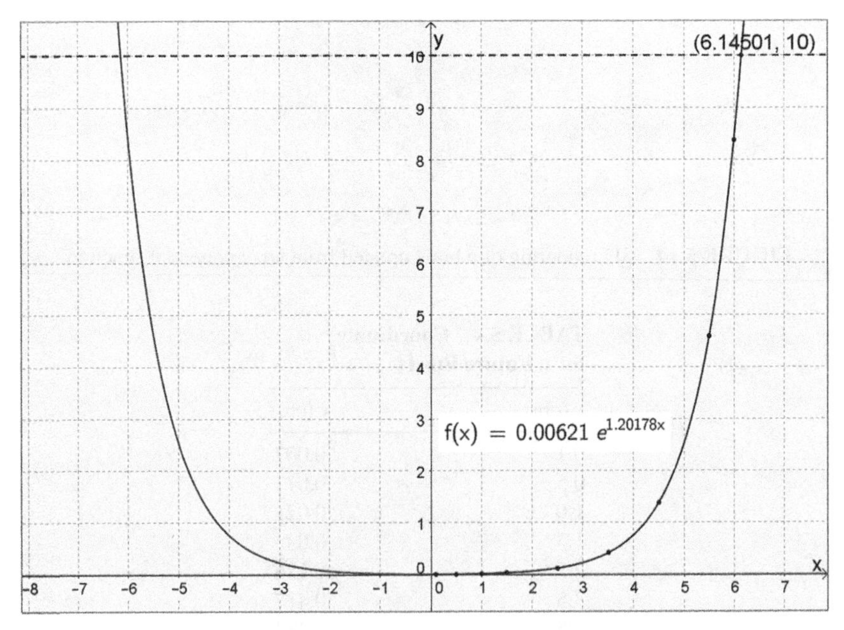

FIGURE 5.15 Mathematical model of a bowl.

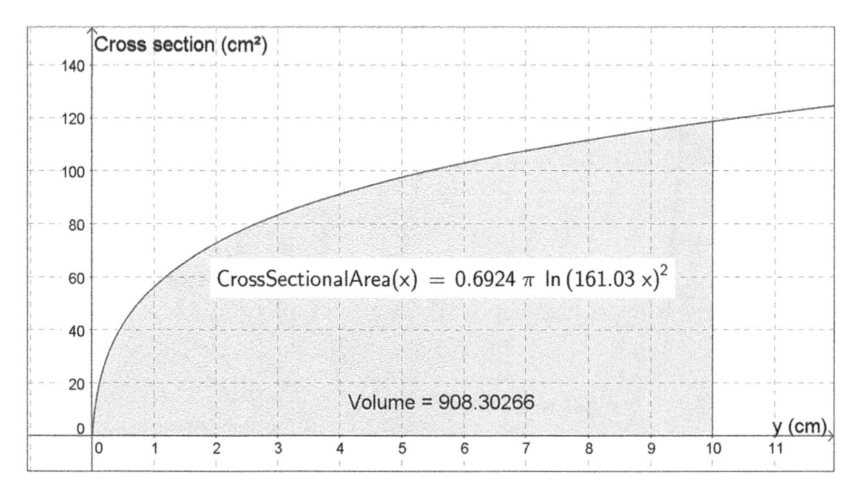

FIGURE 5.16 Cross-sectional area as a function of *y* along the *x*-axis.

In Figure 5.16 you see that the volume = 908.3 cm³. You can also adjust the labels of the axis so that they fit the situation. Right-click on the Graphics area and select Graphics… to change the labels on the *x*-axis and *y*-axis tabs.

The volume is around 0.9 liters. You must now adjust the function so that you get a volume of exactly 1 liter but still retain the same height and width of the bowl.

Given the condition that the outer dimensions remain the same, you use the equation

$$10 = \mathrm{a} \cdot e^{6.145\mathrm{b}}$$

by which you obtain the solution

$$\mathrm{a} = 10e^{-6.145\mathrm{b}}$$

You now have a function with just one parameter to be determined: $f(x) = 10e^{-6.145\mathrm{b}} \cdot e^{\mathrm{b}x}$ or $f(x) = 10e^{\mathrm{b}(x-6.145)}$. You could do this algebraically by solving for b, but since there is just one parameter to solve, you can do this manually with a slider. First, though, you need to redo the construction so that the volume is directly affected by the slider b.

Use the same values, but enter new commands (use **Alt-e** for ***e***):

```
b = 1
g(x) = 10 e^(b(x-6.145))          reflect in the y-axis

NewCrossSectionalArea = π (Invert[g])^2      in Graphics 2

NewVolume = Integral[NewCrossSectionalArea, 0.00621, 10]
```

The command **Invert[g]** produces the inverse of the function g, such that it if you have g(x) = 3x, **Invert[g]** will generate the function h(x) = x/3. Thus you can avoid a bit of tiresome manual work to find the inverse.

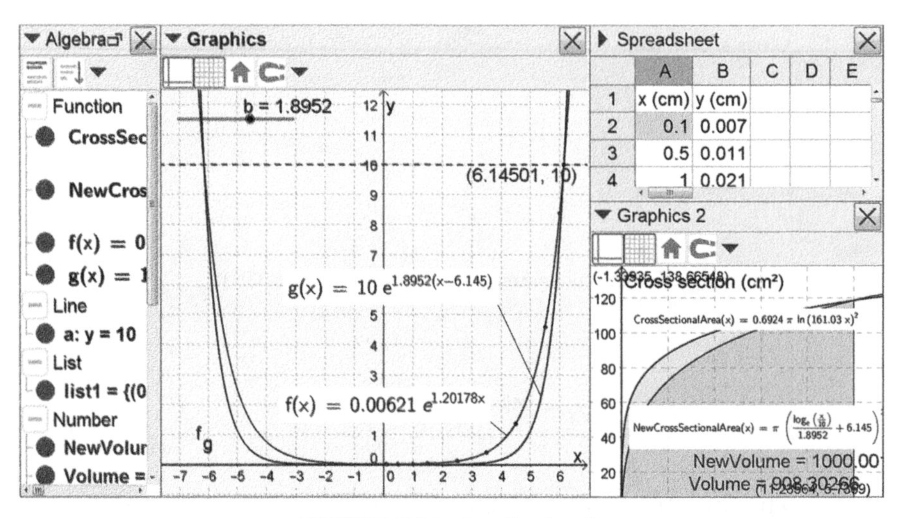

FIGURE 5.17 One liter bowl.

Adjust the value of b. You can see that the new volume is about right if b = 1.9. If you change the increment of the slider (in the properties dialogue of the slider) to 0.001, you will be able to tune it further. You get good conformity when b = 1.8952. Hold down the **Shift** key to make the increment 10 times smaller temporarily. The shape of the bowl has become a little deeper, keeping the outer dimensions. In Figure 5.17 all active windows are visible.

So the function for the one liter bowl is $f(x) = 10 \cdot e^{1.895(x-6.145)}$. You can now start the prototyping of this bowl. You might want to think about a suitable model for the outer surface.

5.4 THE AIRCRAFT WING

Figure 5.18 shows a simple mathematical model of the cross section of a wing for a small remotely controlled aircraft. The model consists of two curves, $f(x)$ and $g(x)$, enclosing a cross section of the wing. The two curves are symmetrical, mirror images of each other in a horizontal symmetry axis.

The group of aircraft scientists who modeled the aircraft wing placed their physical model in a coordinate system and measured the coordinates for the upper curve. These coordinates are found in Table 5.5.

Based on the principles of aero dynamics and approved industrial specifications, this group of scientists knows that a good wing shape is given by

$$f(x) = a\sqrt{x} + bx + cx^2 + dx^3 + ex^4$$

where a, b, c, d, and e, are constants.

Determine accurate values for the constants and then calculate the cross-sectional area of the aircraft wing body profile.

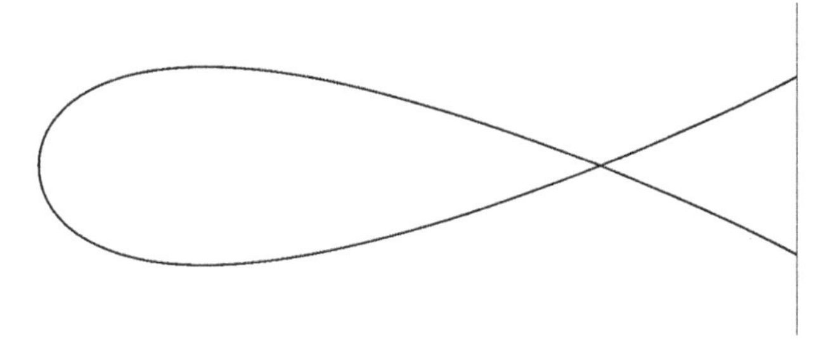

FIGURE 5.18 Schematic image of an aircraft wing.

TABLE 5.5 Coordinates for a Wing

x (m)	y (m)
0.0	0
0.1	0.078
0.2	0.095
0.3	0.100
0.4	0.097
0.5	0.088
0.6	0.076
0.7	0.061
0.9	0.024

Proposed Solution

Enter the given values into the spreadsheet of GeoGebra and create a list of data points. In this problem the suggested model is a linear combination of several elementary functions. You can use GeoGebra to fit such a function to the list of points by typing

```
Fit[list1, {sqrt(x), x, x^2, x^3, x^4}]
```

The regression function works very well together with the data points. The next step is to mirror the function in the x-axis. You can do that easily by clicking on the Reflect tool, then on the function, and finally on the x-axis.

The cross-sectional area is double the integral, but in order to calculate this, you first need to find the upper integration limit, which means the intersection of the function and the x-axis around $x = 1$, with the Intersect tool or by typing

```
Intersect[f,xAxis,1]
```

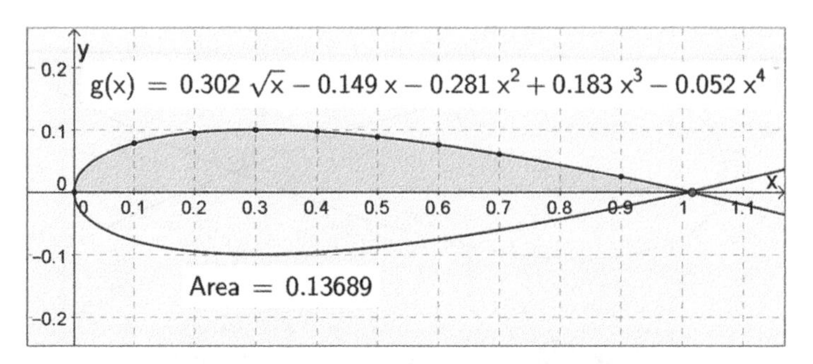

FIGURE 5.19 Model of an aircraft wing.

You need to give GeoGebra the number 1 as a starting point for the search algorithm; otherwise, the algorithm will find the origin. Figure 5.19 shows the result.

The Intersect command creates a new point J. You can find its coordinates in the Algebra window, or you can access the J-point's x coordinate value directly by typing

```
Integral[f, 0, x(J)]
```

Finally find the area with the command

```
Area = 2a
```

where a is the calculated value of the integral. The area is 0.137 m^2.

5.5 THE GATEWAY ARCH IN ST. LOUIS

The arch construction in Figure 5.20 is modeled from the mathematical function

$$y = A - B\cosh\frac{3x}{C}, \quad \text{where } \cosh(x) = \frac{e^x + e^{-x}}{2} \text{ is the } \textit{hyperbolic cosine}$$

The graph of the *hyperbolic cosine* is called a catenary, a term that comes from the Latin word for a chain, *catena*. The shape of the Gateway Arch is in inverted catenary. Any chain or rope with equally distributed weight will form a catenary and all catenaries might be seen as applications of exponential functions.

It is possible to measure the position of the arch. These values are given in Table 5.6. Use these values and determine the arc length of the Gateway Arch in St. Louis and check the accuracy of the result.

FIGURE 5.20 Gateway Arch in St. Louis. Photo: Bev Sykes from Davis, CA, USA (Flickr) [CC BY 2.0 (http://creativecommons.org/licenses/by/2.0)], via Wikimedia Commons. https://upload.wikimedia.org/wikipedia/commons/b/b9/St_Louis_Gateway_Arch.jpg.

TABLE 5.6 Coordinates for the Gateway Arch

Width: x position (m)	Height: y position (m)
–92.5	0.0
–77.5	76.3
–62.5	128.1
–47.5	159.2
–32.5	177.3
–17.5	186.9
–2.5	190.4
2.5	190.4
17.5	186.9
32.5	177.3
47.5	159.2
62.5	128.1
77.5	76.3
92.5	0.0

Proposed Solution 1

Enter the data values into the spreadsheet of GeoGebra, select the points, and right-click on the points and then chose **Create...>list with points**. Next do a general regression with the model function

$$y = A - B\cosh\frac{3x}{C}$$

The input to GeoGebra will be

```
a = 10
b = 10
c = 10
```

These values are only starting values for the regression algorithm:

```
m(x) = a - b cosh(3x/c)
Fit[list1, m]
```

Figure 5.21 shows the result. The model function $m(x)$ is hidden and the function expression has been pulled out from the Algebra window to the graphics area. The arch is almost as broad at the base as it is high.

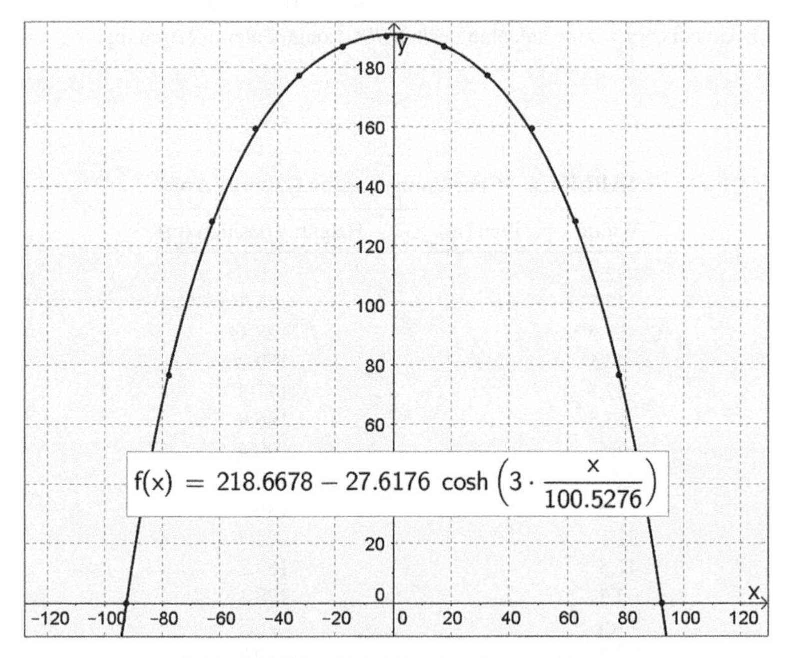

FIGURE 5.21 Model of the Gateway Arch.

The arc length of a continuous curve on an interval [a, b] is given by the expression

$$\int_a^b \sqrt{1+f'(x)^2}\,dx$$

Since GeoGebra is not really constructed for algebra calculations of this level of complexity, you can now use Wolfram Alpha to calculate the arc length. By selecting the function in GeoGebra and pressing **F4**, you can copy the expression to the input field, from where you can copy it and paste it into Wolfram Alpha. Add some information to the input as shown in Figure 5.22 and find the arc length to be some 452.6 m.

When it comes to the question of reliability you have to question the data: Do they show the inside, the outside, or for the middle of the arch? The arch is several meters thick and it contains both an elevator and an observation platform. The English Wikipedia article https://en.wikipedia.org/wiki/Gateway_Arch#Mathematical_elements provides some additional bits of information.

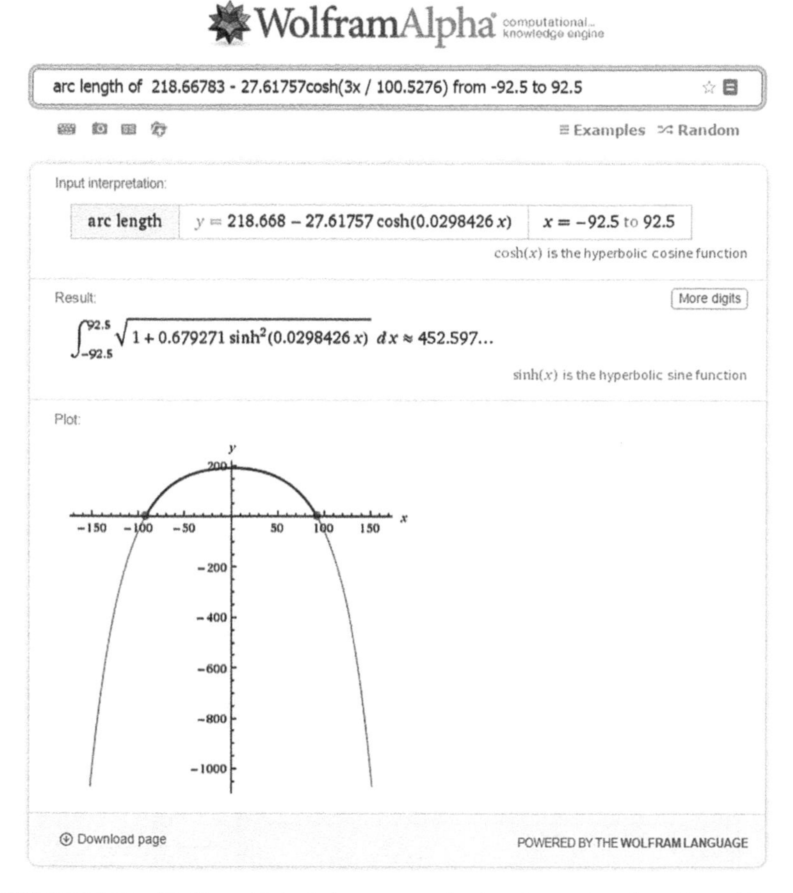

FIGURE 5.22 Wolfram Alpha helps you to calculate the arc length to about 453 m.

Proposed Solution 2

The free version of Curve Expert (www.curveexpert.net) may sometimes be a good alternative when you work a lot with regression. First you enter the data values from the table in the assignment, into Curve Expert, as shown in Figure 5.23.

Afterward you can select **Apply Fit > User model** from the menu to define a user defined model based on the one in the assignment. The dialogue is shown in Figure 5.24.

When Curve Expert asks you about suitable starting values for the regression, you can use a = 200, b = 27, and c = 100.

The results are shown in Figure 5.25.

So from the model $y = \mathrm{A} - \mathrm{B}\cosh(3x/\mathrm{C})$, you have A = 218.6678, B = 27.1757, and C = 100.5276. The arc length can then be found using Wolfram Alpha in the same way as in Solution 1, now that you have the model function.

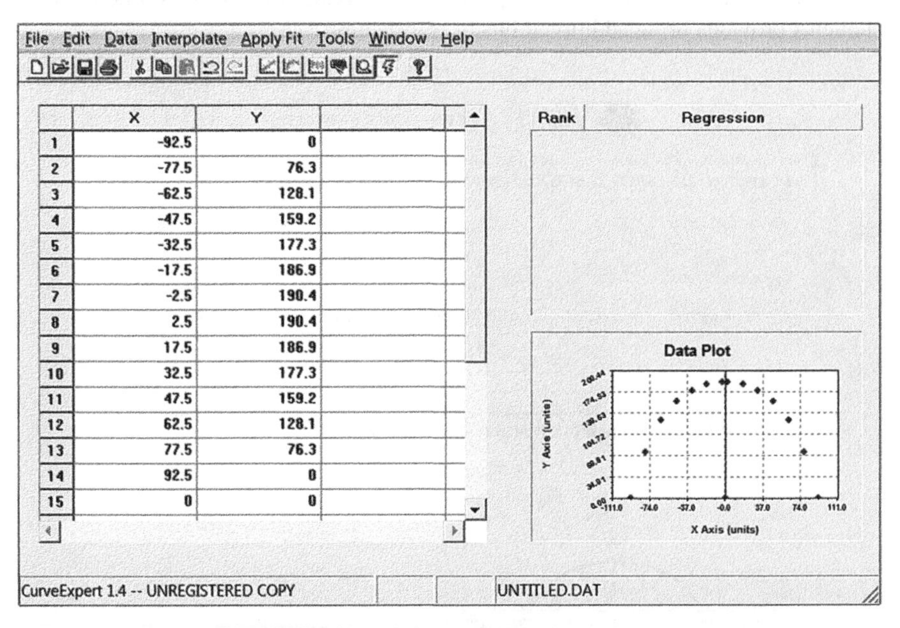

	X	Y
1	-92.5	0
2	-77.5	76.3
3	-62.5	128.1
4	-47.5	159.2
5	-32.5	177.3
6	-17.5	186.9
7	-2.5	190.4
8	2.5	190.4
9	17.5	186.9
10	32.5	177.3
11	47.5	159.2
12	62.5	128.1
13	77.5	76.3
14	92.5	0
15	0	0

FIGURE 5.23 Fitting a curve in Curve Expert.

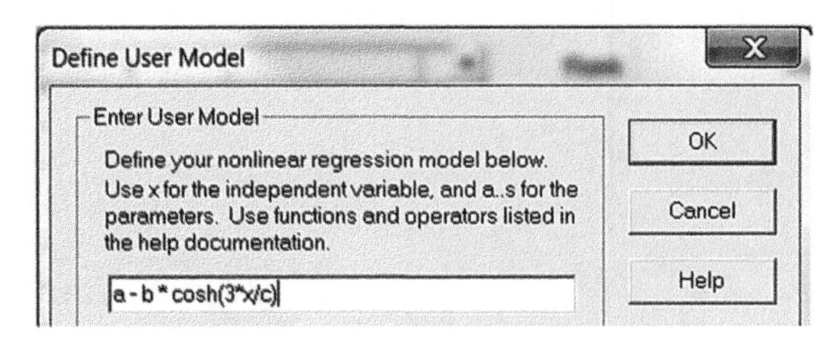

FIGURE 5.24 User-defined model in Curve Expert.

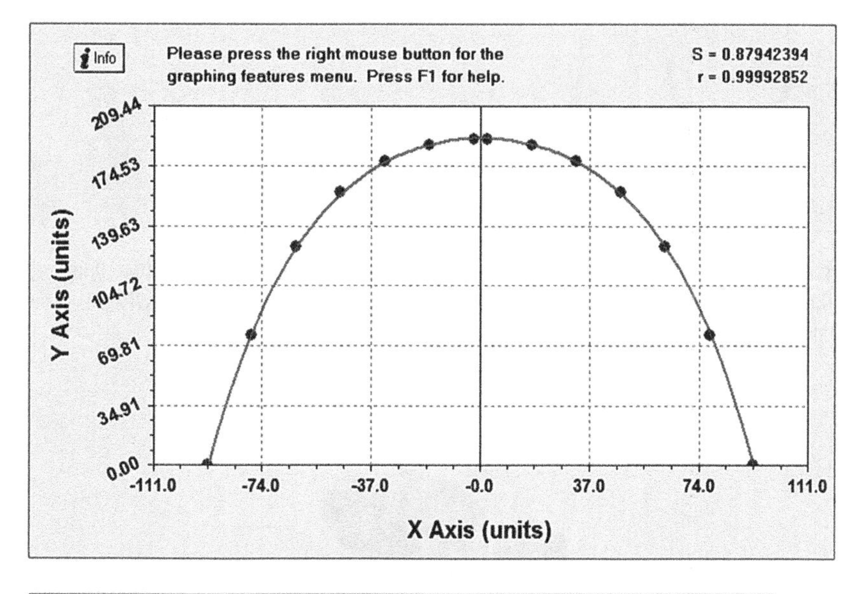

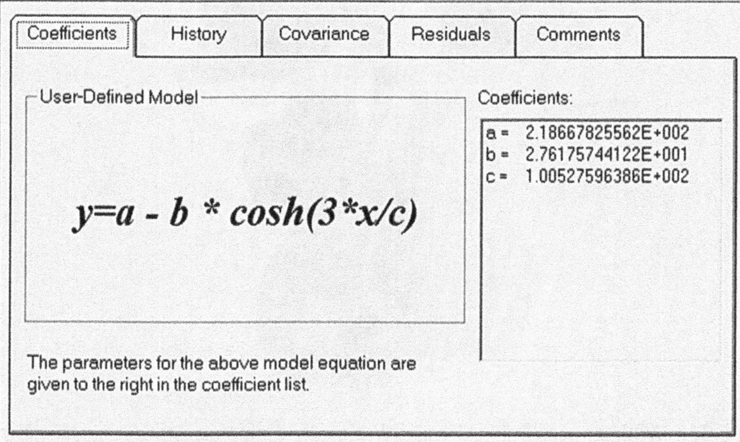

FIGURE 5.25 Result from Curve Expert.

5.6 VOLUME OF A PEAR

In Figure 5.26 you see eight different pears. What is the volume of a pear? Try to calculate the volume in several ways. For instance, you might:

- Put it in a bowl of water and measure the volume increase.
- Take a photo and use your knowledge about rotation volumes.
- Calculate the density from a small cube of pear material. Then measure the weight of the pear and calculate the volume.

FIGURE 5.26 Eight varieties of pears. Photo: Agyle (own work) [Public domain], via Wikimedia Commons. https://upload.wikimedia.org/wikipedia/commons/f/fc/Eight_varieties_of_pears.jpg.

FIGURE 5.27 Williams pear. Public domain/Wikimedia Commons. https://upload.wikimedia.org/wikipedia/commons/0/08/Williams_Bon_Chr%C3%A9tien_1822.png.

Compare your results with each other's:

- Can you define error limits for your results?
- Do the error limits overlap each other's?
- What method seems most reliable?

Combine your results with error limits to a final value with an estimate of the error.

Figure 5.27 shows an 1822 drawing of the Williams, or Bartlett, pear.

Proposed Solution

Figure 5.28 shows a way to calculate rotation volumes in GeoGebra. You can take a photograph of a pear and insert it into the background of GeoGebra. You can also easily take photos of glasses, pots, pitchers, avocados, mangos, and other items. Make sure that you stand at a distance and use the camera zoom to avoid wide-angle distortions. Position a ruler next to the object you photograph in order to calibrate the coordinate system.

Then identify the pear's contours with points and adjusted a sixth-degree polynomial to these points. Why sixth degree? The degree number must reasonably be even, and a second-degree polynomial would not be good enough. A fourth-degree polynomial produces results that are not satisfying. So go with the sixth-degree polynomial.

The intersection points with the x-axis, Q_1 and Q_2, can be found by typing

```
Q = Root[f]
```

or with the Intersect tool after which you can finally calculate the volume by typing

```
Integral[π f(x)^2, x(Q_1), x(Q_2)]
```

The π-sign is accessible from the keyboard shortcut **Alt-p**. GeoGebra gives the volume as 277 cm^3.

You could do this interactively with a pear that you photograph (together with a ruler) during the lesson. After that the students could calculate the volume. It may be determined experimentally by putting the pear in a small bowl or large cup and then adding or taking away water, measuring the volume of the water and thereby to get an approximate volume of the pear. Some students could also make a perfect cubic centimeter of the pear and weigh it, or slice the pear and calculate the volume of each slice.

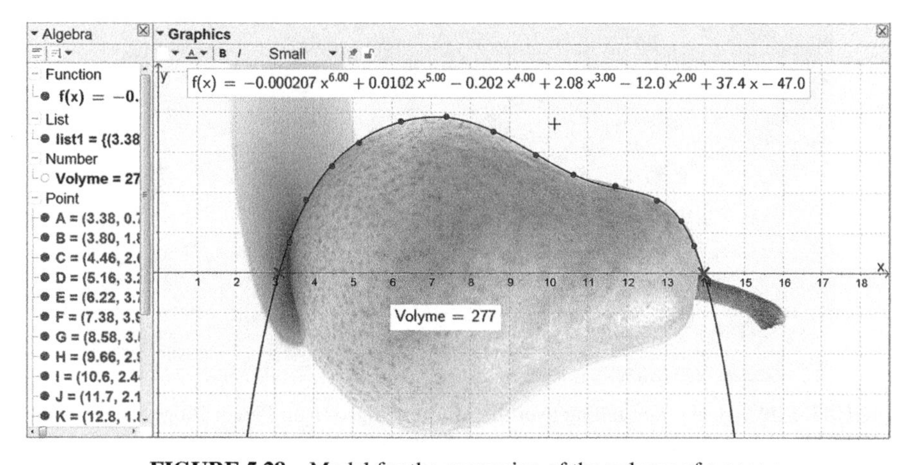

FIGURE 5.28 Model for the measuring of the volume of a pear.

5.7 STORM FLOOD

Figure 5.29 shows a graphical representation of the water flow in m^3/s in a river during a heavy rainstorm that lasted for 12.5 hours. How much rain did the area receive?

Use the information in the figure to construct a mathematical model that describes the water flow $F(t)$ during this time and do as well some error analyses.

If the storm covered a portion of the Murrumbidgee River area, as seen in Figure 5.30, and the average rainfall in this area during the storm was reported to be 130 mm, what area did the rainstorm cover?

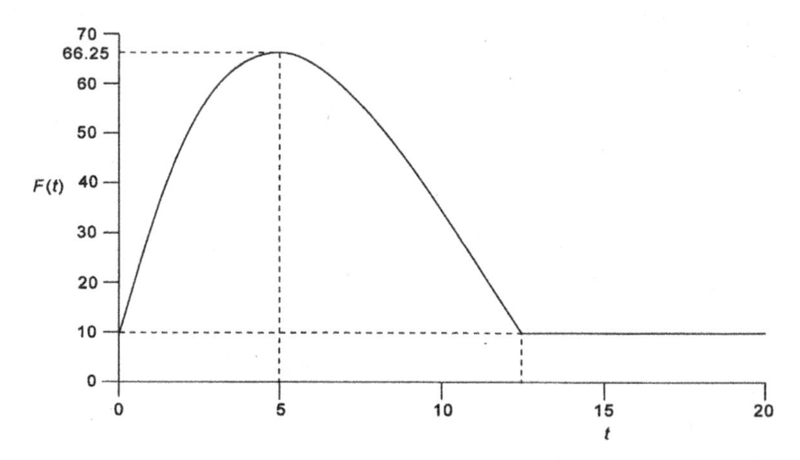

FIGURE 5.29 Water flow in a river.

FIGURE 5.30 Heavy rain falling over the Murrumbidgee river. Photo: Bidgee (own work) [CC BY-SA 2.5 au (http://creativecommons.org/licenses/by-sa/2.5/au/deed.en)], via Wikimedia Commons. https://upload.wikimedia.org/wikipedia/commons/1/1d/Heavy_rain_falling_over_the_Murrumbidgee_River.jpg.

Proposed Solution

Start with the photo of the graphical representation and insert it into GeoGebra as in Figure 5.31 so that you can compare the graph with the model.

In GeoGebra you can insert images using the Image tool. Starting with a screen dump of the diagram, you can also use any skillfully taken photo. Then click in the coordinate system where you want the image's lower left corner (a little below and to the left of the origin) and select the image to insert in the dialogue box.

You can define up to three points, A, B, and C, where you want the image's lower left, lower right, and upper left corner to be. Right-click on the image and select properties to lock the image to these positions, as in Figure 5.32.

Use the image as a *Background image*, in the **Basic** tab. This gives you the possibility to show the grid above the image.

By adjusting the positions for the points A, B, and, C, you have the possibility to move the image around until it matches the coordinate system of GeoGebra. If you want to move two or three points at the same time, you can select all these points and move them with the arrow key.

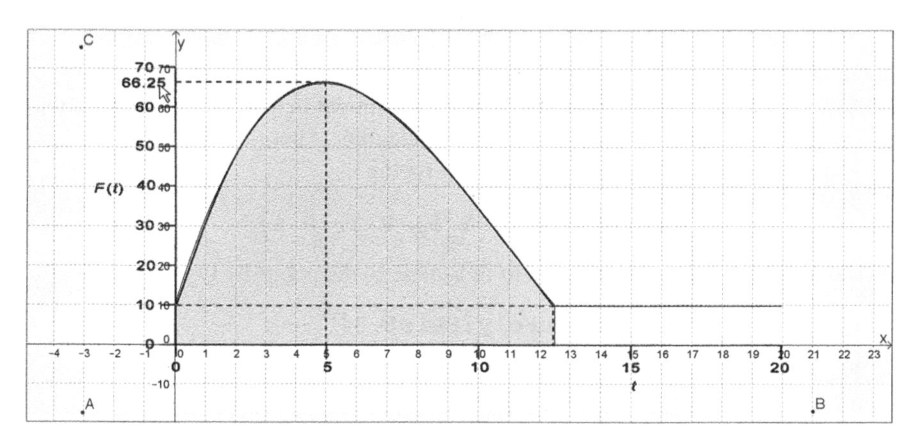

FIGURE 5.31 Background image in GeoGebra.

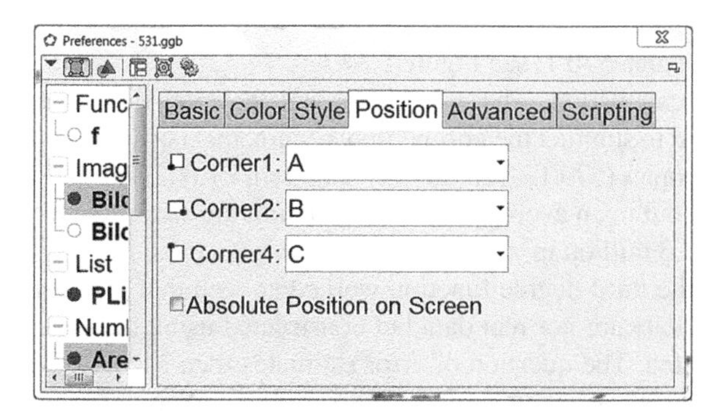

FIGURE 5.32 Positioning a background image in GeoGebra.

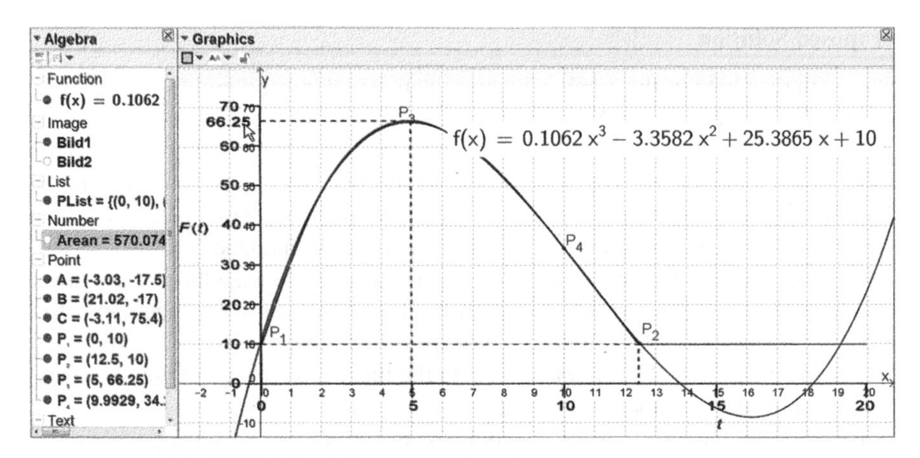

FIGURE 5.33 Measuring a given function and recreating it.

Some points in the diagram are easy to identify. Create these points in GeoGebra by typing, for example,

```
P_1 = (0, 10)
```

in the input field.

A quick glance at the diagram indicates that it might be part of a cubic function. A second-degree function is symmetrical, the diagram is not.

You can create a list of the data points by typing

```
PList={P_1, P_2, P_3, P_4}
```

Having done that, ask GeoGebra to fit a cubic to the points by typing

```
FitPoly[PList,3]
```

The result is seen in Figure 5.33.

It seems as if the agreement is good. The area under the curve can be calculated by the command

```
Area = Integral[f,0,12.4]
```

which yields the area 570.1; see Figure 5.34.

Now to the calculation. The area represents the total rain volume in the unit $m^3/s \cdot h$. You need to subtract the normal flow of rain, that is, $10\ m^3/s \cdot 12.5\ h$. The total volume then becomes $(570.1 - 125)\ m^3/s \cdot h \cdot 3{,}600\ s/h = 1{,}602{,}360\ m^3 \approx 1.60$ million m^3. If this is spread out to an average depth of 130 mm, the area is equal to 1.60 million $m^3/0.130\ m = 12.3$ million $m^2 = 12.3\ km^2$.

In this case the third-degree function worked so well that you might suspect that in this case the data are not real data but constructed using the very cubic function you have recreated. The question of error estimates then becomes somewhat moot, but you could always ask your students to solve the problem differently.

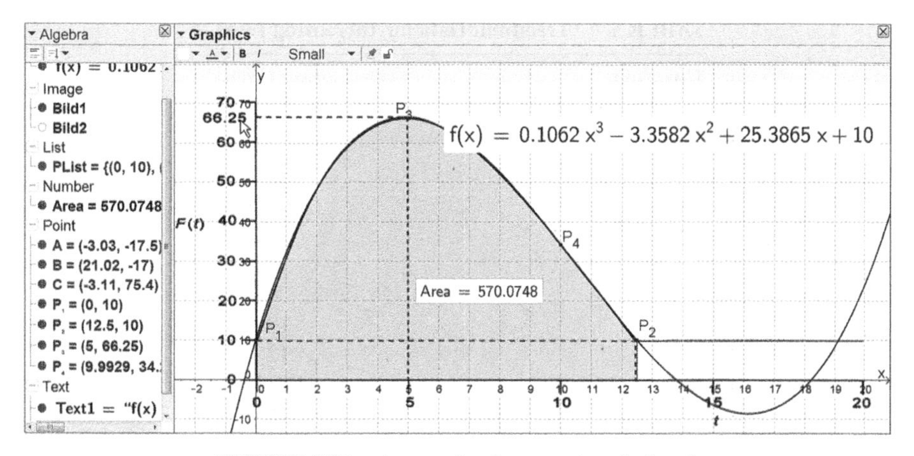

FIGURE 5.34 Area under the curve is calculated.

Alternative models in this case could be:

- Second-degree functions, adjusted so the areas look the same.
- Approximations with upper and lower Riemann sums from some measured values.
- Measure the area directly in GeoGebra by creating a polygon to fit the graph.
- Approximation of a rectangle, with adjusted height so that the areas seems to be the same.

Two different ways to estimate the errors would be to vary some of the parameters within a reasonable interval and see what the effect is for the area, or to work with relatively few errors so that it is easy to estimate the error in the rain amount at the storm's end.

For example, in the model the points $P_1 - P_3$ are well determined, but P_4 depends on the reading. If you read it off the graph at time $t = 10$ h, you can move the point up and down and see how the area varies: 34.7 is clearly above the curve and 33.9 equally clearly below the curve. These values give areas in the intervals 569.1 to 571.4, which correspond to ±0.2%. Note that this is an error limit within the model and the real error most likely is much larger. Comparisons with other types of models could give a better estimation of the real rain amount for an area such as this.

5.8 EXERCISE

Power

The power that is delivered during an exercise run can be seen as a function of time. The data in Table 5.7 come from measuring at a treadmill, where the runner jogs on a rolling carpet. The treadmill used in this test could be run at different speeds and be

TABLE 5.7 Treadmill Data for Increasing Load

Time (min)	Power (Watt)
1	14
2	28
3	45
4	55
5	76
6	80
7	93
8	107
9	126
10	152
11	155
12	183
13	191
14	211
15	222

inclined at different angles. Speed and inclination were increased constantly during this measurement experiment. As a result the runner had to use more and more force to stay on the rolling carpet. The table values are averages for a large group of good runners. Construct a mathematical model to fit the values, and a model that describes the relationship between the running time on the treadmill and the average delivered power.

Oxygen Consumption

Oxygen consumption is a function of the power you are able to generate. Table 5.8 describes the relationship for the same runners that generated Table 5.7. Construct a mathematical model that describes this relationship.

Find the relationship between the time on the treadmill and the oxygen consumption for this group of runners and calculate the total oxygen consumption for these runners during a 15 minute long exercise round.

Michael and Jacob

Two friends, Michael and Jacob usually exercise together for a 5,000 meter race, as illustrated in Figure 5.35. They are roughly the same level of fitness but use different tactics when they race. Michael prefers to start slowly and then increase his speed along the way, whereas Jacob enjoys starting at a high speed and then gradually adjusts to a speed that matches his fatigue level. Their speed outputs are listed in Table 5.9.

You can assume that Michael and Jacob are such good runners that their oxygen consumption can considered to on par with the data given in Table 5.7 and Table 5.8. Determine which of these two strategies consumes the most oxygen. Can you predict who will likely win the race?

TABLE 5.8 Oxygen Uptake

Power (Watt)	Oxygen uptake (liters/min)
10.0	0.35
20.0	0.98
30.0	1.54
40.0	1.82
50.0	2.36
60.0	2.56
70.0	2.80
80.0	3.20
90.0	3.45
100.0	3.85
110.0	3.94
120.0	4.08
130.0	4.21
140.0	4.43
150.0	4.61
160.0	4.72
170.0	4.88
180.0	4.93
190.0	5.04
200.0	5.23
210.0	5.18
220.0	5.31
230.0	5.14
240.0	4.96
250.0	4.89

FIGURE 5.35 Sprint finish. Photo: US Navy, Mass Communication Specialist 1st Class Demetrius Kennon [Public domain], via Wikimedia Commons. https://upload.wikimedia.org/wikipedia/commons/e/e8/US_Navy_100519-N-7367K-001_A_Sailor_and_a_Soldier_based_in_southern_Mississippi_sprint_to_the_finish_line_during_the_Run_for_Relief_5K_Challenge.jpg.

TABLE 5.9 Power during a Race

Time (min)	Michael's power (Watt)	Jacob's power (Watt)
0	30	163
1	31	155
2	35	150
3	39	130
4	43	110
5	47	100
6	50	90
7	60	80
8	70	75
9	80	70
10	90	65
11	105	65
12	120	65
13	140	63
14	148	60
15	150	55

Proposed Solution

Power

Enter the values from the table into the spreadsheet of GeoGebra, which you can show with the keyboard shortcut **Ctrl-Shift-S**. After that, select and right-click on the points and select **Create... > List with points**. **Ctrl-drag** the axis in order to zoom out the window so that it will be large enough for all the points to be visible.

The points seem to be on a straight line, so find the best line through the set of points by typing

```
FitLine[list1]
```

You should get a relatively good fit and errors that show no signs of systematicbe-havior. However, the line does not go through the origin, which it normally would do when the delivered power is zero while the treadmill is standing still. So do another fit to a straight line through the origin, $y = k \cdot x$. This is not a standard model, so you have to use the command

```
Fit[list1, {x}]
```

which creates a linear combination of all functions in the list between the braces, in this case just the function $y = x$. This is a common situation that most graphic calculators have difficulties with. The two different linear functions are shown in Figure 5.36.

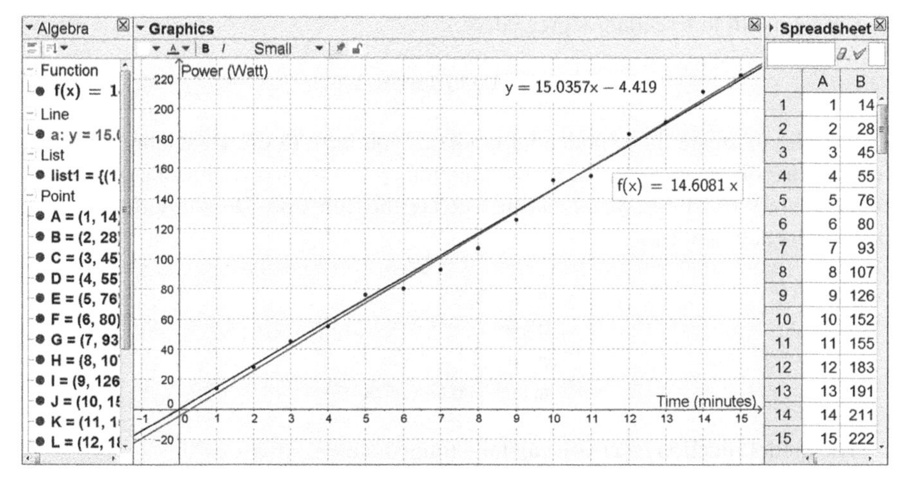

	A	B
1	1	14
2	2	28
3	3	45
4	4	55
5	5	76
6	6	80
7	7	93
8	8	107
9	9	126
10	10	152
11	11	155
12	12	183
13	13	191
14	14	211
15	15	222

FIGURE 5.36 Power that seems to increase linearly over time.

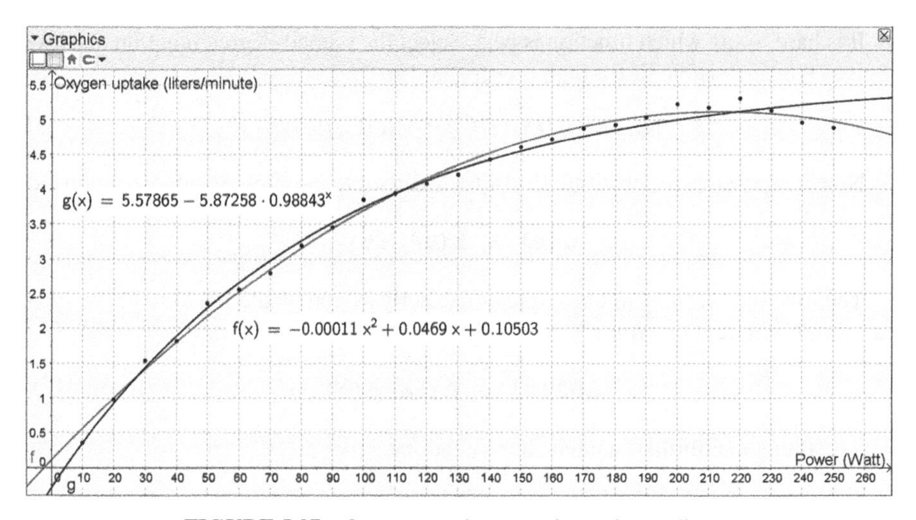

FIGURE 5.37 Oxygen uptake appearing to be nonlinear.

The relationship between the power P(W) and time t(minutes) on the treadmill is

$$P(t)=14.6t$$

Oxygen Uptake

Press **Ctrl-N** to open a new GeoGebra window and repeat the procedure with the data from the oxygen uptake. This time the data seem to follow either a second-degree polynomial or an exponential rise as seen in Figure 5.37.

Create the fit to a second-degree curve:

```
FitPoly[list1,2]
```

In order to create the exponential gradient, you have to use GeoGebra's general regression tools again. Since the model function $y = b - c \cdot a^x$ not is a linear combination of elementary functions, you must first create this function. Do so by defining the parameters a, b, and c as

```
a = 1
b = 1
c = 1
m(x) =b-ca^x
```

The model function $m(x)$ you can hide immediately.

Now type

```
Fit[list1, m]
```

It is hard to say which function is best. Select the second-degree function because it gives the best visible fit for high powers, and so

$$U_{O2} = -0.00011P^2 + 0.0469P + 0.105$$

Now it is possible to substitute $P = 14.6t$ directly into GeoGebra with the command

```
u(t) = f(14.6t)
```

The function $u(t)$ may not be expressed exactly as you would like it to be, but you can simplify it a bit by typing

```
u02(t) = Expand[u]
```

which, after some further manual simplifications, gives

$$U_{O2}(t) = -0.0234t^2 + 0.6847t + 0.105$$

This is a useful, practical application of all that practicing with $f(g(x))$.

Generate table data into cell C1 and C2 by typing 1 and 2. Then select these two cells and copy this selection down to cell C15 by dragging the fill handle. Next in cell D1 type

```
=u02(A1)
```

and copy this down as indicated in Figure 5.38.

In order to show these new points graphically, you need a new coordinate system. Activate Graphics 2 by pressing **Ctrl-Shift-2** or by selecting **Show>Graphics 2** in the menu. When this Graphics window is active, select all the values in column C and D, right-click, and select **Create>list with points.**

▾ Spreadsheet

D1 =u02(C1)

	A	B	C	D
1	10	0.35	1	0.76639
2	20	0.98	2	1.381

FIGURE 5.38 Table created with values for the new function *u02*(*t*).

Location

☑ Graphics ☐ Graphics 2 ☑ 3D Graphics

FIGURE 5.39 Selecting the graphics areas where the function should be visible.

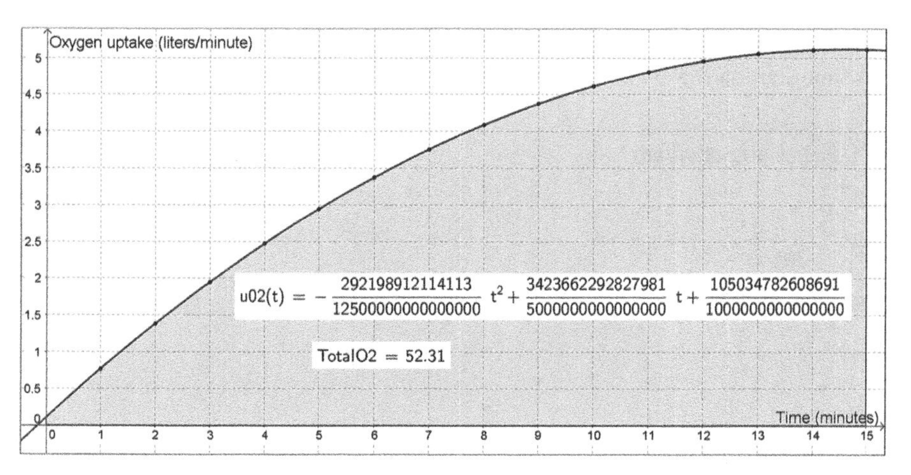

FIGURE 5.40 Calculating the integral that gives the total amount of oxygen taken up by the runner.

The function $U_{02}(t)$ can now be visualized in Graphics 2. Right-click on the function and select the **Advanced** tab. At the bottom of the tab shown in Figure 5.39, you can select the different graphics areas where the function should be visible.

When the function is visible in Graphics 2, the points ought to lie exactly on the graph, since they are generated by the function.

In order to finally calculate the total amount of oxygen during the exercise round, you have to calculate the integral of this function from $t=0$ to $t=15$. You can do this by typing

```
TotalO2 = Integral[u02, 0, 15]
```

The result is represented by a number, **TotalO2**, in the Algebra window. Drag it out to the graphics area as in Figure 5.40. The total amount of oxygen taken up is about 52.3 liter for these runners.

Michael and Jacob

By pressing Ctrl-N for a new GeoGebra window, entering the table values for Michael and Jacob, and creating points for them as in Figure 5.41, you can see that the power output is quite different for the two runners.

Their running styles are not smooth over time, so it will probably be hard to find algebraic functions to fit their running styles. Fortunately, GeoGebra can create artificial functions to a list with values by drawing straight short lines between the points. We will now show you how to do the following:

- Generate oxygen uptake values for the runners.
- Create artificial functions of these values.
- Calculate the total oxygen uptake for the runners.

Go back to the GeoGebra file where you earlier calculated oxygen uptake in Graphics 2. Make the spreadsheet a little bit larger and enter Michaels and Jacobs time and power values in any free columns, a good suggestion is F, G, and H—but this time start on row 3 (we will explain why in the next step); see Figure 5.42.

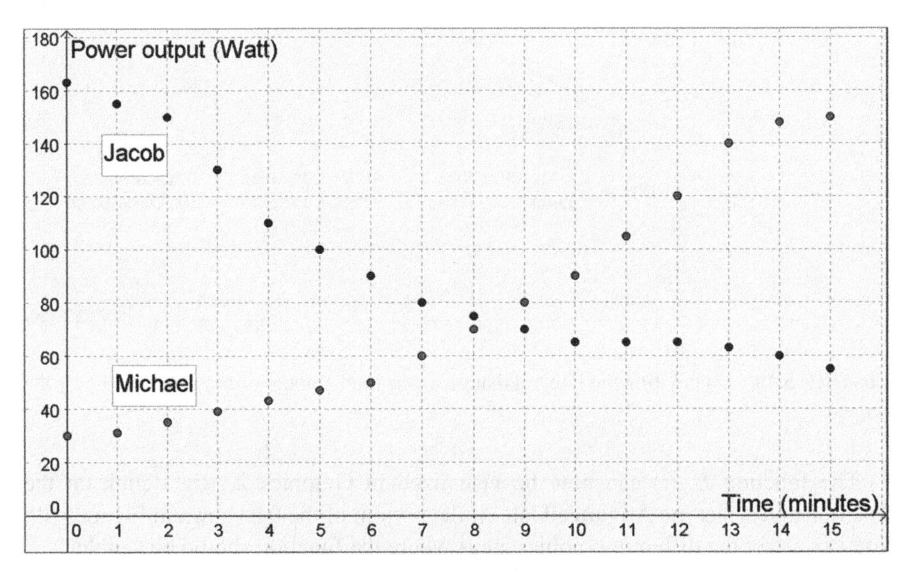

FIGURE 5.41 Two different running styles.

	F	G	H
1			
2	Time	**Michael**	**Jacob**
3	**0**	30	163
4	**1**	31	155
5	**2**	35	150

FIGURE 5.42 Let the values start on row 3.

	F	G	H	I	J
1				0	0
2	Time	Michael	Jacob	15	15
3	0	30	163	1.41332	4.836
4	1	31	155	1.45353	4.73978
5	2	35	150	1.61218	4.67252

FIGURE 5.43 Oxygen uptake values in column I and J.

The oxygen uptake ability as function of the effect *P* is already defined as a second-degree regression function, which in this case we name *f*(*x*). In order to generate Michael's oxygen uptake at the time $t=0$, you need to calculate $f(P(t))$, which you can do in cell I3 by typing

```
=f(G3)
```

By copying this cell one step to the right, J3 will contain the formula **f(H3)**, which is the calculation of Jacob's oxygen uptake ability. Then select both I3 and J3 and copy them downward.

Click one time in Graphics Area 2 to make it active. Then select the time values in column F, drop the mouse button, press and hold **Ctrl** while selecting the values in column I. Right-click on the selected area and select **Create...>list with points**. You should now be able to see Michael's oxygen uptake in Graphics Area 2. Then do the same thing with the values in columns F and J to show Jacob's values; see Figure 5.43.

You are now ready to create the artificial functions for these values. In order to do this, GeoGebra demands that you first create a list with values $\{x_{min}, x_{max}, y_1, y_2, y_3, \ldots, y_n\}$, so enter the values 0 and 15 into the cells I1, I2, J1, and J2 according to Figure 5.43. Then select the cells I1 to and with I18, right-click on the selected region and select **Create...>List** (not **List with points**). Immediately after you have created the list, you can type **listMichael,** which will rename the list to this somewhat more logical name. If you do not rename the list directly, then you can always right-click on it in the Algebra window and select **Rename**. Thereafter you create **listJacob** in the same way. The functions are then created by the commands

```
Michael(x) = Function[listMichael]
Jacob(x) = Function[listJacob]
```

Specifying the names of the functions as you create them makes for good housekeeping. Finally calculate the integrals by typing

```
TotalO2Michael  =  Integral[Michael,0,15]
TotalO2Jacob  =  Integral[Jacob,0,15]
```

The integrals are calculated using the trapezoidal method; see Figure 5.44.

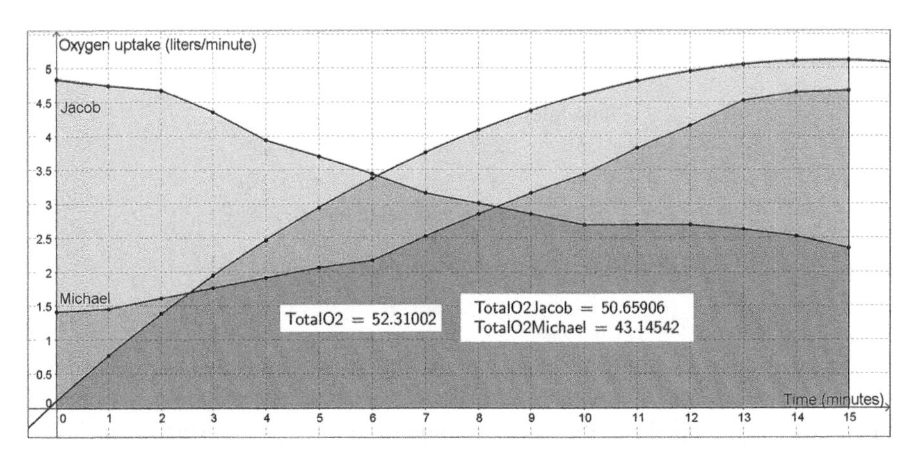

FIGURE 5.44 Jacob will take up more oxygen than Michael.

Jacob seems, in general, to be above Michael in oxygen uptake. Provided that they both can be represented by the earlier results and that they both are well trained, Jacob seems to be more fit and will likely win the race.

5.9 BICYCLE REFLECTORS

In many countries, side reflectors, in orange, yellow, or white, such as shown in Figure 5.45, must be mounted on your bicycle when you ride in traffic. These reflectors are often mounted on the wheel spokes. When you ride the bicycle, the reflector moves up and down on the wheel. What curve will a reflector follow if it is mounted three quarters of the way out from the hub toward the tire?

Students may also look into the question as to how far the reflector moves for every mile that the bicycle is moving.

Proposed Solution

In order to investigate and solve the problem, it helps to have a model of a moving wheel. Introduce the parameters r for the radius of the wheel, in meters, and v, the speed in m/s together with the variable t, for the time in seconds. The radius for an ordinary bicycle wheel is about 0.45 m and a bicycle is most likely run in the speed interval 2–10 m/s. Enter the values

```
r = 0.45
v = 5
```

Also create a *time-slider* for t that runs from 0 to 20 with an increment of 0.01 and a slider a, to represent the relative distance from the hub to the reflector. In this example, set

FIGURE 5.45 Bicycle side reflectors. Photo: No machine-readable author provided. Gerfriedc assumed (based on copyright claims). [GFDL (http://www.gnu.org/copyleft/fdl.html), CC-BY-SA-3.0 (http://creativecommons.org/licenses/by-sa/3.0/) or CC BY-SA 2.5-2.0-1.0 (http://creativecommons.org/licenses/by-sa/2.5-2.0-1.0)], via Wikimedia Commons. https://upload.wikimedia.org/wikipedia/commons/9/94/Hyperbikes.jpg.

```
a = 0.75
```

but allow a to take values from 0 to 2 with the increment of 0.001.

The movement of the reflector consists of several composite movements. You can split the movement into the movement of the hub forward and the movement of the reflector around the hub. It will also become a lot easier if you split the movement into x- and y-components. Assume that the reflector is positioned straight down toward the ground at the time $t=0$ and $x=0$.

The movement of the hub is not too difficult to write down. It moves linearly forward with the speed v, and its position at time t can therefore be described as

$$\begin{cases} x = \mathrm{v} \cdot t \\ y = \mathrm{r} \end{cases}$$

The reflector's movement around the hub can be described with trigonometric function, but which one? The amplitude should be $\mathrm{a} \cdot \mathrm{r}$. In one second the wheel will move v meters, which corresponds to rolling an angle equal to $\mathrm{v}/2\pi\mathrm{r}$ parts of one turn, or v/r radians. This is the angular velocity $\omega = \mathrm{v}/\mathrm{r}$, but should it be $\sin(\omega t)$ or $\cos(\omega t)$, or maybe $\sin(\omega t + 90°)$, or a negative sign as $-\cos(\omega t)$? The advantage of working in GeoGebra is that you can try different options if the algebra becomes tricky. Simply start by creating the points

```
Hub = (v t, r)
C = (a r cos(v t/r), r + a r sin(v t/r))
R = C + (v t, 0)
```

It would probably be easier if you could see the hub and the spoke to which the reflector is mounted. You can create and visualize it with the commands

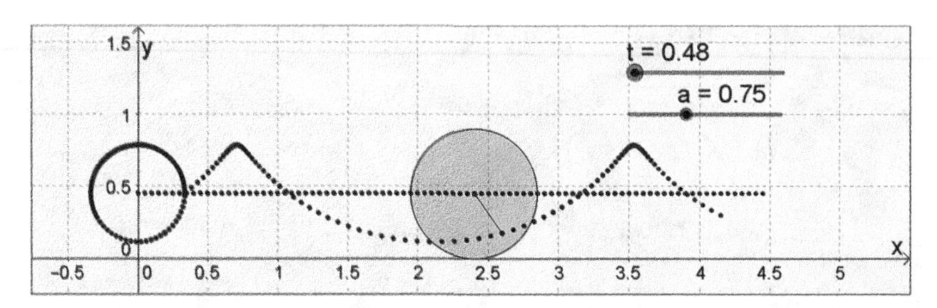

FIGURE 5.46 First attempt with the reflector moved in the wrong direction.

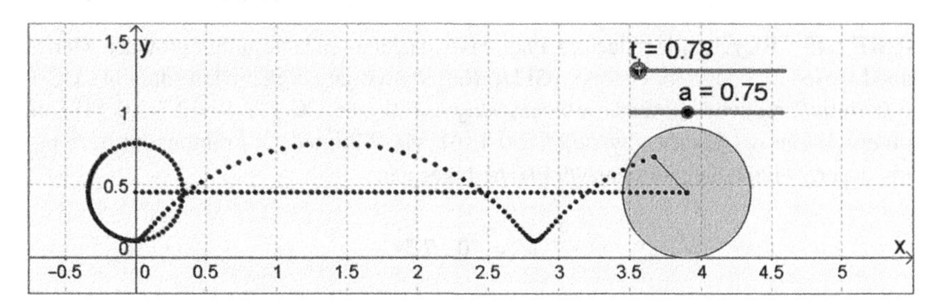

FIGURE 5.47 Correct path for the reflector.

```
Wheel = Circle[Hub, r]
Spoke = Segment[Hub, R]
```

In order to get the wheel really circular, right-click in the graphics area and select **xAxis : yAxis > 1:1**.

Activate Trace on all points by right-clicking on the word **Points** in the Algebra window and select **Trace on**. Then select the slider *t* and use the arrow keys. See if you can get something similar to Figure 5.46.

The reflector seems to be moving in the wrong direction, but it also starts in the wrong position. Try to change the sign for cosine in the *x*-coordinate for T, subtract $90° = \pi/2$ from both *x*- and *y*-coordinates and redefine C to

```
C = (-a r cos(v t/r-π/2), r+a r sin(v t/r-π/2))
```

This works much better; see Figure 5.47. You can now turn off the trace of the points; point C can be hidden. You can also erase the traces with **Ctrl-F**. Instead, create the locus for R when *t* varies by clicking on the locus tool, then on R and finally on the slider *t*. You can now vary the value of ***a*** and see how the curve is changing if you mount the reflector at different distances from the hub.

When $\boldsymbol{a} = 1$ you will get the curve normally called the *cycloid*, seen in Figure 5.48, even if all values for ***a*** actually give different kinds of cycloids. This is a good example of a curve in *parameter form*, where the *x*- and *y*-coordinates can be expressed as functions of another parameter *t*.

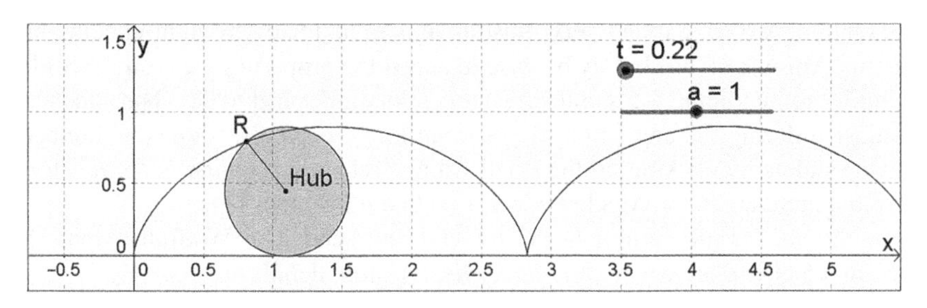

FIGURE 5.48 Cycloid.

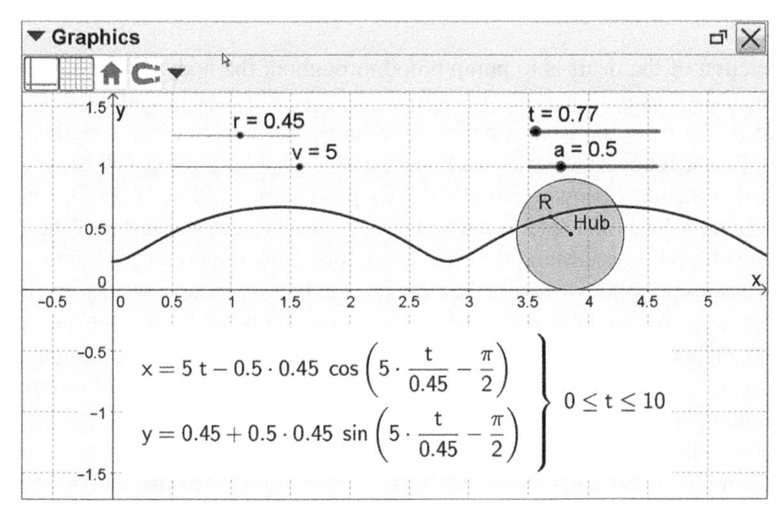

FIGURE 5.49 Parametric curve in GeoGebra.

$$\begin{cases} x = \mathrm{v} \cdot t - \mathrm{a} \cdot \mathrm{r} \cdot \cos\left(\dfrac{\mathrm{v} \cdot t}{\mathrm{r}} - \dfrac{\pi}{2}\right) \\ y = \mathrm{r} + \mathrm{a} \cdot \mathrm{r} \cdot \sin\left(\dfrac{\mathrm{v} \cdot t}{\mathrm{r}} - \dfrac{\pi}{2}\right) \end{cases}$$

Physical movement in two dimensions is often suitable to express in parameter form. One of the most common examples of this is projectile motion, which can be modeled by the following parametrized expression:

$$\begin{cases} x = x_0 + v_{0x} \cdot t \\ y = y_0 + v_{0y} \cdot t - gt^2 \end{cases}$$

In GeoGebra you can sketch the curve directly in parameter form by the command

```
Curve[v t-a r cos(v t/r-π/2),r+a r sin(v t/r-π/2),t,0,10]
```

In Figure 5.49 you see an example of this curve for a=0.5.

Something we have found very illustrative is to animate *t* by right-clicking on *t*, selecting **Animation on**. It may be wise to adjust the properties for *t*, on the **Sliders** tab, limit *t* to maximum 1.5 s and adjust the animation speed to 0.5. The sliders can be adjusted during the animation. If you want to, you can also show the numbers r and v as sliders by clicking on the small white circle, the *visibility indicator*, to the left of the numbers in the Algebra window so that it becomes blue.

For the arc length, formulate the integral and send it to Wolfram Alpha. See Problem 5.5, *The Gateway Arch in St. Louis*, for more details on this.

5.10 CARDIAC OUTPUT

The function of the heart is to pump blood throughout the body. The blood carries oxygen (O_2) from the lungs to the various tissues of the body and carries carbon dioxide (CO_2) from the tissues back to the lungs. In constructing a mathematical model of the circulatory system in a human body, consider it to be a closed loop and assume that the blood flowing around this loop is incompressible. Consequently the total volume V of blood (measured in liters) in the system is constant. The rate at which this blood flows around the circulatory loop is an important parameter.

You can, in principle, measure the flow rate (in liters/minute) past any given point in the system. Attention is ordinarily focused on the heart itself, and the *cardiac output* (*CO*) is the rate at which blood is pumped out of the heart. The cardiac output of the heart is the product of the *stroke volume*—the volume of blood pumped per beat—and *the heart rate*—the number of beats per minute.

The so-called dye solution method (Stewart–Hamiltons method), where you add a small amount of harmless dye to the heart, may be used for the measurement of the heart's *cardiac output*, CO. This is defined as

$$\mathrm{CO} = \frac{\mathrm{I}}{\int_0^{\infty} C(t)\,dt}$$

At the time, for the observation, I milligram dye is injected in the right atrium of the heart. This is at the time t=0. The concentration C(*t*) of the dye that leaves the heart through the aorta at the time *t* seconds is measured by a probe over a certain time interval $0 \le t \le T$. At the end of the time interval, at *T*, the *recirculation* has started to disturb the measurement and the observation is canceled. According to medical praxis, the measurement of the dye concentration is done over equally large time intervals for the length of the entire test [0, *T*].

In one instance, a patient was injected with 5.68 mg dye in the right atrium and the dye concentration in aorta was measured every second for 25 s. The dye concentration reading is plotted on a paper strip where a 55 mm reading corresponds to a concentration of 5 mg/liter. The result is shown in Table 5.10.

Determine the *cardiac output* of this patient. Consider how the recirculation affects the result.

TABLE 5.10 Reading of Dye Concentration in Blood

Time (s)	Reading (mm)
0	0
1	5
2	20
3	50
4	88
5	115
6	122
7	115
8	100
9	80
10	66
11	53
12	41
13	35
14	29
15	24
16	20
17	17
18	15
19	13
20	12
21	13
22	14
23	15
24	16
25	18

Proposed Solution 1

Enter the values in the spreadsheet. Calculate the concentration, dividing by 11 (since 55/5 = 11) and then create the points. The commands used to get Figure 5.50 are

```
C2 = B2/11
D2 = (A2, C2)
```

These are copied down alongside the table. Then an extra point is added in cell D28 by typing

```
D28=(25,0)
```

To determine the area under the curve, which is only known from point to point, use the polygon in Figure 5.50. The polygon was created from measured points using the Polygon tool ▷. In the input field you should type

```
Polygon[D2:D28]
```

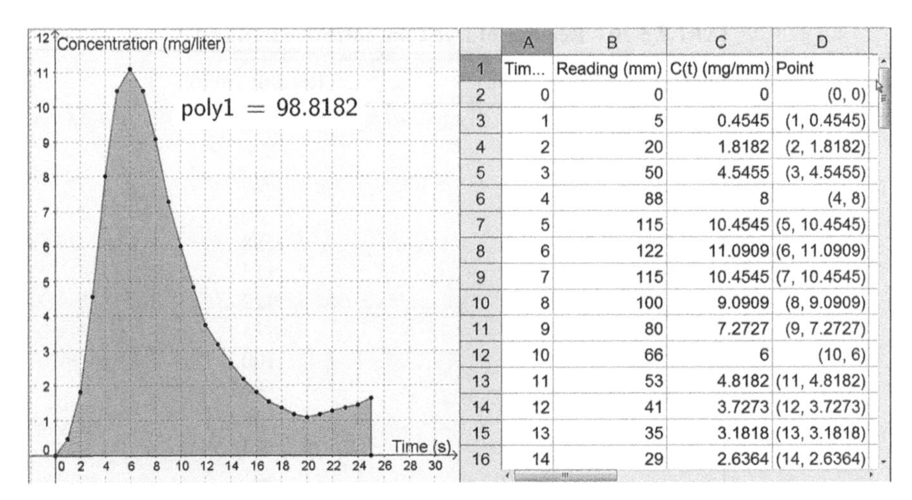

	A	B	C	D
1	Tim...	Reading (mm)	C(t) (mg/mm)	Point
2	0	0	0	(0, 0)
3	1	5	0.4545	(1, 0.4545)
4	2	20	1.8182	(2, 1.8182)
5	3	50	4.5455	(3, 4.5455)
6	4	88	8	(4, 8)
7	5	115	10.4545	(5, 10.4545)
8	6	122	11.0909	(6, 11.0909)
9	7	115	10.4545	(7, 10.4545)
10	8	100	9.0909	(8, 9.0909)
11	9	80	7.2727	(9, 7.2727)
12	10	66	6	(10, 6)
13	11	53	4.8182	(11, 4.8182)
14	12	41	3.7273	(12, 3.7273)
15	13	35	3.1818	(13, 3.1818)
16	14	29	2.6364	(14, 2.6364)

FIGURE 5.50 Area calculated by fitting a polygon to data points.

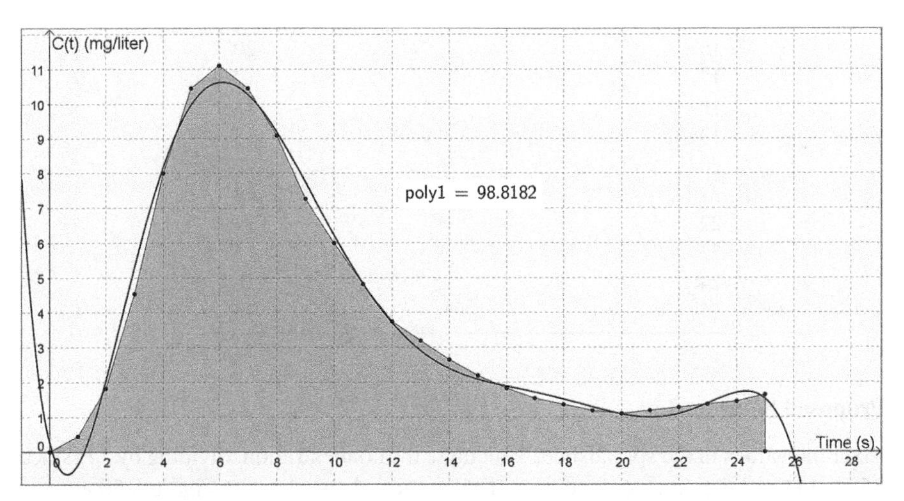

FIGURE 5.51 Seventh-degree polynomial is not a candidate for a good fit but there are worse fits.

The area reported is 98.8. The polygon method is a good method to use when it is hard to find suitable functions for your data.

You could also fit a function to the points. It is, however, quite difficult to know what model function to use in describing a pulse function like this one. Even if you try and do arrive at a seventh-degree polynomial fit, you will see that, as in Figure 5.51, the curve will still deviate somewhat from the measured points at the start, the top, and at the tail of the data.

Going with the data from the polygon, the patient's CO = 5.68 mg/98.8 mg·s/liter = 0.05749 liter/s = 3.45 liter/min. Observe how the units are reduced against each other. Unfortunately, something has been forgotten!

Proposed Solution 2

The mistake in Solution 1 was that blood recirculation was ignored. In all closed systems (the blood system of the body may be viewed as a closed system) everything is recycled as long as it does not get lost to the outside. In this patient's test, this means that part of the curve depends on the fact that the dye was measured two or even three times. How can you solve that? Since the decreasing part of the curve will follow an exponential decrease, you could try to take logarithms of the y-values to see whether they follow a straight line, as they then should.

Observe the fact that you can fit almost any function to the data, which does not mean that using logarithms is invalid. Our eyes can easily spot linear trends, and deviations from them, but it is more difficult with other functions.

Activate Graphics Area 2 with the keyboard shortcut **Ctrl-Shift-2** and then click in it. In the spreadsheet, create logarithms of the concentrations in the E-column (you have to exclude the initial zero value from the points) and create new points in the F-column. These new points will show up in Graphics Area 2.

In Figure 5.52, the straight line is the result of typing the command

```
FitPoly[F10:F14, 1]
```

A spreadsheet region like F10 : F14 can be part of a command just as in Excel, and it is automatically interpreted as a list by GeoGebra. Use **FitPoly[]** with the degree set to 1 instead of **FitLin[]** to get a function that you can make function calls to. Here, the function gets called $q_1(x)$ because all the usual function names are already used for the sides of the polygon.

You now have to adjust the values for recirculation. The values up to and including 10 s are good, but the rest of the values need to be exchanged for values

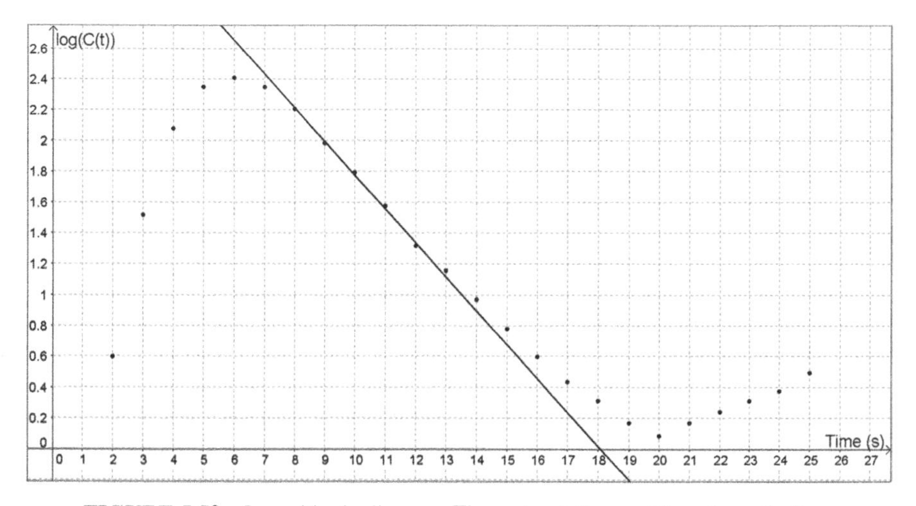

FIGURE 5.52 Logarithmic diagram. The recirculation part lies above the line.

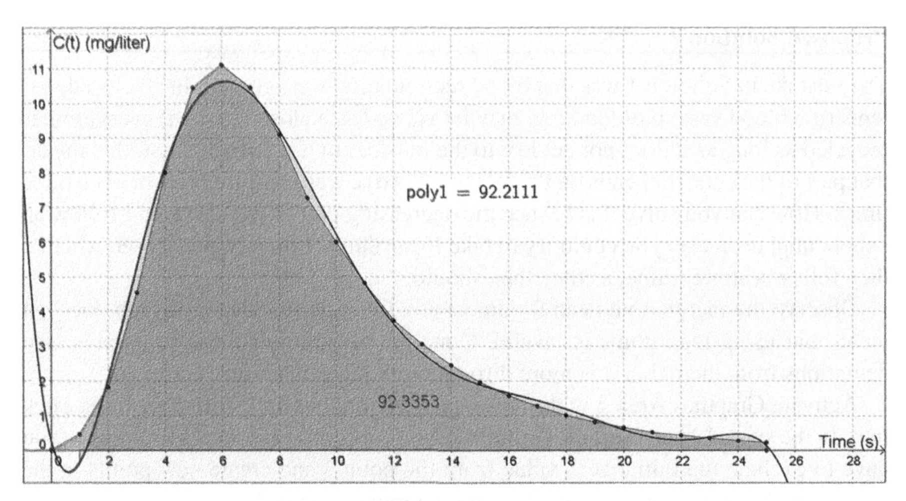

FIGURE 5.53 Diagram adjusted for the recirculation.

equal to the straight line. You can do this by clicking in column C. From C15 and down, type

```
=e^q_1(A15)
```

and copy this down. Use **Alt-e** for ***e***. Since the line depends on values in the rows 10–14 in the spreadsheet, you must start on row 15. The points will now move and you will get a new value for the area = 92.2. The line lies in the logarithmic diagram, which is why you have to take ***e*** raised to the function value to recreate the original values and compensate for recirculation.

You can also try to integrate the seventh-degree polynomial. Integrate from 1 in order to avoid the negative area in the beginning and type

```
Integral[p_1, 1, 25]
```

You now get the 92.3 for the area. These new values from Figure 5.53 agree rather well, so you get a *cardiac output* of 3.69 liters/min, which is a reasonable result.

5.11 MEDICATION

L-DOPA is a substance used in medicines for patients with symptoms from Parkinson's disease. In Table 5.11 are listed the amounts L-DOPA in blood, in nanograms per milliliter, for a typical medication dosage of L-DOPA. The L-DOPA in each patient's blood was measured as a function of time, in minutes, after the medication was

TABLE 5.11 Amount of Medicine in Blood over Time

t (min)	L-DOPA (ng/ml)
0	0
20	300
40	2,700
60	2,950
80	2,600
100	1,550
120	1,100
140	900
160	725
180	600
200	510
220	440
240	300
300	250
360	225

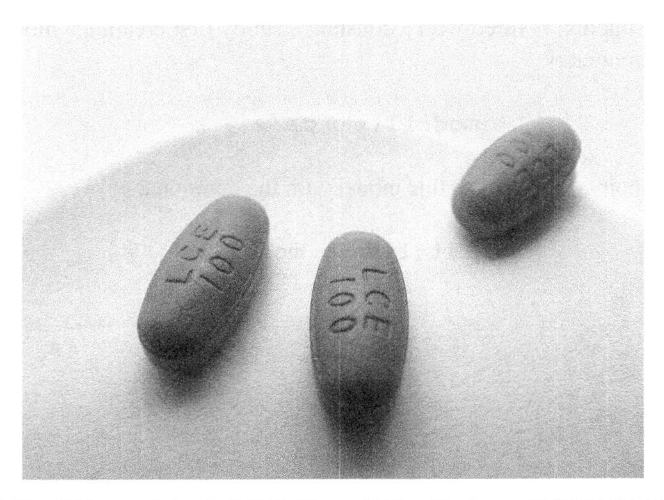

FIGURE 5.54 Pills containing L-DOPA. Photo: Walter Hochauer [CC BY-SA 2.0 de (http://creativecommons.org/licenses/by-sa/2.0/de/deed.en)], via Wikimedia Commons. https://upload.wikimedia.org/wikipedia/commons/0/09/Stalevo.jpg.

started. The medicine is typically dispersed in pills, such as the pills shown in Figure 5.54.

Suggest at least two different functions that you think will fit the data and discuss the relative advantages and disadvantages of your suggestions.

When will the level of L-DOPA in the blood go beneath 10 nanograms per milliliter?

Calculate the area and try to give it a reasonable interpretation.

Proposed Solution

A typical pulse is shown in Figure 5.55. This pulse is only positive, it starts at the origin and approaches zero as time goes to infinity. It is quite possible for the level of L-DOPA to be maintained at a level close to zero for a very long time, but it is reasonable to expect the medication to eventually be voided completely.

The pulse's second derivative seems to change sign from + to – to +. It is not easy to see immediately what kind of function you could use to model these data, but there are several options.

One is to split the function into several parts. Once you have passed the maximum, the rest of the course follows an exponential decay. So the start of the course could be modeled with a polynomial that has the appropriate maximum and minimum points.

Figure 5.56 shows one such possibility. A polynomial without a constant term was used by typing

```
list2 = {A, B, C, D, E, F}
Fit[list2, {x, x², x³}]
```

With this model the first six points are used to model a cubic polynomial. The exponential function is fitted with a constant term by first creating a model function with three parameters:

```
model1(x) =c a^x + b
```

Then the points are fitted to this model with the command

```
Fit[list3, model1]
```

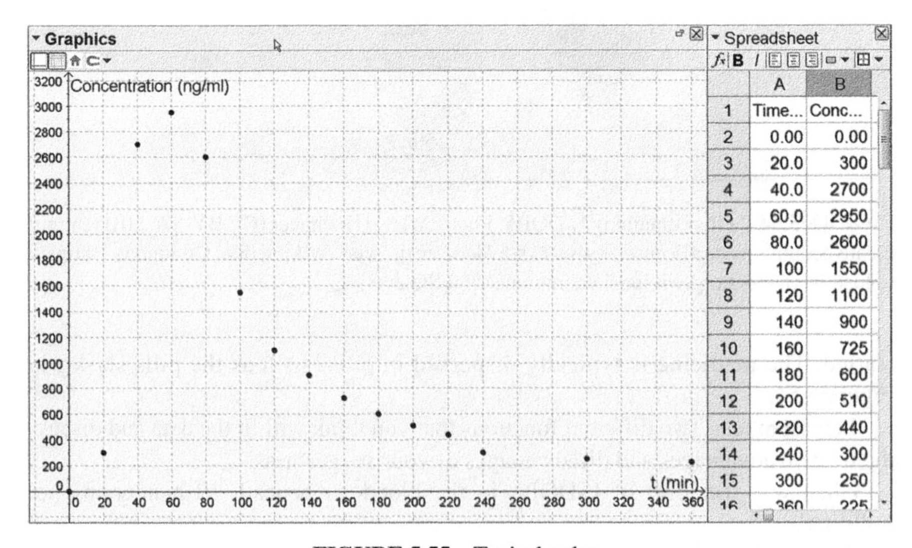

	A	B
1	Time...	Conc...
2	0.00	0.00
3	20.0	300
4	40.0	2700
5	60.0	2950
6	80.0	2600
7	100	1550
8	120	1100
9	140	900
10	160	725
11	180	600
12	200	510
13	220	440
14	240	300
15	300	250
16	360	225

FIGURE 5.55 Typical pulse.

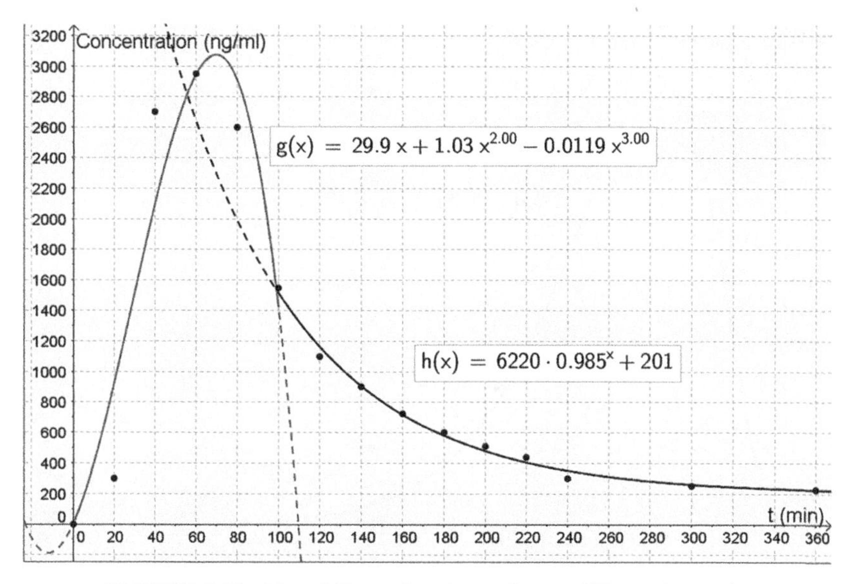

FIGURE 5.56 Two different functions adjust to different intervals.

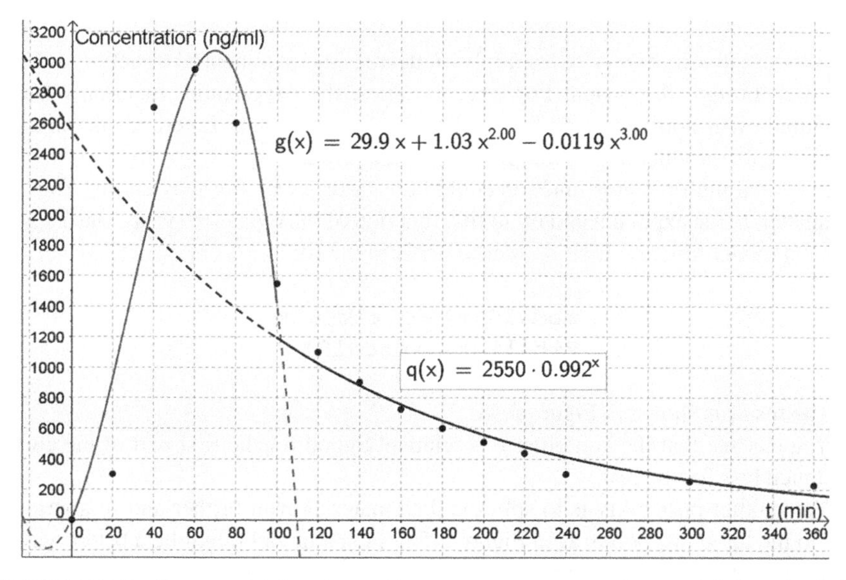

FIGURE 5.57 Exponential function without constant term used to model the tail of the pulse.

where all the point except the first five points are in **list3**. The fit is ok, but unfortunately this concentration will never go down to zero.

The simpler model for the second half of the pulse, shown in Figure 5.57, gives worse compliance with data but is somewhat more realistic.

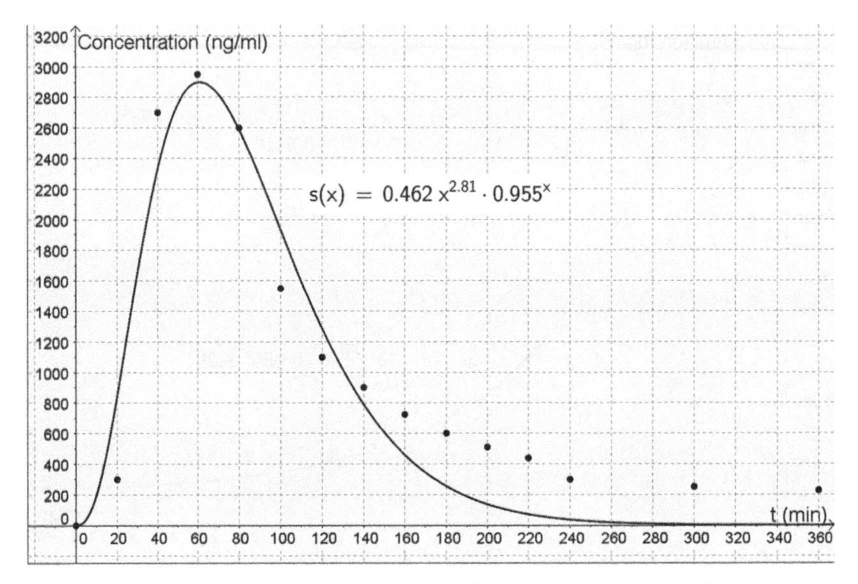

FIGURE 5.58 Fitting a product of a power function and an exponential function.

Another possibility is to multiply a decreasing exponential function with a function through the origin. For large x-values, the exponential function's lesser magnitude will dominate, and for smaller x-values, the other function's dominance will force the product to go through the origin as well.

A typical multiplicator function could be $y=x$ or $y=x^2$. But why chose? You can parameterize the exponent and try to fit a function of the type $y=c \cdot x^k \cdot a^x$. Once again, create a model function first in order to perform the fit:

```
model3(x)=c x^k a^x
Fit[list1, model2]
```

The result is shown in Figure 5.58.

You can see that the compliance is relatively good for the first part of the curve, but much less for the rest.

Yet another possibility is to split the data material even further and fit a straight line to the three last points as in Figure 5.59. Such a model might be a better way to find a suitable value for the time the level of L-DOPA goes beneath 10 nanograms per milliliter.

In order to get some grip on this model, also try to fit a straight line to the data points that are between the top and the last tail.

This extreme ad hoc model, however, lacks a theoretical foundation, though it does give a relatively good functional fit to the data. For a concentration of 10 ng/ml, it gives 697 min, or almost 12 h (713 min until the concentration reaches 0), and the area 374,000 ng/ml min = 374 µg/liter min. You could, of course, have fitted a polygon to the points, which is the same thing as using the trapeze method.

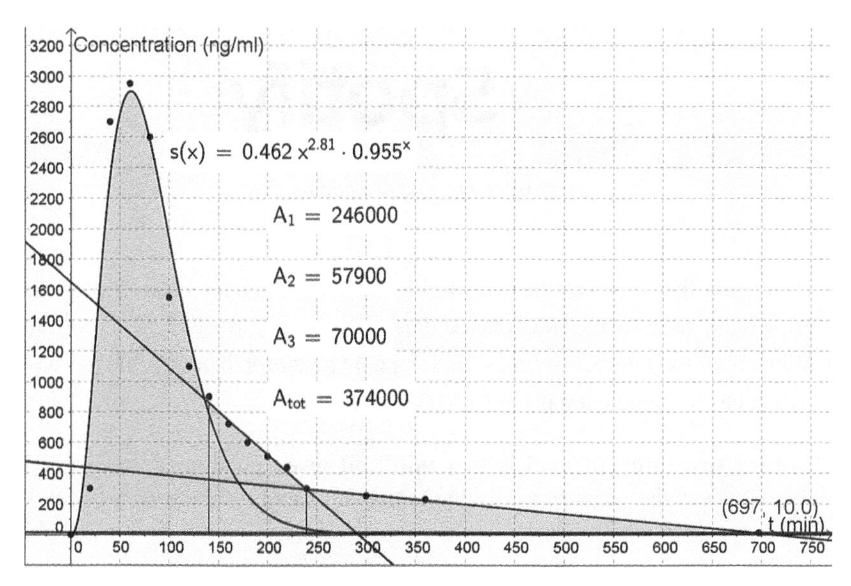

FIGURE 5.59 Model in three parts.

In order to understand the meaning of the area must you first realize that the blood is moving with a speed, or with a flow Φ that can be measured in liters per minute. If you compare these units with the units for the area, then you can see that

$$[\text{ng/liter}] \cdot [\text{min}] \cdot [\text{liter/min}] = [\text{ng}]$$

So the area under the curve multiplied by the blood flow seems to give the total mass of the medication that has passed the probe:

$$m = \Phi \cdot \int \text{konc } dt$$

This is complicated by the fact that is takes such a long time for the medication to be measured by receptors that the blood will have gone several rounds in the body. This is called recirculation and may possibly be compensated with the same methodology used in Problem 5.10, *Cardiac Output*.

5.12 NEW SONG ON SPOTIFY

Figure 5.60 shows the by now well-known Spotify logotype. On Spotify, new music comes and goes. A particular song called "Temporarily Popular" is played on Spotify 12,500 times the first week after its release. During the tenth week the song hits its maximum, and then the number of times the song is played per week decreases. After 25 weeks the song is more or less forgotten and almost not played at all. In order to pay a fair compensation to the songwriters, Spotify employees has asked you to find some additional information:

- How many times was the song played in total?
- How many times was the song played on average during these 25 weeks?

FIGURE 5.60 Spotify logotype.

- How many listeners play the song during week 10, when it reaches a maximum?
- How many times is the song played during the first 12 weeks?
- When does the number of times the song is played drop below 9,000 per week?
- When has the song been played 100,000 times?

Construct two different models that fulfill all given conditions and use these models to answer the questions. Discuss similarities and differences between you models.

Proposed Solution 1

In this problem you will mainly use GeoGebra as a graphing tool.

You want to construct a function $S(t)$ that describes the number of times the song is played, up to and including week t, which fulfill the following conditions:

$$S(0) = 0$$
$$S(1) = 12{,}500$$
$$S'(25) = 0$$
$$S'(10) = 0 \quad \text{and} \quad S''(10) < 0 \qquad \text{(maximum playing rate)}$$

You also want this function to be increasing in the interval $0 \leq t \leq 25$ and to have a maximum rise at $t = 10$.

The function $S(t) = \text{A} \sin(\text{B}t + \text{C}) + \text{D}$ can, for suitable values on the parameters A, B, C, and D, probably describe the total number of playings during $0 \leq t \leq 25$. This also makes it convenient to think about the properties its derivative – the playing rate – has. The playing rate can be described by $S'(t) = \text{AB} \cos(\text{B}t + \text{C})$, defined in the same interval as $S(t)$.

The playing rate $S'(t)$ goes from a maximum down to zero during weeks 10 to 25. This means that a quarter of a period is 15 weeks and a whole period is 60 weeks, which in turn means that $\text{B} = 2\pi/60 = \pi/30$, giving $S'(t) = \text{A}\pi/30 \cos(\pi t/30 + \text{C})$.

If $\text{C} = 0$ this function will have its maximum at $t = 0$ and its zero at $t = 15$. Obviously you have to displace the function 10 weeks to the right. Ten weeks is 1/6 of a period, which gives $\text{C} = -\pi/3$. The function may therefore be written as $S'(t) = \text{A}\pi/30 \cos(\pi t/30 - \pi/3)$. If you ignore the multiplicative constant, the cosine function will look like the one in the Figure 5.61.

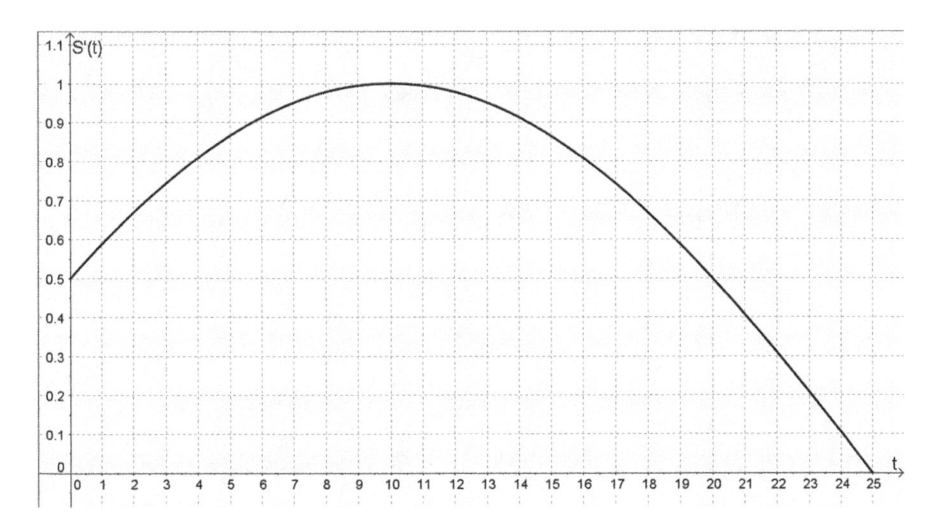

FIGURE 5.61 Having determined the period and phase, you now need to determine the amplitude.

What you have left to do is to calculate the amplitude A in order to complete the playing rate function $S'(t)$. You know that the song was played 12,500 times in first week, so set

$$\int_0^1 S'(t)\,dt = 12.5$$

where you switch to counting in thousands.

The calculation gives

$$\left[A\sin\left(\frac{\pi}{30}t-\frac{\pi}{3}\right)\right]_0^1 = A\left[\left(\sin\left(\frac{-9\pi}{30}\right)\right)-\left(\sin\left(\frac{-10\pi}{30}\right)\right)\right] = A\left[\sin\left(\frac{\pi}{3}\right)-\sin\left(\frac{3\pi}{10}\right)\right] = 12.5$$

which gives A = 219.3. So the playing rate can now be written

$$S'(t) = \frac{219.3\pi}{30}\cos\left(\frac{\pi}{30}t-\frac{\pi}{3}\right), \qquad 0 \le t \le 25$$

as is shown in Figure 5.62.

You must next determine D in $S(t) = A\sin(Bt+C)+D$. The condition $S(0)=0$ immediately suggests that $D = 219.3\cdot\sin(\pi/3) = 189.9$. The model for the total number of played songs is therefore

$$S(t) = 219.3\cdot\sin\left(\frac{\pi}{30}t-\frac{\pi}{3}\right)+189.9$$

which is shown in Figure 5.63

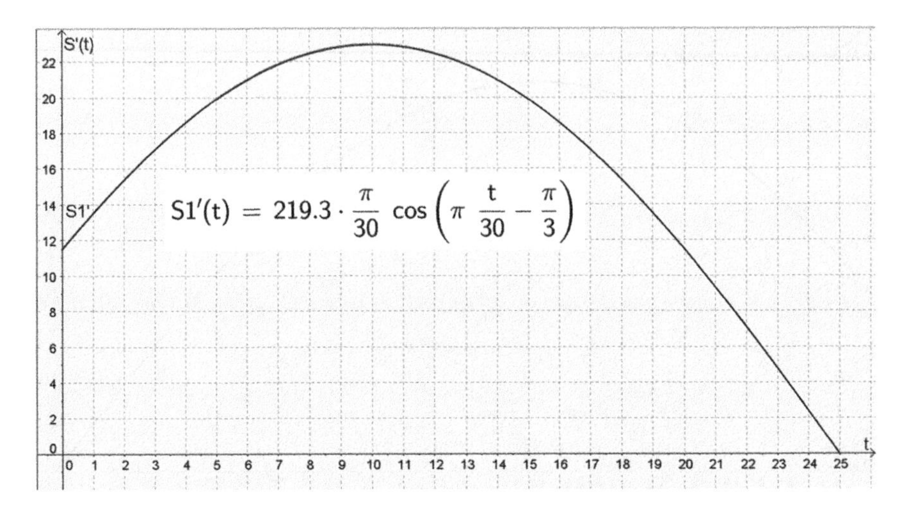

FIGURE 5.62 Complete model for the playing rate.

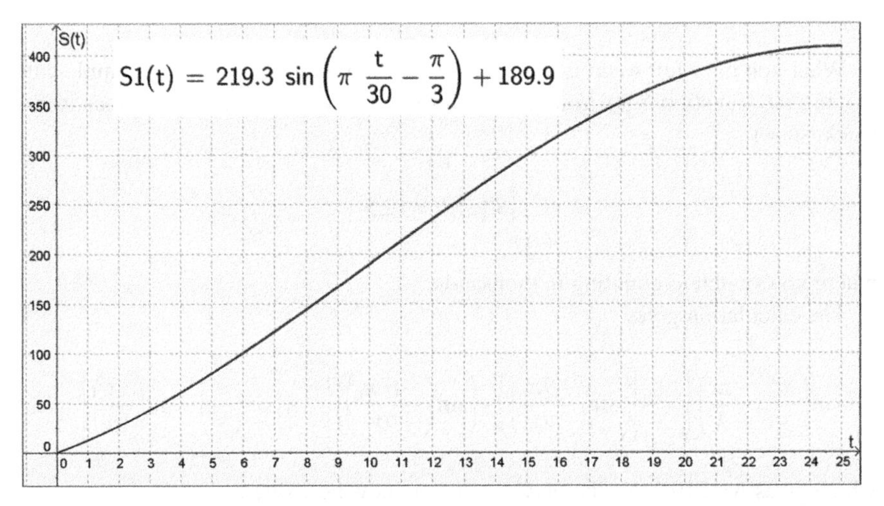

FIGURE 5.63 Complete model for the total number of played songs.

You can now use the model to answer the questions:

- The total number of times the song was played $= S(25) = 409{,}000$ times.
- The average number of times the song was played per week $= 409{,}000/25 = 16{,}360$ times/week.
- The number of times the song is played reaches its top week 10 and is then played $S'(10) = 22{,}970$ times/week
- The total times the song was played after 12 weeks $= S(12) = 235{,}470$ times
- The playing rate dips below 9,000 times/week in week 21.
- The song was played 100,000 times at $S(t) = 100$, which is in week 6.

Proposed Solution 2

As you have seen, the playing rate describes a function with its maximum point at $t=10$. Now assume that this function is a quadratic function. The total number of times the song was played will then be given by a cubic polynomial.

Assuming that the playing rate has a maximum point and is symmetrical around $t=10$, you can write

$$S'(t) = -a(t-10)^2 + b$$

The condition $S'(25)=0$ means that $b=225a$. Substituting this gives

$$S'(t) = -at^2 + 20at + 125a$$

You know that the song was played 12,500 times in first week, so set

$$\int_0^1 S'(t)\,dt = 12.5$$

Once again this represents thousands of times the song is played. The calculation gives $a=0.09224$ and $b=20.79$, so the function $S'(t)$ is given by

$$S'(t) = -0.0924(t-10)^2 + 20.79 = -0.0924t^2 + 1.848t + 11.55, \quad 0 \le t \le 25$$

which is shown in Figure 5.64.

$S(t)$ is acquired from integration:

$$S(t) = -0.0308t^3 + 0.924t^2 + 11.55t + C$$

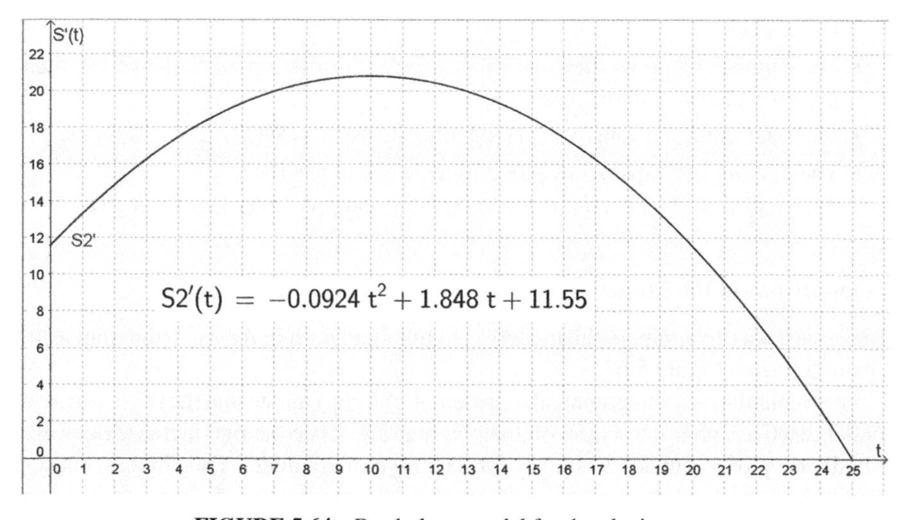

FIGURE 5.64 Parabola as model for the playing rate.

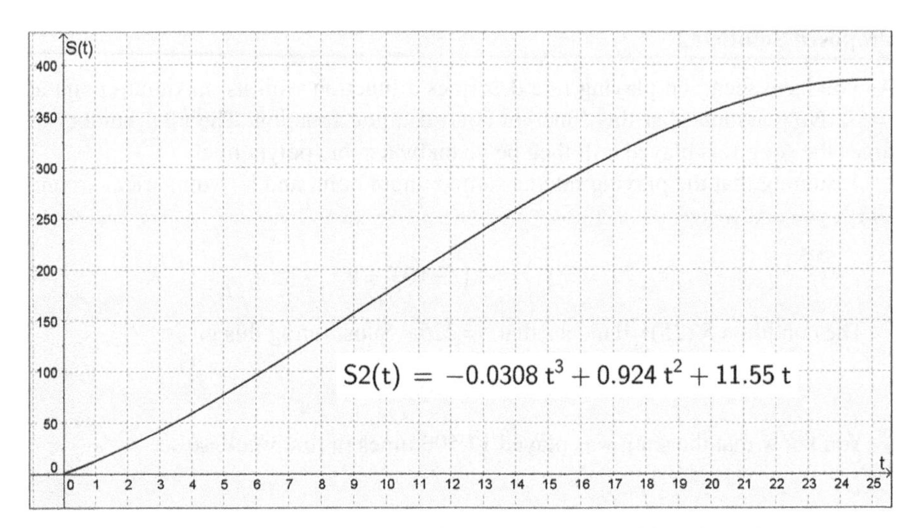

FIGURE 5.65 Total number of played songs as a cubic polynomial.

From condition $S(25)=0$, we learn, somewhat surprisingly, that C=0. The final model is therefore

$$S(t) = -0.0308t^3 + 0.924t^2 + 11.55t, \quad 0 \leq t \leq 25,$$

shown in Figure 5.65.

You can now once again use the model to answer the questions:

- The total number of times the song was played $= S(25) = 385{,}000$ times.
- The average number of times the song was played per week $= 385{,}000/25 = 15{,}400$ times/week.
- The number of times the song is played reaches its top week 10 and is then played $S'(10) = 21{,}690$ times/week.
- The total times the song was played after 12 weeks $= S(12) = 218{,}430$ times.
- The playing rate dips below 9,000 times/week in week 22.
- The song was played a total of 100,000 times at $S(t) = 100$, that is, in week 7.

Comparisons of the Models

The easiest way to compare the models is to put them both in the same diagram as in Figure 5.66 and Figure 5.67.

From calculations and graphical representations you can see that the trigonometric model reach a higher maximum on the playing rate, since the rate increase does not initially drop of as quickly as it does for the polynomial model. This, in turn, causes the total times the song was played, which is the integral of the rate, to become higher.

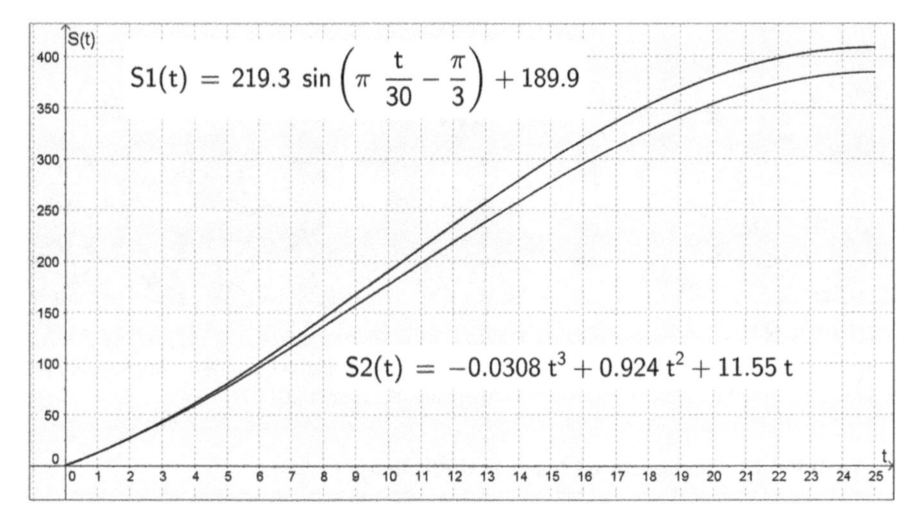

FIGURE 5.66 Comparison of the total times the song was played.

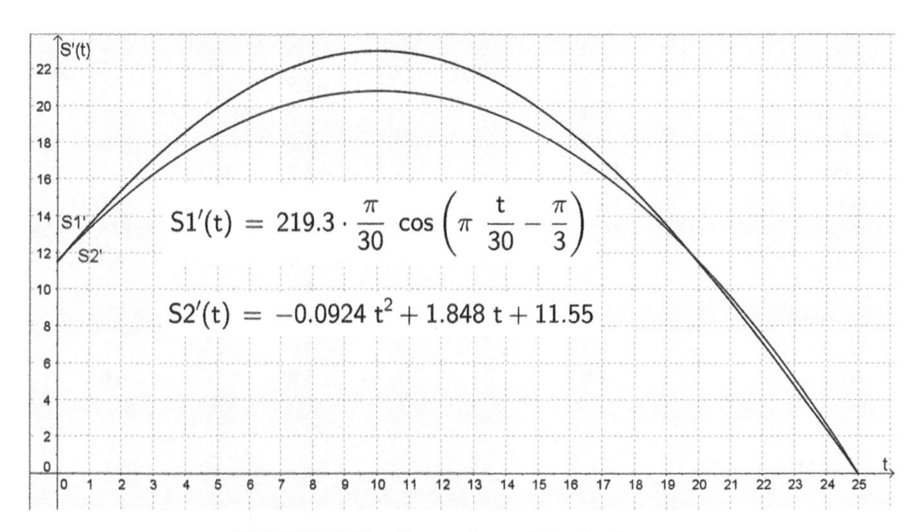

FIGURE 5.67 Comparison of the playing rate.

5.13 TEMPERATURE CHANGE

Figure 5.68 shows a panorama of Gothenburg, Sweden's second largest city. The average temperature around and inside Gothenburg has changed noticeably over the last 20 years. Table 5.12 gives data from *Statistics Sweden*, the Swedish central agency for statistics, which is found online at http://www.scb.se/en_/.

Represent the temperatures in a diagram and illustrate the changes in some suitable way. Construct a model that fits the data and that describes the changes in temperature.

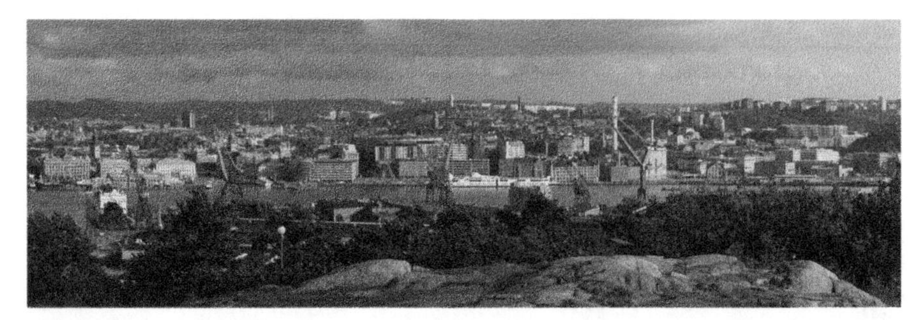

FIGURE 5.68 Gothenburg panorama. Photo: David Lindecrantz (own work) [CC BY-SA 3.0 (http://creativecommons.org/licenses/by-sa/3.0)], via Wikimedia Commons. https://upload.wikimedia.org/wikipedia/commons/f/fb/Gothenburg-Panorama-20110911.jpg.

TABLE 5.12 Temperature Change in Gothenburg

Month	Average temperature (°C) 1991–2005	Average temperature (°C) 1961–1990	Change
January	−4.0	−6.5	+2.5
February	−4.3	−6.0	+1.7
March	−1.2	−2.5	+1.3
April	3.3	2.2	+1.1
May	8.5	8.2	+0.3
June	12.8	13.2	−0.3
July	15.8	14.8	+1.0
August	14.8	13.6	+1.2
September	10.1	9.4	+0.7
October	4.7	5.0	−0.3
November	−0.1	−0.7	+0.6
December	−3.4	−4.6	+1.2
All year	**4.8**	**3.9**	**+0.9**

Proposed Solution

You need to think the problem through before you select a model. Temperature models for temperate countries tend to be periodic. GeoGebra has a built-in trigonometric regression model that fits a function of the type $y = \mathrm{a} + \mathrm{b} \cdot \sin(\mathrm{c} \cdot x + \mathrm{d})$ to the data. It gives an acceptable result for both series of temperature, but the difference between the functions is not at all the same as the difference displayed by the data.

One problem here is that GeoGebra tries to fit a value to c, corresponding to the angular frequency ω. But you already know this. It is reasonable to believe that the temperatures have fluctuations where the period is one whole year and thus $\omega = \mathrm{c} = 2\pi/12$.

Therefore make another attempt with a general regression and with the model function

```
m(x) = a + b sin(2πx / 12 + c)
```

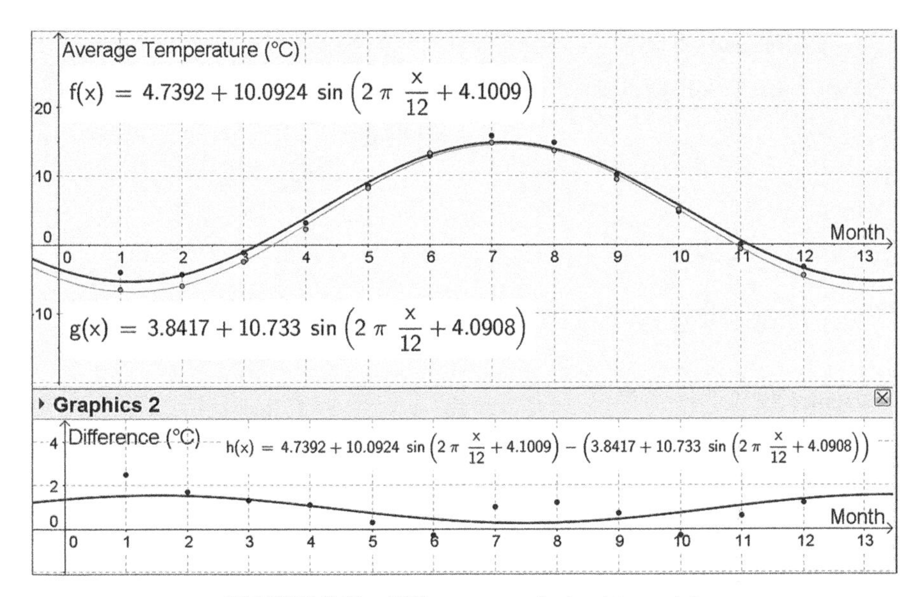

FIGURE 5.69 Differences can be hard to model.

You then fit the functions by typing

```
f(x) = Fit[list1, m]
g(x) = Fit[list2, m]
```

where **list1** and **list2** contain the temperature data.

As you can see in Figure 5.69, you still have a really bad fit for the temperature differences, even though the average temperatures are in phase and seem to be adequately modeled. The model's precision is simply too weak to model differences.

How can you improve the precision? A polynomial can be made to fit the data better and better the higher the temperature degree, and this idea can be used in trigonometric functions as well. In the polynomial case, this idea leads to the concept of the Taylor and McLaurin series, and in the trigonometric case, this idea will lead to the concept of the Fourier series. In this sense then, is it possible for us to compare the trigonometric model to a linear polynomial with just one constant term and one first-degree term? The comparison between a polynomial and a trigonometric function in fact shows the general similarities between these different structures. In the trigonometric case, you create new, higher degree terms by adding overtones. In Figure 5.70 the first overtone has been added with a frequency twice as high as the frequency for the fundamental tones:

```
m(x) = a + b sin(2π x/12 + c) + p sin(4π x/12 + q)
```

It is not unreasonable to do this. The temperature depends on several, quite different phenomena, each of them periodic, but with different phases (parameters c and q).

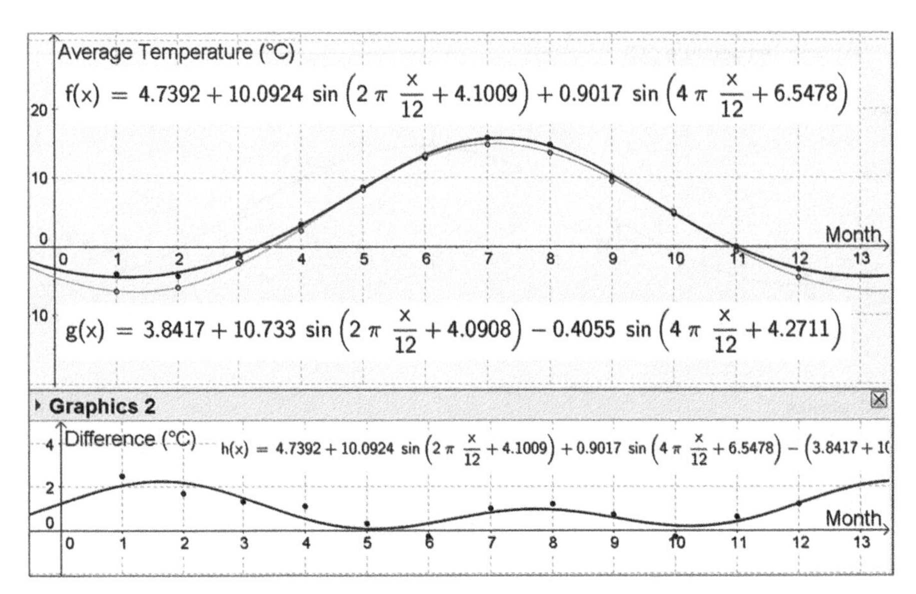

FIGURE 5.70 Two sine functions are better than one.

The Earth's distance to the Sun varies, and that in turn affects the incoming radiation, the insolation. It would be perhaps be best if the data could be modeled with the same phase, which hasn't been done here. In particular, one of the term's phases differs from the others, which lowers the reliability of the results. In order to partly improve this anomaly, you could try a new model function that locks the phases to each other. Obviously there is room for improvement here.

5.14 TAR

The poisonous substances in cigarette smoke are generally called *tar*. Many cigarettes have filters that absorb tar, but the tar cannot be completely absorbed. A smoked cigarette will give of tar in a nonlinear way. The closer to the filter you get when smoking the cigarette, the more tar you will inhale. The tobacco closest to the filter acts like a small filter in itself, and some of the tar will accumulate there during the smoking. If this tobacco is smoked as well, it will release the accumulated tar into the smoke.

A typical cigarette has about 8 centimeters of tobacco and a filter. It is possible to verify that tar is inhaled according to the relation $r(s)=Ae^{0.025s}-Be^{0.02s}$, where s is the distance in cm along the cigarette measured from where the cigarette is lit, r is the amount of inhaled tar in mg/cm, and A and B are positive constants. The total amount of inhaled tar I during smoking of a cigarette from s_1 to s_2 ($s_1<s_2$) is then given by

$$I=\int_{s_1}^{s_2} r(s)\,ds$$

TABLE 5.13 Amount of Tar Inhaled in Different Sections of the Cigarette

$I(s_i, s_j)$ ($s_{i,j}$ in cm)	Amount of tar (mg)
$I(0, 1)$	61.4
$I(1, 2)$	64.2
$I(2, 3)$	67.0
$I(3, 4)$	—
$I(4, 5)$	73.1
$I(5, 6)$	—
$I(6, 7)$	79.6
$I(7, 8)$	—

In order to determine the constants A and B and thereby get a model that can calculate the amount of inhaled tar for arbitrary values for s_1 and s_2, a research team has measured the inhaled tar by a mechanism constructed to mimic a human respiratory system. The data shown in Table 5.13 were compounded during such a laboratory study.

Calculate with as close accuracy as possible the values for the constants A and B. Then use these values to compute the amount of tar that is inhaled from the first to the last of a cigarette's centimeters during the smoke. How much more tar will the last centimeter give compared to the first?

Proposed Solution 1

You know that $r(s) = Ae^{0.025s} - Be^{0.02s}$ and that the inhaled tar from a cigarette from s_1 to s_2 ($s_1 < s_2$) is given by

$$I = \int_{s_1}^{s_2} r(s)\,ds$$

When you integrate, you will find that

$$I = \left[A\frac{e^{0.025s}}{0.025} - B\frac{e^{0.02s}}{0.02} \right]_{s_1}^{s_s}$$

According to Table 5.13, integration from 0 to 1 cm will give the amount of tar as 61.4 mg, and integration from 1 to 2 will give the amount 64.2 mg, and so forth.

So the equation system is

$$\left[A\frac{e^{0.025}}{0.025} - B\frac{e^{0.02}}{0.02} \right] - \left[A\frac{1}{0.025} - B\frac{2}{0.02} \right] = 61.4$$

$$\left[A\frac{e^{0.05}}{0.025}-B\frac{e^{0.04}}{0.02}\right]-\left[A\frac{e^{0.025}}{0.025}-B\frac{e^{0.02}}{0.02}\right]=64.2$$

To avoid manual calculation errors, you could solve this with Wolfram Alpha by sending the following command:

```
(A*exp(0.025)/0.025-B*exp(0.02)/0.02)-(A*1/0.025-B*1/0.02)
= 61.4;
(A*exp(0.05)/0.025-B*exp(0.04)/0.02)-(A*exp(0.025)/0.025-
B*exp(0.02)/0.02) = 64.2;
```

Wolfram Alpha will respond with A=301.191 and B=241.159. The model will consequently be $r(s)=301.191e^{0.025s} - 241.159\ e^{0.02s}$. This function can also be graphed in GeoGebra. You can use the variable **s** and start by typing

```
r(s) = ...
```

You have to change the axis and adjust the numbers of the shown decimals. Imagine a cigarette laid in the **s**-direction with the mg/cm amount of tar in the **r**-direction. This model, shown in Figure 5.71, is only applicable in a limited interval.

You can calculate the values of the model and then how they stand up against the values you already have in the table. The command

```
integrate (301.191*exp(0.025*x)-241.159*exp(0.02*x) dx
from x=0 to 1
```

gives the first value in Table 5.14 from Wolfram Alpha.

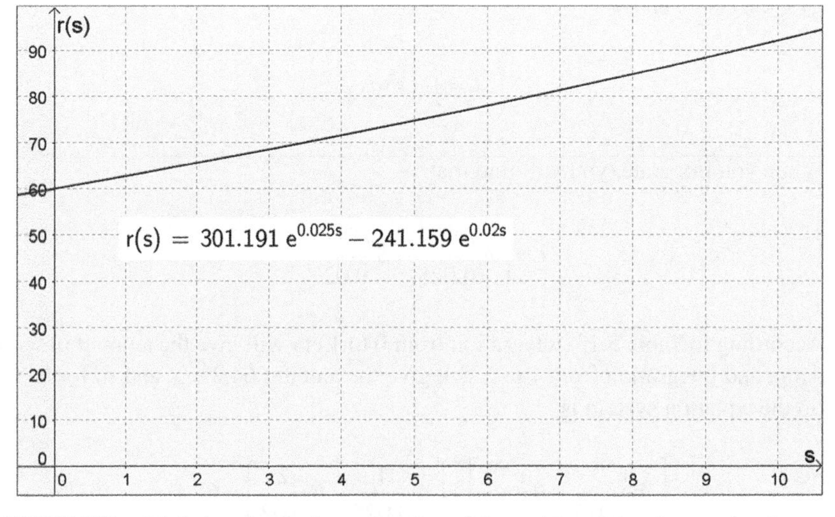

FIGURE 5.71 Inhaled tar (mg/cm) as a function of the position of the glow on the cigarette.

TABLE 5.14 Comparison of Measured and Calculated Values of Tar Amounts

$I(s_i, s_j)$ ($s_{i,j}$ in cm)	Measured tar (mg)	Calculated tar from integration of $r(s)$ (mg)
$I(0, 1)$	61.4	61.40
$I(1, 2)$	64.2	64.20
$I(2, 3)$	67.0	67.10
$I(3, 4)$		70.09
$I(4, 5)$	73.1	73.19
$I(5, 6)$		76.39
$I(6, 7)$	79.6	79.70
$I(7, 8)$		83.12

So the model delivers almost the same values as in Table 5.14 but you can also see that the amount of tar increase a lot during the 8 cm as cigarette can be smoked. The last centimeter gives 21.7 mg more tar than the first, an increase of 35 percent! Likely this is something for all smokers to think about.

Proposed Solution 2

In the first solution you used Wolfram Alpha in order to solve an equation system of the first degree. In a way that was a waste of information. The numerical answers that Wolfram Alpha gave you were not all you received. You also received the simplified equations $1.0126\,A - 1.01007\,B = 61.4$ and $1.03824\,A - 1.03047\,B = 64.2$.

Since you have five values in the table, it should be possible to set up five such equations. But this equation system, having two variables and five equations, is overdetermined and has no unambiguous solution. How can you work around this problem?

One way is not to consider the relations as an equations system, but as points in a coordinate system. Every equation has three numerical values and the parameters A and B. The numerical values could represent three-dimensional points and the equations should then represent different planes in space. For given x- and y-coordinates, every equation then shows how to calculate the plane's z-coordinate. A completely free plane should have four parameters, as in $Ax + By + Cz + D = 0$, but in this case C is already -1 and $D = 0$. It is, in principle, quite possible to fit a plane to these points with some higher order least square method, but that requires better tools.

What you can do instead is to systematically investigate what solutions each pair of equations gives. To be able to deal with overdetermined equation system can be useful in many practical situations. First use Wolfram Alpha to simplify the equations. They are displayed in Table 5.15.

You could continue with Wolfram Alpha and two of these equations for all 10 combinations that are possible, but it is actually more convenient to do this in GeoGebra by replacing A with x and B with y. You will then get five straight lines. It is quickly done to use the Intersect tool and click on two lines at a time to create the 10 intersection points.

TABLE 5.15 Equations from Wolfram Alpha

$I(s_i, s_j)$	Equation
$I(0, 1)$	1.0126 A – 1.01007 B=61.4
$I(1, 2)$	1.03824 A – 1.03047 B=64.2
$I(2, 3)$	1.06452 A – 1.05129 B=67.0
$I(4, 5)$	1.1191 A – 1.09419 B=73.1
$I(6, 7)$	1.17648 A – 1.13885 B=79.6

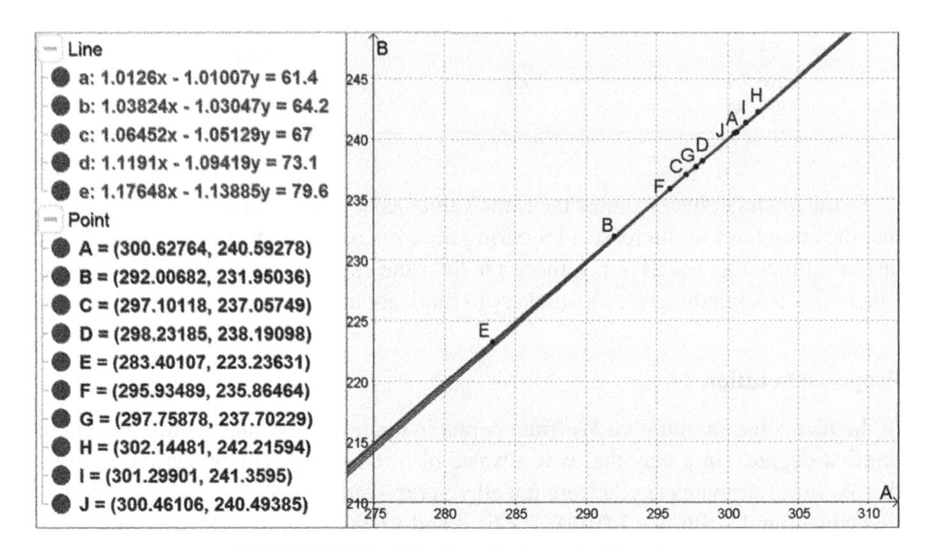

FIGURE 5.72 Five lines intersect in 10 points.

Note in Figure 5.72 that the lines are as good as parallel and the solutions have therefore small, but not insignificant, distributions. You see that solution A, which corresponds to the combination of equations 1 and 2, does not give the same answer as before once the numbers have been rounded off during the simplification.

Theoretically could you make a new table now and recalculate the integrals, but only do this for the solution of E, which is the one furthest away from solution A. Ask Wolfram Alpha to

```
integrate (283.4011*exp(0.025*x)-223.2363*exp(0.02*x) dx
from x=0 to 1
```

and so on, to get the results in Table 5.16.

With this last solution method you get a lower level of agreement with the data, especially with the two last centimeters. The last centimeter now delivers 21.0 mg more tar than the first centimeter, a 34% increase. If you use the differences in final results as the final error limits, the results seem to be accurate ±1 mg/1%.

TABLE 5.16 Comparing Different Solutions to Estimate Errors

$I(s_i, s_j)$ ($s_{i,j}$ in cm)	Measured tar (mg)	Calculated tar solution A (mg)	Calculated tar solution E (mg)
$I(0, 1)$	61.4	61.40	61.49
$I(1, 2)$	64.2	64.20	64.20
$I(2, 3)$	67.0	67.10	67.00
$I(3, 4)$		70.09	69.90
$I(4, 5)$	73.1	73.19	72.89
$I(5, 6)$		76.39	75.99
$I(6, 7)$	79.6	79.70	79.18
$I(7, 8)$		83.12	82.49

5.15 BICYCLE REFLECTORS REVISITED

In Problem 5.9, *Bicycle Reflectors*, the distance traveled by the reflector was never explicitly calculated. Find the distance traveled by the reflector relative to the distance the bicycle travels as a function of the distance from the hub. Is it possible to find a symbolic expression for this? Where should you place the reflector if you want it to go 25% farther than the distance cycled?

5.16 GAS PRESSURE

For a given quantity of gas, Boyle's law states that the pressure exerted by the gas and the volume it takes up are inversely proportional to each other; that is, $P = C/V$, or equivalently, $PV = C$, for some constant C. Assume that the volume changes over time, so that V is a function of t. Find the rate of change of the pressure with respect to time in two different ways: in terms of V and dV/dt only, and in terms of P and dV/dt only.

Assume that the pressure exerted by the gas is modeled by the equation

$$P(t) = \sqrt{1+t}$$

Determine the rates at which the pressure and the volume are changing, with respect to time.

A volume of gas occupies 6.52 liters at a pressure of 0.92 atmospheres. Determine the volume if the pressure is increased to 1.44 atm. If the pressure change is linear and occurs over a period of 5 s, what is the instantaneous rate of change of the volume with respect to time at 2 s?

5.17 AIRBORNE ATTACKS

Here is a very simple model for airborne attacks and the defense of the targets. First assume that all targets have the same value, all bombs have the same strength, all missiles the same probability to hit, and so on. Also assume that one bomb hitting the target will be enough to destroy it. The targets are spaced so that an anti-aircraft gun at one target cannot defend other targets.

There are m defensive missiles that can hit enemy attackers and the probability for a missile to destroy an attacker is k.

If there are n missiles against one attacker, then the attacker is destroyed with probability

$$A = 1-(1-k)^n$$

This is the complementary event of not being destroyed.

If there are b attackers against a single target, the probability that all attackers are destroyed is A^b, so the target is lost with probability $1 - A^b$.

The best defending strategy is to divide all missiles equally among the b attackers. This means that the probability that the target is destroyed is

$$P = 1-\left(1-(1-k)^{m/b}\right)^b$$

What is the best strategy for the attackers? For a numerical example, assume there are 400 attackers attacking 50 targets, each defended with $m = 20$ missiles, each missile capable of bringing down an attacker with a probability $k = 0.4$. How many targets should the attackers "go for" in order to maximize the number of destroyed targets?

5.18 RAILROAD TRACKS

To avoid the effects of thermal expansion, railroad tracks are laid down with a small, but decisive, gap between them. Assume for a minute that they didn't, and lad down a stretch of 100 m of solid track, fixed to the ground only at the ends, in the midst of winter. How high would the middle of the track rise on a hot summer day if it takes the shape of a circular segment? How high would it rise if instead, it broke in the middle to form a flat triangle?

Find a relation between the temperature and the track rise for both scenarios. Also find and plot the derivative, that is, how many meters (cm, mm, etc.) the tracks will rise for each degree the temperature rises.

5.19 COBB–DOUGLAS PRODUCTION FUNCTIONS

You decide to go into business, so you set up a company that does gardening jobs in a small town where you supply the tools and pay fixed wages. According to basic economic theory, you need two things to do this; capital K (e.g., cost of tools) and labor L (e.g., worker-hours per year). You rent the tools at the yearly rate r = 5%. The wage for your employees is w = 12 USD per worker-hour.

Given the amount of capital K and Labor L, the output, or number of job hours is assumed to follow a Cobb–Douglas type production function: $F(K, L)=A\cdot K^{\alpha}\cdot L^{\beta}$, where $A=2.5$, $\alpha=0.4$, and $\beta=0.6$ for this problem. Often, but by no means always, $\alpha+\beta=1$. Suppose further that $p=20$ USD is the price per job-hour that you charge. Then the profit is a function of L:

$$Profit(L)=p\cdot A\cdot K^{0.4}\cdot L^{0.6}-r\cdot K-w\cdot L$$

where each term has the units USD/year.

We assume that you have a fixed number of tools, that is, the capital $K=1000$ USD. You can then choose how many employees L you need to hire to maximize your profit.

Find this number, using the data given in the problem. Then investigate the effect each parameter has on the number of employees needed to maximize your profit. Which parameters have a linear effect and which don't? Can you solve the problem algebraically? What interpretation can you give to the parameter A?

Suppose instead that you have a tightly knit gang of friends that you employ and so the labor is fixed at 8 full-time employees. How much money should you invest in tools to maximize your profit this time?

5.20 FUTURE CARBON DIOXIDE EMISSIONS

Projections for yearly carbon dioxide emissions worldwide for the next 700 years are listed in Table 5.17. The time variable t corresponds to the year starting in 2000.

Construct a mathematical model to fit these data and use it to estimate how much carbon dioxide will be emitted in total over the next 400 years.

What is the total amount of carbon dioxide emitted over the next 700 years, according to this model?

TABLE 5.17 Predicted Carbon Dioxide Emissions

Year (since 2000)	Carbon dioxide emission (billion tons/year)
0	5.5
100	10
200	13.7
300	12
400	7
500	3.5
600	1.5
700	0.5

5.21 OVERTAKING

A particular car is said to accelerate from 60 km/h to 100 km/h in 10 s in fourth gear.

If the acceleration could be kept constant, then the speed would increase linearly, but since the acceleration cannot be maintained, we will use another model:

$$v(t) = \mathrm{a}\sqrt{t} + \mathrm{b}$$

where a and b are constants and t is the number of seconds the car has accelerated.

Construct a model of this in GeoGebra and find the values of the constants a and b. What distance would be needed to overtake another vehicle in 10 s, if your initial speed is 60 km/h? How far would you move in relation to the other vehicle? Is this dependent on the speed of the other vehicle? Investigate!

If you change down to third gear before overtaking, you lose a second for the changing but you could finish the overtaking sooner. Assume that in third gear you could accelerate from 60 km/h to 100 km/h in just 6 s. Under what circumstances would it be beneficial to change down to third gear to overtake another vehicle?

5.22 POPULATION DYNAMICS OF INDIA

In the Wolfram Mathworld article on life expectancy at http://mathworld.wolfram.com/LifeExpectancy.html

You find detailed information on how to calculate the remaining life expectancy from a population distribution. Use the Indian age distribution table found at https://en.wikipedia.org/wiki/Demographics_of_India#Structure_of_the_population_.5B62.5D to calculate the remaining life expectancy data of India. Plot this data against age. The calculations are mostly differences between table values. Can you set up the corresponding relations for the continuous versions of the variables? What variables are derivatives of other what other variables?

The Population dynamics evolve over time. See a good example of this at http://www.gapminder.org/population/tool/ How does this affect the remaining life expectancies?

5.23 DRAG RACING

Drag racing is a sport where you sit in a racing car and try to accelerate as hard as possible to finish a quarter-mile (402.34 m) as quickly as possible, reaching a speed as high as possible. Afterward, you are presented with a time slip, such as the one in Figure 5.73, where your data are recorded. These data show how long it took you to travel various distances. The R/T values represent your reaction times and correspond to a distance of 0 ft. Then times for 30 ft, 330 ft, 1/8 mi, 1,000 ft, and 1/4 mi are recorded, together with your speed at 1/8 mi and 1/4 mi.

```
Car #...          54

Class ...

DIAL  ...
R/T   ...        .734
60'   ...       2.063
330   ...       5.387
1/8   ...       8.025
MPH   ...       94.23
1000  ...      10.312

 E.T. ...      12.259

MPH   ...      116.14
```

FIGURE 5.73 Drag racing time slip sample.

If you search the Internet for photographs of "Drag strip time slip," you will find lots of examples of these time slips. Use the data provided in Figure 5.73, or any other time slip, to calculate either the 0–60 time (i.e., how long it took to reach 60 mph) or the 0–100 time (i.e., how long it took to reach 100 km/h). Your model should take into account that the speed at two different times and the total distance are known. Compare your results from your final model with results from a simple polynomial fit to the time and distance data only. How much more accuracy can you get by taking the known speeds and distances into account?

5.24 SUPER EGGS

Assuming a circle has a radius of r and is placed in the origin of a coordinate system, then the equation describing this circle is

$$\left(\frac{x}{r}\right)^2 + \left(\frac{y}{r}\right)^2 = 1$$

An ellipse with semi-axes a and b has the equation

$$\left(\frac{x}{a}\right)^2 + \left(\frac{y}{b}\right)^2 = 1$$

A curve known as the *super-ellipse* may be thought of as a generalization of the ellipse, and it has nearly the same equation

$$\left(\frac{x}{a}\right)^n + \left(\frac{y}{b}\right)^n = 1$$

where normally, $n > 2$ but this is not strict.

If this curve is rotated around the x-axis, a super-ellipsoid is formed. Investigate the volume of a super-ellipsoid with respect to n for different values of b. You can assume that $a = 1$ without loss of generality.

5.25 MEASURING STICKS

A cylindrical tank has a flat bottom so that its cross section is a circle that has "lost a segment." The owner wants to have a measuring stick that is able to measure the volume of the content directly by lowering it down from the top opening until it hits the bottom of the tank.

Create the scale that should be engraved on the tank. That is, make a table that converts from cm on the stick to liters in the tank. The tank measures 160 cm in diameter but only 130 cm from top to bottom and has a capacity of 2 m^3.

What would the scale look like for different shapes of the tank, such as a flattened sphere, an octagon, or a frustum?

5.26 THE LECTURE HALL

A large and sloping lecture hall is completely full when you arrive. The hall is 15 m wide and you have to stand at one of the side walls in order to get a reasonable view of the 5 m wide screen at the front wall of the room. The stage extends 6 m from the front wall before it starts to climb 15 cm every 60 cm. Where along the side wall should you be standing to maximize the viewing angle of the screen? You can attempt to solve this problem as if the room were flat first, before attempting the three-dimensional case.

5.27 PROGRESSIVE BRAKING DISTANCES

When learning to drive, you are taught *progressive breaking*, meaning you should brake gently, slowly increasing the pedal force until you reach a maximum force and maximum deceleration, and then ease off and release the breaks just before coming to a standstill to avoid a jerk. Assuming that this results in a pulse-shaped deceleration and that the maximum comfortable deceleration is 5 m/s^2, find the stopping distance as a function of speed for this kind of breaking. There may be other parameters to take into consideration as well as other observations. For instance, the stopping distances should be longer than the emergence break distances that result from breaking with maximal pedal force. Also note that the stopping distance = reaction distance + braking distance. For a choice of pulse-like functions, see Appendix B.

5.28 CYLINDER IN A CONE

A cylinder with radius r and height h is to be inscribed in a cone with radius R and height H. Find the maximum volume of the cylinder and its dimensions.

This typical optimization problem can be seen as the intersection of a surface

$$v(r,h) = \pi r^2 h$$

and a plane corresponding to the condition that the cylinder is inscribed in the cone. The local maximum of the intersecting curve of the surface and the plane will give you the maximum volume of the inscribed cylinder. Use GeoGebra's 3D window to solve this, and other typical optimization problems.

6

USING DIFFERENTIAL EQUATIONS

Suppose that you know how a certain quantity, for instance, the temperature of coffee in a coffee cup, the number of people infected with a virus, or the concentration of carbon dioxide in the atmosphere changes over time. The rate of change of this quantity is the derivative, so you can work out how fast the temperature changes, how fast the number of infected people changes, or how fast the concentration of carbon dioxide changes.

Suppose that you actually know the value of the quantity right now and you wish to predict its value in the future. To do this, you must know how fast the quantity is changing. But the rate of change will often depend on the quantity itself: this gives rise to what is called a *differential equation*—an equation relating the derivative of a quantity to its value.

Now suppose, for example, that $N(t)$ is the number of bacteria growing in a petri dish. At the start of the experiment, suppose there are 1,000 bacteria, so $N(0) = 1000$. This is known as an *initial condition* or an *initial value*. The rate of change of N will be proportional to N itself: if there are twice as many bacteria, then N will grow twice as rapidly. So

$$\mathrm{d}N/\mathrm{d}t = \mathrm{k}\,N$$

where k is a constant, and $\mathrm{d}N/\mathrm{d}t$ is the derivative (rate of change) of N with respect to time. You would have to do further experiments to find out the value of the parameter k.

Mathematical Modeling: Applications with GeoGebra™, First Edition. Jonas Hall and Thomas Lingefjärd.

Companion website: www.wiley.com/go/Hall/MathematicalModeling

It is possible to verify that

$$N(t)=1000e^{kt}$$

is a solution of this differential equation with the given initial condition. To do this, first calculate $N(0)$ and verify that it satisfies the initial condition:

$$N(0)=1000e^{0}=1000$$

The next step is to calculate dN/dt and verify that it satisfies the differential equation

$$dN/dt=1000ke^{kt}=k(1000e^{kt})=kN$$

as required.

Differential equations lie, in some ways, at the very heart of modeling. With the help of differential equations, you can handle realistic and complicated situations of change. When doing so, you need to formulate problems, solve problems symbolically, analytically or numerically, and analyze solutions with respect to parameters. Furthermore you need to handle stability issues; you need to get the rate of change from tabular values and infer differential equations from these. There are many mathematical and cognitive issues you need to handle when dealing with mathematical modeling and differential equations.

Differential equations can be modeled either in a spreadsheet like Excel or GeoGebra or directly in GeoGebra, thereby getting dynamic graphical solutions. GeoGebra will also allow you to solve systems of differential equations graphically, which will enable you to formulate and solve quite complex real world problems.

6.1 COOLING III

A small amount of water is cooling after it has been poured from a kettle to a cup. The time t in seconds and the temperature T in degrees Celsius are in Table 6.1. At the time of the experiment, the room temperature was 22.3°C.

Construct a mathematical model that fits *all* these data and use the model to determine how long it will take until the water cools down to 26°C. Start by using Newton's cooling law but also try to find other models that may work.

Proposed Solution

You already tried fitting ordinary exponential functions to these data in Problem 3.6, *Cooling I*. You then experienced some difficulty in making the exponential function fit well with the initial values. It comes naturally to see this as an exercise about exponential functions. Since Newton's cooling law says that the rate of heat loss for a body is proportional to the difference in temperature between the body and its surroundings, exponential functions are likely solutions.

TABLE 6.1 Cooling Data

t (s)	T (°C)	t (s)	T (°C)
0	69.58	510	39.40
30	66.11	540	38.70
60	61.41	570	38.00
90	58.07	600	37.32
120	55.60	630	36.67
150	53.58	660	36.08
180	51.66	690	35.50
210	50.05	720	35.00
240	48.52	750	34.53
270	47.24	780	34.04
300	46.00	810	33.59
330	44.96	840	33.20
360	43.96	870	32.76
390	42.92	900	32.37
420	41.95	930	32.00
450	41.05	960	31.64
480	40.18	990	31.30

Newton's cooling law can be stated as a differential equation $T' = \mathrm{k} \cdot (T - \mathrm{T}_0) = \mathrm{k} \cdot \Delta T$, but this is not the only possible model. A more careful look at the situation indicates that you might try $T' = \mathrm{k} \cdot \Delta T^{\mathrm{p}}$, where $\mathrm{p} = 5/4$ or $\mathrm{p} = 4/3$ depending on the type of flow you have at the contact surface. Unfortunately, this differential equation is so hard to solve that neither GeoGebra nor Wolfram Alpha manages to solve it symbolically for an unspecifed value of p. Yet both GeoGebra and Wolfram Alpha can solve the differential equation $T' = \mathrm{k} \cdot \Delta T^{5/4}$. In GeoGebra you first open the CAS window from the menu with, **Show>CAS** or using the keyboard command **Ctrl-Shift-K**. You then type

```
SolveODE[y' = k y^(5/4)]
```

Wolfram Alpha's answer is somewhat simpler, but equivalent. It is a rational function:

$$T(t) = 256 / (\mathrm{c} - \mathrm{k} \cdot t)^4$$

Having done this, try to fit the data to a model function based on

```
m(x) = 256/(c - k x)^4 + 22.3
```

where you have switched to using x rather than t. Allow the sliders c and k to be created, and set them to c = 1.5 and k = 0.001 in order to have suitable start values for the algorithm. Finally hide the model function m and type

```
Fit[list1, m]
```

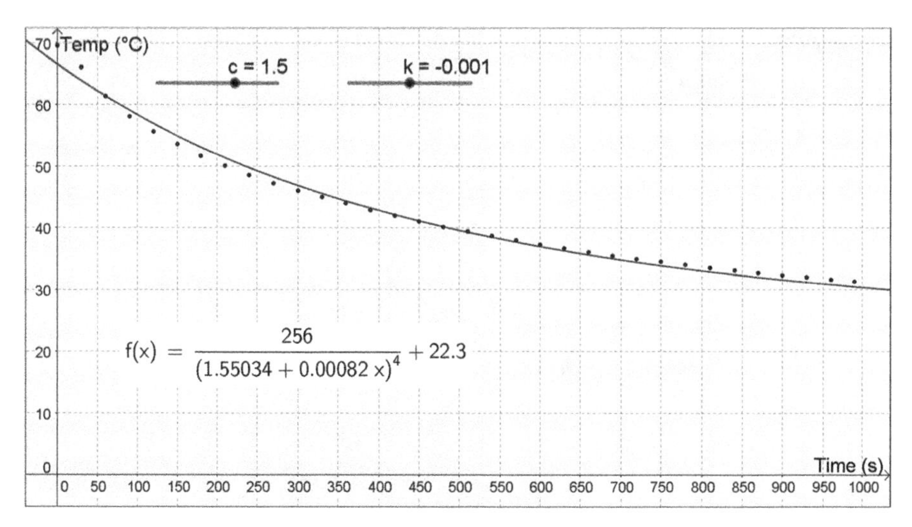

FIGURE 6.1 Fitting a rational function to cooling data.

There are, however, still systematic errors. You can see that the fitted function in Figure 6.1 lies above the data points in the middle and below the data points at the beginning and at the end of the interval. Will the exponent 4/3 work better? In this case you get $T(t)=27/(\mathrm{c}+\mathrm{k}\cdot t)^3$ as the solution for the differential equation and then create

```
m2(x) = 27/(c + k x)^3 + 22.3
```

and after that a new function by typing

```
Fit[list1, m2]
```

The results in Figure 6.2 are better but might become even better. Maybe you can ask GeoGebra to find the best exponent in this situation? Try the model function

```
m3(x) = a/(c + k x)^p + 22.3
```

and accept to create the new sliders for a and p. Set a = 22.4, c = 0.6, k = 0.00113, and p = 1.4 in order to get reasonable starting values and then type

```
Fit[list1, m3]
```

This time you get a very good fit to the data values (shame on you otherwise, since you have introduced two more parameters). The disadvantage is that you no longer have your function clearly linked to a differential equation. So you are on a rather insecure theoretical ground.

You can answer the question by plotting the line $y=26$ in Figure 6.3 and finding the intersection between the line and your function by typing

```
Intersect[a, h]
```

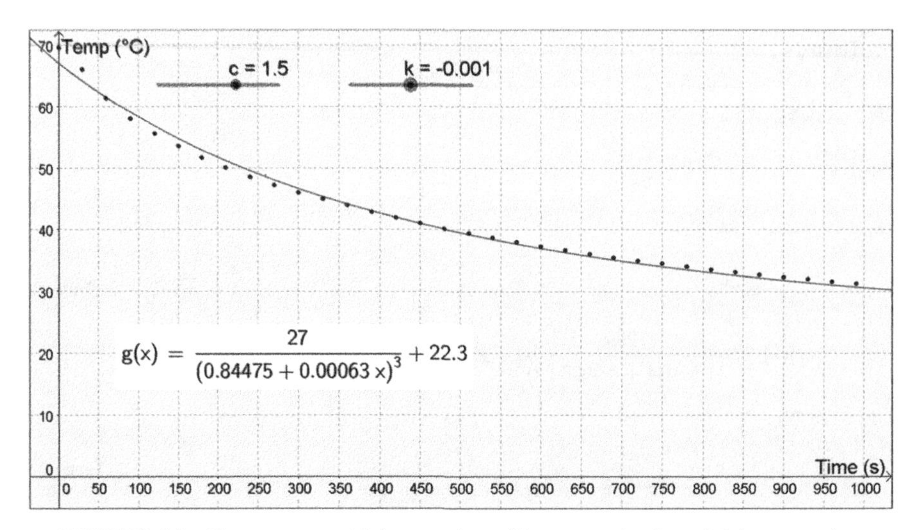

FIGURE 6.2 New attempt with better—but still not completely satisfying—results.

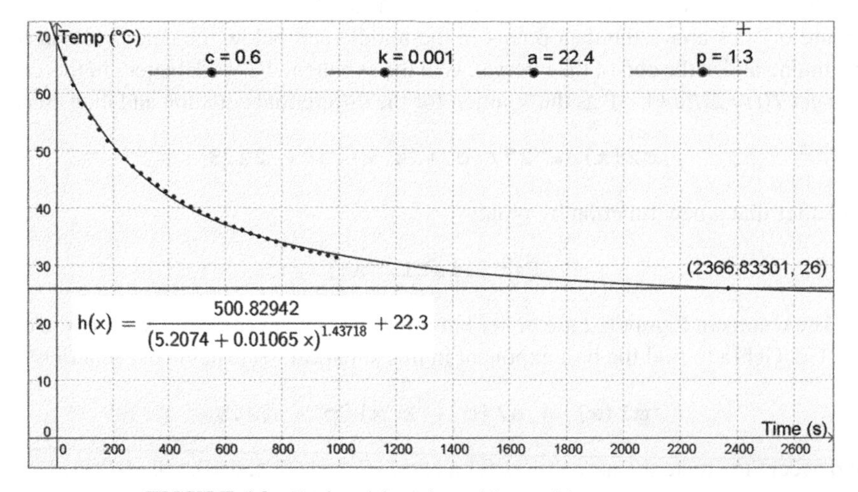

FIGURE 6.3 Rational function with an arbitrary exponent.

You get an answer quite different from than the one you got when you solved Problem 3.6, *Cooling I*. There you got around 26 min, but now you get 39.5 min.

Why is there such a huge difference? Generally speaking, it is always hard to extrapolate data, to draw conclusions from the model about something that will happen outside the data interval. Interpolation, to draw conclusions inside the data interval, is so much easier. In this case you have measured the temperature for 16.5 min and you are asked to find a time value at least 10 min, or even up to 25 min, away.

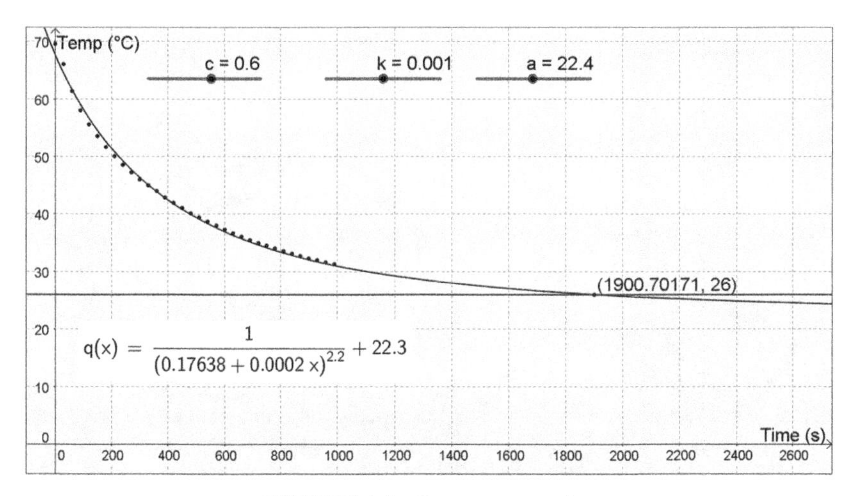

FIGURE 6.4 Best answer so far.

If you try different values with a fixed parameter p and use the model function of the type

```
m4(x) = 1/(c + k x)^2.2 + 22.3
```

where the value 2.2 is varied manually, you will get different answers for each new value of the exponent. Precisely around $p = 2.2$ the resulting function seems to follow the data at the end of the interval exceptionally well; see Figure 6.4. This approach will yield an answer of about 31.5 min. Is this the "best" answer?

Newton's cooling law is easy to understand, but creating solutions with different methods can bring students to understand that mathematics is much more than just one single solution to check against the answer key at the end of the book.

6.2 MOOSE HUNTING

Figure 6.5 shows a female moose with calves. There are currently about 300,000 moose in Sweden in summertime, and about 90,000 of them are shot during annual hunts. In the 1980s the animal herd was twice as large as today, and caused many more traffic accidents as well as forest damage. The Environmental Protection Agency in Sweden decided to halve the number of moose, which was done by issuing more hunting licenses. The herd is now considered stable, meaning around 90,000 new calves are born every year. Some moose obviously die of natural causes, but we will ignore that for the time being.

For most wild animals, their environmental growth is not without limits. In general, it follows Pierre Francois Verhulst's model from the mid-1800s, whereby growth is not just proportional to the number of individuals—which usually leads to *exponential growth*—but also to the remaining environmental growth space.

FIGURE 6.5 Moose cow with calves. Photo: Denali National Park and Preserve (Moose Cow and Calves Uploaded by AlbertHerring) [CC BY 2.0 (http://creativecommons.org/licenses/by/2.0)], via Wikimedia Commons. https://upload.wikimedia.org/wikipedia/commons/1/1f/Moose_Cow_and_Calves_%286187093400%29.jpg.

The theory has a parameter, K, called the species *carrying capacity*, which is the maximum number of individuals that can live in the area. If the number of individuals is y, then the remaining environmental growth space is $(\mathrm{K} - y)$. Verhulst's model then says that $y' = \mathrm{k} \cdot y \cdot (\mathrm{K} - y)$, whose solutions are called *logistic functions*.

Solve this equation and fit the solution to known facts about the moose. Then try to adjust the model to take the annual hunt into account. How many animals can be shot yearly for the moose herd to still remain stable and not be at risk of extermination?

Proposed Solution

There are several ways to solve a differential equation in GeoGebra. In this chapter we will show you the common ones, but you can also construct your own numerical solutions in the spreadsheet or use the command

```
SolveODE[]
```

in several different ways.

Read more about this and about other commands in the user handbook at http://wiki.geogebra.org/en/SolveODE_Command.

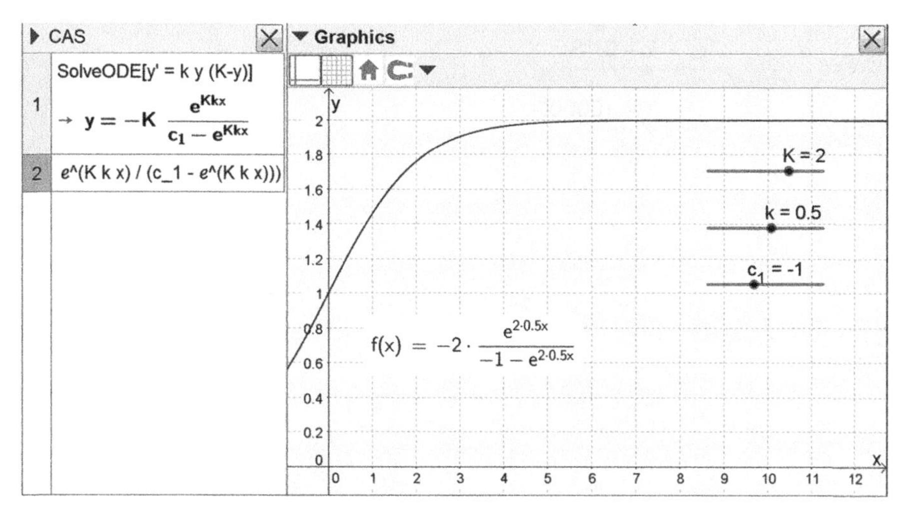

FIGURE 6.6 Symbolic solution of a differential equation in GeoGebra.

A Symbolic Solution

Start GeoGebra and open the CAS window by from the menu by selecting **View > CAS.** Then type

```
SolveODE[y' = k y (K-y)]
```

GeoGebra will give a symbolic answer with an integration constant c_1.

With the cursor in row 2 of the CAS window, you can type a right parenthesis, ")", which will copy the output in an editable format to row 2. When you copy this into the ordinary input field and press **Enter**, GeoGebra will ask if you want to create sliders for k and K. Answer yes, and the graph is drawn complete with sliders. But the graph is flat, and you are never asked if you wanted to create a slider for the integration constant.

That depends on the fact that it is already created but defined as being = 0. Integration constants are normally defined as help objects that are often hidden. To show the integration constant, click in the Algebra window on the Auxiliary Objects button to the left in the Style Bar of the Algebra window. Show the number c_1 as a slider by clicking in the white visibility button to the left of c_1 in the Algebra window. If you change the value, GeoGebra will show you the shape of the graph. In Figure 6.6 the CAS window and the Graphics window are shown but not the Algebra window.

A Numeric Solution

Open a new GeoGebra window by pressing **Ctrl-N**. Type the following command into the input field:

```
NSolveODE[{k y (K - y)}, 0, {210000}, 5]
```

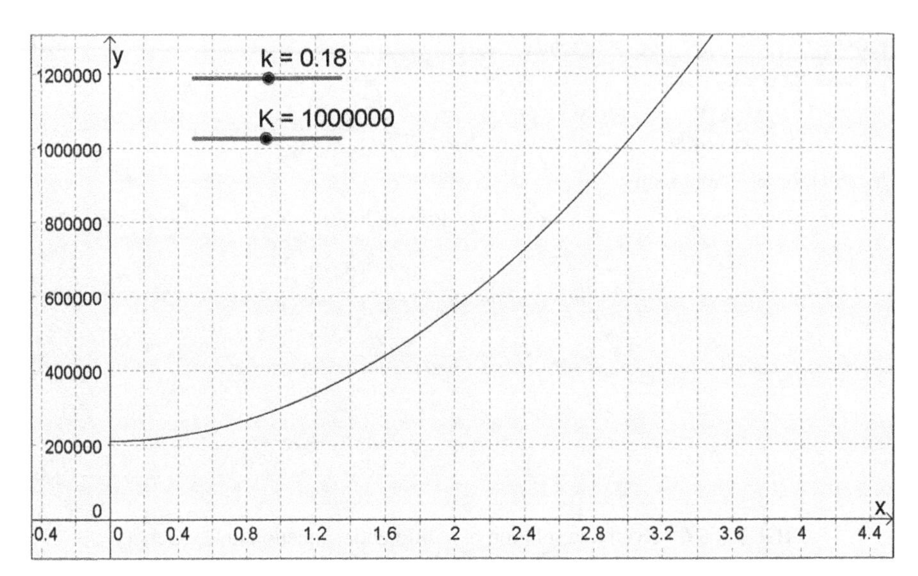

FIGURE 6.7 Numerical solution of a differential equation in GeoGebra.

Press **Enter** and answer yes to the question if you want to create new sliders.

In this command, do not write "$y'=$" first! Both the derivative and the initial y-value should be inside the list braces, since the command can solve several equations at the same time. The result, shown in Figure 6.7, will not be an algebraic function, but a curve, or *path*, that GeoGebra calculated numerically. This means, for instance, that GeoGeobra cannot find the derivative, or its extreme points.

A Direct Solution

Open yet a new window by pressing **Ctrl-N**. Create a point A on the y-axis with the point tool •A. Then type the following two commands into the input field:

```
SolveODE[k y (K-y), A]        (accept to create a sliders)
SlopeField[k y (K-y)]
```

When the slope field is shown as in Figure 6.8, it is harder to see the sliders. You can hide or show the sliders by clicking on the round blue visibility indicator buttons to the left of the numbers in the Algebra window. In order to change the value of a number, select the number in the Algebra window and use the arrow keys. You can also right-click on the number and select **Properties>Sliders**, and change Minimum, Maximum, and Increment values for the sliders.

Rest of the Problem

Continue from the direct solution. The first thing to do is to zoom and move A to (0, 210,000). The most convenient way to do this is to select the points' properties,

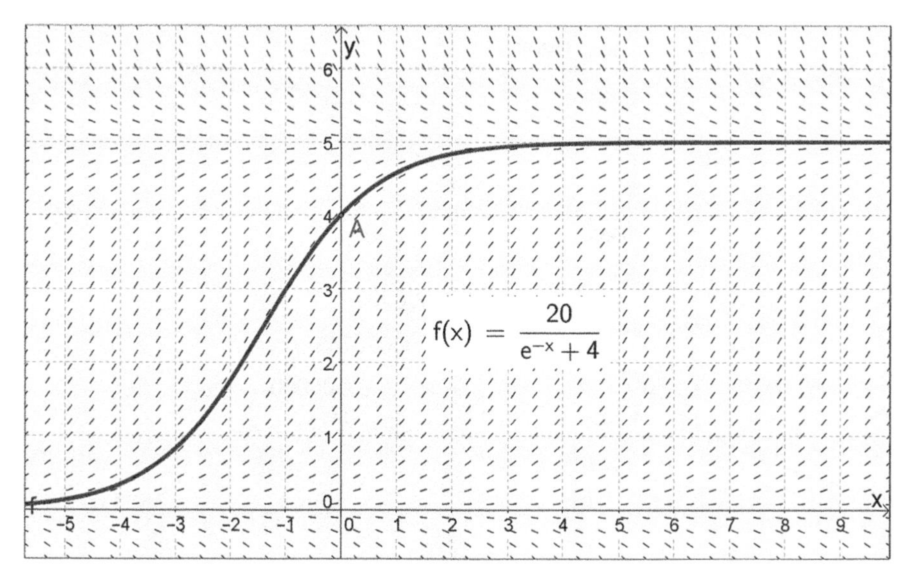

FIGURE 6.8 Direct solution of a differential equation in GeoGebra with a slope field.

the **Algebra** tab, and change the step length to 10,000. Once the point has been selected, you can move it with the arrow keys.

Next you need to adjust the sliders. Set K to go from 0 to 1,000,000 with an increment of 10,000 and k to have an increment of 0.00000001. Remember that K is the maximum number of animals and that there were as many as 600,000 animals in Sweden during the 1980s. So set K = 1,000,000 as a possible maximal limit and adjust k so that the number of animals grows from 210,000 to 300,000 animals in one year. You will find that k = 0.00000047. Setting K = 500,000 animals as in Figure 6.9, you will get k = 0.00000143. There appears to be several different solutions to the problem. You need to know more about the parameters from direct field studies before you can determine exactly what is happening.

In order to take the yearly hunt into account, consider that the hunt decreases the growth speed by taking away a number of individuals every year. You can currently ignore details such as whether these are bulls, cows, or calves. The differential equation becomes $y' = \mathrm{k} \cdot y \cdot (\mathrm{K} - y) - \mathrm{H}$, where H stands for the number of individual moose shot in the yearly hunt, at present some 90,000 animals per year.

Now go back to K = 1,000,000 and k = 0.00000047.

```
H = 90000

SolveODE[k y (K-y) - H, A]
```

To better see the solution, hide the slope field and your earlier solution. The new solution in Figure 6.10 is not, as you might have expected, a straight horizontal line, which would indicate a stable moose population, but a decrease that will lead to a

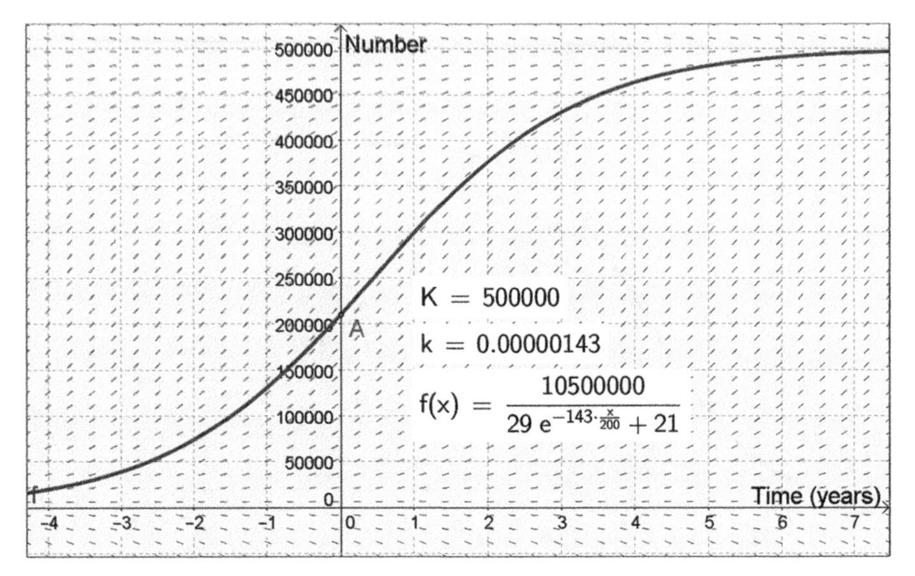

FIGURE 6.9 Possible future for the moose without hunting.

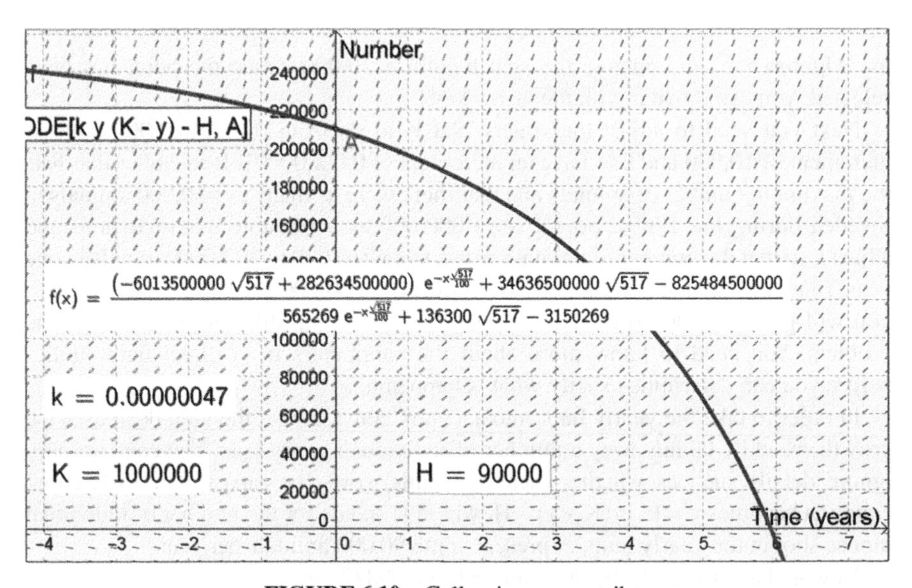

FIGURE 6.10 Collapsing moose tribe.

collapse in about six years. Why? We know that we can hunt 90,000 animals during the year and still have a stable population.

The answer is that the model is too limited. The hunt is not going on all through the year, and moose do not breed all through the year either. The model does not take these important factors into consideration at all. In an attempt to try to adjust the

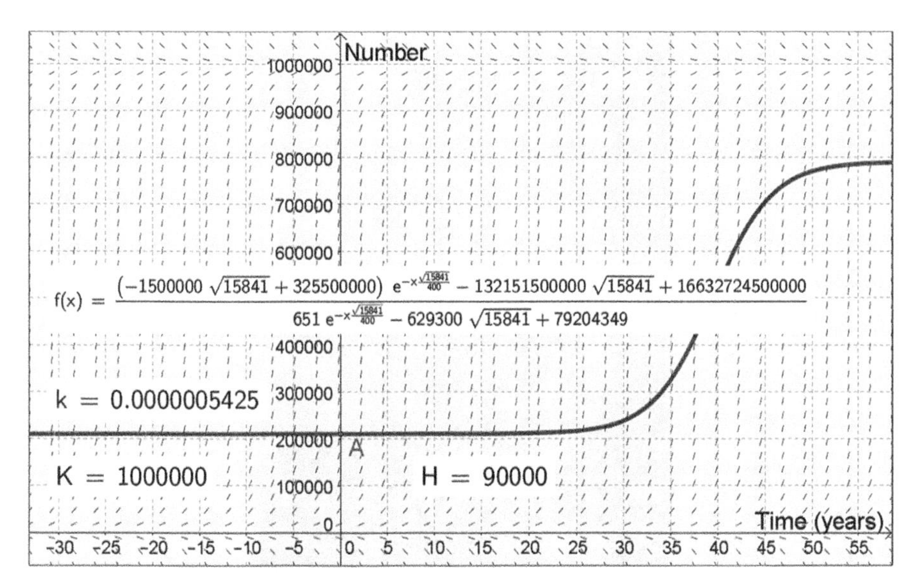

FIGURE 6.11 Moderate hunt to ensure a stable population.

model to reality, change k until you get a stable population, which happens around k=0.0000005425. It is obvious that the model is highly sensitive to k and not stable.

Adjusting H a small amount up or down, you get what might be called asymmetric results. If H falls just a little bit below 90,000 shot animals a year, you get a steady population of about 800,000 animals in a few years, as shown in Figure 6.11. If you instead increase J a little bit above 90,000 shot animals a year, the moose tribe will collapse.

Threshold effects and instability are important concepts that are well worth discussing when you study differential equations. These types of threshold effects are real and even if the moose will survive by constantly adjusting the number of animals shot, some animal populations have, or have almost, been eradicated, such as the North Sea cod, the East African Ningu, and the North American Passenger Pigeon, the last one dead in the early 1900s. Swedish hunters often point to the risk of a collapse of the Swedish moose population as the main argument against protecting a wolf population that mainly eats moose calves. Whether the wolves should be decimated in order for us to hunt moose for our own food and pleasure, or whether we should build a society where moose, humans, and wolves can live side by side is a complicated and difficult political and ethical question. Obviously humans do not profit from the elimination of either moose or wolves in the long run.

6.3 THE WATER CONTAINER

A water dispenser in the school cafeteria, shown in Figure 6.12, contains about 20 liters of water and empties from the bottom, as you fill your glass. It takes about 2.1 s to fill a glass when the dispenser is full. How long time will it take to fill a glass when the dispenser only have 25% of its water left?

FIGURE 6.12 Water dispenser. Photo: Monika Wahi (Own work) [CC BY-SA 3.0 (http://creativecommons.org/licenses/by-sa/3.0)], via Wikimedia Commons. https://upload.wikimedia.org/wikipedia/commons/5/54/Water_dispenser_with_lemons.JPG.

You may assume that the water level in the dispenser labeled y can be modeled by the differential equation $y' = -k\sqrt{y}$ for some constant k. Try finding out why?

Proposed Solution

Start with illustrating the differential equation and its solutions. Create a point

```
A = (0, h)
```

and accept to create the slider h. This point will serve as the initial value for the height of the water. Next create the slope field by typing the command

```
SlopeField[-k sqrt(y)]
```

Now accept the new slider k, and create a solution through the point A with

```
SolveODE[-k sqrt(y), A]
```

After that, move the sliders and drag the function expression from the Algebra window to the graphics area as in Figure 6.13. It is only the decreasing part of the quadratic function that is the solution. When the water dispenser is empty, it is empty.

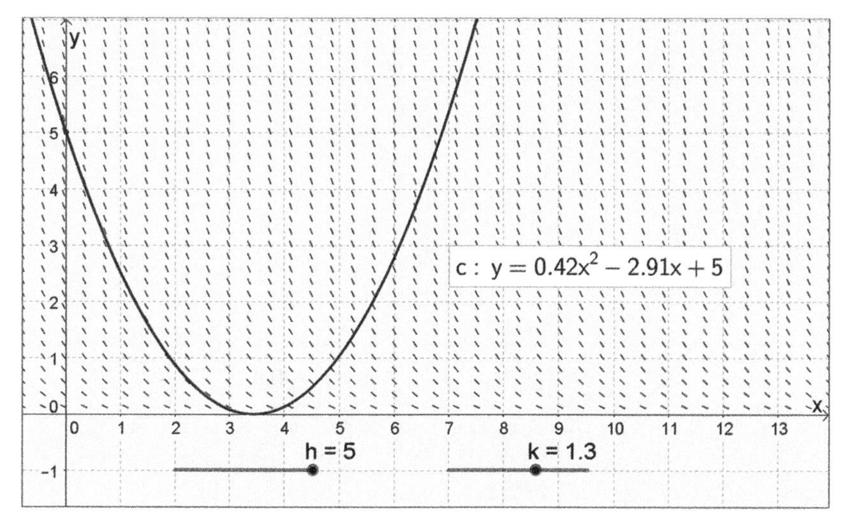

FIGURE 6.13 Slope field with a solution through a given point.

By changing the sliders around, you will get a feeling for how the water level in the water dispenser is changing over time, depending on the parameter value.

Now create a horizontal line

```
y = 0.25 h
```

and an intersection point between the line and the graph by using the Point tool and clicking at the intersection. The point's x-coordinate tells you when the height reaches 25% of the original height, if the water is allowed to just pour out. Observe that k has to do with the speed of the water.

You now want to use the function, but GeoGebra has not created a function but a conic section, which you cannot make function calls to. You are therefore forced to reconstruct a part of the function with a trick. Create an arbitrary point C on the conic section, the graph. You now have three points; A, B, and C, that all are on the graph. Typing

```
FitPoly[A, B, C, 2]
```

will reconstruct part of the graph as a function you can use.

Create the derivative

```
v(x) = f'(x)
```

and dash it, using the Style Bar. The commands

```
v0 = v(0)
```

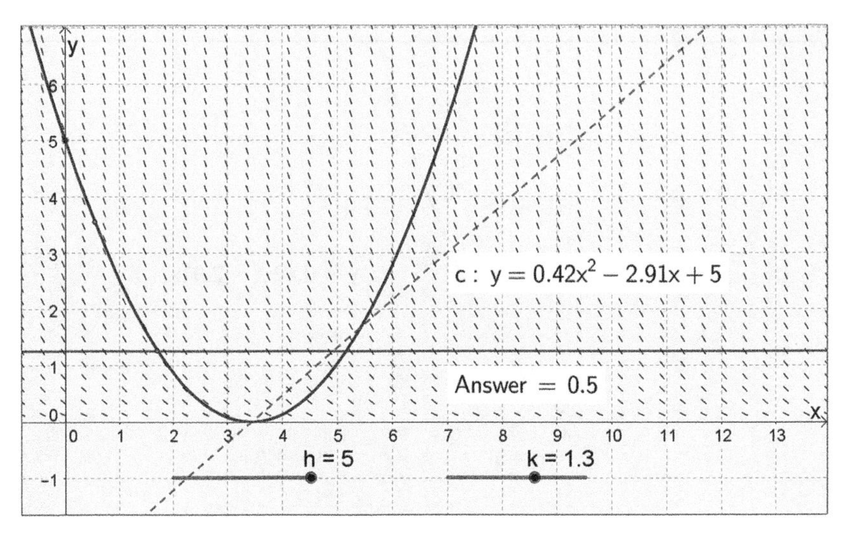

FIGURE 6.14 Complete solution.

```
v25 = v(x(B))
```

create the speeds for the full and the quarter full dispenser. Then

```
Answer = v25/v0
```

gives you the result shown in Figure 6.14. Regardless of the values of h and k, the glass will always take twice as long time to fill when the dispenser is a quarter full compared to when it is full.

6.4 SKYDIVING

Figure 6.15 shows 16 skydivers at an event over Peterborough, England. Those who skydive as an activity or as a sport usually do it for the experience of the free fall, without the parachute open. The parachute is just there so that they can jump again.

During the initial part of the dive, the speed increases very rapidly due to Earth's gravity. Soon though, the air resistance becomes so large that the speed no longer increases. The skydiver has reached the *terminal velocity*. For a typical skydiver falling belly down and arms outstretched the terminal velocity is almost 200 km/h or some 120 mph.

To avoid the use of oxygen tubes, the skydivers normally don't jump from heights above 4,000 m, or 13,000 ft. What will be the longest jump possible (in time) if you assume that the skydiver also must have enough time to release the parachute and lower the speed to secure a safe landing, with time to maneuver to the intended landing area? How long will the skydive last if you assume that the skydiver

FIGURE 6.15 Skydivers quickly reach their terminal velocity. Photo: Alistair Christie (Template:Stuart Meacock) [CC BY-SA 4.0 (http://creativecommons.org/licenses/by-sa/4.0)], via Wikimedia Commons. https://upload.wikimedia.org/wikipedia/commons/2/22/Skydiving_16_way_Over_Peterborough.jpg.

immediately after the jump dives head first with arms closed to the body, and therefore achieves a terminal velocity of 320 km/h, some 200 mph?

A hunting peregrine can also reach this impressive terminal velocity, but the peregrine is able to start the dive from a maximal height of about 4,500 m. The dive will be discontinued when the peregrine hits one of the prey's wings at about the height of 150 m; the peregrine will then experience rough breaking with an acceleration corresponding to 25 g. How long time will it take before the peregrine reaches 90% of its terminal velocity?

Proposed Solution

The diver experiences a downward constant force of gravity that you can write as

$$F_g = \mathrm{m} \cdot g$$

where m is the mass of the skydiver and $g = 9.81$ m/s^2. This value varies from approximately 9.79 m/s^2 at the equator to 9.83 m/s^2 at the poles with 9.81 m/s^2 as an average value.

You also have to consider the air resistance, that you can assume is proportional to the square of the velocity, or $F_{air} = \mathrm{k} \cdot v^2$. The gravity and the air resistance have opposite direction; the resulting force F downward may therefore be written $F = \mathrm{m} \cdot g - \mathrm{k} \cdot v^2$. The acceleration the skydiver experiences can be calculated as

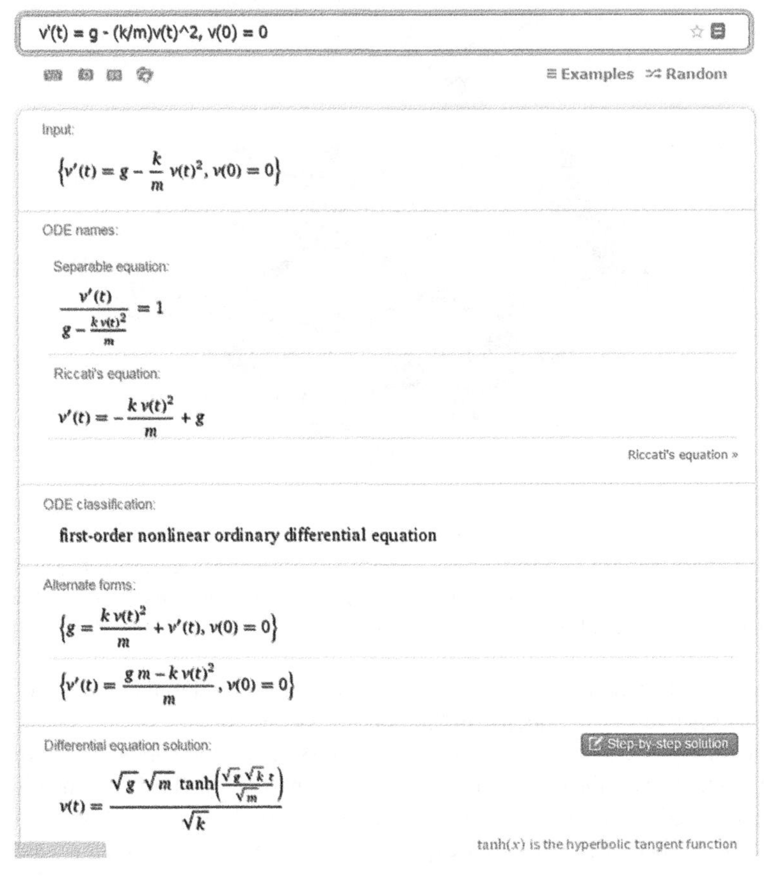

FIGURE 6.16 Wolfram Alpha used to solve a differential equation based on initial conditions.

$a = F/\mathrm{m}$. Observe that this acceleration is the time derivative of the velocity, that is, $a = v'$, and that you therefore can write the differential equation as

$$v' = g - \frac{\mathrm{k}}{\mathrm{m}} v^2$$

where g, k, and m are constants. As initial conditions, the vertical velocity should be zero when you start the dive, $v(0) = 0$. GeoGebra is not able to solve this equation symbolically, but Wolfram Alpha is as seen in Figure 6.16. The solution contains a function, $\tanh(x)$, also called a *hyperbolic tangent* and defined as

$$\tanh(x) = \frac{e^x - e^{-x}}{e^x + e^{-x}}$$

The hyperbolic functions are nice applications of exponential functions.

A simple online search for skydiving discloses that most skydivers release their parachute at about the height of 1,000 m, or at least no later than at the height of 800 m. The vertical velocity before landing can, with modern parachutes, be as high as 10 m/s, only to be slowed drastically just before touchdown.

Start with an estimate of the problem. Assume that initially the velocity is increasing linearly $v = g \cdot t$ as if no air resistance existed. 200 km/h is about 55 m/s, which will take $t = v/g = 55/9.82 = 5.6$ s. During that time the average velocity is 27.5 m/s and the skydiver falls 27.5 m/s · 5.6 s = 154 m. If the total height of fall is 3,200 m before the skydiver must release the parachute, the skydiver still has 3,046 m to fall, which will take an additional 3,046 m/55 m/s = 55.4 s for a total time in free fall of about 61 s. In reality it should take a little longer, since the acceleration gradually decreases in the initial phase.

In GeoGebra, enter the solution from Wolfram Alpha as a function where you have merged the square roots somewhat. You may use t as an independent variable in GeoGebra if you clearly define the function as $v(t) = \ldots$

```
v(t) = sqrt(g m/k) tanh(sqrt(g k/m) t)
```

Accept to create the sliders g, m, and k and enter the values for $g = 9.81$ m/s^2 and m = 80 kg, a reasonable enough total weight of a young skydiver, including clothing, coverall, helmet, and parachute. By varying the value of k, you can change the final speed. Place a point A on the curve somewhere in the area where the terminal velocity has been reached, just so that you can read its coordinates in the Algebra window, and adjust k to get a final velocity of 55 m/s. This procedure gives k = 0.260. You can now calculate the area under the curve, from 0 to A′s x-coordinate by typing

```
FallDistance = Integral[v, 0, x(A)]
```

By adjusting A until you get a total falling distance of 3,200 m, you get the total fall time from the x-coordinate of A to 62.1 s. This is surprisingly close to the previous estimate of 61 s; see Figure 6.17.

If you now vary the mass, the terminal velocity changes, which in turn affects the total fall time. A heavier jumper reaches a higher final velocity and will result in a quicker jump. By first adjusting the mass, and then the position for point A, you can get some new data for the relation between mass and time of flight. Enter these values into the spreadsheet that you open by pressing **Ctrl-Shift-S**. Also open up Graphics Area 2 with **Ctrl-Shift-2**, and make sure that it is selected and active. Select the data values, right-click, and select **Create... > list with points**. From the way the points lie you probably want to do a power regression

```
FitPow[list1]
```

getting the results shown in Figure 6.18.

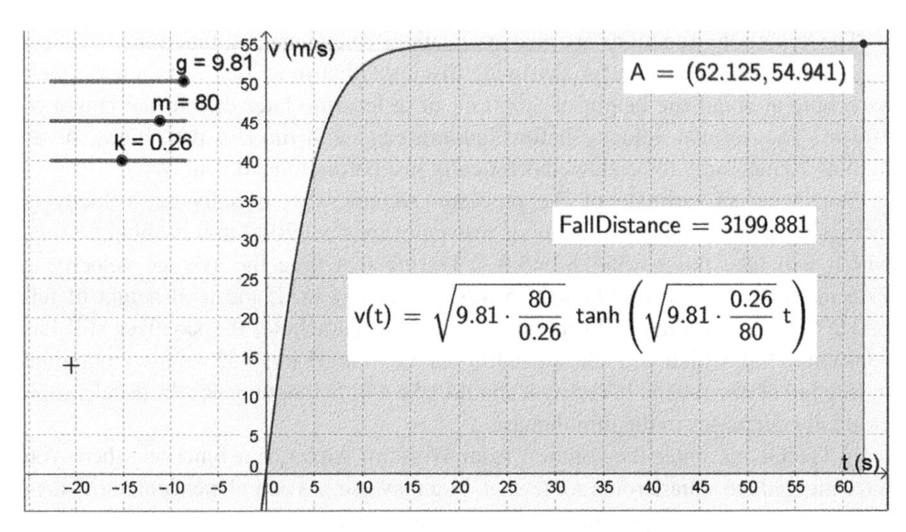

FIGURE 6.17 *vt*-Diagram for a parachute jumper.

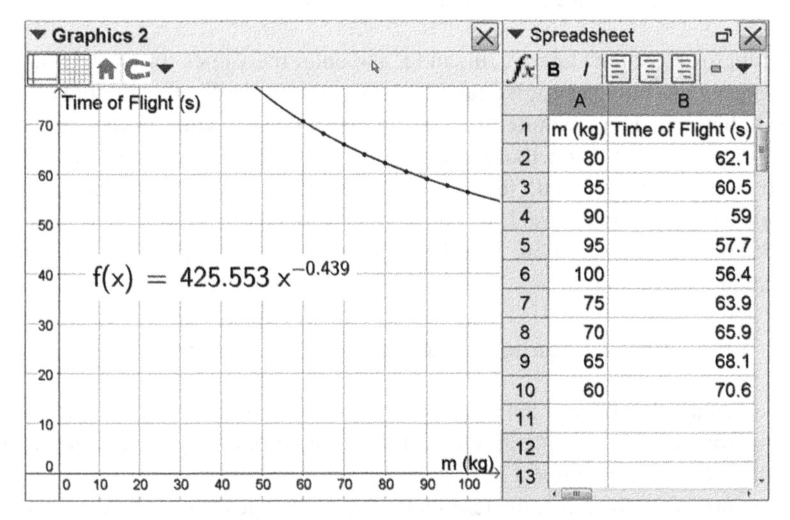

	A	B
1	m (kg)	Time of Flight (s)
2	80	62.1
3	85	60.5
4	90	59
5	95	57.7
6	100	56.4
7	75	63.9
8	70	65.9
9	65	68.1
10	60	70.6
11		
12		
13		

FIGURE 6.18 Time of flight almost inversely proportional to the square root of the mass.

A skydiver can reach velocities around 320 km/h = 89 m/s. That velocity indicates heavier jumpers, so adjust the mass to 100 kg and re-adjust k until the *y*-coordinate of A is about 89 m/s. Thereafter, just as before, adjust A:s position until the total falling distance = 3,200 m, and find the time of flight to be about 42 s, as seen in Figure 6.19.

Female peregrines are heavier than the males. The current observed peregrine diving record was set by a female with a top speed of 389 km/h = 108 m/s. Female peregrines can have a weight of 1.5 kg; therefore set m = 1.5 kg and k = 0.00126 so that the terminal velocity becomes 108 m/s. This time adjust A to get a total falling

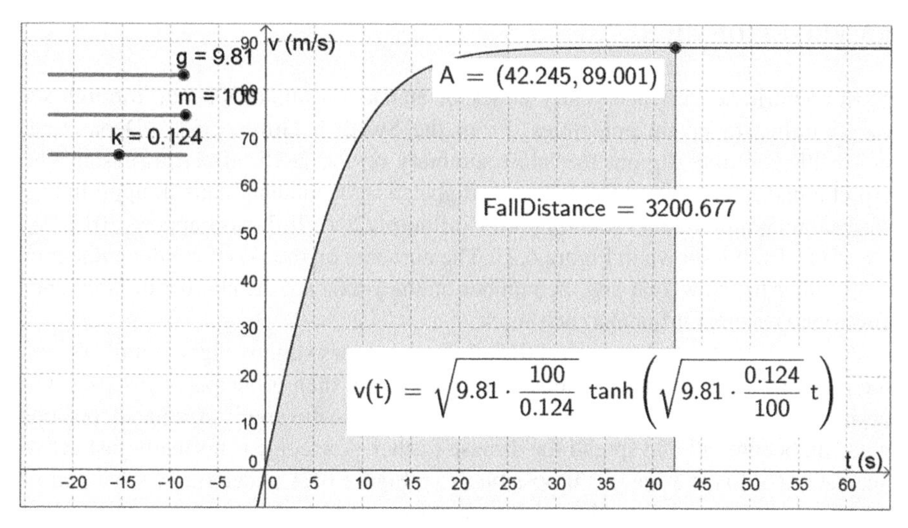

FIGURE 6.19 *vt*-Diagram for a skydiving human.

FIGURE 6.20 Peregrine in flight. Photo: Kevin Cole from Pacific Coast, USA (Peregrine Falcon (Falco peregrinus)) [CC BY 2.0 (http://creativecommons.org/licenses/by/2.0)], via Wikimedia Commons. https://upload.wikimedia.org/wikipedia/commons/1/1e/Peregrine_Falcon_in_flight.jpg.

distance of 4,350 m. It is not likely that the bird maintains this speed during the entire fall, but should it do so, it will have about 48 s to find its prey.

The peregrine reaches 90% of 108 m/s, which is about 97 m/s, already after about 16 s and 975 m. You can see this by dragging A until the y-coordinate equals 80. Sometimes the prey discovers the peregrine and also dives to escape. Should this happen high enough, the peregrine is able to pass the prey, turn around, aerobreak, and close the distance so that the prey helplessly falls into the claws of the peregrine.

In Figure 6.20 you see a peregrine flying against a backdrop of cliffs in California.

6.5 FLU EPIDEMICS

Most countries experience yearly attack or attacks on human's health through so-called influenza or flu epidemics. From the Swedish Government's Department of Health you may dig out the latest summary on the 2015 influenza season. The report found at http://www.folkhalsomyndigheten.se/documents/statistik-uppfoljning/smittsamma-sjukdomar/Veckorapporter-influensa/2014/Influensarapport_2015-20_v4_150601.pdfis shown in Figure 6.21. The numbers on the x-axis are the number of weeks after the New Year and the number on the y-axis are the number of laboratory confirmed cases of influenza each week.

A reasonably typical year was 2015, though it shows slightly higher numbers than the previous years. If enough people are vaccinated, then influenza or diseases will never spread, since the virus will not seek out persons who aren't vaccinated, persons who can become ill and spread the disease further. A vaccine is normally just active against one virus at a time. Is it possible from Figure 6.21 to determine what part of the population must be vaccinated for the current virus to be stopped so that it does not spread?

Here a so-called SIR model could be used to organize differential equations that describe how the number of susceptible individuals, the number of infected, and the number of immune/recovered individuals varies over time. Assume that there were 9 million susceptible individuals in Sweden and that only one of them arrived in Sweden with the influenza from a journey aboard.

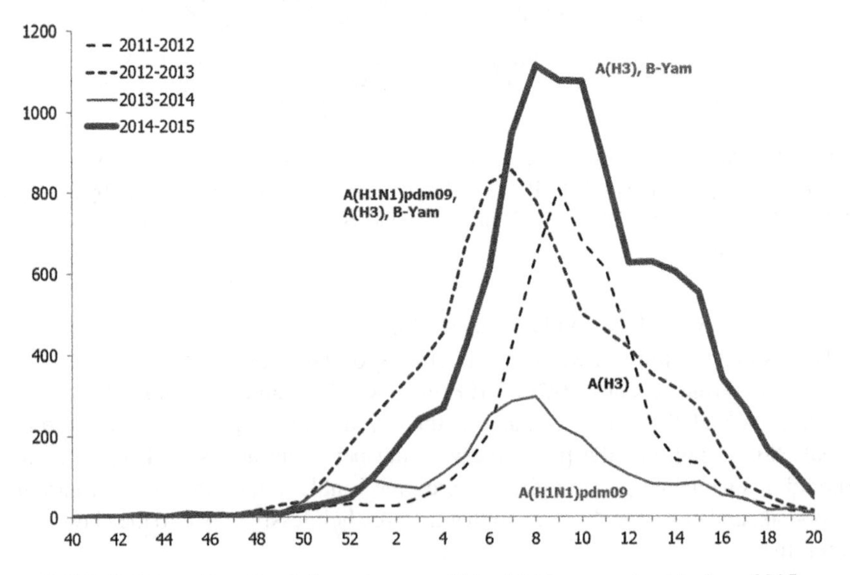

FIGURE 6.21 Swedish Influenza season 2014–2015 summarized in June 2015.

Proposed Solution 1

The basic SIR model looks like:

$$\begin{cases} \frac{dS}{dt} = -\beta IS \\ \frac{dI}{dt} = \beta IS - \gamma I \\ \frac{dR}{dt} = \gamma I \end{cases}$$

where $S(t)$ is the number of *susceptible individuals*, $I(t)$ is the number of *infected individuals,* and $R(t)$ is the number of immune or *recovered individuals.*

These differential equations build on at least the following assumptions:

- The population is constant.
- The number of infected individuals is proportional to the number of individuals who are susceptible and to the number of individuals who are infected—if 20 infected individuals are walking around among 300 uninfected, then that will results in six times as many new infected cases as opposed to 10 infected individuals walking around among 100 uninfected. This proportionality constant is called β.
- The number of recovered individuals is proportional to the total number of infected. This proportionality constant is called γ.

Fortunately, you do not need to solve this system of differential equations algebraically. Everything is possible to solve numerically, and a good start is to rewrite your differential equations as difference equations:

$$\begin{cases} \Delta S = -\beta IS \cdot \Delta t \\ \Delta I = (\beta IS - \gamma I) \cdot \Delta t \\ \Delta R = \gamma I \cdot \Delta t \end{cases}$$

In GeoGebra you can start by creating some numbers that you will need:

`S_0 = 9000` (in thousands of individuals)

`I_0 = 1`

`β = 0.001`

`γ = 1`

`Δt = 0.1`

The Greek letters can be found in the symbol table, which you reach by clicking on the α-symbol in the very right of the input field: Inmatningsfält: .

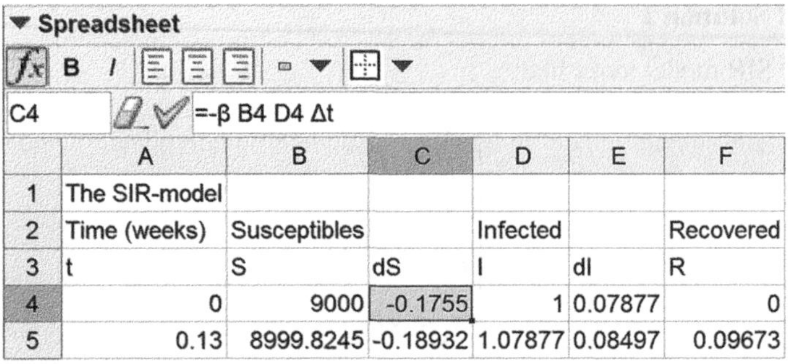

▾ Spreadsheet

C4 =-β B4 D4 Δt

	A	B	C	D	E	F
1	The SIR-model					
2	Time (weeks)	Susceptibles		Infected		Recovered
3	t	S	dS	I	dI	R
4	0	9000	-0.1755	1	0.07877	0
5	0.13	8999.8245	-0.18932	1.07877	0.08497	0.09673

FIGURE 6.22 Building a spreadsheet model of epidemics.

By selecting a number in the Algebra window, you can use the arrow keys and change the value of the number. Default value for the increment is 0.1, but if you right-click on the number, select **Object Properties...**, **Slider** tab, you can change the increment to something more suitable, for instance, 100 for S_0 or 0.00001 for β.

With these numbers in place, you can start building a model for epidemics in the spreadsheet that you open with the keyboard shortcut **Ctrl-Shift-S**.

In Figure 6.22 you can see the how to start building the model. There is one column for time and two columns for every dependent variable: one for the variable and one for its derivative. Now type:

In A4: **0**
In A5: **=Δt**

Select these two cells and copy them down, by dragging the fill handle, so that you get about 100 cells filled with time values. Once this is done, you continue with

In B4: **=S_0**
In D4: **=I_0**
In F4: **0**

These are the initial conditions for the differential equations. Thereafter enter the difference equations:

In C4: **=-β B4 D4 Δt**
In E4: **=(β B4 D4 - γ D4) Δt**
In G4: **= γ D4 Δt**

The new values for the variables are the sum of the old value and the difference:

In B5: **=B4+C4**
In D5: **=D4+E4**
In F5: **=F4+G4**

An important step is to complete row 5 by copying C4 down to C5, E4 down to E5, and G4 down to G5.

Spreadsheet

H4 =(A4, B4)

	E	F	G	H	I	J
1						
2		Recovered				
3	dI	R	dR			
4	0.07877	0	0....	(0, 9000)	(0, 1)	(
5	0.08497	0.09673	0....	(0.13, 8999.82...	(0.13, 1.078...	(0.13, 0.C

FIGURE 6.23 Creating points in the spreadsheet.

You have now created the beginnings of a table with values, but you want to see this graphically as well. Therefore create data points:

```
In H4:    =(A4,B4)
In I4:    =(A4,D4)
In J4:    =(A4,F4)
```

Right-click on these points—either in the graphics area or in the spreadsheet—select **Object Properties** and set their size to 2 and their colors to three distinctly different colors. Having done that, select the cells H4, I4, and J4 and copy them down to row 5. Figure 6.23 shows the points created in rows 4 and 5 of the spreadsheet.

Now select all used cells in row 5 except A5, from B5 to J5, and copy them all down at once by dragging the fill handle for the selection, all the way to row 100 or so. After zooming, adding text labels, and playing around with the parameters, this should result in something like the graphic display shown in Figure 6.24.

To get the small segments that connect the points and help illustrate the graph more clearly, you can type

```
In K4:    =Segment[H4, H5]
In L4:    =Segment[I4, I5]
In M4:    =Segment[J4, J5]
```

as in Figure 6.25. As you can see, the GeoGebra spreadsheet can contain all kinds of objects. The number shown is the same as would be shown in the Algebra window.

If you have remembered to copy everything down, then the model is now ready and you can investigate how the solutions depend on the parameters β and γ. As expected, β controls how fast the disease is spreading, higher values giving a faster transmission rate and higher values for γ a faster decay. If you increase both β and γ, you get both a faster spread and a faster decay. The graph that shows the number of infected persons will be more compressed over time but will give roughly the same maximum values.

In order to be able to use the model to solve the problem, insert Figure 6.21 as a background image in GeoGebra by using the Image tool and then clicking at a point in the third quadrant. The image is anchored in two points, but since you

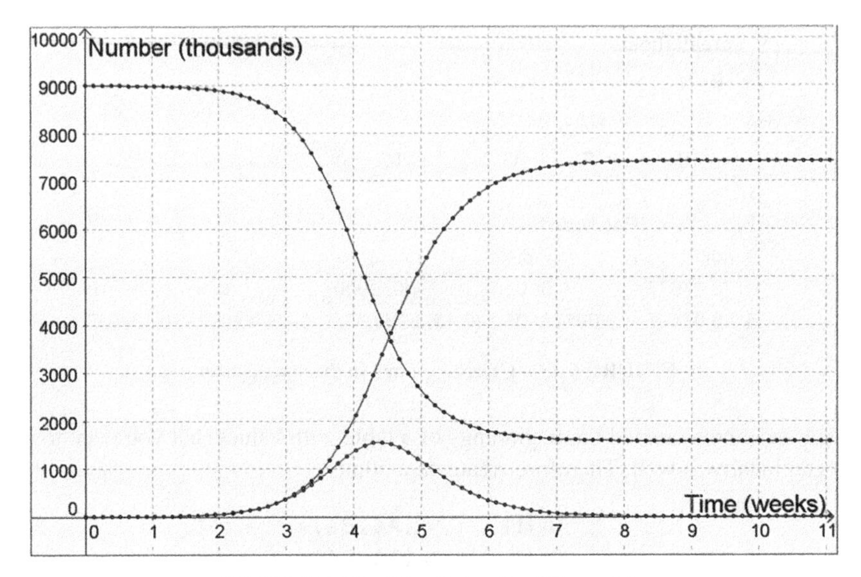

FIGURE 6.24 Visualizing the SIR model.

▼ Spreadsheet

Drag or select objects

K4 =Segment[H4, H5]

		I	J	K	L	M	N
1							
2							
3							
4	)	(0, 1)	(0, 0)	0.2184	0.152	0.16204	
5	..	(0.13, 1.078...	(0.13, 0.096...	0.22966	0.1553	0.1667	

FIGURE 6.25 Connecting the dots.

probably have very different scales on the axes, the picture may initially appear as a thin line. Create a third point C in the second quadrant and go into the properties of the image. On the **Position** tab you can anchor the upper left corner of the image to C. In the **Basic** tab of the image's properties, you select **Background image**.

Now you can drag the anchor points to adjust the image exactly to the coordinate system. This is a little tricky, and you may want to turn off the point capture options in the Style Bar so that the points aren't captured by the grid intersections. An interesting but quite different mathematical modeling problem is to calculate the coordinates for the image's vertices from the coordinate systems position in the image.

You can now adjust the parameters β and γ so that the epidemic is matched by the model. You can also adjust the starting point of the infection, something you do by selecting all points A, B, and C simultaneously and moving the image horizontally. It is a complicated, tricky process because you need to adjust the values of β and γ

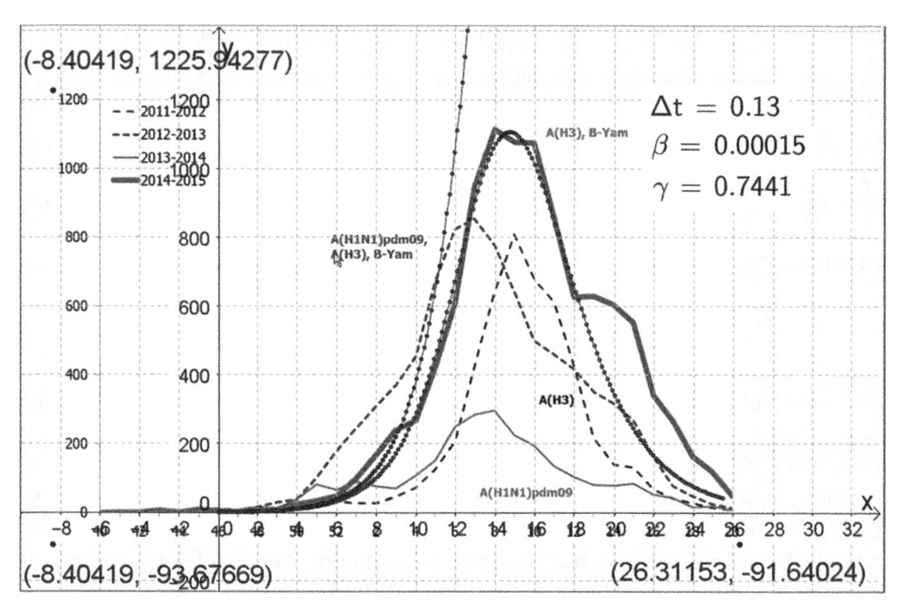

FIGURE 6.26 Adjust the parameters so that the model fits with the epidemic.

and the image's position manually, with arrow keys, forward and back until you find some values that fit reasonably well. Nevertheless, this works relatively well because β affects the model's maximum horizontally and γ affects the maximum value mostly vertically and also affects the width of the pulse. The result is shown in Figure 6.26.

In the SIR model you can now define a parameter $R_0 = S_0 \cdot \beta/\gamma$ that represents the number of average individuals that an infected person will infect before the person becomes immune or recovers. If $R_0 < 1$, the epidemic will never pick up speed but just die out. By vaccination of a population, the number of initially susceptible individuals S_0 can be reduced and consequently also R_0.

The part of the population that needs the vaccine, v, is derived from the SIR model in the following way: R_0 is the number of infectious contacts when the whole population is susceptible and v is the part of the population that is vaccinated. After a part of the population is vaccinated, the number of infections per person will be $R_0 \cdot (1 - v)$. In order for the epidemic to die out, this expression must be < 1, which gives you that $v > 1 - 1/R_0$. Open a new GeoGebra window, plot this function and place a point at the curve. In this case $R_0 = S_0 \cdot \beta/\gamma = 1.81$, and you can see in Figure 6.27 that $v = 0.45$. This means that 45% of the population needs to be vaccinated for this specific epidemic to stop.

In reality, not everyone in a population is susceptible. Some individuals in a population are immune from previous vaccinations; some might be naturally immune, and so forth. Perhaps only half the population is susceptible?

You can test this hypothesis by changing the population parameter S_0 to 4,500, half of Sweden's population. You have to adjust β and γ again, whereupon you can find reasonable values at $\beta = 0.000467$ and $\gamma = 1.33$, which gives $R_0 = 1.58$ and in turn

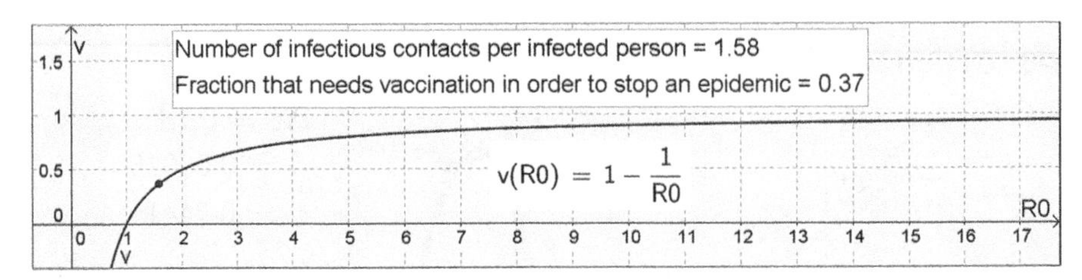

FIGURE 6.27 Relatively low value for R_0 still demands that more than every fourth person in the population must be vaccinated to stop the epidemic.

$v = 37\%$ of the susceptible part of the population. This result is not the same that you got earlier but very roughly of the same magnitude. If you try with $S_0 < 4{,}500$, the shape of the curve agree less with the actual course of the influenza, which makes it likely that $S_0 \geq 4{,}500$.

Different years bring different influenzas, but for a normal healthy person, catching the flu is not dangerous. Therefore it is a generally a problem to convince enough individuals to vaccinate themselves every year. During the pandemic warning of 2009 the disease was considered to be far more dangerous and a large vaccination venture was launched; this in turn caused the pandemic to become very limited that year—at least in Sweden.

Proposed Solution 2

This solution is perhaps more elegant and less time-consuming, but there are educational advantages in letting the students build their own solutions in the spreadsheet.

Instead for building a numeric solution in the spreadsheet, you will now use GeoGebra to solve the problem numerically behind the scenes. The following sequence of commands will enable a numerical solution of a system of differential equations in GeoGebra:

Start a new window by pressing **Ctrl-N**. Then define the start values and parameters you need. Look at the previous solution to get good starting values for the parameters and type

```
S_0 = 9000
I_0 = 1
β = 0.000467
γ = 1.33
```

Next define the derivatives as multivariable functions:

```
S'(x,S,I,R)  = - β S I
I'(x,S,I,R)  = β S I - γ I
R'(x,S,I,R)  = γ I
```

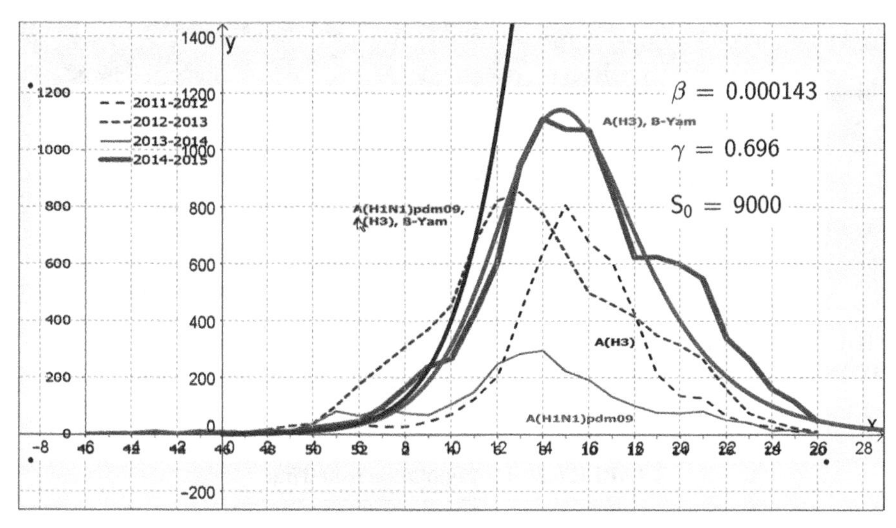

FIGURE 6.28 GeoGebra's own numerical solution.

Finally tell GeoGebra to solve the system of differential equations. The derivatives and the starting values will be passed to GeoGebra in separate lists:

```
NSolveODE[{S', I', R'}, 0, {S_0, I_0, 0}, 30]
```

This command will give you numerical solutions in the interval $0<x<30$. These solutions deviate somewhat compared to the previous solutions, possibly because the algorithm is more accurate and the step size is smaller. Insert the image again, color the individual solutions, and adjust the parameters so that the solutions will fit better. The result is shown in Figure 6.28.

The new values for β and γ give you $R_0 = 1.85$, nearly the same as your first attempt, which in turn gives the fraction that needs to be vaccinated to prevent an epidemic to 45%.

6.6 USA'S POPULATION

How many people will we become? Figure 6.29 tickles the imagination in an ever important question. Today, population models attract large interest. Population projections for the future are needed in order to plan for future needs and resource allocations. Table 6.2 shows how the population of the United States has changed from 1790 to 2010 as measured on December 31 every tenth year.

Use these data to make several different population projections for 2020 and 2050. In particular, study the change of the rate of change over time and make some different projections from these observations.

FIGURE 6.29 People—How many will we become? Photo: Greg Solomon via Flickr, CC BY-SA 2.0. https://www.flickr.com/photos/thisparticulargreg/362937046/sizes/o/.

TABLE 6.2 US Population over Time

Year	Population (millions)
1790	3.929
1800	5.308
1810	7.240
1820	9.638
1830	12.87
1840	17.07
1850	23.19
1860	31.44
1870	38.56
1880	50.19
1890	62.98
1900	76.21
1910	92.23
1920	106.0
1930	123.2
1940	132.2
1950	151.3
1960	179.3
1970	203.3
1980	226.5
1990	248.7
2000	281.4
2010	308.7

Proposed Solution 1

Start by entering the data into GeoGebra's spreadsheet. Using years as a variable is seldom a good idea. Instead, introduce a variable t for time that represents decades (tenths of years) from 1800. Also introduce P for the population in millions.

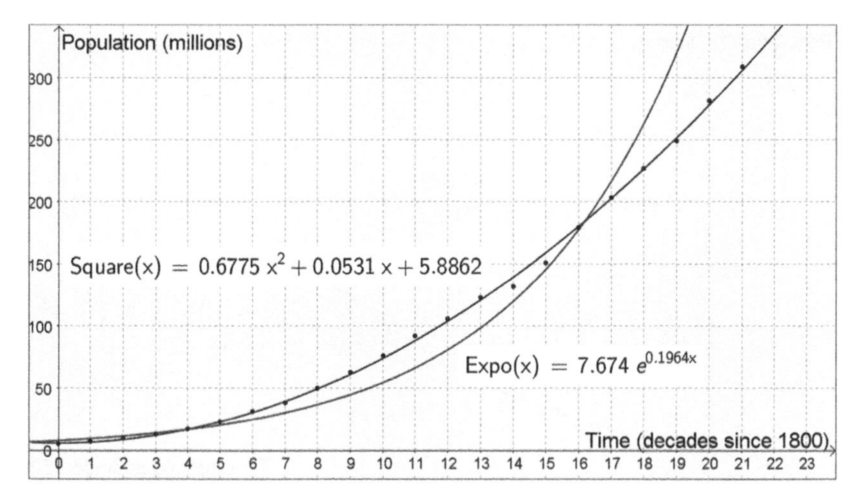

FIGURE 6.30 US population growth over the last 200 years.

Enter years in column A, t in column B, the number of million residents in column C, and in column D, a definition of the data points. In cell D2 type

```
=(B2, C2)
```

and copy this down. Then select these points in column D, right-click and select **Create...>list** to create **list1**.

At a first glance the points seem to follow either a quadratic or an exponential function, so try fitting these functions to the data with the following commands:

```
Square(x)  = FitPoly[list1, 2]
Expo(x)  = FitExp[list1]
```

It is always a good habit to label your functions with clear names.

It is quite obvious from Figure 6.30 that a quadratic function fits the data much better than an exponential function. The prognoses for years 2020 and 2050 can be calculated by typing

```
ProgSq2020 = Square(22)
ProgSq2050 = Square(25)
```

According to this model, USA is going to have about 335.0 million residents in 2020 and 430.6 million residents in 2050.

Proposed Solution 2

Population growth does not normally follow quadratic relations. But quadratic relations often fit well as a first approximation to many other relations. If the data do not follow an exponential function, a logistic function may work better.

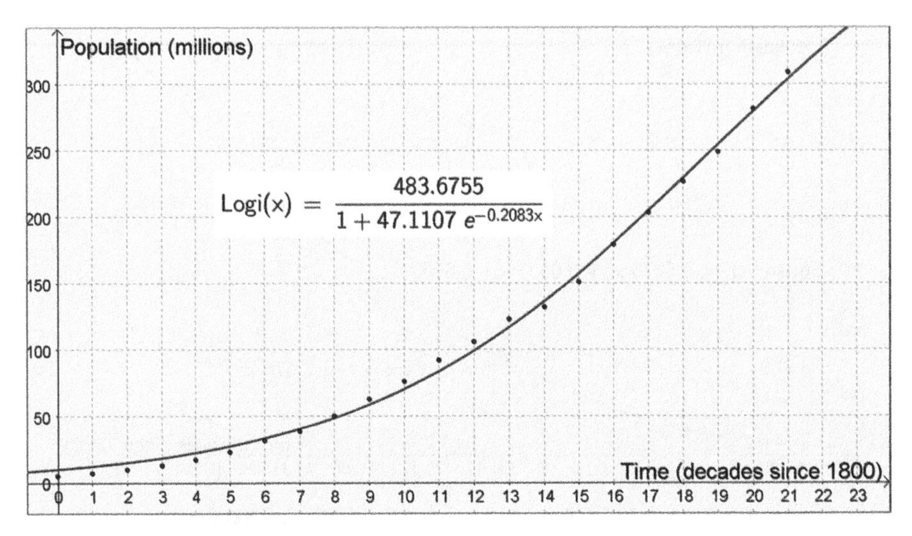

FIGURE 6.31 Logistic function that fits the data well.

	A	B	C	D	E	F
1	Year	t (years)	P (millions)	Point	Error (logistic)	Sqared error
2	1790	-1	3.929	(-1, 3.929)	-4.2658	18.1968
3	1800	0	5.308	(0, 5.308)	-4.7454	22.5187
4	1810	1	7.24	(1, 7.24)	-5.0826	25.8331

FIGURE 6.32 Spreadsheet so far.

The basic differential equation for population growth proportional to the population can be written $P' = k \cdot P$. The solutions to this equation are exponential functions. When resources are limited, the population can only grow to a certain value K. It is common to also allow P' to be proportional to the factor $(1 - P/K)$, which for small values for P is close to 1 but decreases toward 0 as y goes to K. The differential equation

$$P' = k \cdot (1 - P/K) \cdot P$$

has the logistic function as a solution. In GeoGebra the logistic function has a command all to itself, so you can type

```
Logi(x) = FitLogistic[list1]
```

Figure 6.31 shows a much better fit for the logistic model than for the purely exponential model. Once again, you can calculate the prognoses to find 326.5 million residents for year 2020 and 384.5 million residents for year 2050.

In Figure 6.32 is shown how to also calculate the sum of the squares of the errors in order to compare the models.

In E2: `= C2 - Logi(B2)`
In F2: `= E2²`

The square can be created pressing **Alt-2**. These two formulas have to be copied down. Note that in the GeoGebra spreadsheet, the equals symbol "=" is optional when entering formulas.

Cell **F25** is the right place to calculate the sum by typing

```
= Sum[F2:F24]
```

Note that "**F2:F24**" is a valid way of typing a list in GeoGebra when the cells contain numbers. The sum of squared error's, SSE=519.8, for the logistic model, while the same value for the quadratic fit is SSE=183.0. The model with more theoretical strength actually works less well in this case. You probably need a new theoretical explanation and understanding of population growth.

Proposed Solution 3

In the instructions to this problem you were asked to study the rate of change and draw conclusions from what you saw. The first model came from the differential equation $P' = \mathrm{k} \cdot P$, which gives exponential equations as solutions. This is built on the assumption that the population growth is proportional against the population P. In the logistic case you changed the proportionality constant k to the expression $\mathrm{k}(1 - P/\mathrm{K})$, which in turn is a function of P.

Now assume that $P' = r(t) \cdot P$, where $r(t)$ is some function of t, but not P. Since you know the values of P, you can calculate P' for every point (except the end points) and also $r(t) = P'/P$, where r is the *relative rate of change*. In general, r could be a function of both t and P, but in this case you can limit the complexity to consider only when r is a function of t.

Calculate an approximate value for the derivative using the central difference quotient

$$P' \approx \frac{P(t+\Delta t) - P(t-\Delta t)}{2\Delta t} \quad \text{giving} \quad r(t) \approx \frac{P(t+\Delta t) - P(t-\Delta t)}{2\Delta t \cdot P}$$

In this case $\Delta t = 1$. $P(t+\Delta t)$ corresponds to the value of P one step down in the table and $P(t-\Delta t)$ corresponds to one step up in the table.

You want to sketch this diagram in a separate window, so open Graphics Area 2 with **Show>Graphics 2** or by **Ctrl-Shift-2**.

In I3: `= (C4-C2)/(2C3)`
In J3: `= (A3,I3)`

These formulas are copied down, and in Figure 6.33 you can see that the function $r(t)$ is not constant but decreasing over time.

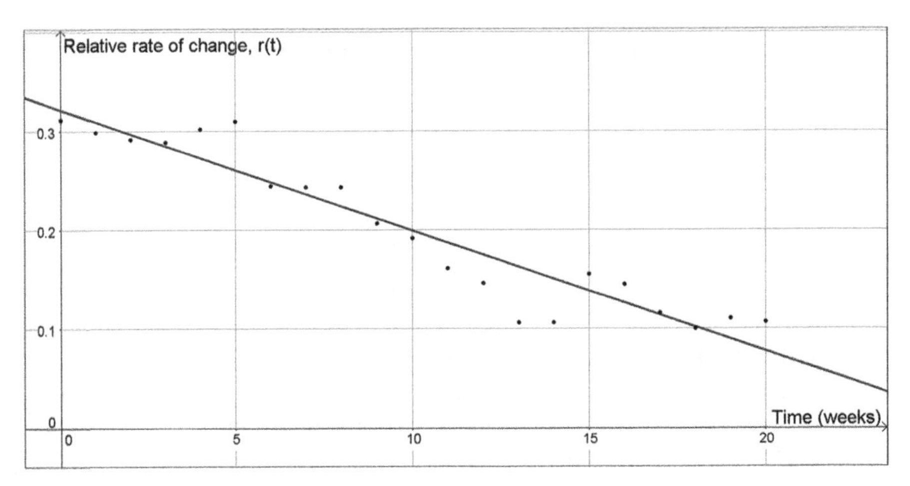

FIGURE 6.33 Graphics Area 2 showing $r(t)$ decreasing over time.

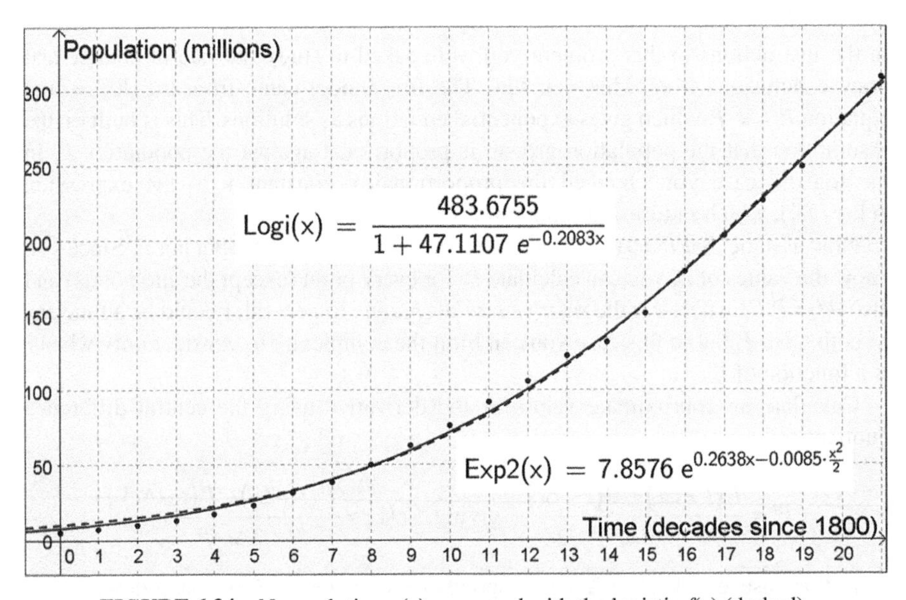

FIGURE 6.34 New solution $g(x)$ compared with the logistic $f(x)$ (dashed).

First assume that the relative rate of change, $r(t)$, decreases linearly over time. By selecting the cells H3 to H23, right-clicking on the selection, and choosing **Create...>list**, you can place these points in **list2**. Create the line you see in Figure 6.34 by typing

```
FitLin[list2]
```

You can now formulate the differential equation $P' = (a - b \cdot t) \cdot P$. This differential equation can be solved in different ways such as by finding the correct integrating factor, but you can also find the form of the equation with Wolfram Alpha or with GeoGebra. In GeoGebra, you can create a new window and type

```
SolveODE[a - b x]
```

to see the form of the solution, and in Wolfram Alpha, you can enter the whole equation as it stands, **P′ = (a – b · t) · P**, to get the general solution

$$P(t) = C \cdot e^{a \cdot t - (b \cdot t^2/2)}$$

Now, when you have the form of the solution, you can do a general fit to find a specific function of this form to model the data. Start by creating the model function $m(x)$ by typing

```
m(x) = c e^(a x + b x²/2)
```

Use **Alt-e** for the base for natural logarithms. When asked, answer yes to the question about sliders for a, b, and c. Then hide the model m and the sliders and type

```
Exp2(x) = Fit[list1, m]
```

In case you get an undefined or unsatisfactory function, it is because the starting values for a, b, and c are too far away from the real values for GeoGebra to be able to find the solution. If so, show $m(x)$ again and change the values of the sliders until a solution appears.

You can then clearly see, from Figure 6.34, that this model fits the data better than the logistic solution. You can calculate the error's square sum SSE in the same way as before by entering

In K2:	`= C2 - Exp2(B2)`
In L2:	`= K2²`

and then copying these formulas down and summing. You get SSE = 309.0, which is better than the logistic function but not better than the quadratic function. You can calculate the prognoses for 2020 to 331.2 million residents and for 2050 to 400.7 million residents. But note, in Figure 6.35, that the long-term behavior of this function is unacceptable for a population model.

Proposed Solution 4

Looking at the prognoses from Solution 3 from a long-term perspective, you might already have observed that the predictions are catastrophic. After about 2060 the function $r(t)$ becomes negative, which causes the population to have a peak soon after

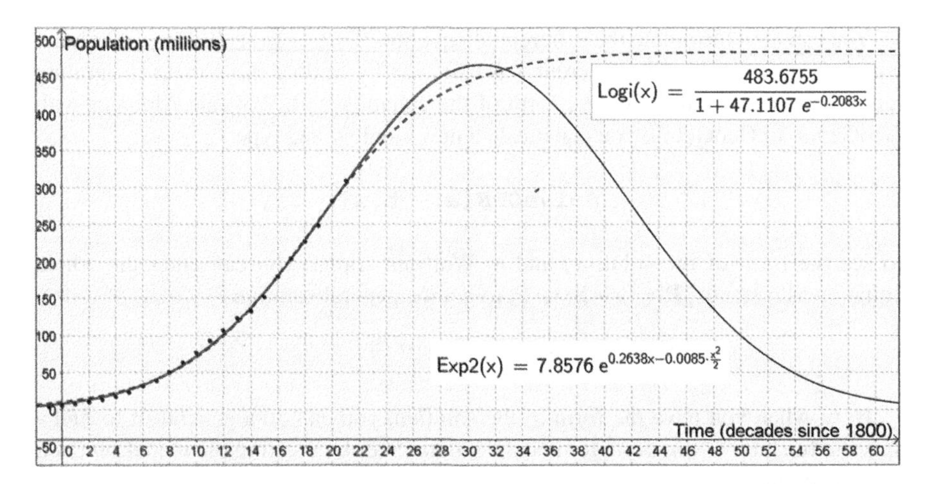

FIGURE 6.35 Long-term prognoses differences.

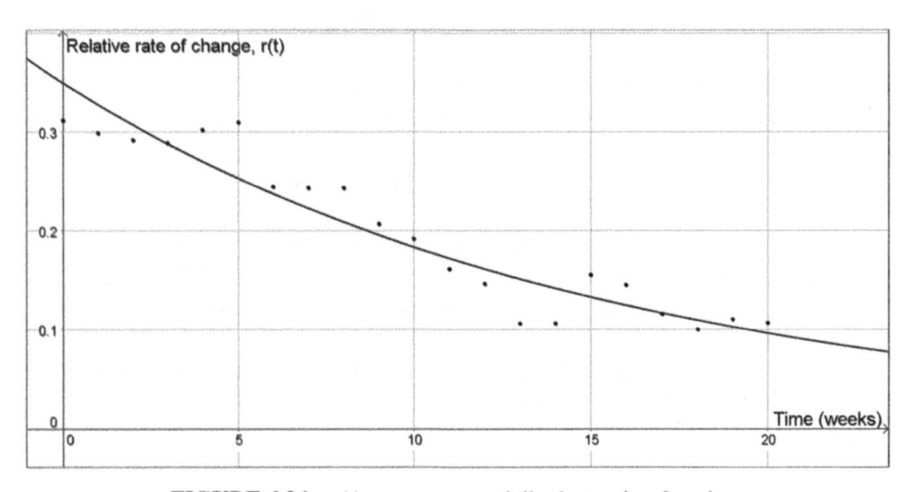

FIGURE 6.36 $r(t)$ as an exponentially decreasing function.

2100, and then decrease again. This is nothing you aimed at building into the model; nevertheless, it comes as a natural consequence of the assumption that $r(t)$ is linear and decreasing.

To amend this, now assume that the $r(t)$ decreases exponentially so that $r(t)>0$ but that it approaches zero as t increases. Figure 6.36 shows an exponential fit to $r(t)$.

In this case assume that the differential equation can be written as

$$P' = \left(\text{C} \cdot e^{\text{k} \cdot t}\right) \cdot P$$

which will give a solution of the form

$$P(t) = \text{C}_1 \cdot e^{\left(\text{C}_2 \cdot e^{\text{k} \cdot t}\right)/\text{k}}$$

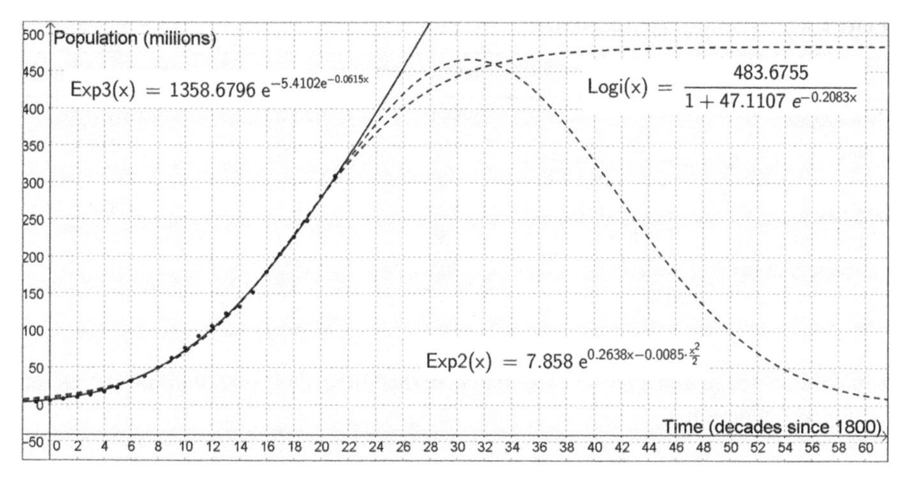

FIGURE 6.37 Another function that can help you understand the data.

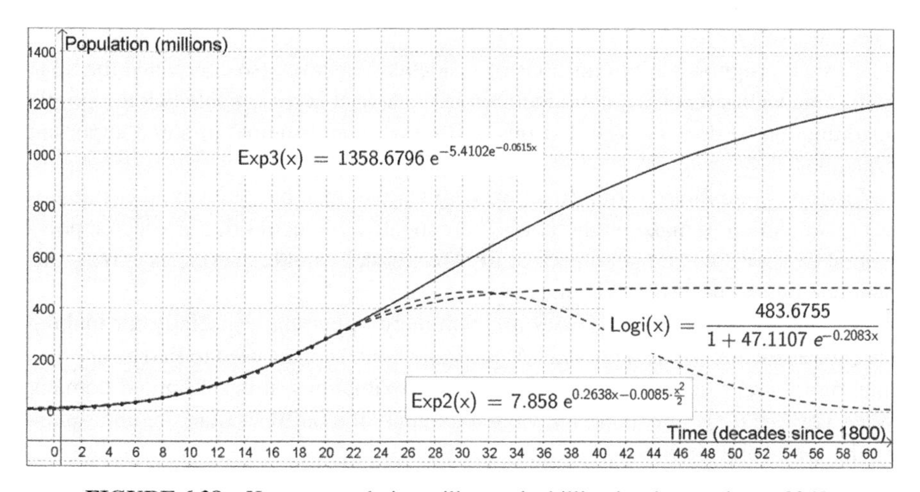

FIGURE 6.38 Human population will pass the billion level around year 2260.

In the same way as in Solution 3, create a model function

```
m2(x) = c e^(a e^(b x))
```

which you hide and then

```
Exp3(x) = Fit[list1,m2]
```

For this new function, $h(x)$, solid in Figure 6.37, SSE=205. The prognoses for 2020 and 2050 are 335.2and 424.2 million residents, respectively. The long-term behavior of this function, shown in Figure 6.38, resembles a logistic function with

TABLE 6.3 Comparing the Models

Model	$r = P'/P$	SSE	2020 prognosis	2050 prognosis
Exponential	K	48791	576.9	1040
Square	$\frac{2a \cdot t + b}{a \cdot t^2 + b \cdot t + c}$	183	335.0	430.6
Logistic	$k \cdot (1 - P/K)$	520	326.5	384.5
r Linear in t	$a - b \cdot t$	309	331.2	400.7
r Exponential in t	$C \cdot e^{k \cdot t}$	205	335.2	424.2

a maximum level, but this level is much higher than the corresponding level for the logistic function.

Comparison and Comments

Table 6.3 shows all scenarios for the US population $P(t)$ that you have derived and calculated so far.

Here all prognoses are in millions of residents. We have also completed the table with some values that we did not mention in the text. It is remarkable that the simple quadratic model does so well, but this will serve as an example of some important principles.

First, it is one thing to fit a function with known data, but quite another thing to use this function to make predictions. Theoretically, you could fit a 22-degree polynomial to your data and get a perfect fit. That model would nevertheless give completely wild and useless predictions.

Second, the underlying theories are extremely important, especially for making prognoses outside the data interval. A quadratic model can be used to find an approximation to a real situation, but as with all approximations, it has a limited domain. What you can do is to expand the valid data interval a bit by exchanging the square model for a cubic model.

Third, and most important, there is nothing that tells you that the mechanisms that controlled the development of society and population growth in 1790 are the same in 2010. The underlying economic, political, and demographical processes that lead to population growth look different in different eras. All you can hope for with these simple analyses is to find a reasonable model that you can use for a short-term prognosis.

In order not to give up on this problem yet, dig deeper into how polynomials of higher degree could be used to manage these data. To get these values quickly, use the *Two Variable Regression Analysis* tool shown in Figure 6.39. To use this tool, you must first select your data in the spreadsheet, and then click the tool button.

The *Two Variable Regression Analysis* tool makes it convenient to get a feel for the effect of increasing the polynomial degree. The population change between $t = 13$ and 14, in other words, between year 1930 and 1940, is relatively low because of the

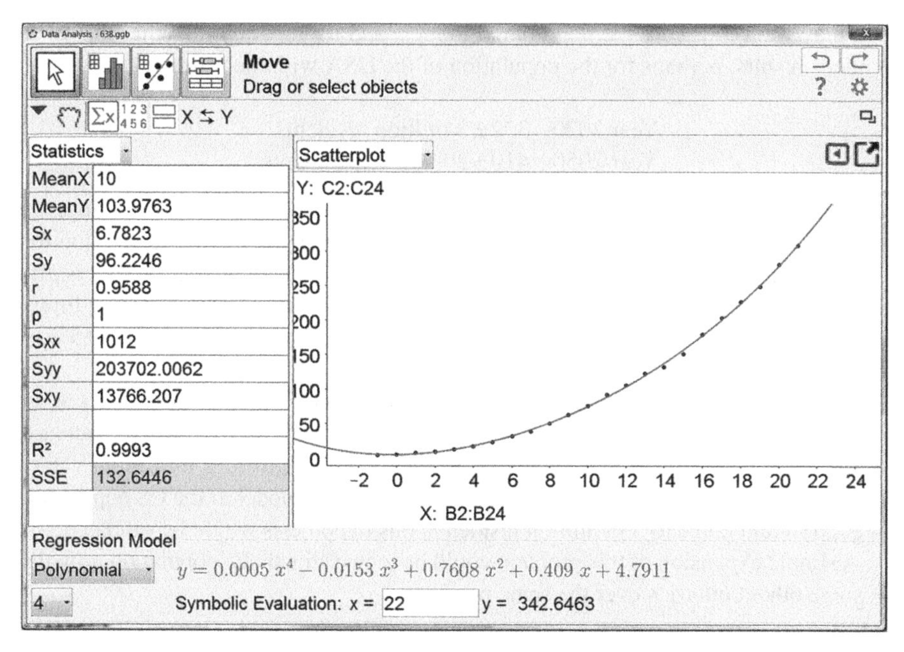

FIGURE 6.39 Two Variable Regression Analysis tool gives quick information when comparing different standard models.

TABLE 6.4 Comparing Polynomial Models

Model	SSE	2020 prognosis	2050 prognosis
Square	183	335.0	430.6
Cubic	143	339.5	442.3
4th degree	133	342.6	457.5
5th degree	132	343.8	465.6
6th degree	107	334.3	360.0
7th degree	106	331.7	315.2
8th degree	91	348.4	751.6
9th degree	85	364.7	1 388

Second World War. The algorithms used by GeoGebra try to adjust for this in order to minimize the sum of squared errors. One effect, among others, is an increasingly strong and unrealistic hook to the left of the data. It is safe to suspect that the values and prognoses to the right of the data are as unrealistic as the ones to the left. You can also see giant leaps in the prognosis structure in Table 6.4, for example, between degrees 5 and 6, then between degrees 7 and 8.

Our conclusion must be that the error's squared sum, SSE, alone is not enough to determine if a model is useful for predictions and prognoses. In your final prognoses you may want to ignore the polynomial models and try to weigh together the other model's results: perhaps for the population of the USA with the estimates

Year 2020: 332 ± 3 million residents
Year 2050: 410 ± 20 million residents

The error limits do not stretch out entirely down to the prognosis from the logistic model, a choice defended by remembering that this model had the largest value for the SSE. Furthermore these estimates are from several different models, a so it is difficult to estimate a confidence interval. What you can observe is that the error limits corresponds to relative errors of a magnitude of about ±1% to year 2020 and ±5% to year 2050.

Though the model now has some connection to theory, the model still treats r as an exponential function of t. If that model were accurate, the birth rates would decrease slowly down to zero and the population would slowly adjust to a maximum value. This model had a SSE value not far from that of the square model. If the US population in 2020 should come to 335 million residents, this hypothesis would be validated.

A simple expansion of this problem could be to investigate the population growths of some other countries over the same period.

6.7 PREDATORS AND PREY

In a relatively restricted area in the African savannah, game wardens have counted the number of cheetahs and their main prey, the antelopes, every third year. The results are represented in Table 6.5. As you can see, the number of antelopes has decreased strongly for a long time. The game wardens are worried that the antelopes will disappear completely from the area, and consequently that the cheetahs too will disappear.

This type of relation between a predator and its prey is often modeled by Lotka–Volterra's equations, which describe the dynamics between two species. Strictly speaking, herbivores are also considered to be *predators* in that the herb or grass they eat is their *prey*.

$$\begin{cases} \dfrac{dx}{dt} = \text{a}x - \text{b}xy \\ \dfrac{dy}{dt} = -\text{c}y + \text{d}xy \end{cases}$$

TABLE 6.5 Are Antelopes Endangered?

Year	Antelopes	Cheetahs
2005	700	300
2008	600	500
2011	400	600
2014	200	500

Here t is the time, x the number of prey, y the number of predators, and dx/dt and dy/dt are the rates of change in the two species. The parameters a, b, c, and d are positive numbers and these parameters affect the solutions to the differential equations.

If $y=0$, the first equation represents exponential growth, indicating that there is plenty of food for the prey. If, however, $x=0$, the second equation represents a decreasing exponential function that corresponds to the predators' natural extinction, as clearly there is nothing for them to eat.

At the same time the xy-terms represent the fact that predators eat the prey. This will lead to a decrease in the number of prey and an increase in the number of predators. The consumption rate and the birth rate are both proportional to the number of prey and the number of predators, $-\mathrm{b}\cdot xy$ and, respectively, $+\mathrm{d}\cdot xy$.

Make a numerical model of Lotka–Volterra's equations and try to adjust the parameters to the known values. What will, according to your model, happen to the number of animals?

Proposed Solution 1

All first-order differential equations with initial conditions can be solved numerically by relatively simple numerical methods. If you organize the data in a spreadsheet, the solution will not be too difficult to manage.

Euler's method for solving differential equations is built on the idea that you start at a point, you follow the tangent to the function at this point to the next point, and then you repeat this procedure until you arrive at the solution. Using t as the independent variable, you have

$$y'(t) \approx \frac{\Delta y}{\Delta t}$$

or

$$y(t+\Delta t) \approx y(t) + y'(t)\cdot \Delta t$$

and the corresponding equation for x. These are known as *difference equations*, rather than differential equations.

Since you already have equations that express the derivatives as function of x and y, you are now ready to start building the model in the spreadsheet.

Start GeoGebra and first create sliders for a, b, c, and d. You need to have short step lengths: 0.0001 or 0.00001. Also create a slider for Δt with the step length 0.01.

Place points in the graphics area in Figure 6.40 that show the table values. Let the x-axis represent the number of antelopes and the y-axis represent the number of cheetahs.

Now show the spreadsheet with **Show > Spreadsheet**, or **Ctrl-Shift-S** and also show Graphics Area 2 with **Show > Grahics 2** or **Ctrl-Shift-2**. Enter the values into the spreadsheet as in Figure 6.41.

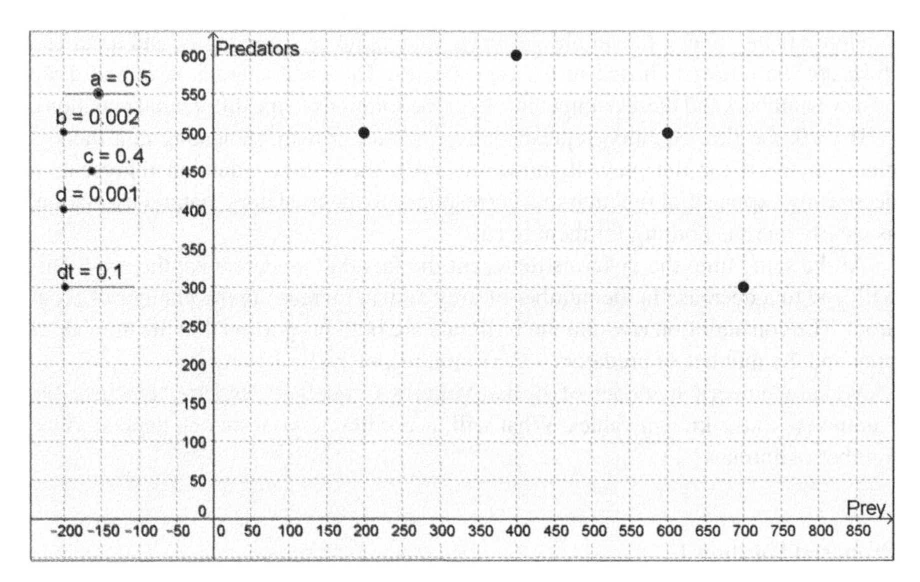

FIGURE 6.40 Table values in a phase diagram.

	A	B	C	D	E	F
1	t	x	y	Fas	Pray	Predators
2	0	700	300	(700, 300)	(0, 700)	(0, 300)
3	1	693.08	309	(693.08, 309)	(1, 693.08)	(1, 309)

FIGURE 6.41 Model under construction.

Click in the Graphics Area 1 before entering

In D2 `= (B2, C2)`

In the same way, click in Graphics Area 2 before entering

In E2: `=(A2, B2)`
In F2: `=(A2, C2)`

Color these points by right-clicking on them, select **Properties**, the **Color** tab.

Figure 6.42 shows Graphics 2 with labeled and scaled axes. You can now continue to build the model:

In A3: `1`

In B3 and C3 you can calculate the evolution of new individuals through the difference equations. Enter

In B3: `=B2 + dt (a B2 - b B2 C2)`
In C3: `=C2 + dt (d B2 C2 - c C2)`

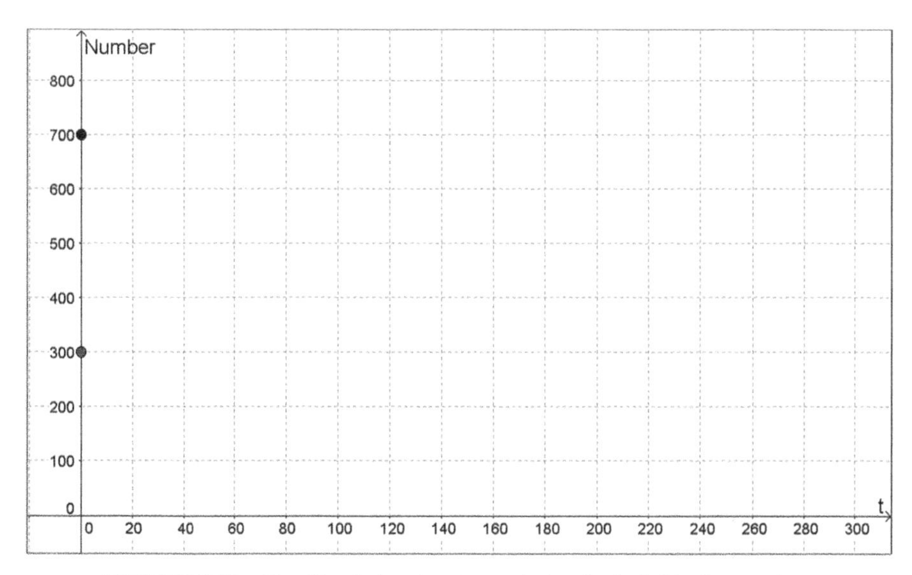

FIGURE 6.42 Graphics 2 showing the evolution for individuals over time.

The expressions in the parenthesis correspond to *dx/dt* and *dy/dt*. Either enter the contents of D3:F3 manually, in the same manner as previously, or select the cells D2:F2 and copy them down to row 3.

You need to construct one more row manually. Select both cells A2 and A3, and copy these down to A4, which will then contain the formula =2A3 – A2, causing the series to repeat. The cells B4:F4 should be copied down from B3:F3. Finally select all cells A4:F4 and copy them down to row 300.

What you see might, to begin with, look a little bit messy, but by adjusting the parameters to, for instance, a=0.5, b=0.002, c=0.4, d=0.001, and d*t*=0.1, you get something that begins to be understandable, as seen in Figure 6.43.

Obviously the populations are cyclic. When the number of prey goes down, the number of predators decreases, which stabilizes the population of prey. Once the prey recover, the number of predators increases, until they once again start to decimate the prey.

By varying *dt,* you should see that the solutions represent closed curves in the phase diagram, but since you are working with numerical solutions, errors will make the solutions spiral out. You can now try to fit the solution to the data by changing the parameters. It can be well worth the time to play with the parameters for a while and observe what they do. But to simultaneously adjust four different parameters manually to fit a graph to data can just about drive you insane. So take it slow, be systematic, and it might just work.

- a and b seem to affect the extension in *y*-direction
- c and d seem to affect the extension in *x*-direction

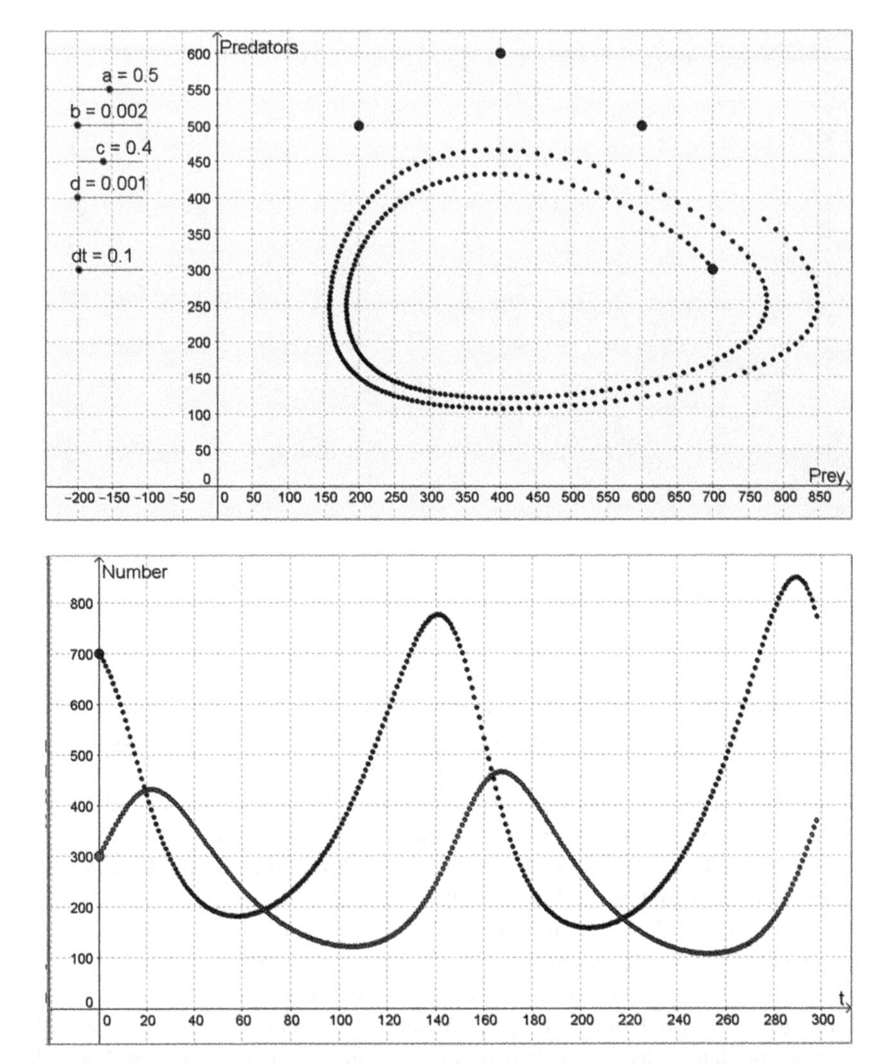

FIGURE 6.43 Typical phase-and-time diagram for Lotka–Volterras equations.

- b and d seem to have a greater effect than a and c and also seem to adjust the shape around the maximum points more than a and c.

Since you are about to adjust four parameters to three measured values (the solution curve is already locked to the first points as an initial condition), there should be infinitely many solutions. The following method should finally work:

Start by adjusting d until the maximum value for the curve fits rather well with the measured value from 2011 when there were 600 cheetahs in the area. Here d is about 0.001. Then adjust b, so that the curve goes through the measured value from 2008.

TABLE 6.6 Four Different Solution Attempts

Attempt	a	b	c	d	Prey min	Predator min
1	0.183	0.00059	0.185	0.00050	155	130
2	0.478	0.00147	0.419	0.00113	160	147
3	0.396	0.00125	0.375	0.00102	155	135
4	0.095	0.00030	0.084	0.00023	153	142

Adjust c, so that the curve reaches the measured value from 2014. Adjust a, so that the curve will reach the measured value from 2008.

You have to adjust the values for b, c, and a several times in this manner. Try to get the measured values from 2008 and 2011 perfected, and note how the solution curve behaves around the measured value from 2014. If the solution curve goes outside the measured value, it needs to have a sharper curvature. If so, decrease b and adjust the shape back to fit the measured value with a. If the solution curve goes inside the measured value, then it needs a less sharp curvature. If so, increase the value of b and then re-adjust a. If needed, adjust the solution curve somewhat with c, and then with a again. Four different attempts using this method produced the values in Table 6.6.

You can see that the minimum values for the populations are about the same in all four cases. Accordingly, this model shows that there is no imminent risk that the antelope will become extinct in the area. The number of cheetahs in the area has been decreasing since 2011, and according to the model, the number of antelope will be decreased to about at least 150 beasts.

The key phrase here is "*according to the model.*" There are many different assumptions in the model, and it does not take into consideration that the number of beasts must be a whole number or that births and deaths occur unevenly through the year. Random events might decrease the numbers under the normal minimum and create huge damage to the population.

Proposed Solution 2

Lotka–Volterra's equations are well known for modeling the relationships between predator and prey, but they not the only such equations. In the late 1980s Arditi and Ginzburg suggested that the consumption and birth rates in these models should not be proportional to the number of prey x but instead be functions of the relative number prey per predator x/y. In their model it is not the total number of prey but the number of prey per predator that is important. Arditi and Ginzburg claim that data that show real interactions in nature can be modeled better by their model than by the Lotka–Volterra model.

The consumption rate $g(x/y)$ should be so constructed that g is proportional to x/y for small values of x/y, but goes to some constant for large values of x/y. Such a function could be, for instance,

$$g(x/y) = \frac{K \cdot (x/y)}{R + (x/y)}$$

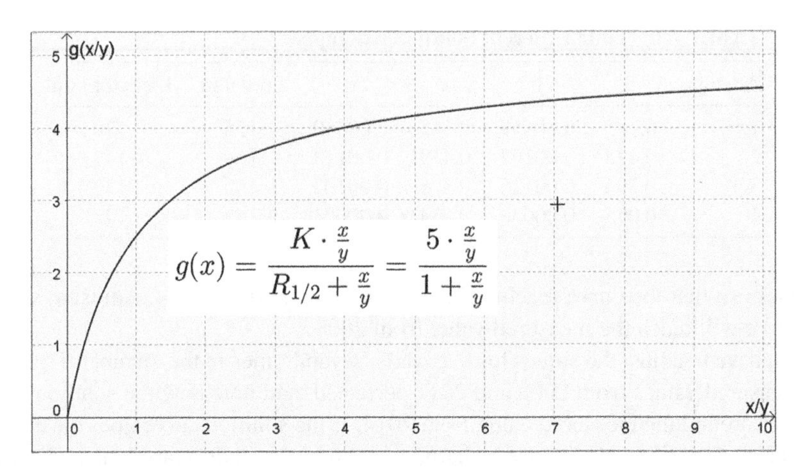

FIGURE 6.44 Function for the food intake relation. Maximum value $K=5$. Half this value is reached when $x/y=1$.

where K is the maximal value and R is the value for x/y when $g=K/2$. An example is shown in Figure 6.44

The function has its derivative value $g'(0)=K/R$ in the origin, corresponding to b in the Lotka–Volterra model. Furthermore Arditi and Ginzburg assume (based on observations) that the birth rate h for predators is proportional to the food intake, that is, $h=\mathrm{e}\cdot g$, where e is the conversion efficiency that measures the number of new predators born for each eaten prey.

You can still assume that the food for the antelopes is enough to not limit their growth, so you still can let the first term be $\mathrm{a}\cdot x$, which gives an exponential increase of antelopes if they were not hunted by the cheetahs.

If you add all this up and in the equations of Lotka–Volterra exchange $-\mathrm{b}\cdot x$ for g and $+\mathrm{d}\cdot x$ for h, you will get

$$\begin{cases} \dfrac{dx}{dt} = \mathrm{a}x - \dfrac{\mathrm{K}\cdot x}{\mathrm{R}+(x/y)} \\ \dfrac{dy}{dt} = -\mathrm{c}y + \dfrac{\mathrm{e}\cdot\mathrm{K}\cdot x}{\mathrm{R}+(x/y)} \end{cases}$$

To be able to model this, you have to create new sliders for K, R, and e that will replace the sliders for b and d, which you do not need any more. Then change the formulas in B3 and C3:

In B3: `=B2 + dt (a B2 - K B2/(R + B2/C2))`
In C3: `=C2 + dt (e K B2/(R + B2/C2) - c C2)`

After this you should copy down B3 and C3. One again, you will encounter solution curves that seem to go here and there, but after playing with the parameters

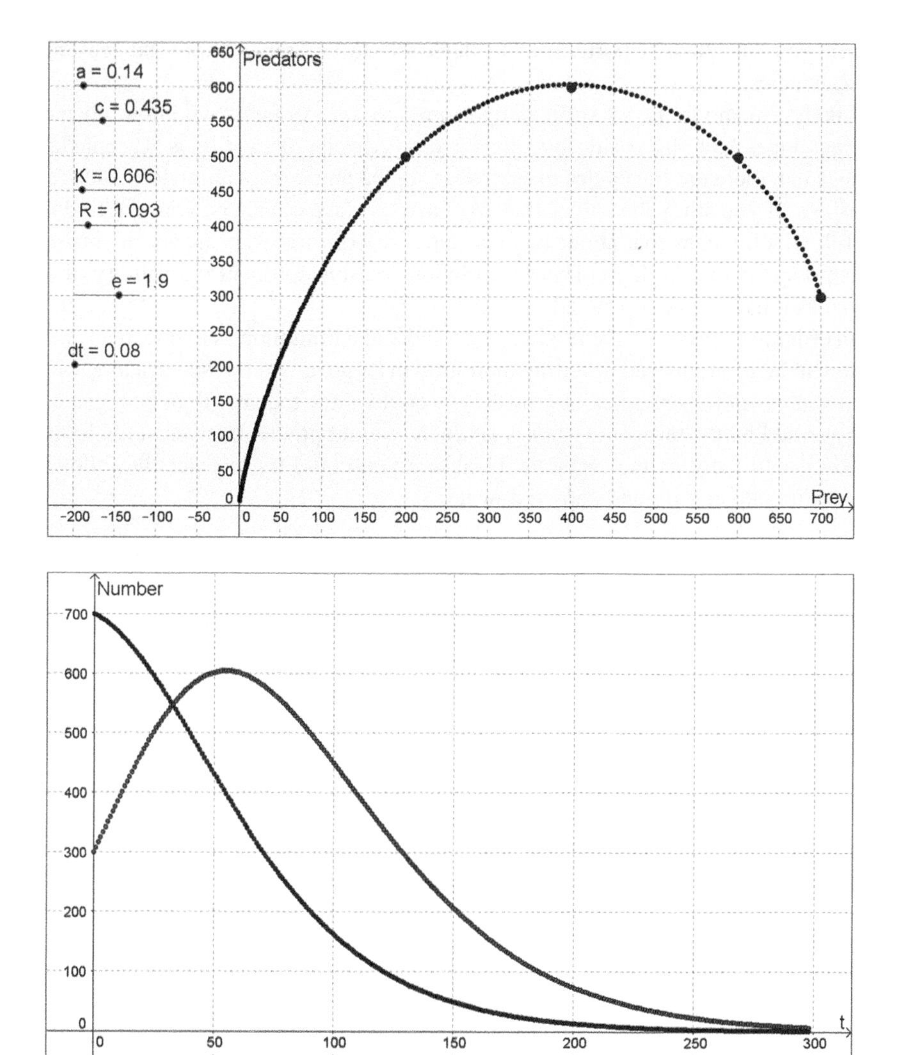

FIGURE 6.45 Catastrophic situation where both spices becomes extinct.

for a while, you will soon find that this model has some problems with life cycles. Either, on the one hand, the predators die out immediately or they grow too strong, decimate their prey, and then die from starvation later on, or on the other hand, both prey and predator grow to infinity—since you did not add a mechanism that limits the growth of the prey.

You can nevertheless adjust the parameters so that the solution curve matches the data for this model too. If you adjust the parameters so that a=0.14, c=0.435, K=0.606, R=1.093, and e=1.9, the solution curve will pass neatly through the table points and then down to the origin as in Figure 6.45. In this model both antelopes and cheetahs die out. Trying out a method that fails is no failure for the students. Instead, it will give excellent opportunities to demonstrate communication and reasoning competencies.

Likely this model is not sufficiently correct as yet. As far as we know, cheetahs and antelopes have lived side by side for long periods of time. It is reasonable to believe that there are life cycles that can explain the numbers of animals in such relationships. If you study the values that you have for the parameters, you will find that e = 1.9, which means that about two cheetahs are born for every newborn antelope, which does seem a bit high. There are obvious reasons to doubt the validity of this model as it has been constructed.

In order to improve on the model, you could add a limitation as to how many antelopes can be born. Instead of letting their growth be governed by the term $a \cdot x$, which results in unlimited exponential growth, you could allow the growth to be logistic, to be governed by the term $a \cdot (1 - x/k) \cdot x$, which without predators gives you a logistic solution with the maximal value k. Therefore you could try to enter the following expression into cell B3 and copy it down:

```
= B2 + dt (a B2 (1 - B2/k) - K B2/(R + B2/C2))
```

Unfortunately, this improvement of the model still gives no life cycles similar to those we found in the Lotka–Volterra model, at least no life cycles passing through the data.

Proposed Solution 3

Lotka–Volterra's equations contain continuous variables for the number of predators and the number of prey. In reality these will, of course, be integers. It is meaningless to talk about decimal antelope. Besides the growth process occurs stepwise, since births normally happen once per year. Also the number of newborn and the number of deaths are randomized. You are now ready to create a new modified model to reflect some of this insight

To start with, let's decide that dt = 1 year, rather than any arbitrary time unit. Then implement normally distributed randomized variation factors for a, b, c, and d, constructed so that the variation factors have an average value of 1, with standard deviations that you can control with parameters σ_a, σ_b, and so on. Instead of the term $a \cdot x$ that in the previous formulas were represented by **a B2**, there will be a term

```
floor(a x RandomNormal[1, σ_a/sqrt(x)])
```

where the function **floor()** rounds off down to whole numbers. The other functions will be changed in the same manner.

First erase all rows where $t > 60$ and enter **dt = 1**. Zoom to the x-axis in Graphics 2 so that you see the remaining points clearly. Create four new sliders σ_a, σ_b, σ_c, and σ_d and label them σ_a, σ_b, σ_c, and σ_d, allowing them to go from 0 to 1 as illustrated in Figure 6.46.

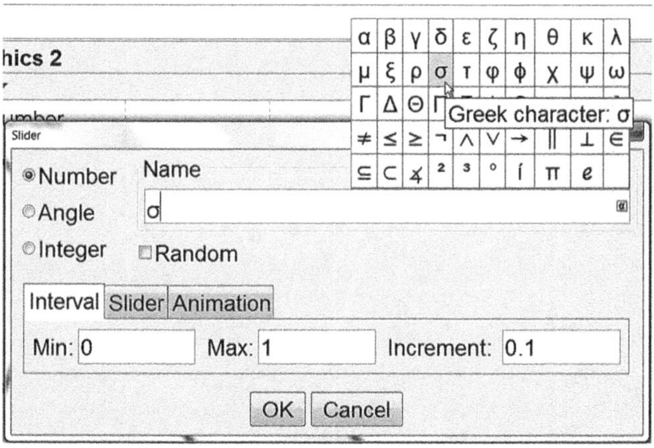

FIGURE 6.46 Greek letters accessed by clicking the α-sign in the input field.

In B3 type:

```
B2 + dt (floor(a B2 RandomNormaldistribution[1,
σ_a/sqrt(B2)]) - floor(b B2 C2 RandomNormaldistribution [1,
σ_b/sqrt(B2)]))
```

And in C3 type:

```
C2 + dt (floor(d B2 C2 RandomNormaldistribution[1,
σ_d/sqrt(B2)]) -floor(c c2 RandomNormaldistribution[1,
σ_c/sqrt(B2)]))
```

Then copy these down and try to make sense of the result. Figure 6.47 shows a reasonable attempt.

After having limited the standard deviations to 0 and played with the parameters for a while, you should find a solution where a=0.13, b=0.0005, c=0.25, and d=0.00062.

Unfortunately, there are difficulties with the time aspects. If you adjust the parameters so that you have three years between the first and the second measuring point, the model forces you to have two years between the second and the third. If you succeed to get the distance between the second and the third point to three years, the distance between the third and the fourth point easily become five or six years instead of three.

If you increase the standard deviations and repeatedly press F9 or Ctrl-R, you will see the effects of the built-in randomness. It seems as if the solution is not substantially changing in the short run but it is not producing a closed graph either. It seems more as if the solution is casting itself between extremes.

According to this model, you get a minimum value of 154 antelopes for 2019. The number of cheetahs will then be reduced drastically, to less than 100 in the 2025

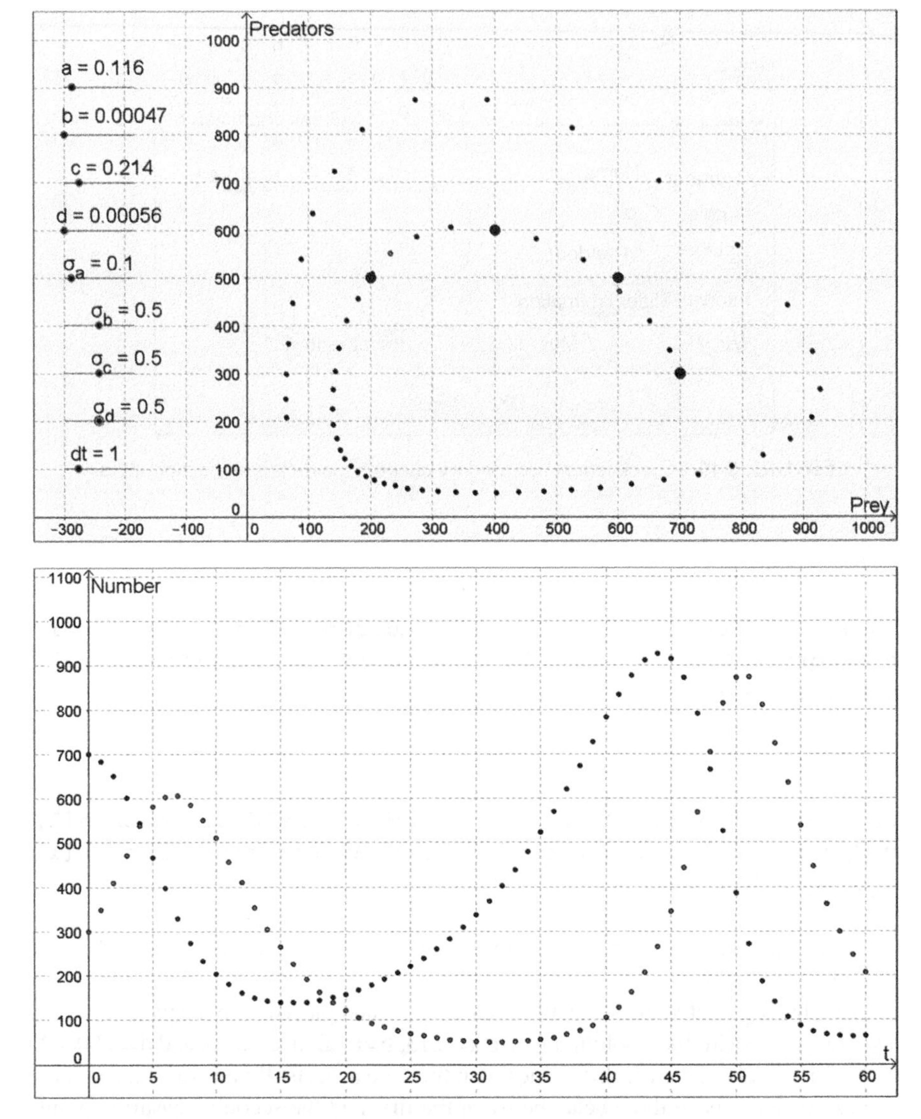

FIGURE 6.47 Stochastic model affected by randomness.

and will then have smaller and smaller minimum values each cycle. The antelope herd will grow and reach ever higher peaks.

If this model were real, the extremes would become larger and larger, and at the end the antelope herd would probably become so small that they would die out. It is unclear just exactly when that would happen and we must once again remind ourselves that this is just a model and that there are many other factors to look into that this model does not consider.

6.8 SMOKE

In a room measuring 4 m × 3 m × 2.5 m, and initially filled with fresh air, a group of people have started to smoke. Together they produce 0.04 m^3 smoke per minute. The smoking goes on during a lunch break from 12:00 to 13:00 p.m. The smoke contains 0.1 mg/m^3 of the carcinogenic substance benzo pyrene.

Well-mixed smoke-filled air is leaving the room at a speed of 0.02 m^3/min and is replaced with the same amount of fresh air. Construct a mathematical model that describes how the amount of smoke is changing in the room over time.

Examine your model. Will the legal airborne *permissible exposure limit* (PEL) for benzo pyrene be exceeded in the room? If so, when will this happen? When, after 13:00 p.m., can you consider the air as free from smoke? What will the air in the same room look like after a working week if the same group of smokers produces the same amount of smoke every lunch break? What will it be like after four weeks? In what way will the results change if the smokers smoke more or less?

Assume that the legal airborne PEL for benzo pyrene is 0.3 mg/m^3 in the country where the room is situated.

Proposed Solution 1

Let $y(t)$ denote the total amount of benzo pyrene in the room in mg at the time t, and let t denote the time after 12:00 p.m. the first day, measured in minutes. During the first hour the following *flow rates* can be identified:

Incoming:	$0.04\ m^3/min \cdot 0.1\ mg/m^3 = 0.004\ mg/min$
Outgoing:	$0.02\ m^3/min \cdot (y\ mg/30\ m^3) = 0.002 \cdot y/3\ mg/min$

The difference between these two is the total change of the amount of smoke:

$$y' = 0.004 - \frac{0.002}{3}y$$

This differential equation must be solved. With the initial condition $y(0)=0$, you can also determine the integration constant. You need to know how much smoke there is after 60 min, so you need to calculate $y(60)$. Send the following command to Wolfram Alpha:

```
y' = 0.004 - 0.002y/3; y(0) = 0; y(60)
```

getting back

$$y(t) = 6 - 6e^{-0.002t/3} \quad \text{and} \quad y(60) = 0.235263\,\text{mg}.$$

This equation is, of course, also possible to solve by hand. During the rest of the first day there is another situation where the incoming amount is 0 mg/min. You therefore get another differential equation with another initial condition, leading to another solution.

$$y' = -\frac{0,002}{3}y, \quad y(60)=0.235263\,\text{mg}$$

Sending

```
y' = - 0.002y/3; y(60) = 0.235263; y(1440)
```

to Wolfram Alpha will, surprisingly, not give any results. Removing the last part and sending

```
y' = - 0.002y/3; y(60) = 0.235263
```

gives the result

$$y(t) = 0.244864e^{-0.2t/3}$$

Wolfram Alpha refuses to calculate $y(1440)$, claiming the calculation time is exceeded but Wolfram Alpha will happily calculate

$$0.244864e^{-0.002\cdot 1440/3} = 0.0937\,\text{mg}$$

This is the amount of benzo pyrene in the room when the smokers return on the second day. You can see that you have to repeat the whole procedure and adjust the initial conditions to the values you get from the previous step all the time. You could use Wolfram Alpha for this step by sending sequences of commands with the same structure as above. This solution might feel a little desperate because, if you need to change a parameter, you are forced to start all over again.

Proposed Solution 2

In GeoGebra you can do a fully dynamical model and the construction of this model will also show some very powerful techniques.

Start by showing Graphics Area 2 with **Show... Graphics 2**, and in there create sliders for all parameters you need. Create them in Graphics 2 to get the parameters organized and out of the way for the results that you want to show in the primary Graphics window.

Open the **Properties** dialogue for the sliders and select suitable values for each parameter, namely the smoke production speed, defined to 0.04 m^3/min, can be labeled "Smoke" and have the properties **Min = 0**, **Max = 0.1** and **Increment = 0.001**. Select short names for the sliders:

```
Step = 0.5

Smoke = 0.04
```

```
Vol = 30
STime = 60
Benz = 0.1
Vent = 0.02
```

Step is a parameter that will represent the step length in the numerical solutions of differential equations later on.

Next define the functions for incoming and outgoing flows as functions of x and y with the commands:

```
In(x, y) = Smoke Benz
Out(x, y) = Vent/Vol y
```

It will now be easier to define the two different functions for the first derivatives:

```
f1(x, y) = In(x, y) - Out(x, y)
f2(x, y) = - Out(x, y)
```

Activate the spreadsheet with **Ctrl-Shift-S**. Check to see that Graphics 1 is active and create a point

In A1: `(0,0)`

Then solve the first differential equation by typing

```
B1 = SolveODE[f1, x(A1), y(A1), STime, Step]
```

The whole solution for the differential equation is in cell B1 in the spreadsheet. Observe that the solution is dynamic. You can move the point A1 representing the initial condition or change any of the parameters and see the result immediately. Since you have labeled the solution B1, it is automatically saved in the spreadsheet, which you will use later when you will copy it down.

But now create a point as far right as possible on the solution

In C1: `=Point[B1,0.999999]`

The number 0.999999 places the point as far the right as possible. If instead you were to use 0.5, the point would be positioned at $t=30$ min. See the manual at http://wiki.geogebra.org/en/Point_Command for more information. You then execute the same procedure for the time of the day when no one is smoking in the room:

```
D1 = SolveODE[f2, x(C1), y(C1), 1440, Step]
```

And in E1: `=Point[D1,0.999999]`

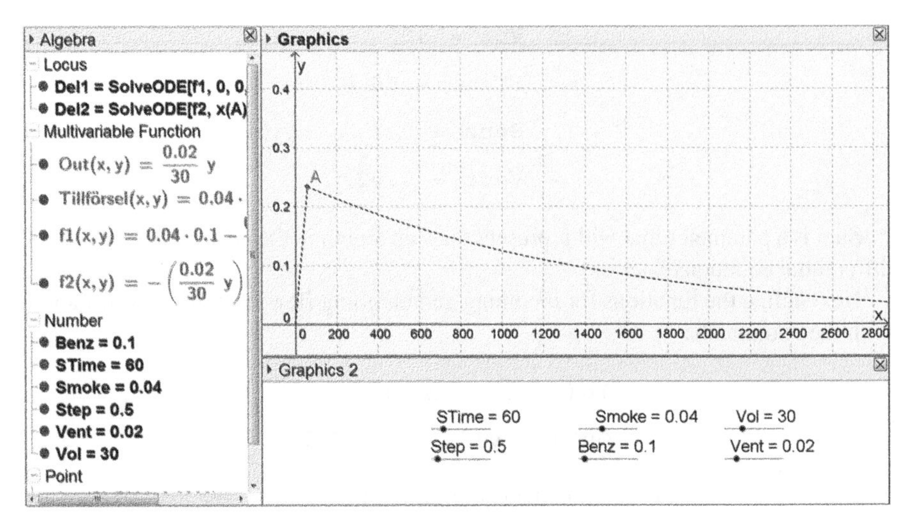

FIGURE 6.48 Dynamic solution of a differential equation.

The solution for the first 24 hours is now ready and is totally dynamic; see Figure 6.48.

You will use the spreadsheet to repeat this solution day by day. Start by copying E to $t=0$ by typing

In A2: `=(0, y(E1))`

After that, select B1 to E1 and copy them one step down, to row 2. You now have the solution for day 2. Then select the cells A2 to E2 and copy them down to row 5.

If this is a room that is only used for smoking from Monday to Friday, then you need to make some small adjustments to the model. Open up the **Properties** dialogue for cells B6 and B7 and change f1 to f2 in both cells. That means that the ventilation is up and running and that no one smokes in the room during lunch break on Saturday or on Sunday. In Figure 6.49 the Saturday and Sunday solutions are shown at the bottom. The amount of benzo pyrene will be about 0.0221 mg/m^3 before lunch on Monday morning again.

To see what has happens after four weeks, you need to copy the first cells A5 to E5 to row 8. Then you can copy the whole block A2:E8 with **Ctrl-C** and paste it into cells A9, A6, and A23. It might be little bit thickly to see what happens in the somewhat crowded Graphics window, but you can read the values for Monday 12:00 p.m. from cells A1, A8, A15, A22, and A29. A1 is, of course, 0; all the others have 0.0221 mg benzo pyrene in the entire room. That they are the same indicates that the solution converges quickly; see Figure 6.50.

Observe that in total, the solution consists of iterated solutions of 56 differential equations, each of them depending on a value from the previous solution. So the solution is fully dynamic, and it is possible to adjust any parameter you have defined. You can easily experiment with this model to find suitable airflow for ventilation if the amount of smoke should increase.

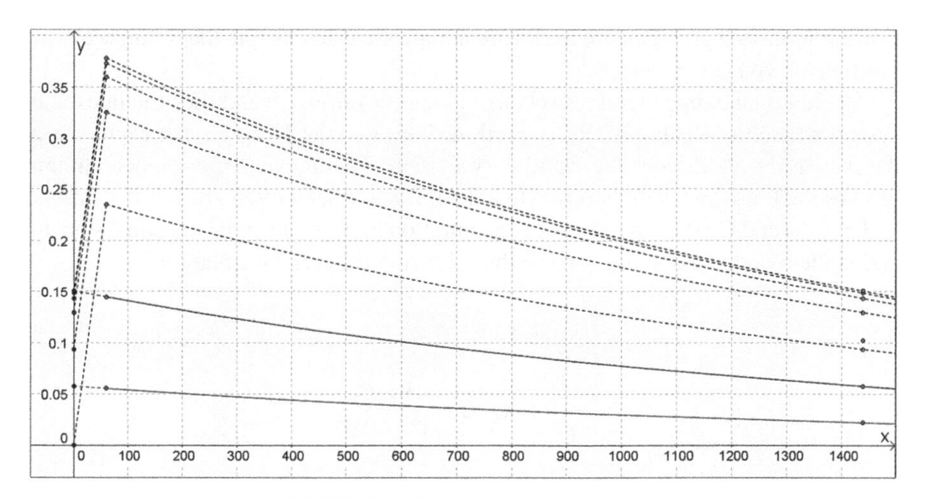

FIGURE 6.49 Seven-day solution.

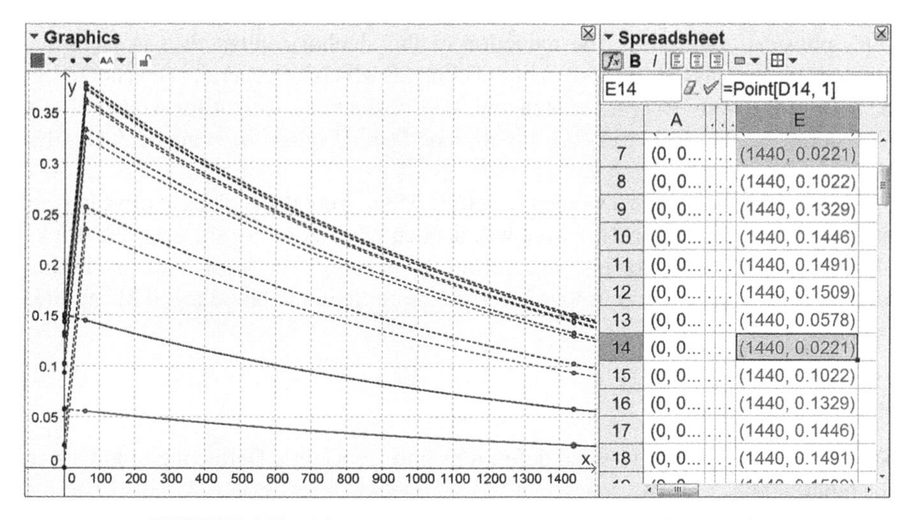

FIGURE 6.50 Final value does not change from week to week.

6.9 ALCOHOL CONSUMPTION

On many occasions in life it is important to know the fine line between being sober and drunk. It is especially important if you plan to drive a car after drinking. At many dinner parties you may even hear someone saying: *I will just take only one glass of wine because I am driving.*

There are, of course, many factors that influence how much a person can drink without getting drunk. Your body weight, the amount of alcohol you normally drink, how much food you eat and when you eat it, your sex, the time of alcohol consumption, and your general health are among the usual factors that influence how much drink you can handle.

Alcohol intake, in general, travels from mouth to stomach and then the intestinal system where the alcohol is slowly portioned out into the blood stream from which the liver removes it from the circulatory system. The alcohol in the blood stream reaches the brain and is that portion that determines whether you are sober or drunk.

Let $C_i(t)$ and $C_b(t)$ be the alcohol concentration in *promille* in the intestinal system and in the blood stream. You can use the following differential equations:

$$\begin{cases} \dfrac{dC_i}{dt} = I - k_1 \cdot C_i \\ \dfrac{dC_b}{dt} = k_2 \cdot C_i - \dfrac{k_3 \cdot C_b}{C_b + 0.05} \end{cases}$$

Here I is the alcohol intake in promille (tenth's of %) per hour, while the constants k_1 to k_3 control the speed of the process. If you drink on an empty stomach, $k_2 = k_1 = 6$, but if you eat at the same time, then $k_2 = k_1/2$. k_3 is controlled by the liver, reducing the alcohol amount with approximately 8 g alcohol per hour regardless of the person's physical conditions. The reduction of the alcohol content then is $k_3 = 8/m_V$, where m_V = the mass of the body's liquid content, which, in general, is considered to be 87% of the body weight for men and 71% of the body weight for women. Let's also define a standard drink as 1 bottle of beer = 1 glass of wine = 2.5 cl 40% alcohol = 14 g of 40% alcohol.

How much alcohol can a women weighing 55 kg drink if she plans to drive her car back home 6 h later? Consider what will happen whether or not she eats, she drinks all the alcohol directly, or just sips during the evening. The legal limit for driving with alcohol in your blood in Sweden is 0.2 promille in the blood and 0.21 mg/liter in exhaled air.

Proposed Solution

Start by entering parameters and definitions into GeoGebra. Define the following in the input field:

```
Bodyweight = 55

LiquidFactor = 0.71

LiquidVolumeMass = Bodyweight LiquidFactor

k_3 = 8/LiquidVolumeMass

k_1 = 6

k_2 = 6

I = 0
c_s = 14/LiquidVolumeMass
```

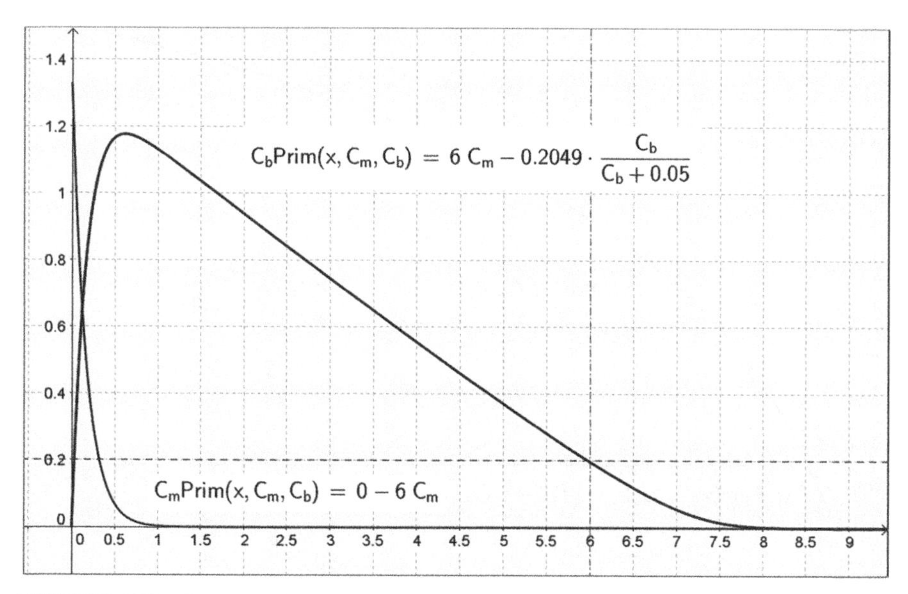

FIGURE 6.51 Amount of alcohol in the blood decreases almost linearly over time if taken at the start of the evening.

```
n = 3
c_0 = n c_s
```

Next define the derivatives as multivariable functions:

```
C_i'(x, C_i, C_b) = I - k_1 C_i
C_b'(x, C_i, C_b) = k_2 C_i - k_3 C_b/(C_b + 0.05)
```

When you define functions this way, you can automatically solve a system of first-order differential equations without the need for manual calculation in the spreadsheet. The command to solve the system is

```
NSolveODE[{C_i', C_b'}, 0, {c_0, 0}, 10]
```

You should first give a list of functions that are derivatives, then that you want to solve from $x = 0$, then a list of initial values corresponding to the functions, and last that you want to solve these equations to where $x = 10$. The complete syntax for this command can be found at http://wiki.geogebra.org/en/SolveODE_Command.

When you execute the command, the solution curves will be drawn in GeoGebra. You can add a line **y=0.2** in order to identify and visualize the legal limit and another line **x=6** in order to identify and visualize the time limit in this particular problem. You can change the increment for n to 0.01 in the **Properties** dialogue and then adjust the value until the curve for the alcohol in the blood goes down to 0.2 promille in six hours, as shown in Figure 6.51.

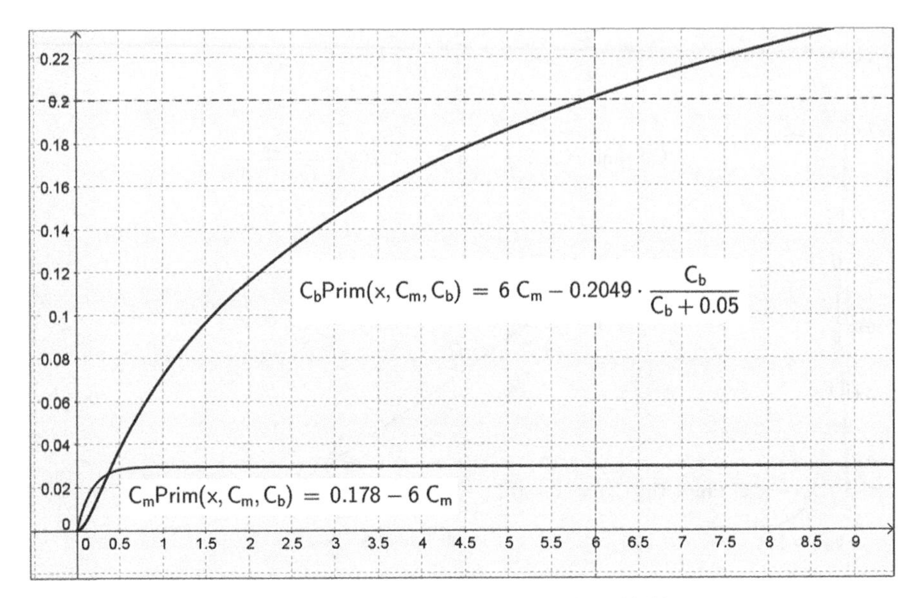

FIGURE 6.52 Alcohol intake for the same person when drinking at a constant rate.

For this part of the problem, where you study the effect of alcohol in a woman whose weight is 55 kg and who drank some alcohol in the early evening without eating, you will find that she can drink 3.7 standard size drinks and still legally drive her car back home after six hours. The model is built on the assumption that all alcohol will be absorbed in the body the first minute, which perhaps is not so realistic.

With some simple modifications you can model how much alcohol the same woman can drink in other situations. To construct a model that pays attention to the intake of food, you can set k_2=k_1/2. Adjust n once again and get 7.4 drinks. This is indeed a rather simple modeling of the effects of alcohol when eating, but it still gives some indication of the woman's sobriety.

If you instead want to model the effect of a continuous alcohol intake, you would set $n=0$ and vary I. Without food, you will get I=0.18 drinks/h for 6 h ≈ only one drink for the whole time, indicated in Figure 6.52. With food you will get slightly more than two drinks. You have to be more careful with drinking if the drinking was continuous, since some of the last drinks could be taken close to the time limit. To eat continuously is, of course, not realistic in most situations.

6.10 WHO KILLED THE MATHEMATICS TEACHER

You arrive at the crime scene. A mathematics teacher has been killed on the day of the math finals. All students and staff are present. The ordinary police officers are busy interviewing people in order to determine exactly who were where during the time of death and finding out who has an alibi and who has not. Your mission is to examine the dead person and the crime scene. You are a professional crime scene investigator.

The day started out well enough, being sunny and warm with 25 degrees early on in the morning. Later on it started to cloud over and now, at 10:00 a.m., the air temperature reads 22.0°C. You also measure the inner temperature of the body to 30.3°C. In order to establish the time of death of the mathematics teacher, you will need another measurement later on. Meanwhile you closely observe the surroundings.

The mathematics teacher seem to have been stabbed in the back, and is lying in a small back part of the schoolyard where normally no one spends much time other than entering and exiting the school. Was the mathematics teacher going into the building when the attack occurred? A bicycle is parked here, so that might be a reasonable guess. The weapon is still in the wound: A papercutter type of knife was used that can be found in almost any office. Anyone could get hold on such a papercutter. You spray it for prints, but no luck. Wiped clean after the deed, and most likely it was premeditated, so you note to yourself.

It must be difficult to creep up on someone in this yard, you think. Maybe the mathematics teacher heard someone coming up from behind, but proceeded walking without giving it a second thought. It could have been someone known to the teacher, or maybe even a parent or some student. On an exam day like this there are many unknown people on school grounds. People that are not normally here. Parents, grandparents, relatives, friends, all are ready to congratulate or comfort the overtested children …

What could be the motive for such a violent killing? An former student the teacher had previously failed? A jealous colleague? A parent to a former student who didn't get into the university because of a bad math grade? Someone from the catering firm who was annoyed because the teacher always took two glasses of milk to the food tray despite being told not to? Right now, everyone was a suspect, except for the janitor. Everyone wants to get in touch with the janitor on a day like this, so it would be difficult for the janitor to get away after killing someone. It would more likely be someone who could fly under the radar, so to speak. Someone no one would be interested to talk to during this day. But who could that be?

Right then your captain approaches you, saying:

> "Thanks to our new digital interrogation system, Comprehensive Hearings External Additive Tally, we have managed to identify some individuals without alibi for the early morning. Here is the list. Let me know as soon as you have an answer. I will be at the principal's office for the time being. The principal has helped us to organize all witness hearings and is offering coffee, though without donuts. And be quick about it."

You look at the printout and you read:

Persons without alibi (time):

Cally Crook	**Parent in T14f, 48 years**	**7:23 – 8:15**
Sy Smoke	**Catering assistant, 41 years**	**8:38 – 8:48**
Charlie Bal	**PE teacher 30 years**	**8:35 – 8:55**
Nicci Vulture	**Language teacher, 77 years**	**8:50 – 9:28**
Alex Game	**Student in Sc13d, 17 years**	**9:25 – 9:48**

Right as usual, you think to yourself while you put on a sweater. Time to measure the body temperature again. Now, at 11:00 a.m., the body temperature is measured to 26.6°C and you start doing the arithmetic to find the precise time when the mathematics teacher died. Based on the information available from other teachers and the principal, the mathematics teacher was well and healthy and could therefore be expected to have a body temperature of about 36.8°C—at least while alive.

So back to the main question: Who killed the math teacher?

Proposed Solution 1

Select the variables so that you measure the time from 10:00 a.m. and the surrounding temperature, which is 22°C. Newton's law of cooling will yield the following system of equations:

$$\begin{cases} 8.3 = C \cdot a^0 \\ 4.6 = C \cdot a^1 \end{cases}$$

This is a system you can solve in different ways. In GeoGebra you create the points

```
A = (30.3, 0)
B = (26.6, 3)
```

Then create the model function

```
m(x) = p q^x + 22
```

With the use of sliders, you adjust the model function slightly so that it goes though or close to points A and B. Hide the model function and finally fit a function of this type by typing

```
Fit[{A,B}, m]
```

Here the temperatures are measured from 0°C again.

In Figure 6.53 lines have been added that indicate the temperature of the surroundings and the body temperature. You can find the intersection between the body temperature and the model, seeing that the mathematics teacher died 0.98 h before 10:00 o'clock, which is 9:01. You learn that the prime suspect is the language teacher Nicci Vulture, 77 years old, ripe for retirement, who could have stabbed the mathematics teacher from behind with a papercutter ...? Is that even possible? With a nagging feeling that something is not completely right, you slowly walk back into the school building and to the principal's office to report your conclusions to your chief. It is a little bit warmer inside than outside and a shiver runs down your back. You take a few deep breaths and try to calm down. Your conclusions are surely on target...

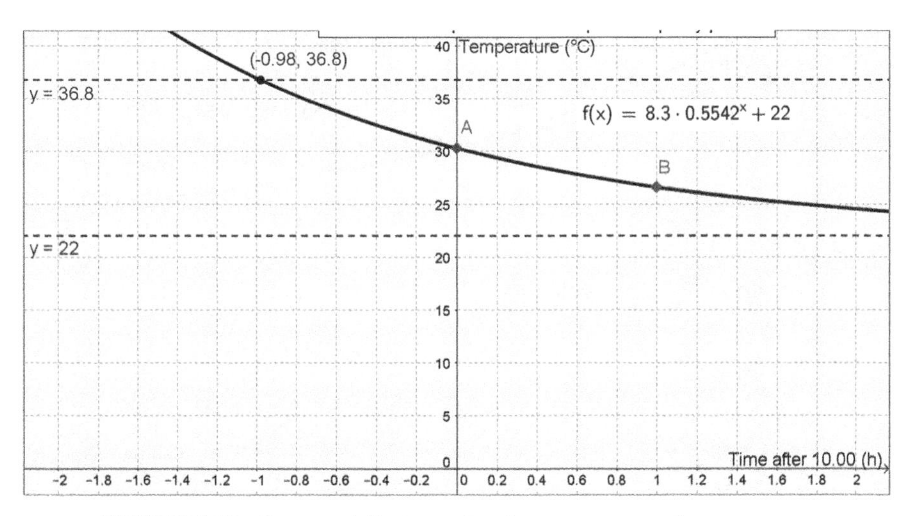

FIGURE 6.53 Exponentially decreasing fit to two measured temperatures.

The Insight …

On your way to the principal's office, it suddenly strikes you. The temperature outdoors is falling! At 9 o'clock it was 25°C, at 10 o'clock it was just 22°C, and now—at almost 11 o'clock—the temperature is down to about 20.5°C. Your model is wrong, but how do you find a model that takes into account an outdoor temperature that is falling? You will have to solve the basic differential equation again. Of course, it wasn't the language teacher Nicci Vulture, 77 years old, who killed the mathematics teacher, you think while you are finishing off your new calculations. It must have been … You search in the list of the suspects. How strange… Who did kill the mathematics teacher?

Proposed Solution 2

The temperature is falling at 3°C per hour since at least 8:30 that morning. Newton's cooling law claims that the speed of cooling is proportional to the difference in temperature. You therefore set up the differential equation $T' = \mathrm{k} \cdot (T - (22 - 3t))$, where the time t once again is measured in hours after 10:00. Open up the GeoGebra CAS window from the menu **Show > CAS** and type

```
SolveODE[y' = k(y - (22 - 3x))]
```

You find that $y = (\mathrm{c}_1\ \mathrm{k}\ e^{\mathrm{k}x} - 3\mathrm{k}x + 22\mathrm{k} - 3)/\mathrm{k} = \mathrm{c}_1 e^{\mathrm{k}x} - 3x + 22 - 3/\mathrm{k}$. Now you create a new model function

```
m2(x) = c e^(k x) - 3x + 22 - 3/k
```

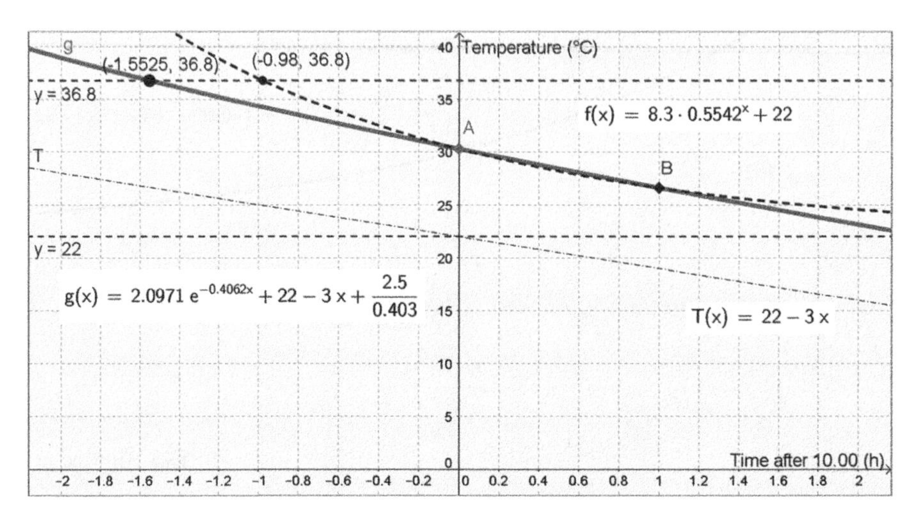

FIGURE 6.54 Solid curve showing body temperature, which decreases when the outdoor temperature is falling.

You allow GeoGebra to create the necessary sliders and then fit a function of this type to the data, using

```
Fit[{A,B}, m2]
```

According to this model, shown in Figure 6.54, the mathematics teacher was killed as early as 1.5525 h before 10:00 or at 8:27. But not one of the suspects lacks an alibi for that time …

Right as Usual

"… is really good. Everyone enters their information and the system correlates and compares, always finds all the false information, and then print a list of all the suspects," says the Captain to the principal when you enter the office. They both look up at you.

"Aha, here is our technician," says the chief. "Well, which one of the suspects is it? I have them all in custody, in a classroom in the building with a guard outside and I would like to release those innocent as soon as possible. So …?"

"You can release all of them," you say in a clear voice. "None of them is guilty. They all have an alibi for the time of death."

"Excuse me? How is that possible? You must have made a mistake as usual …" the Captains starts, but you interrupt him and say:

"The guilty party is sitting right there," you say, pointing at the principal who seems to shrink." The principal helped organize all the interviews, right? That gave him the opportunity to avoid giving any information that could be correlated to the other interviews. The principal could, so to speak, fly under the radar. I'm guessing that the motive is the school's economy, since all schools have less financing these days. And mathematics teachers have the highest salaries because they are better educated." You look at the principal. "Did you try to improve the school's finances?"

"Substitutes are so much cheaper ...," the principal snivels, while police officers cuff him and take him away.

Later, back at the station, the Captain has a chat with you. Again, it is not the first time this conversation unfolds.

"I re-did the calculations when I found that the temperature had fallen ...," you try to explain, but the chief is just smiling ironically.

"I knew you were wrong as usual," says the chief.

"To make an error does not mean that you cannot yet be right," you answer him as you spot some donuts on the desk. "The big error would be to give up. I did not give up, and in the end I was right."

...as usual, you think privately, and take a donut.

6.11 RIVER CLAMS

River clams live in river mouths where tidal changes mean they will be exposed to a surrounding where the salinity will vary cyclically, as a sine function. The salt will wander in and out from the clam with a speed that can be assumed to be proportional to the difference in salinity (compare this to Newton's cooling law). Set up a differential equation and solve it both algebraically and numerically. Describe how the solutions will vary, depending on your parameter values.

6.12 CONTAMINATION

An oil spill in a lake contaminates both the lake water and the sea bottom. From the sea bottom the contamination slowly leaks back to the seawater. The water in the lake is regularly replaced through streams. How long time will it take before 90% of the contamination is removed from the lake?

6.13 DAMPED OSCILLATION

Differential equations for damped oscillations, or for swinging motions with friction, can be written $m \cdot y'' + c \cdot y' = -k \cdot y$. Here m is the mass of the object, k is a spring constant, and c is a friction coefficient. Set up the differential equation and solve it both algebraically and numerically. Let m and c be constant and find the value for k that will produce a *critically damped* solution. Do this several times for different values for c, and find a relation between c and k that guarantees critical dampening.

6.14 THE POTASSIUM–ARGON METHOD

The carbon-14 method is useful for determining the age of biological material no older than 50,000 years. For geological minerals another method is better, a method built on the decay of potassium. When a potassium atom decays, it can do this in two different ways, either decay to argon or to calcium. The decay constants for these reactions are $k_1 = 5.76 \cdot 10^{-11}$/year and $k_2 = 4.85 \cdot 10^{-10}$/year respectively. The decay process can be described as a system of differential equations

$$\begin{cases} K' = -(k_1 + k_2)K \\ A' = k_1 K \\ C' = k_2 K \end{cases}$$

The age of the material can be determined by measuring the amount of argon and the amount of potassium, A/K. Solve the system numerically and try to determine A/K for a mineral with the age 2.5 million years. Can you find a relation between the ratio A/K and the age? How old is a bit of material whose A/K ratio is half of what the 2.5 million year old material was?

6.15 BARIUM, LANTHANUM, AND CERIUM

Barium-140 is a fission product with a half-life of 12.8 days that decays with beta radiation to lanthanum-140 which also is radioactive, decaying with beta radiation with a half-life of 40.5 h to the stable isotope cerium-140. Create a system of differential equation that describe how the amount of the three different substances vary over time and solve this numerically. Starting with a bit of pure barium, when is the lanthanum–barium ratio exactly 0.01?

6.16 IODINE

At accidents where radioactivity is involved it is not unusual that different isotopes of radioactive iodine are released into nature. At the Fukushima accident in 2011 a considerable amount of iodine-131 was released, and small traces of this can be found even today all over the world. Iodine that enters the body will be assimilated by the thyroid, a sensitive organ that can be protected by keeping the iodine at normal levels with normal iodine at all times. In many countries, iodine is added to table salt, not because of the danger of a nuclear accidents but because it counteracts a goiter disorder.

Another iodine isotope, iodine-133 decays to xenon-133, which decays to cesium-133, which is stable. Find the half-time for these substances, and how to convert them to decay constants. Use the system of differential equation from Problem 6.14, *The Potassium–Argon Method*, and determine how much radioactive xenon remains in your body 24 h after you have assimilated 1 g of iodine-133.

6.17 ENDEMIC EPIDEMICS

Read through Problem 6.5 on flu epidemics. Create a new system of differential equations where children are born all the time and where adults die all the time. The newborn are normally uninfected, and susceptible (S) while the deaths rates are all the same from the different parts of the model—susceptible, infected, and recovered. For reason of simplicity, can the birth and death rates be set equal? Show that this model gives raise to new outbreaks of disease when the number of susceptible in the population has sufficiently increased.

6.18 WAR

A very simple model of a war might be constructed like this: assume that there are two armies, red and blue, and that one army's soldiers kill/wound the other army's soldiers with a speed that is proportional against the number of remaining soldiers, *R* and *B*, respectively. You can formulate this as

$$\begin{cases} R' = -a_B B \\ B' = -a_R R \end{cases}$$

Solve the system numerically and investigate its properties for the different initial conditions *R* and *B*.

Expand the model in some way and investigate the situation further. You might, for instance:

- Allow one army to use randomized firing as is typical when fighting hidden troops, namely guerrilla fighting. The guerilla soldiers will then have their numbers reduced with a speed proportional to both the enemy soldiers and their own numbers.
- Allow both armies to use randomized firing, as in trench warfare.
- Let the constant a = firing velocity multiplied by the probability of shooting an enemy soldier, and vary the value for these parameters.
- Let one or both armies get reinforcements or become decimated by disease.
- Divide one army into one camp and one fighting contingent. Only the fighting contingent is actually fighting, but they are supported from the camp according to a logistic model. This variant can be used to model the Battle of Thermopylae (seen in the movie *The 300*).

6.19 FARMERS, BANDITS, AND RULERS

China's history is very much about different dynasties. But between these large, well-organized and stable dynasties were periods of anarchy that were characterized by large groups of bandits that terrorized the peasants. A model for describing this has been created, a model that split the residents in three parts, Farmers (F), bandits (B), and rulers and soldiers (R). The model consists of the following system of differential equations:

$$\begin{cases} F' = rF\left(1-\dfrac{F}{K}\right)-\dfrac{afB}{b+F}-hFR \\ B' = \dfrac{eaFB}{b+F}-mB-\dfrac{cBR}{d+B} \\ R' = \dfrac{faFB}{b+F}-gR \end{cases}$$

The model resembles the Lotka–Volterras equations (see Problem 6.7, *Predators and Prey*) with bandits as predators and the peasants as prey. The soldiers and rulers both obtain feed from the peasants and bandits. The peasants will grow logistically if there are no soldiers or bandits.

Try to interpret each term and every parameter in the equations and explain their meanings.

Set r=1. b=0.17, m=0.4, g=0.009, K=1, h=0.1, c=0.4, e=1.2, a=1, d=0.42, and f=0.1. Solve the equation system numerically for initial conditions $F(0)=0.7$, $B(0)=0.1$, and $R(0)=0.2$.

Try to vary the parameters and to interpret what these changes mean in reality. One suggestion is to change the parameter c to 0.8 or bring h up to 2.

6.20 EPIDEMICS WITHOUT IMMUNITY

Create a system of differential equations that describes an epidemic where no one stays immune but instead becomes susceptible over and over again. What happens then?

6.21 ZOMBIE APOCALYPSE I

A village with 6,000 residents is terrorized by zombies. Initially, 4 out of every 10 survivors are attacked and converted to zombies at night. Model the situation as a logistic growth of zombies. How long time will it take before 90% of the residents are zombies? Round up.

6.22 ZOMBIE APOCALYPSE II

Use the SIR model but allow the infected to have the possibility to enter a second-step infection whereby they become zombies. From this state it is possible to recover, but with low probability, and the presence of zombies heavily affects the probability of being infected. Try to set reasonable values for all parameters. What fraction of the population survives?

7

GEOMETRICAL MODELS

The study of geometry is also the study of how we humans see the world. Part of how we communicate the images we see around us is through basic geometrical properties and names of geometrical objects. The molecules of our DNA, the cornea of our eye, snowflakes in wintertime, pinecones, flower petals, diamond crystals, the branching of trees, a nautilus shell, the star we spin around, the galaxy we spiral within, the air we breathe, and all life forms as we know them emerge out of timeless geometric representations. Our languages are often built around geometrical concepts: *Circle of life, Level with me, Get to the point, Social circle, Fair and square, Pyramid scheme*, and so on.

We also receive a broader and richer understanding of geometrical concepts if we are allowed to investigate the unexpected applied contexts in which they appear. In applied settings geometrical concepts take on new and different interpretations; for instance, a finite geometrical construction may become a graph and a knot can become a description of a quantum mechanical operator. There is a fascination in seeing familiar theorems and principles show their values in different roles. Without geometrical investigations, there is a risk that a student never gets to see the whole picture of geometry.

The study of triangles is sometimes known as triangle geometry, and this is a rich area of geometry filled with beautiful results and unexpected connections. In 1816, while studying the *Brocard points* of a triangle, Crelle exclaimed, "It is indeed wonderful that so simple a figure as the triangle is so inexhaustible in properties. How many as yet unknown properties of other figures may there not be?"

Mathematical Modeling: Applications with GeoGebra™, First Edition. Jonas Hall and Thomas Lingefjärd.

Companion website: www.wiley.com/go/Hall/MathematicalModeling

FIGURE 7.1 Supported pens. Photo: R. M. Koske CC BY-SA 2.0 via Flickr. https://www.flickr.com/photos/67146024@N00/5076817972/sizes/o/.

We will guide you in some initial but still intriguing investigations of triangles in this chapter.

Making a geometrical model of a real situation allows you to make measurements directly in the model, which can generate interesting relations that you may investigate further. But "pure" geometrical constructions—without any connection to the "real" world—are also an abundant source for investigative possibilities.

7.1 THE LOOPING PEN

Figure 7.1 shows several pens and pencils standing in a mug. If you lean a pen on a book, cup, or any other support so that the tip of the pen rests on the table, then the top of the pen will be above the support. If you now allow the tip of the pen to slide on the desk (or on a paper; after all, you don't want to draw on the desk, do you?), then the end of the pen will trace a loop in the air. Will that loop be part of a circle?

Proposed Solution

Assume that the pen is sliding on a line passing directly below the support point. In GeoGebra you can now start building a model of this situation using two sliders, which you can place in Graphics Area 2 so that they do not disturb the rest of the construction.

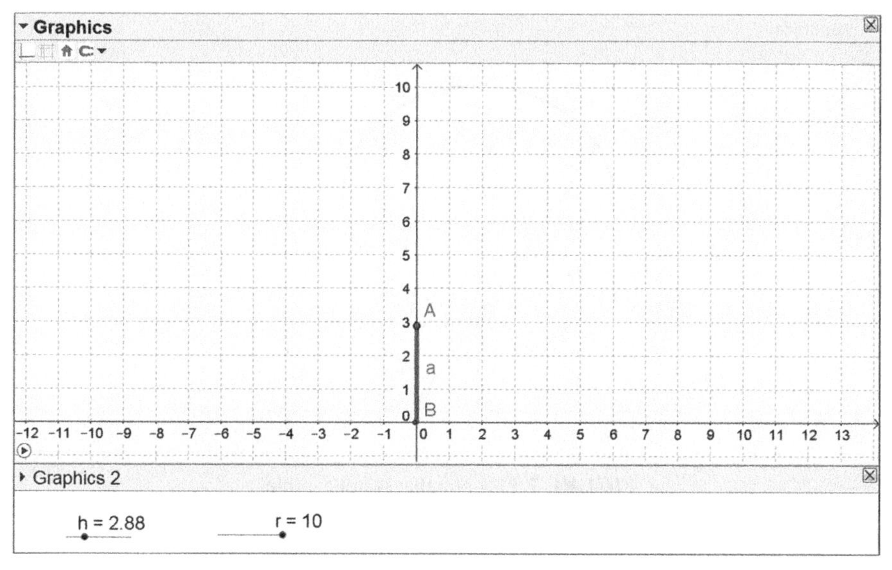

FIGURE 7.2 Building a model of a sliding pen.

Press **Ctrl-Shift-2** or select **View>Graphics 2** from the menu. Drag the Graphics 2 window so that it is directly below the Graphics 1 window, and hide its axes and grid. Click on the Slider tool, and then click in Graphics 2. Name the sliders h and r, and set their properties so that they run from 0 to 10.

Then create a point A on the y-axis to represent the support point by typing

```
A = (0,h)
```

in the input field. Create another point on the x-axis that will represent the tip of the pen. This point must be able to slide on the table (the paper). In order to create B, you can use the Point tool •A and click once on the x-axis.

You can now make the support visible by typing

```
Segment[A,(0,0)]
```

which will create a segment between A and the origin. Using the Style Bar, add some extra color and thickness, as in Figure 7.2.

You are now ready to create a representation of the pen of length r. You need a segment that goes from B toward A and has the length r. You can construct it with the following sequence of commands in the command field:

`Circle[B,r]`	Create the circle c
`Line[A,B]`	Create the line a
`Intersect[a,c]`	Create the points C and D (assuming that point D is interesting)
`Segment[B,D]`	Create a representation of the pen

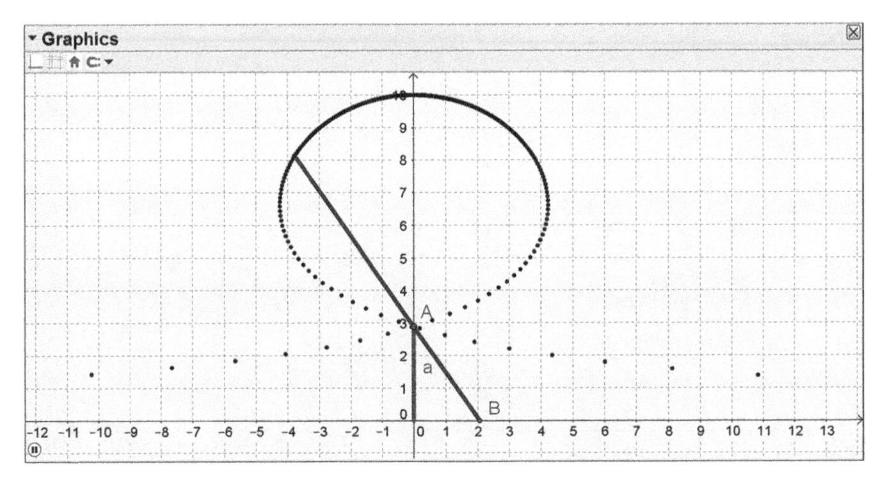

FIGURE 7.3 Loop that is not a circle.

Allow the pen a bit of color and extra thickness from the Style Bar.

You could also have done this with the standard tools for Circle, Line, Intersection, and Segment.

You now have the model ready for exploration. By dragging B, you can make the pen to slide along the support and you can also get a sense of the shape of the loop. In order to see the loop better, you can right-click on D at the top of the pen and select **Trace on**. After that you right-click on B and select **Animation on**.

This curve is known as a *strophoid*. The whole curve can be generated by typing

```
Locus[D,B]
```

The curve, seen in Figure 7.3, is obviously not a circle.

Some similarly interesting and historically famous curves are the *lemniscate*, the *cardioid*, the *cycloid*, the *tractrix*, and the *cissoid* among many others.

7.2 COMPARING AREAS

In Figure 7.4 you see a rectangle that has been divided into four different areas: A1, A2, A3, and a central area. How do these areas relate to each other? Can you show this?

Make your constructions and do your investigations with the help of GeoGebra. Argue for your conclusions and attempt a proof.

Proposed Solution

Create the rectangle using perpendicular lines to the axes, constructed with the Perpendicular Line tool. Next you can create points for the intersections between these lines and between these lines and the axes. Then create the rectangle with the

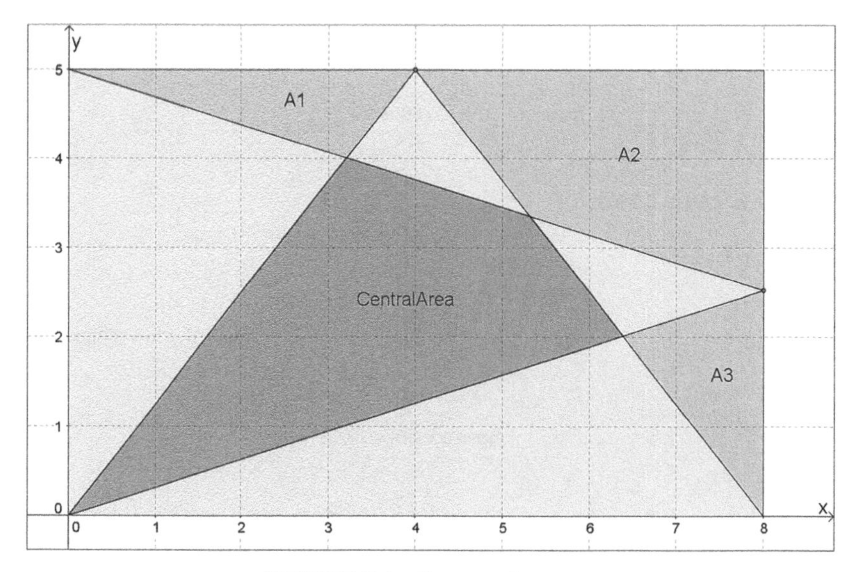

FIGURE 7.4 Compare the areas.

Polygon tool. After this the lines can be hidden. You can add points on the side of the rectangle and also the segments. Now create the areas A1, A2, A3, and the central area with the Polygon tool.

After some experimentation, you might conclude that the sum A1+A2+A3 is equal to the central area. But how can you prove this? By using the Area and Distance tools, you can experiment to find data to build your arguments, shown in Figure 7.5. You might investigate when the two triangles you are not currently using have the same area. Is it when H is on the diagonal?

What happens if the situation is simplified, for instance,when F coincides with A or C? Is it easier to show something then?

In Figure 7.6 you can see one such special case, and here is an outline for a proof for this case.

Let h and b be the sides of the rectangle. Let g be the height in ΔBEG, and let G be the height in ΔADG. You should see that ΔADG ~ ΔBEG (why?), but the base in the small triangle is f = BE/BC of the large triangle. If so, the heights must also relate to each other the same way so that $G \cdot f = g$. The sum of the heights $= h$, so $G = h - g$.

The area of ΔBEG can be written $(g \cdot f) \cdot b/2$. In the same fashion, ΔACE's area can be written $h \cdot (1 - f) \cdot b/2$. The area for ΔBEG and ΔACE together then becomes $g \cdot f \cdot b/2 + h \cdot (1-f) \cdot b/2 = g \cdot f \cdot b/2 + h \cdot b/2 - h \cdot f \cdot b/2 = h \cdot b/2 - (h \cdot f \cdot b/2 - g \cdot f \cdot b/2) = h \cdot b/2 - G \cdot f \cdot b/2 = h \cdot b/2 - g \cdot b/2 = (h - g) \cdot b/2 = G \cdot b/2 =$ the area for ΔADG. Q.E.D.

Similarly, you can do the proof for some other special cases and finally make an attempt to prove the general case. In this case, as in many others, a change of perspective gives us a simpler proof. Looking at the rectangle both vertically and horizontally, and concluding that the "big triangles" have are half the size of the

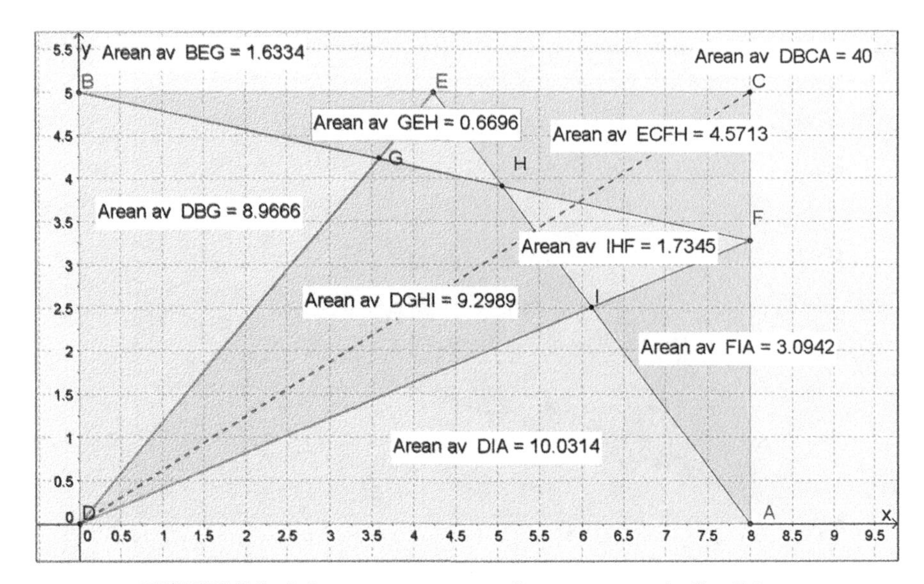

FIGURE 7.5 It is easy to measure and compare areas in GeoGebra.

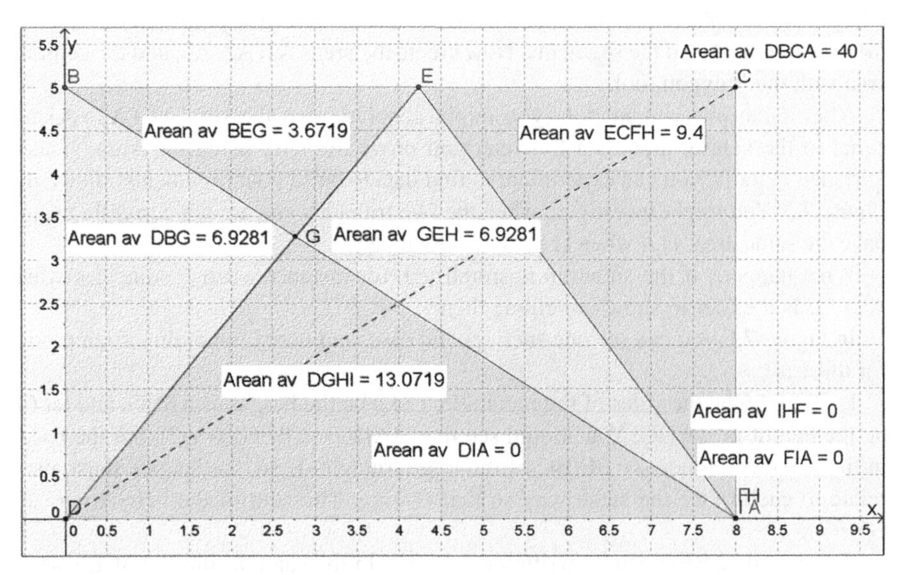

FIGURE 7.6 Special case that might be easier to prove.

rectangles we see that ΔDBG + A1 + A2 + ΔIHF + A3 = ΔGEH + central area + ΔDIA, and that A1 + ΔGEH + A2 + A3 + ΔDIA = ΔDBG + central area + ΔIHF. Adding these together and cancelling like terms gives us immediately that A1 + A2 + A3 = central area.

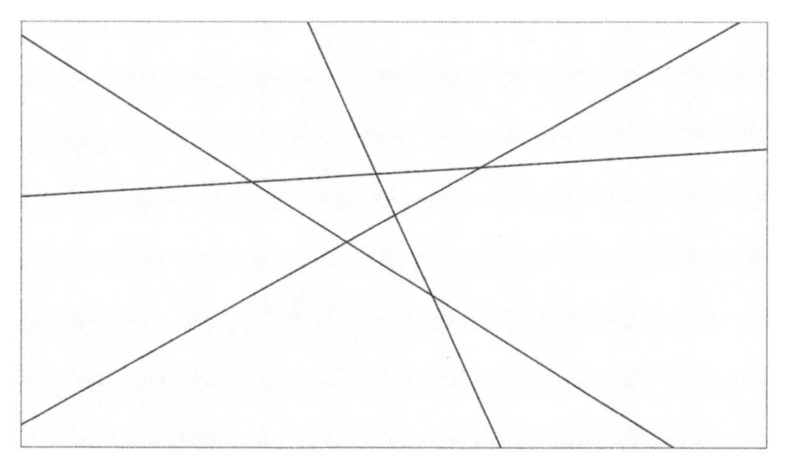

FIGURE 7.7 Four crossing lines form four triangles.

7.3 CROSSING LINES

Four crossing lines form four triangles as in Figure 7.7. Investigate the relationships and properties between the triangles with respect to medians, heights, and the circumscribed and inscribed circles.

Proposed Solution

If you are asked to create circumscribed and inscribed circles as well as other geometrical constructions several times over, it might be well worth simplifying the process. In GeoGebra you have the opportunity to create your own custom tools for constructions you already have done once.

You start by creating one triangle with the Polygon tool and then you create its bisectors with the Angle Bisector tool. After this you create the intersection point of the bisectors and add a line through this point, perpendicular to one of the sides. Last you use the Circle tool with the bisector's intersection point as center and add the intersection point between the perpendicular line and one side as the radius point to complete the construction of the inscribed circle, shown in Figure 7.8. Then hide all extra construction lines.

Since you are going to use this procedure several times, you will now create a tool that will do all this automatically for you. Select the triangle and the circle. In the menu select **Tools > Create New Tool...**, and in the dialogue box select **Circle k** in the List as in Figure 7.9. Then press **Next** or select the **Input objects** tab.

In the Input Objects tab, GeoGebra suggest the points A, B, and C. Accept this for the input choice and select **Next** or select the **Name and icon** tab. Enter "Inscribed circle" and add a help text to help you remember what objects to click on when using this tool. Finally click **Finish**, as in Figure 7.10.

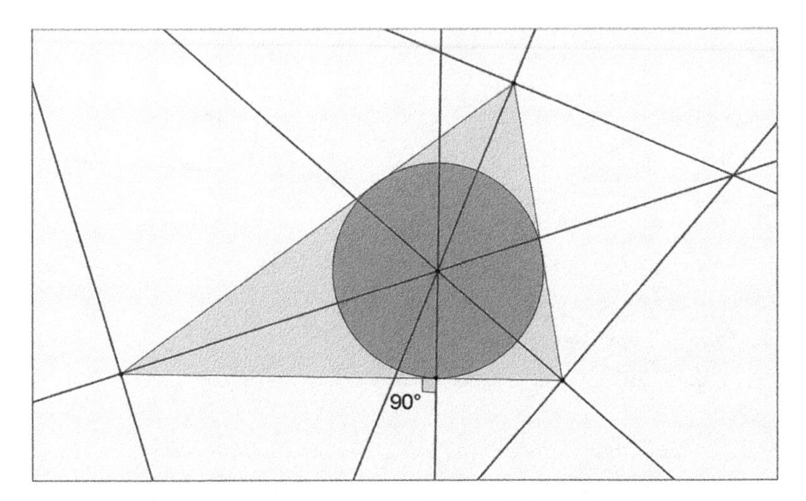

FIGURE 7.8 Bisectors and the inscribed circle of a triangle.

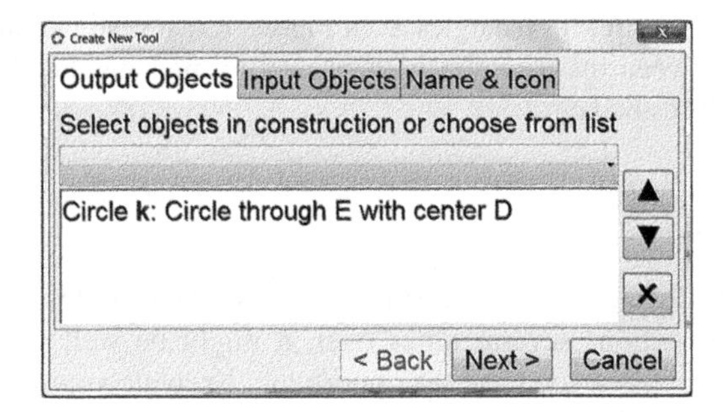

FIGURE 7.9 Create New Tool Dialogue.

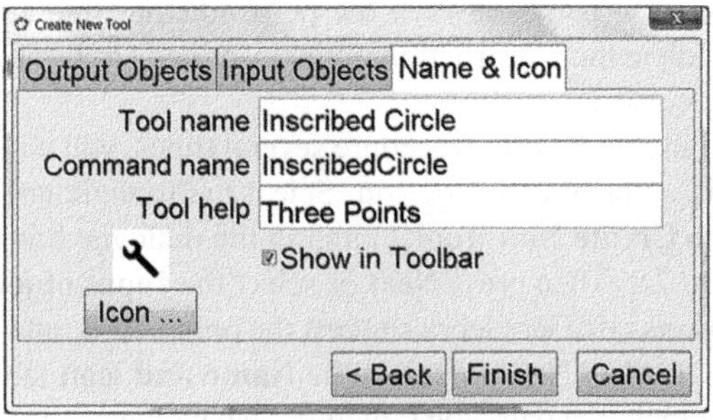

Create New Tool
Output Objects
Input Objects
Name & Icon
Tool name Inscribed Circle
Command name InscribedCircle
Tool help Three Points
Show in Toolbar
Icon ...
< Back
Finish
Cancel

FIGURE 7.10 Remember to write a short help text.

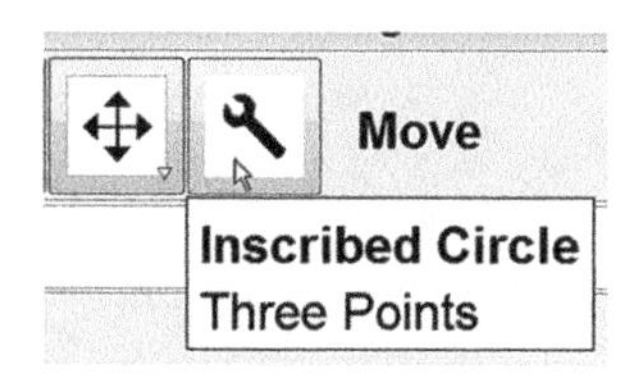

FIGURE 7.11 New tool created.

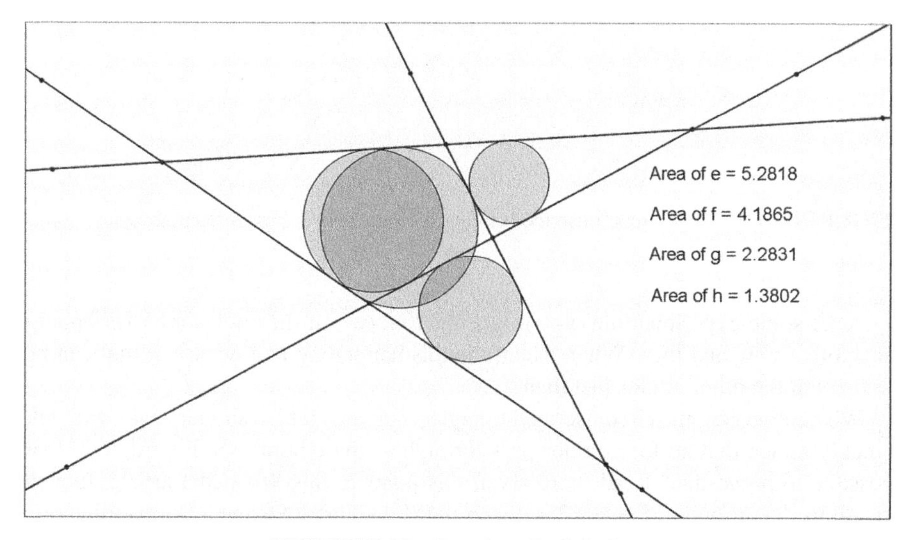

FIGURE 7.12 Four inscribed circles.

You now have a new tool in the tool field and a new command that you can use in the input field: **InscribedCircle[<Point>, <Point>,<Point>]**

This command works just as GeoGebra's other inbuilt commands work.

Your new tool works just like the polygon tool and lies in the toolbar to the right of the other tools; see Figure 7.11. You might click on existing points, or points will be created on the fly where you click. In the same way you might now create tools for the circumscribed triangle or tools for all heights, bisectors, and so forth, in any triangle. Even if you erase all your objects, you will still have your new tools. Read more about how to handle tools in GeoGebra in *Custom Tools*: http://wiki.geogebra.org/en/Custom_Tools.

You can now start investigating the four lines and the triangles they define. First erase all objects you have defined so far. The lines must be constructed with the Line tool and you should place the two control points that define each line close to the edge of the window. Then use the newly created tool to create inscribed circles for each of these triangles. In Figure 7.12 these circles areas have also been measured with the Area tool.

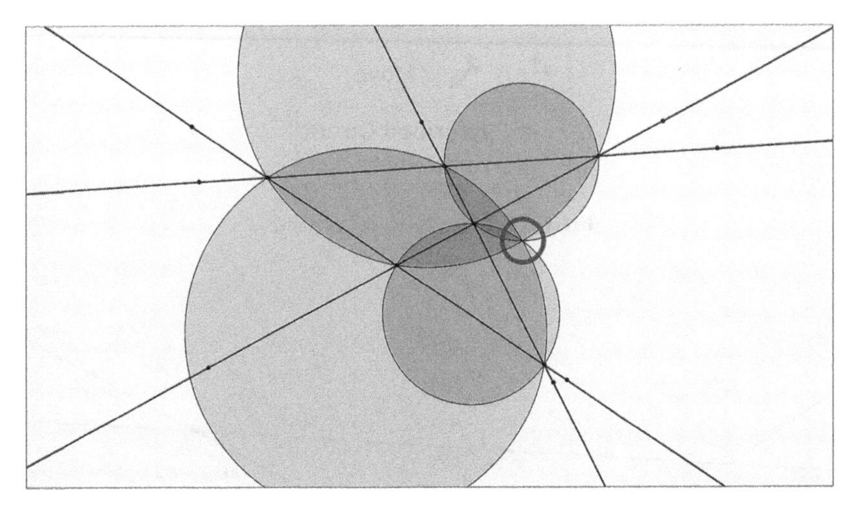

FIGURE 7.13 With four circumscribed circles, all four seem to pass through the same point.

After some experimenting, you might see that two of the circles are overlapping each other now and then. When exactly is this happening and what is it that can be said about the other circles just then?

When observing the circumscribed triangles—created with a similar tool—you will quickly notice that all four circles pass through a single point; see Figure 7.13. Is it possible to prove this? Read more about this point in this Wikipedia article: https://en.wikipedia.org/wiki/Miquel%27s_theorem#Miquel_and_Steiner.27s_quadrilateral_theorem.

In the same way, medians, heights, bisectors, and orthonormals can be investigated. Students can formulate hypotheses about discovered relationships and try to verify or dismiss these by further constructions. As a hallmark of their work, they might just be able to prove the relationships found, or at least look them up, to understand and present them.

7.4 POINTS IN A TRIANGLE

The *centroid*, G (or T) of a triangle, is the common intersection of the three medians. A median of a triangle is the segment from a vertex to the midpoint of the opposite side.

The *orthocenter*, H of a triangle, is the common intersection of the three lines containing the altitudes. An altitude is a perpendicular segment from a vertex to the line of the opposite side. Note that the foot point of the perpendicular may lie on the extension of the side of the triangle. It should be clear that H does not have to lie on the segments that are the altitudes. Rather, H lies on the lines extended along the altitudes.

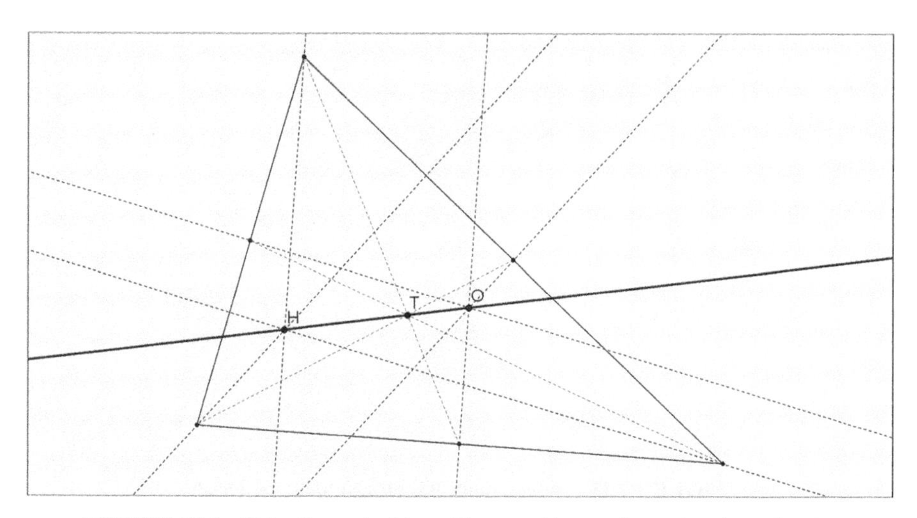

FIGURE 7.14 Euler line coincides with several interesting points in a triangle.

The *circumcenter*, O of a triangle, is the point in the plane equidistant from the three vertices of the triangle. Since a point equidistant from two points lies on the perpendicular bisector of the segment determined by the two points, O is on the perpendicular bisector of each side of the triangle. Note that O may lie outside the triangle. Construct the circumcenter O and explore its location for various shapes of triangles. It is the center of the circumcircle (the circumscribed circle) of the triangle.

The *incenter*, I of a triangle, is the point on the interior of the triangle that is equidistant from the three sides. Since a point interior to an angle that is equidistant from the two sides of the angle lies on the angle bisector, then I must lie on the angle bisector of each angle of the triangle. I is the center of the incircle (the inscribed circle) of the triangle.

Use GeoGebra to construct G, H, O, and I for the same triangle. What relationships can you find among these points? Explore this for many shapes of triangles.

These four points are just the beginning of a long list of points called *triangle centers*. Read more about these at *Encyclopedia of Triangle Centers* at http://faculty.evansWelle.edu/ck6/encyclopedia/ETC.html. Several of these points are associated with different lines and sometimes circles that can be constructed from an arbitrary triangle. One such example is the Euler line, shown in Figure 7.14, which passes through G (or T), H, and O. Also HG = 2 GO.

Now consider Figure 7.15, where an arbitrary triangle ΔABC and a point P are placed so that the sum of the distances from the triangle's vertices to P is as small as possible. Does P lie on the Euler line? What other properties does this point have?

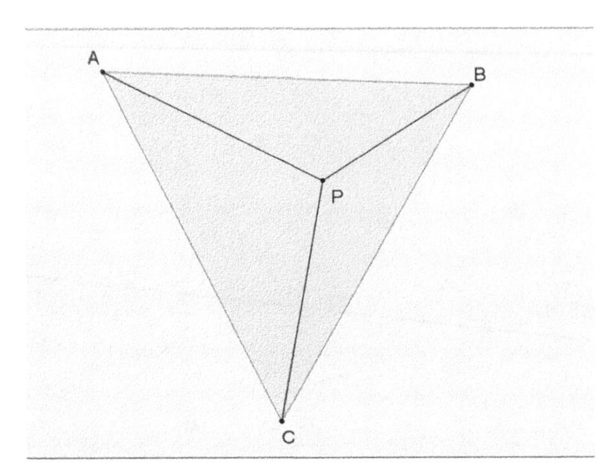

FIGURE 7.15 Which point P is closest to all vertices?

Proposed Solution

Start by creating a triangle with the Polygon tool by clicking on the tool, next on three optional points, and then on the first point once again. Clicking the starting point is the signal that tells GeoGebra that the polygon is finished. After this, create an arbitrary point P inside the triangle with the Point tool. Last create the segments PA, PB, and PC with the Segment tool.

In the Algebra window you can see the lengths of the created segments, but you also need the sum of lengths. So find the sum by typing

```
Sum = d + e + f
```

Now use a text box to create a dynamic text that will show you each segment's length as well as the sum of these lengths. Click on the Text tool, then inside the graphics area where you want the text box to be, and the Text dialogue seen in Figure 7.16 opens.

In the **Edit** box at the top you can type

```
Sum =
```

and then retrieve the objects d, e, f, and the sum from the **Objects** dropdown list. Close the dialogue by clicking **OK**, and the text appears in the graphics view as in Figure 7.17. Dynamic texts can be powerful learning tools because they can contain a mixture of static and dynamic texts that reflect and clarify the properties of other dynamic objects.

You could now move P and see how the sum changes depending on P's position. In order to search for the best position, we advise you to use the arrow keys once P has been selected. Holding down **Shift** while using the arrow keys allows you to increase the precision tenfold by reducing the step size. It might also be a good idea to show more decimals than just the two shown by default in GeoGebra.

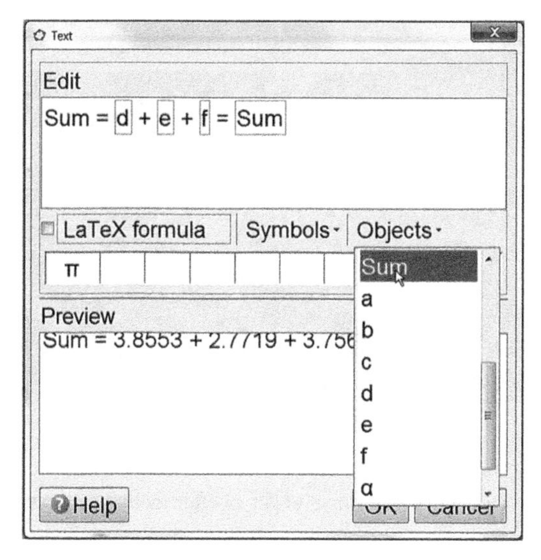

FIGURE 7.16 Creating a dynamic text box.

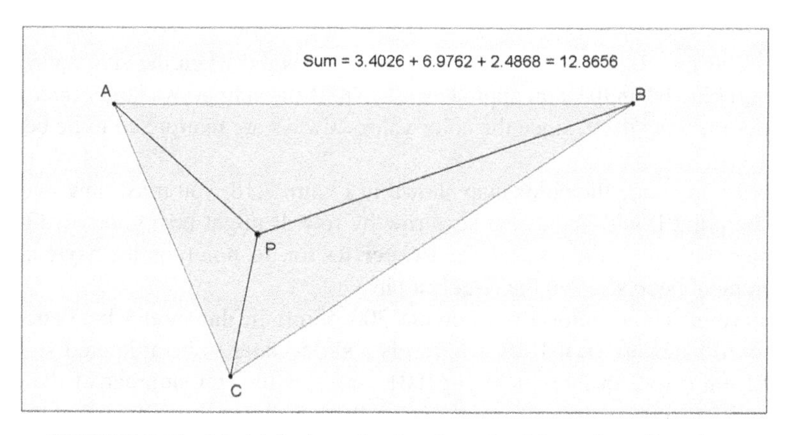

FIGURE 7.17 Model for investigating the point "closest" to all vertices.

When you find the minimal value for P from all vertices, you should try to determine P's properties. From inspection it is clear that P is not the same as any of the previously mentioned points.

A method that, unfortunately, won't help you in this particular case, but which might be valuable in other situations, is to make a *color map*. By right-clicking P and selecting **Trace On**, you allow the point to leave a trace as it moves. In this case it will be hard for you to connect this color to what you are looking for. For this specific triangle the minimum value was close to 12.86. Right-click on P and select **Properties,** then the **Advanced** tab, and type

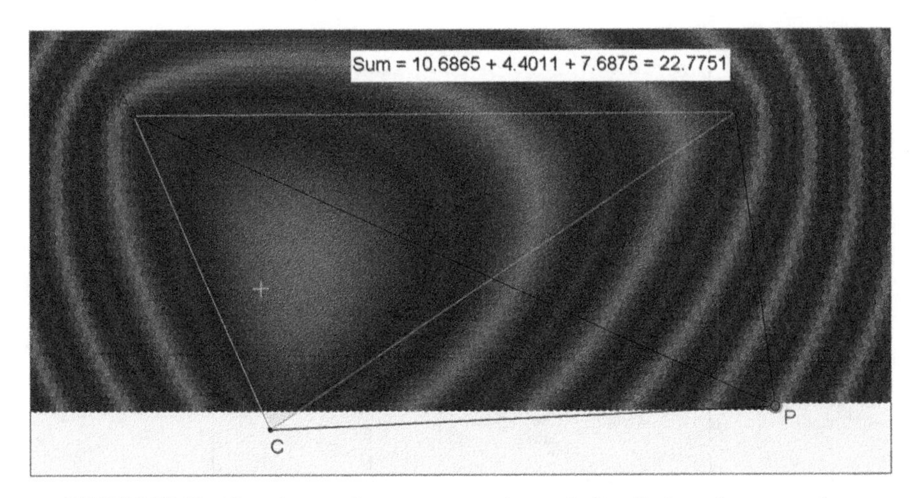

FIGURE 7.18 Drawing a color map to get a better feeling for how the sum varies.

`13.86 - Sum`	in the **Red** field
`0`	in the **Green** field, and
`Sum - 12.86`	in the **Blue** field.

This will give P a *dynamic color* that approaches red when the sum approaches 11.78 and blue when the sum approaches 12.78. If the sum grows larger than 12.78, the pattern repeats itself, since the color values always are interpreted to lie between 0 and 1.

In order to create the color map shown in Figure 7.18, you must now manually move the point P with the arrow keys row by row. It might be a good idea to first determine the size of the point in the **Properties** for the point, on the **Style** tab and the increment (step size) on the **Algebra** tab.

A more efficient method is to create 200 points in the spreadsheet such that their coordinates are (t, n/100), where t is a slider, later to be animated so that it moves from 0 to 2 in increments of 0.01, and n is the row number of the point. This will generate a row of points from 0 to 2 along the y-axis that will all move to the right simultaneously from 0 to 2 along the x-axis as you animate t. If the original point has the correct dynamic color, then so will the rest of the copied points.

Whether the students manage to find the best position experimentally or by Googling probably depends on what kind of students they are. But you can probably expect that some of the students might see that from P, you seem to view the other points 120° from each other. Students should be able to verify this dynamically in GeoGebra as in Figure 7.19 where the Angle tool has been used to measure the angles.

It is relatively simple to show that P does not lie on the Euler line. You could construct the Euler line through the coincident point for the heights and for the bisectors, and so forth, but GeoGebra also has a command that create these points directly.

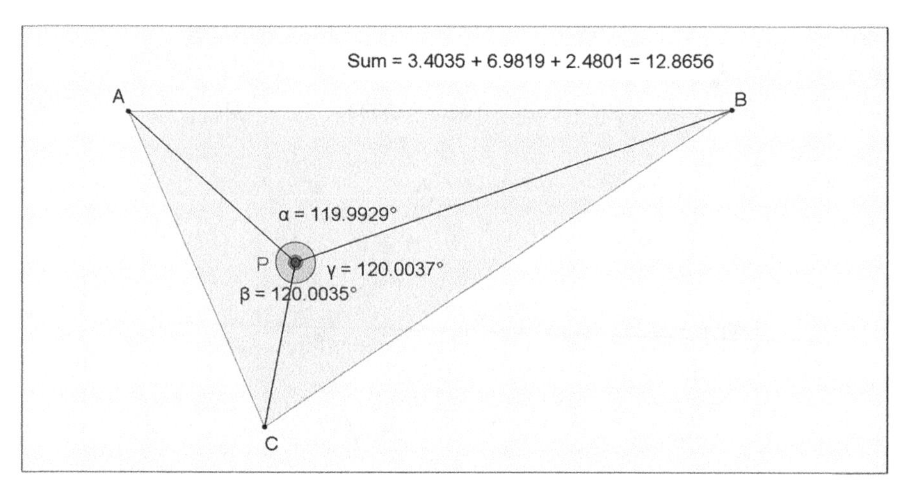

FIGURE 7.19 Angles between PA, PB, and PC that seem to be 120°.

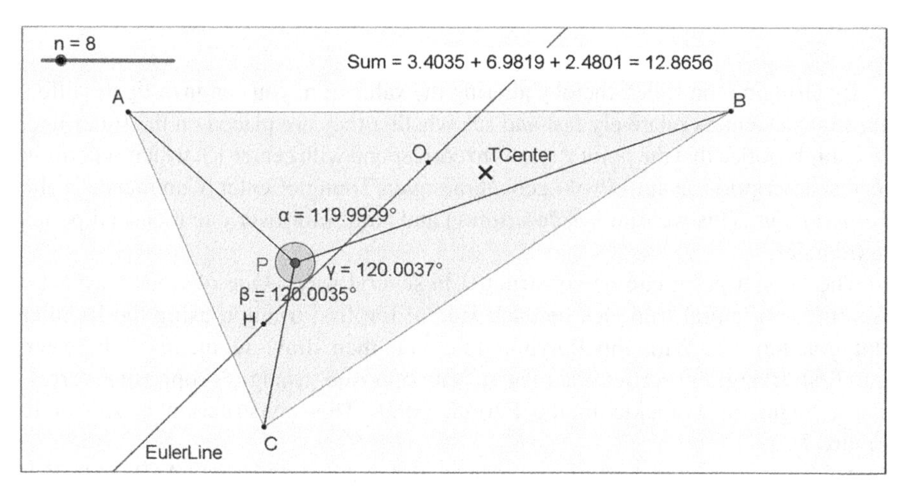

FIGURE 7.20 P does not lie on the Euler line.

Typing the following sequence of commands is probably the fastest way to create the Euler line:

```
O = TriangleCenter[A, B, C, 3]
H = TriangleCenter[A, B, C, 4]
Line[O,H]
```

In Figure 7.20 you also see an integer slider n and a point TCenter. TCenter is defined as

```
TCenter = TriangleCenter[A, B, C, n]
```

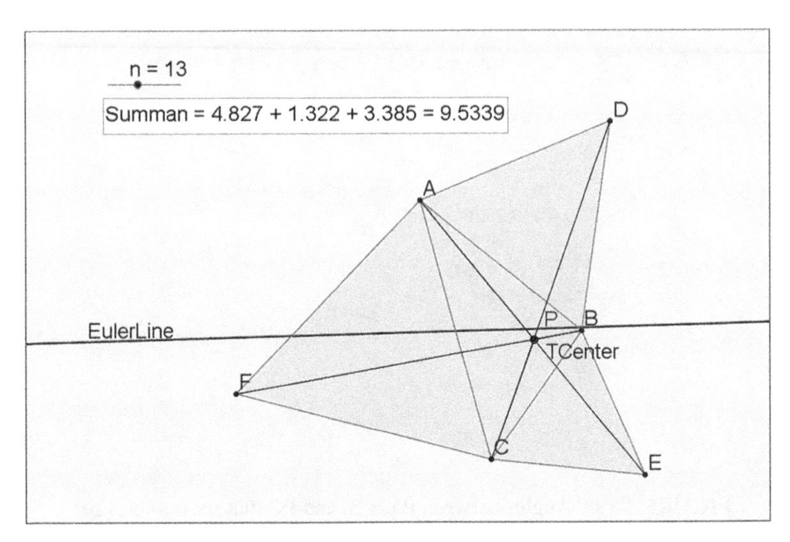

FIGURE 7.21 Constructing the Fermat point.

By changing the slider, thereby altering the value of n, you can investigate different triangle centers relatively fast and see whether they are placed on the Euler line. You might notice that the point P seems to correspond with center 13, which according to the description at http://wiki.geogebra.org/en/TriangleCenter_Command, is the *Fermat point*. This website has description and names for over one thousand points in triangles.

The Fermat point can be constructed in several ways. One obvious way is by creating equilateral triangles on each side of the first triangle using the Regular Polygon tool . With the Polygon tool you then draw segments between the first triangle's vertices across to the opposite triangle's opposite vertex. These segments coincide in the Fermat point. This construction is shown in Figure 7.21.

This construction is dynamic and adjusts if you move any vertex A, B, or C. You also have the foundations for a proof of the Fermat point properties, or at least you can guide your students through some of the proofs at http://en.wikipedia.org/wiki/Fermat_point.

7.5 TRISECTED AREA

Partitioning geometric shapes in different ways has interested mathematicians for a long time, and one of the many, many thought-provoking results can be seen in Figure 7.22. As an introduction to this particular branch of mathematics, we will start by dissecting the area of an arbitrary triangle ΔABC into three parts with equal areas.

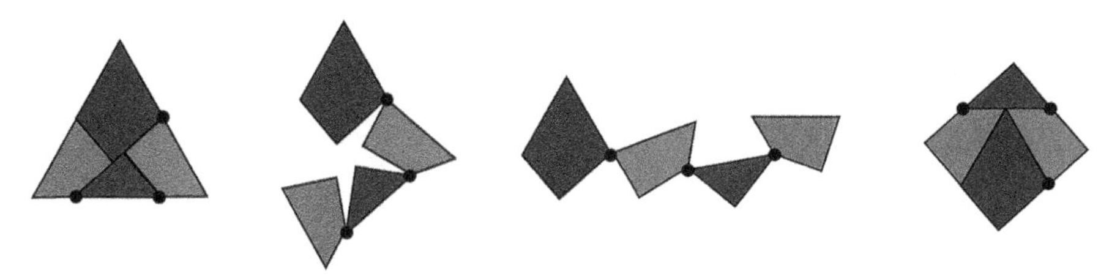

FIGURE 7.22 Dudney's dissection where a triangle can be rearranged to form a square. Photo: Yoni Toker (henry ernest dudeny) [GFDL (http://www.gnu.org/copyleft/fdl.html) or CC-BY-SA-3.0 (http://creativecommons.org/licenses/by-sa/3.0/)], via Wikimedia Commons. https://upload.wikimedia.org/wikipedia/commons/2/2b/Hinged_haberdasher.svg.

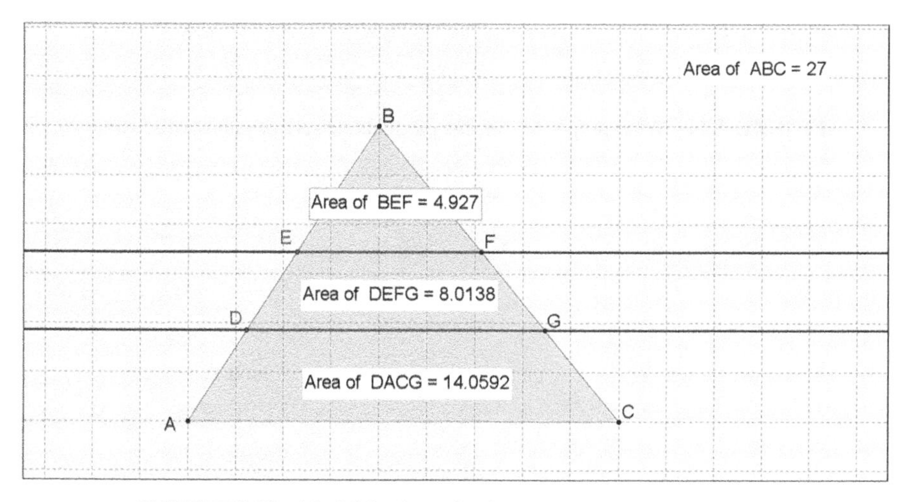

FIGURE 7.23 Model for investigating the trisection of a triangle.

Proposed Solution 1

Start by constructing triangle ABC using the Polygon tool. Next create D and E on the side AB with the Point tool. Using the Parallel Lines tool, draw lines through D and E, parallel to AC. These lines intersect the side BC, and you can then create the intersection points F and G with the Intersect tool.

Now use the Polygon tool again to create the polygons ADGC, DEFG, and EBF. Finally use the Area tool to measure the area of the polygons and the initial triangle (hide one of the polygons temporarily so that you can select the initial triangle).

You can now move the points D and E in Figure 7.23 until you have found the positions that trisect the initial triangle's area. In the properties for the points, on the **Algebra** tab, you change the increment to, for example, 0.001. You can then use the arrow keys on a selected point with good precision. Measure the distances for AD, DE, and EB with the Distance tool and calculate their relation to see if you might discover something.

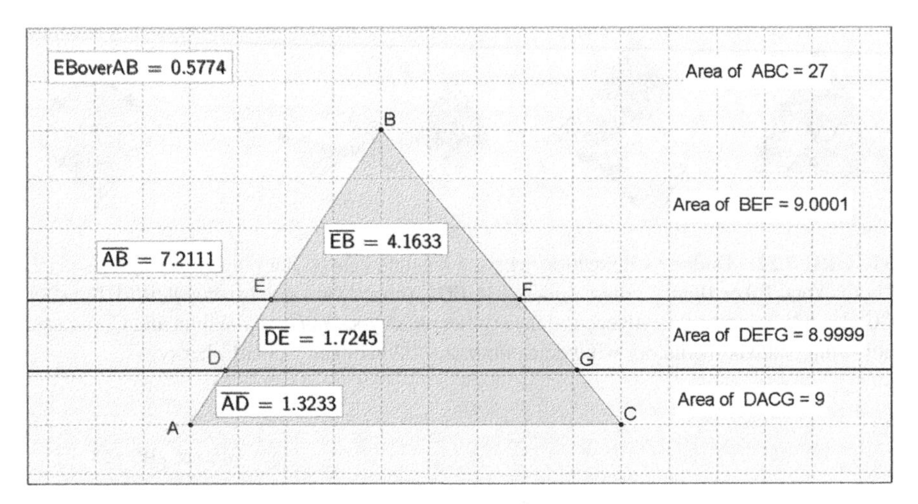

FIGURE 7.24 What properties for this particular trisection can we discover?

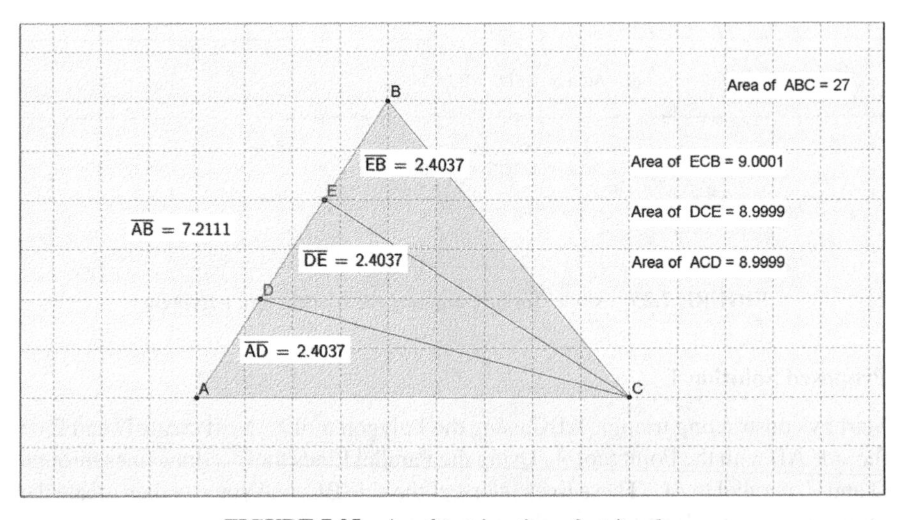

FIGURE 7.25 Another trisection of a triangle.

In the Figure 7.24 the ratio of EB to AB has been measured to 0.5774. If you consider that the problem is about area and that the number 3 or 1/3 probably is involved, it shouldn't take long for you to discover that $\sqrt{1/3} \approx 0.577350\ldots$. You might even verify this conclusion by calculating the ratio DB/AB, which should be $\sqrt{2/3} \approx 0.8165\ldots$. This distance is supported by direct measurement in the construction.

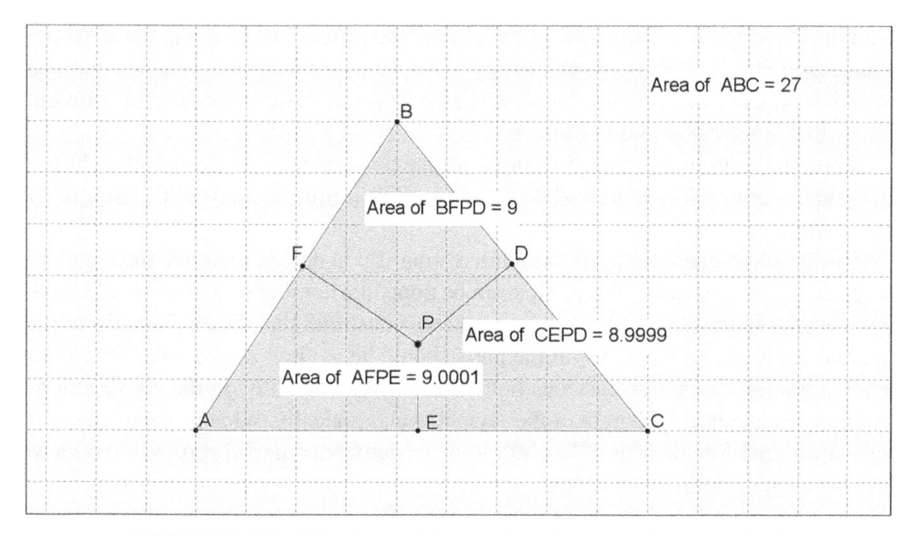

FIGURE 7.26 What is special about P, and what are its properties?

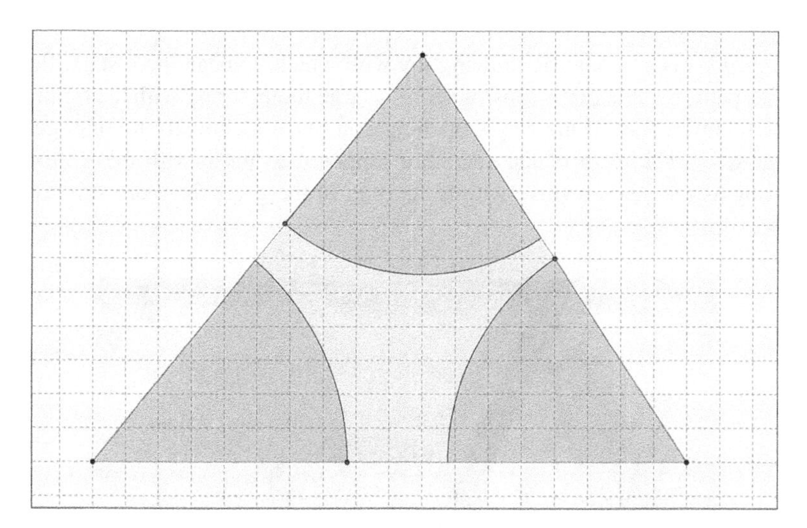

FIGURE 7.27 One of many different ways to divide a triangle into four parts of equal area.

Proposed Solution 2

Another way to partition the triangle is to draw segments from the points D and E directly to the vertex C, then repeating the other steps of the construction from solution 1. You can see in Figure 7.25 that the segments AD, DE, and EB in this case are proportional to the area parts, and since these are equal, so are the segments.

There are, of course, several other ways to divide the area. Select, for instance, a point P in the interior of the triangle and draw lines from P perpendicular to each side with the Perpendicular Line tool . Determine the position of P so that the areas are equal, as in Figure 7.26. Is P equal to any of the other points inside triangles that you have investigated in Problem 7.4, *Points in a Triangle?* Are there several different points that correspond to this solution?

More suggestions on how a triangle might be trisected can be found at http://jwilson.coe.uga.edu/EMT668/EMAT6680.2000/Lehman/emat6690/trisecttri's/trisect.html.

This problem becomes even more interesting if you decide to divide the area into four equal parts instead of three. This can be done in a number of ways, one of which is shown in Figure 7.27; you could even add the constraint that the total lengths of the internal segments you use to split the parts should be as short as possible. This gives many opportunities for discussions both about how to actually do the constructions and how to calculate the length of the segments algebraically. What group of students will win this math competition? See http://www.baumanneduard.ch/EqAreaOverview.htm for a list of optimal solutions.

7.6 SPIROGRAPH

A *spirograph*is is a geometric drawing toy with which you can draw stars, flowers, and other patterns. It is often constructed as a large toothed ring with teeth lining the inside circumference. In the ring's opening you place identically toothed gears of different sizes. The teeth of the gears are supposed to mesh with the teeth of the inside ring as the gears revolve within the ring. In the gears there are several holes

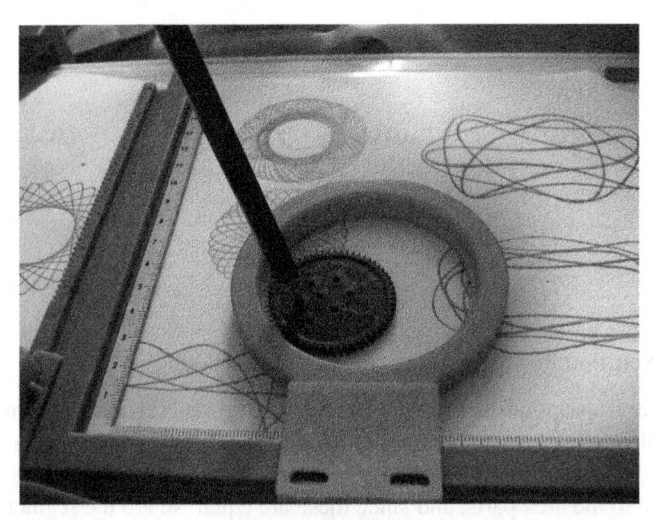

FIGURE 7.28 One example of a spirograph. Photo: Kungfuman (own work) [GFDL (http://www.gnu.org/copyleft/fdl.html) or CC-BY-SA-3.0 (http://creativecommons.org/licenses/by-sa/3.0/)], via Wikimedia Commons. https://upload.wikimedia.org/wikipedia/commons/9/97/Spirograph3.jpg.

drilled at different positions. In one of the holes, you place a pen. Moving the pen will force the gear to spin around at the same time as the pen draws a specific pattern, which will be different for each combination of hole and gear. An example of this toy can be seen in Figure 7.28 and some typical patterns are shown in Figure 7.29.

FIGURE 7.29 Patterns produced with a spirograph.

Construct a model of a spirograph in GeoGebra. Use the model to investigate the mathematical properties of the patterns. For instance, you can investigate what properties of the gear decide the number of flower petals or number of star tips of the pattern, or what radii the inscribed or circumscribed circles have.

Proposed Solution

The spirograph may be constructed using only geometry, without any algebra or trigonometry. Since the construction consists of many steps, we will show them all at once in the spirograph's *construction protocol*, shown in Figure 7.30 in order to make the construction clearer.

Construction Protocol - 2_15_autospirograf.ggb

No.	Name	Toolbar Ic...	Definition	Command	Value
1	Number t	a=2			t = 10.48
2	Number R	a=2			R = 5
3	Number r	a=2			r = 2.4
4	Number a	a=2			a = 0.7
5	Point A		Intersection point of xAxis, yAxis	Intersect[xAxis, yAxis]	A = (0, 0)
6	Circle c		Circle with center A and radius R	Circle[A, R]	c: $x^2 + y^2 = 25$
7	Point B		Intersection point of c, xAxis	Intersect[c, xAxis, 2]	B = (5, 0)
8	Point B'		B rotated by angle t	Rotate[B, t, A]	B' = (-2.47, -4.35)
9	Circle d		Circle with center B' and radius r	Circle[B', r]	d: $(x + 2.47)^2 + (y + 4.35)$...
10	Point B"		B' mirrored at A	Reflect[B', A]	B" = (2.47, 4.35)
11	Segment b		Segment [B", B']	Segment[B", B']	b = 10
12	Point C		Intersection point of d, b	Intersect[d, b, 1]	C = (-1.28, -2.26)
13	Circle e		Circle with center C and radius r	Circle[C, r]	e: $(x + 1.28)^2 + (y + 2.26)$...
14	Point B"$_1$		B' rotated by angle -(t R / r)	Rotate[B', -(t R / r), C]	B"$_1$ = (-0.44, -0.01)
15	Ray f		Ray through C, B"$_1$	Ray[C, B"$_1$]	f: -2.25x + 0.84y = 0.98
16	Circle g		Circle with center C and radius a	Circle[C, a]	g: $(x + 1.28)^2 + (y + 2.26)$...
17	Point D		Intersection point of g, f	Intersect[g, f, 1]	D = (-1.04, -1.61)
18	Segment h		Segment [C, B"$_1$]	Segment[C, B"$_1$]	h = 2.4
19	Locus Geometri...		Locus[D, t]	Locus[D, t]	GeometriskOrt1 = Locus[...

FIGURE 7.30 Spirograph's construction protocol.

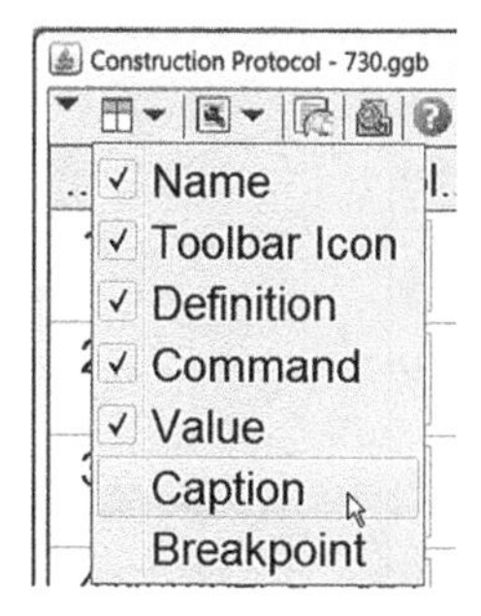

FIGURE 7.31 Selecting which columns to show.

GeoGebra always creates a construction protocol while you are working, and you can always show it with **Ctrl-Shift-L** or by selecting **View > Construction protocol** from the menu. The protocol shows what objects you have created but not the properties of the objects, such as the minimum and maximum values for sliders, or if objects are hidden or visible. In this construction protocol all available columns are shown except for the **Caption** column. This is done from the tool box inside the construction protocol as shown in Figure 7.31. This is an excellent way to study how a construction has been built. The **Caption** column is initially empty but can be filled with your own notes.

As you can see, you can create almost anything with tools and commands. The tools used display their tool icons. We will now explain the different steps in this particular construction.

In **steps 1–4** we create four sliders (t, R, r, a), where *t* represents the time but also the angle that the contact point between the gear and the inside ring has moved. R is the radius for the ring while r is the radius for the gear and finally the slider a represents the distance from the center of the gear to the pen. The time (or angle) *t* is set to vary from 0 to 100 (radians), while the other sliders are set to vary from 0 to 10.

In **steps 5–6** we have used the Circle with Center and Radius tool to construct the circle that represents the circular frame of the spirograph. Our first click on the origin creates the center A as the intersection between the coordinate axes. When the dialogue box asks for the radius for the circle, we enter "**R**."

In **step 7** we create a point where the circle intersect the positive *x*-axis.

In **step 8** we rotate this point an angle *t* around the circle's center. This will create the point B′. If we change the value for *t*, the point B′ will move.

In **step 9** we create a help circle by typing

```
Circle[B′,r]
```

In **steps 10** and **11** we create a diameter in the large circle.

In **step 12** we find the center of the gear as the intersection between the diameter and the help circle.

In **step 13** we create the circle that represents the gear and hide the help circle. Remember that hiding objects or changing their properties does not show in the construction protocol. The construction up to this point is shown in Figure 7.32.

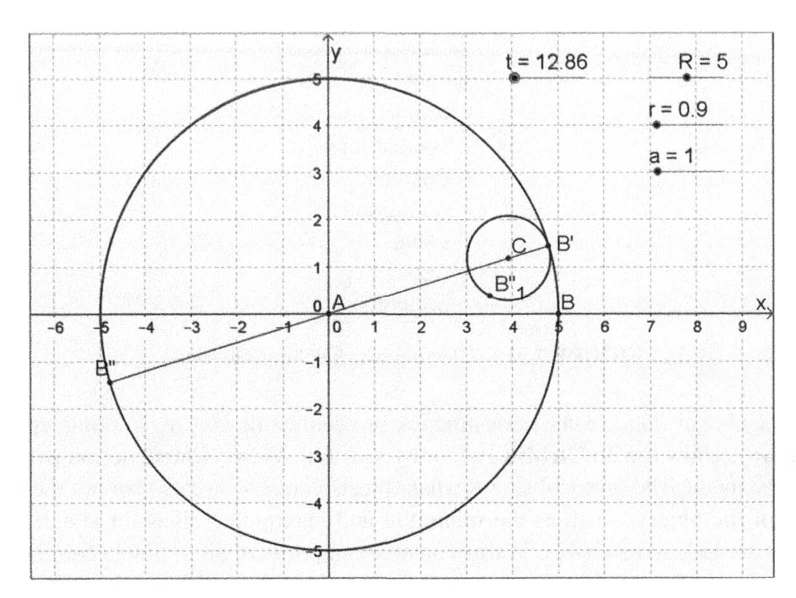

FIGURE 7.32 Spirograph model halfway finished.

Step 14 demands some thought. To create the point B''_1, representing the initial contact point's new position on the gear, we must imagine that the gear is rolling without sliding inside the ring. This new point should be rotating clockwise from B′, but with what angle? We see that the angle is larger than *t*. When the gear has moved one lap around within the ring and point B′ has gone exactly one lap around the gear, then the point B''_1 has rotated a number of laps equal to the relation between the ring's and the gear's perimeters or equally, between their radii, so B''_1 will rotate the angle $(t \cdot \mathrm{R})/\mathrm{r}$.

We must now find the point where the pen's tip will be drawing:

In **step 15** we will construct a ray from the gear's center through the point B''_1.

In **step 16** we will create a circle with the same center as the gear, but with the radius a.

In **step 17** we finally create point D as the tip of the pen.

We will now clean up a bit. We hide the ray and the circle from steps 15–16, and we color the pen, or point D, red. We right-click on D and select **Trace on**. Last we select the slider *t* and adjust the value with help of the arrow keys. In Figure 7.33, we show the pattern being drawn. If we want to erase the trace, we press **Ctrl-F**.

Even if the spirograph now is working, it will be uncomfortable to manually draw the pattern every time. So we finally create, as in **step 19** the locus for point D when *t* varies. We elect to show the locus as a solid green line. We also clean up a little more by hiding everything in the construction except the sliders and the locus.

The construction is now completely finished as in Figure 7.34, and you may start analyzing it. The number of flower petals does not vary with a, so let r to vary from 0.1 to 5 when R=5 and count the number of flower petals. Some of the values are listed in Table 7.1.

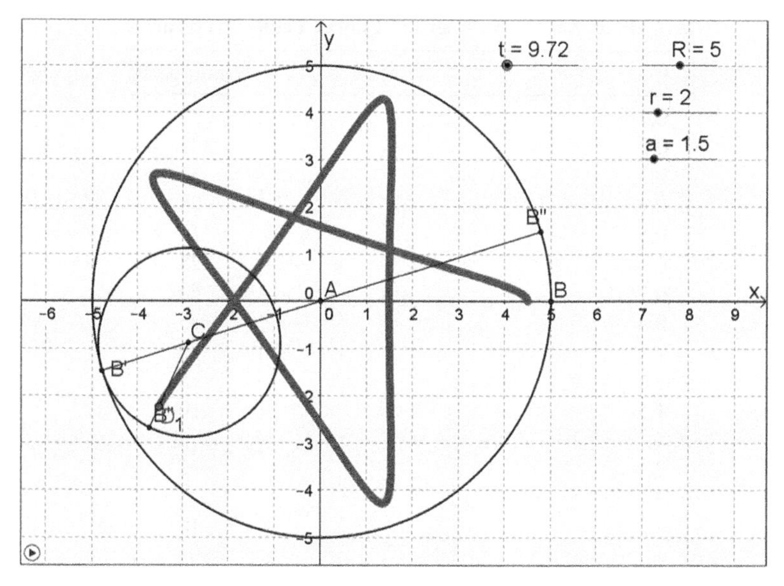

FIGURE 7.33 Spirograph in action.

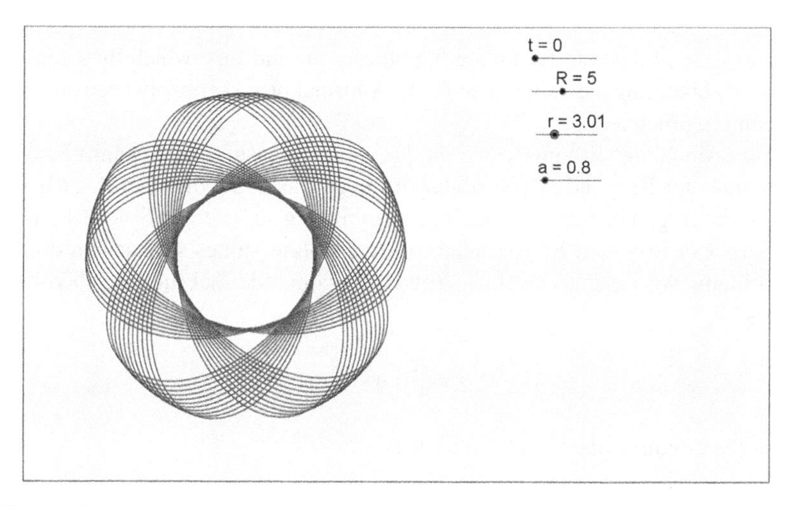

FIGURE 7.34 Auto-spirograph generating a pattern directly when the sliders are updated.

The values oscillate greatly both up and down, but they all seem to be divisible by 5. It is probably no use trying to find a function that describes the number of flower petals as r varies. Instead, by experimenting a little more, you may discover that the number of flower petals depends on the divisibility between circles' radii. More precisely; the number of flower petals is equal to the denominator in the ratio r/R when it is expressed as a rational number in its simplest form.

TABLE 7.1 Number of "Flower Petals" as r Varies

r	Flower petals
0.1	50
0.2	25
0.3	50
0.4	25
0.5	10
0.6	25
0.7	50
0.8	25
0.9	50
1.0	5
1.5	10
2	5
2.5	2
3	5
3.5	10
4	5
4.5	10
4.6	25

Allow your students to formulate hypotheses around this, which they can verify later and by changing the values also for R. A formal proof probably needs to include angles and perimeters.

The inscribed and circumscribed circles can be measured and examined for different values for R, r, and a. The center of the gear is at the distance R – r from the center of the ring. The pen is at distance a from the gear's center. Since r + a may be either larger or less than R, you need to use absolute values so that you don't run into problems with negative values. You could conclude that the inscribed circle's radius is

$$r_i = ||R - r| - a|$$

and that the circumscribed circle's radius is

$$r_c = |R - r| + a$$

7.7 CONNECTED LP PLAYERS

At http://roberthowsare.com/rational-aesthetics/drawing-apparatus/ there is a video showing two LP players connected by a linkage to create a homemade harmonograph. The drawing apparatus is pictured in Figure 7.35.

Produce a model of this system in GeoGebra that will recreate the pattern of the pen.

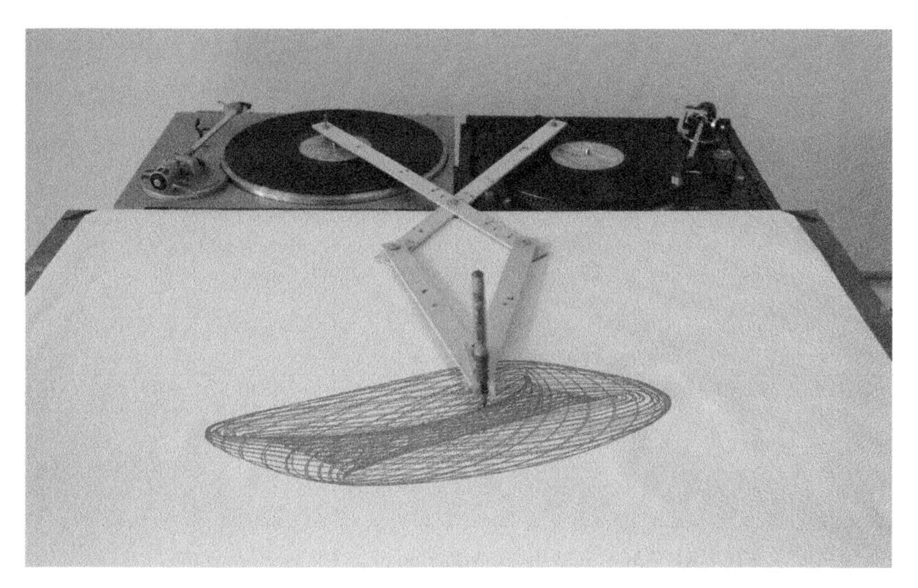

FIGURE 7.35 Mechanical harmonograph. Photo: © Robert Howsare. Reprinted with permission of the Photographer.

For simplicity you may set the lever's lengths in the model to be 3 and 1 LP radii, respectively, and let the long levers be connected at a distance of 2 LP radii from their anchor points on the LP players. Also position the two LP players so that their centers lie at a distance of 2.5 LP radii from each other.

The patterns that will occur are partly based on the fact that the LP players are rotating at slightly different speeds, so your model must allow these rotation speeds to be finely tuned.

Proposed Solution

This is a similar to the spirograph in Problem 7.6 but built on a linkage system. In order to create these links in GeoGebra, you need to understand the important nature of a link: it always has the same length. So to create links in GeoGebra, you use circles where the radii represent the links' lengths. A point that slides on the circle's circumference and a point in the circle's center will be the link's end points.

Even if you won't be showing the axes and grid, when the construction is finished, you should keep them visible now so that points will be easier to place exactly. Start with the construction of the two LP players, letting both LP players have the radius = 2.

First use the with Center and Radius tool to create two circles, both with radius = 2 and with centers in (0, 4) and (5, 4). Next use the point tool to create two points E and F that will slide on each circle, as in Figure 7.36. As long as you place these points on the circles, they will be locked to each circle.

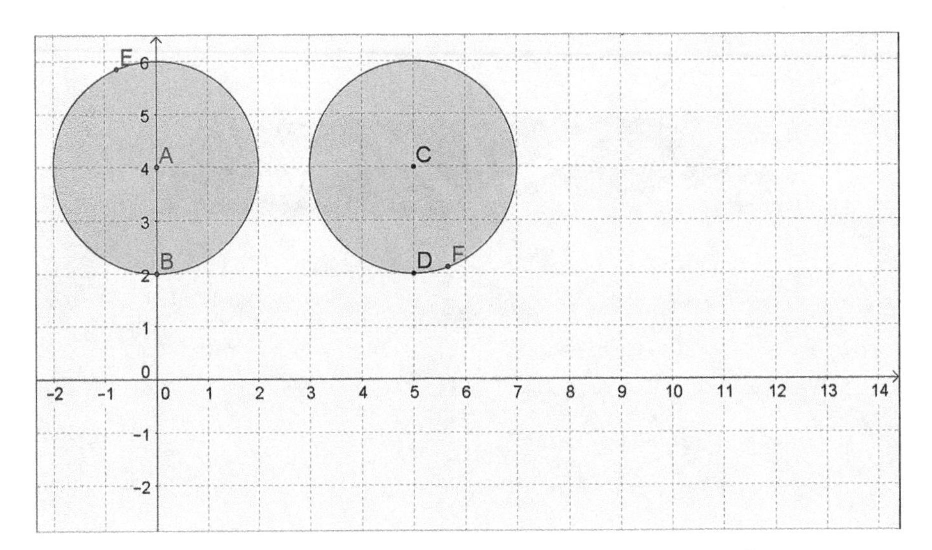

FIGURE 7.36 Start the construction of an LP harmonograph.

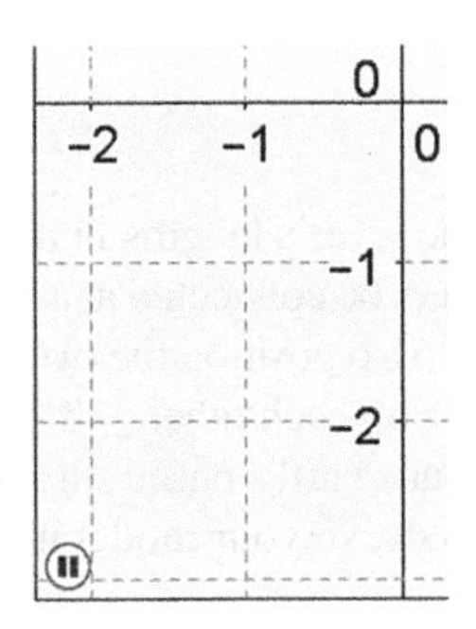

FIGURE 7.37 When something is animated, a pause/play-button appears in the lower left of the Graphics window.

You might already try to animate the points by right-clicking the points E and F one at a time, selecting **Animation on**. They will now rotate around the circles. In order to carry on with the construction, you must stop the animation, which may be done in two different ways: either right-click on the point's representation in the Algebra window and turn the animation off or use the pause button that has appeared in the lower left of the Graphics window, seen in Figure 7.37.

In the video, the links are connected to the LP players at different distances from the center. Therefore create a segment from the center to the animated point for each LP player and create a point for each LP player on these two segments.

With these two new points as centers, you now create two circles with radius 4 using the Circle with Center and Radius tool. Also create these two circles' intersection point in Figure 7.38. The circles you can hide later on. Right now this intersection corresponds to the first intersection of the linkage.

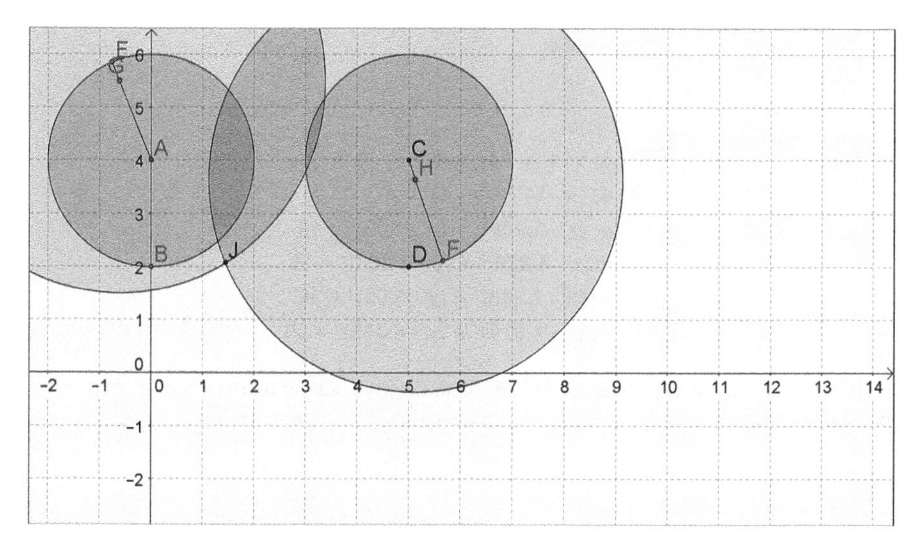

FIGURE 7.38 Point J as equally far away from G and H.

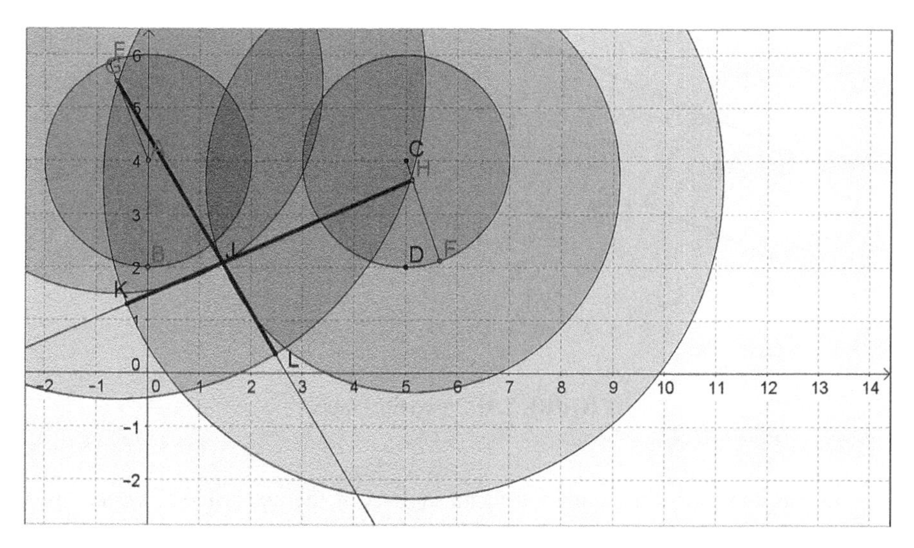

FIGURE 7.39 First links finished.

These two first links should have the length=6. In order to be able to construct their end points, you need two circles with the same center as before, points G and H, and you need two rays from G and H through J.

The intersection points between the rays and the circles with radius=6 are the links' end points. Represent the links with segments between G and L and between H and K. Color them and make them thicker as in Figure 7.39.

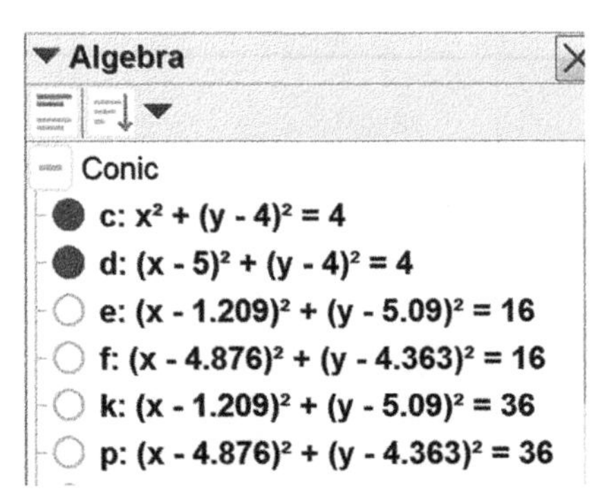

FIGURE 7.40 Each object can be hidden or shown by clicking on the visibility buttons in the Algebra window.

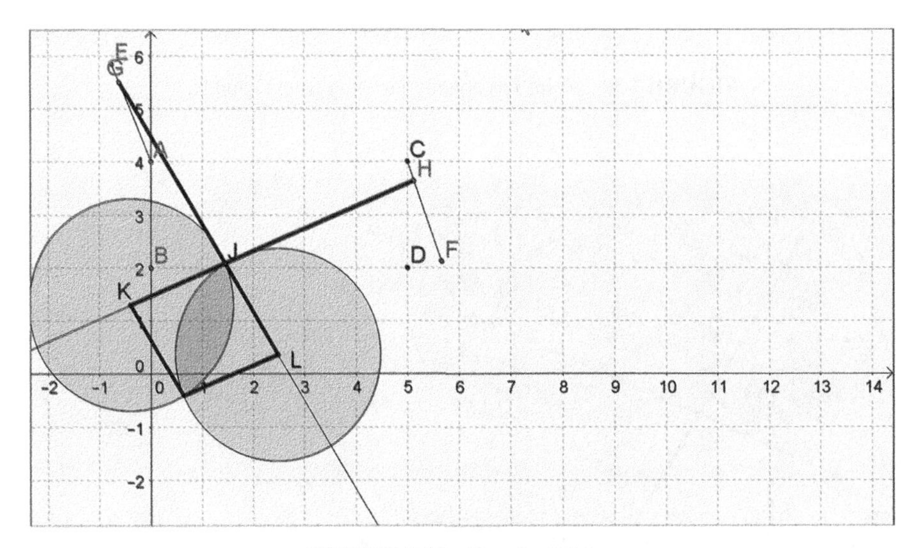

FIGURE 7.41 Two final links.

You can now start the animation again and see if the construction works; as a matter of fact it makes a lot of sense to do this quite often during the construction to check your progress.

Now hide the help circles by clicking on the blue visibility buttons in the Algebra window, shown in Figure 7.40, so that the buttons become white.

The two final links are created in the same way as before. You want them to have the lengths equal to the LP radius, so create two circles with centers in K and L with radius=2: the pen is positioned where these two circles intersect, at N in Figure 7.41.

Hide the last help circles and any points that you do not need to see. Transform N to a pencil by right-clicking on it and selecting **Trace on**. Start the animation with the

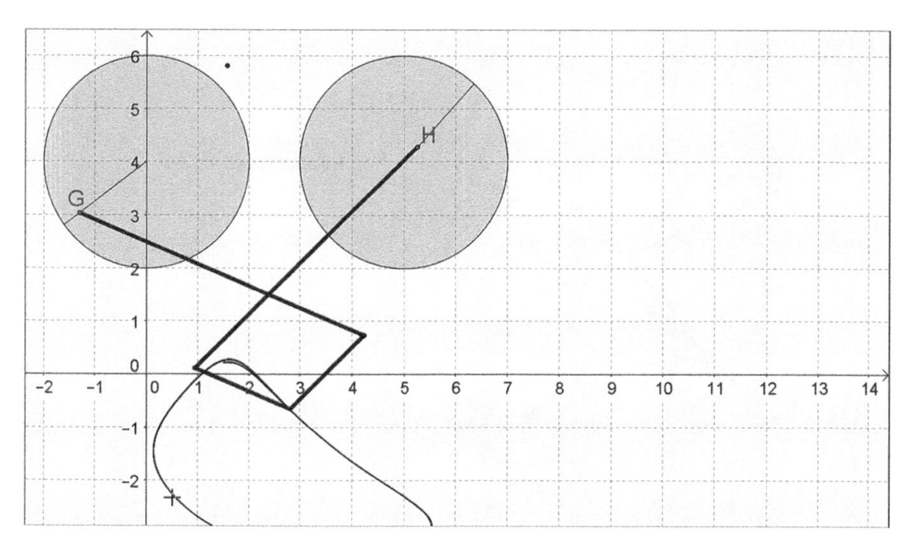

FIGURE 7.42 Model of the linkage that is almost finished.

start button down to the left in the Graphics window; now the construction, shown in Figure 7.42 seems to work. Some adjustments need to be done, however.

You need to adjust the LP players so that their speed is slightly different. You do this by creating two sliders: **Speed** that goes from 0 to 10 with an increment of 0.1 and **RelativeSpeed** that goes from 1 to 2 with an increment of 0.001. Then open the properties dialogue for the points E and F, which are the points that have been animated.

On the **Algebra** tab, in the **Animation Speed** field, type

`Speed` for point E
`Speed RelativeSpeed` for point F

Now you can adjust the speed of the LP players in detail with the sliders. Observe that when the animation speed can be controlled by sliders and stopped by adjusting the sliders to 0, the stop/play-button in the lower left disappears in the Graphics window. Several command fields accept variables, including min, max, and increment values for sliders.

Adjust the thickness of the pen, point N, to 1. Next select both E and F in the Algebra window and start the harmonograph by right-clicking them, selecting **Animation on**, and see the pattern that appears as shown in Figure 7.43.

With the Locus tool you can make different patterns by creating the *locus* of N when E and F move along their path.Click the Locus tool, next N, E, and then N again and finally point F. Moving points are easiest to click in the Algebra window. The result is seen in Figure 7.44.

Try to change the end point distances from the LP players' centers and see how the pattern changes.

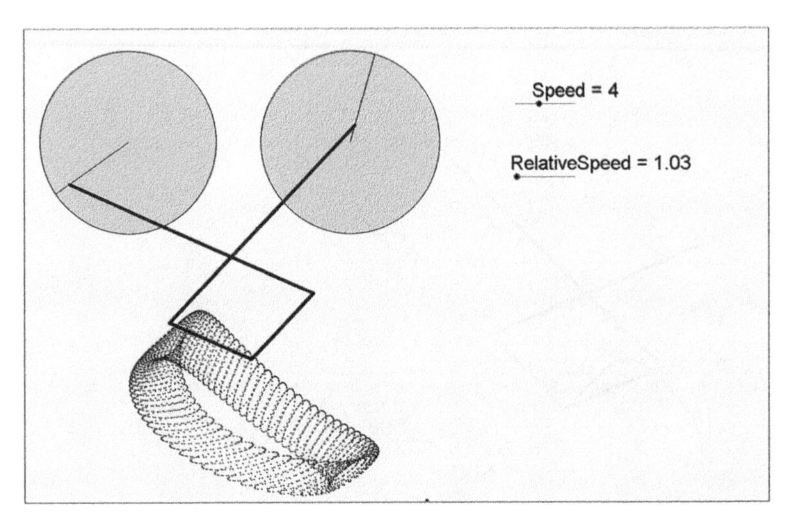

FIGURE 7.43 One of several possible patterns created by the LP-harmonograph simulator.

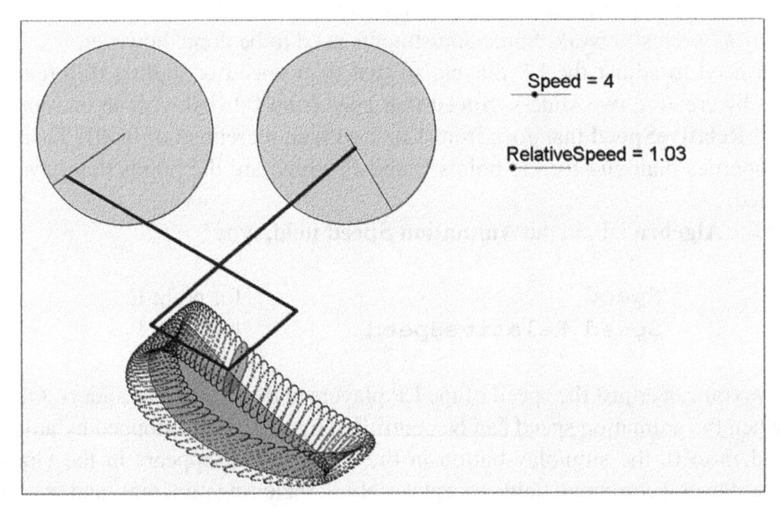

FIGURE 7.44 Locus demonstrates the patterns that would occur if one of the LP players were standing still.

7.8 FOLDING PAPER

An A4-size paper is folded so that one corner lies on the long edge of the paper. A small right-angled triangle is formed in the corner as seen in Figure 7.45. How can you maximize the area of this triangle?

FIGURE 7.45 Folding paper. Photo: Jonas Hall.

Proposed Solution 1

Give each student their own paper and have them select their own way to fold the paper to maximize the area of the triangle. Then ask everyone to carefully measure the base and the altitude of their triangle and calculate the area. Ask them to share their results in a spreadsheet and make a scatterplot of these data. What size base seems to give the maximum area?

Presenting the problem in this way will make it hands-on for the students and will also give some preliminary results against which you can compare later investigations.

Proposed Solution 2

GeoGebra can be used both to create a geometric model of the paper to be folded and to make measurements in this model and then analyze the generated data. You begin by creating the model in the graphics area, using the coordinate axes to line up the paper. Typing

```
x = 29.7
y = 21
```

you create an area representing the A4 sheet of paper, with dimensions 297 by 210 mm.

Using the Polygon tool, you click in turn on the intersections where the four vertices lie. Click once more on the first intersection where you started to tell GeoGebra that you have finished. You may now, if you wish, hide lines and coordinate axes.

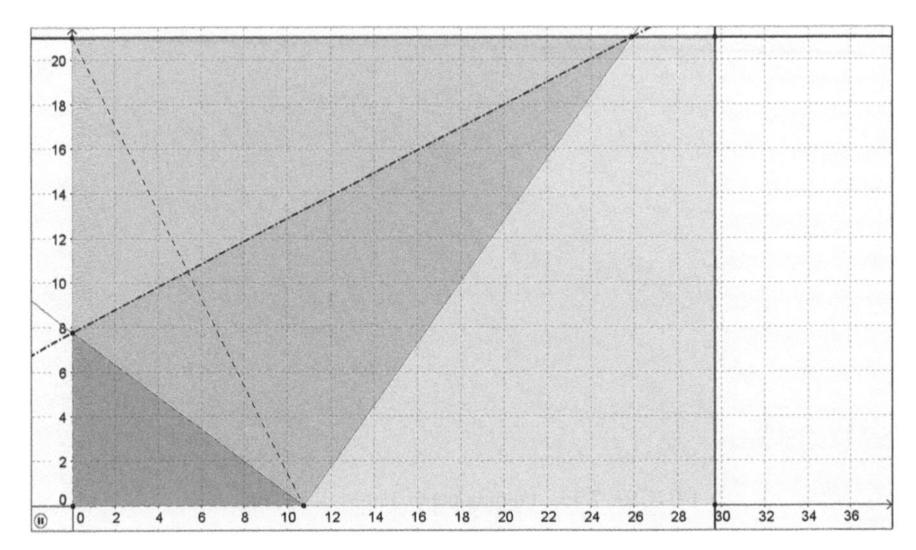

FIGURE 7.46 Our model of the folded paper.

Next use the Point tool on the lower edge to represent the top left corner's position after the fold. Connect the top left corner with this point using the Segment tool. The crease, or fold, itself will be a bisector of this segment, so select the Perpendicular Bisector tool and then click the connecting segment to create the bisector.

You need to be able to use the point where the fold meets the short edge of the paper, that is, where the bisector intersects the edge. To make the intersection so that it can move round all edges of the paper, you need to create the intersection with the polygon, rather than with just one of its edges. Therefore type:

```
Intersect[f, Polygon1]
```

This will generate the two points that you need. Use these points to generate a new polygon to represent the part of the paper that is being turned over, which is the top left triangle of the paper. Then select the Reflect about Line tool, and click this polygon and the bisector fold line to create the illusion of actually folding the paper. Color your polygons so that the different parts can easily be discerned; see Figure 7.46 for an example.

To measure the area of the lower right triangle, you could create a polygon for this area as well. The value given for Polygon3 in the Algebra window is the area you need, and the model is now finished. Make sure it behaves well, and like a real piece of paper, when sliding the point along the lower edge.

Now open Graphics Area 2 by pressing **Ctrl-Shift-2** or selecting **View > Graphics 2** from the menu. Click once in Graphics 2 to make sure it is active and then create a point there by typing

```
(x(E), Polygon3)
```

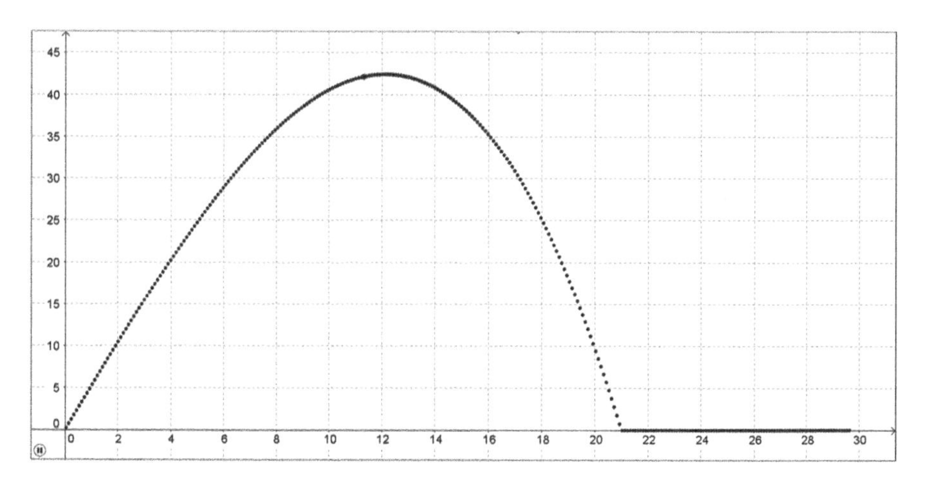

FIGURE 7.47 Trace of our measurements showing a clear maximum.

in the input field, where E is the moving point representing the folded down corner, and Polygon3 is the area of the lower left triangle you are asked to maximize. Right-click on this point in Graphics 2 and select **Trace On**. Then right-click the corner moving on the edge, point E in Graphics 1, and select **Animation On**. This is a typical example of how to build a geometrical model in the regular Graphics window and then analyze it in Graphics Area 2. The trace is shown in Figure 7.47.

You can find a maximum when the lower left triangle has a base of approximately 12 cm. The area is then some 42 cm^2. To get a more accurate value, you have to make a model of the results. The trace is just a set of visual points not remembered by GeoGebra, so the points cannot be used for calculations at all. Create a locus that will at least allow you to place points on it. A locus can be viewed as an automatic trace that GeoGebra remembers, if you wish. Unfortunately, a locus in GeoGebra is not a function but rather a graph drawn with limited resolution, so you cannot automatically find the maximum of a locus either. But there is a work-around:

Turn off the trace of the point in Graphics 2 by right-clicking on it and selecting **Trace On** again. Clear the old traces with **Ctrl-F**. Create a locus in Graphics 2 (make sure it is active) by typing the following command:

```
Locus[H, E]
```

where H is the measurement point in Graphics Area 2 and E is your moving point in Graphics Area 1, at the bottom of the papers edge.

Now place at least four points on the top part of this locus. Then create a cubic function through these points by typing

```
FitPoly[{I, J, K, L, M, N}, 3]
```

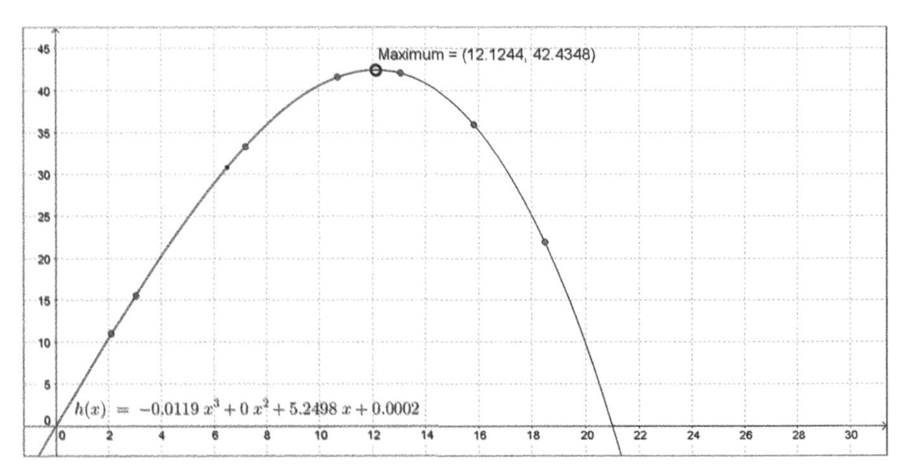

FIGURE 7.48 How to find the maximum of a locus.

You may find the maximum of this function in the interesting interval by typing

`Extremum[h, 10, 14]`	(US English) or
`TurningPoint[h, 10, 14]`	(UK & AU English)

In Figure 7.48 you can see a maximum of 42.43 cm^2 at a base length of 12.12 cm.

A polynomial of degree three was selected partly because the shape—an asymmetric maximum—suggests that it would work and partly because areas are being analyzed, and analyzes often add one degree of complexity.

It is possible to solve this problem algebraically too, which offers a different perspective of the problem. You can then observe that the maximum is achieved as the triangle attains a certain shape. Which shape is this, and why?

7.9 THE LOCOMOTIVE

The old classic steam locomotive like the one in Figure 7.49 was driven by boiling water, letting the steam pressure drive a piston in a cylinder back and forth. This piston was in turn connected to the wheels with rods.

The piston's stroke length, and thus the cylinder's minimum depth, was equal to the diameter of the circle that the rods' connection points traced on the wheels as they turned. When the train moved at a constant speed, these connection points moved around their wheels' hubs with a constant speed, just like points on a unit circle, and the wheels' motion might therefore be said to be capable of generating trigonometric functions.

Assuming that the cylinder is horizontal, will the piston's horizontal component generate a sine or a cosine wave? Will the piston have its highest velocity at the center of the cylinder? Create a model in GeoGebra and analyze it.

FIGURE 7.49 Steam locomotive, or steam engine. Photo: Tobias Johansson/Haldol (own work) [Public domain], via Wikimedia Commons. https://upload.wikimedia.org/wikipedia/commons/b/bf/%C3%85nglok_SJ_E10_1742_2006_G%C3%A4vle.JPG.

Proposed Solution

Since no exact measurements for any specific locomotive are given, you can only create a simplified general model. The driving wheel can then be represented by a circle, the piston by a point moving on a straight line, and the rod by a segment of constant length.

Create a circle c, centered in the origin and with a radius of 3 with either of the circle tools or . You could, of course, use either of the commands

```
Circle[(0,0),(0,3)]
Circle[(0,0),3]
```

as well.

If you connect the rod to the circle's perimeter, the strike length should be equal to the diameter of the circle. So create a segment a, representing the cylinder, between (0, 5) and (0, 11) with either the Segment tool or the command

```
Segment[(0,5),(0,11)]
```

To create the rod connecting the wheel and the piston, first create a point E on the circle. Then create a new circle d, having E as its center but with a radius of 8, using the second of the circle tools, Circle with Center and Radius . This circle intersects the line segment, and at this intersection you can create a point F with the Point tool (if you click close enough to the intersection) or the Intersect tool or the command

```
Intersect[a,d]
```

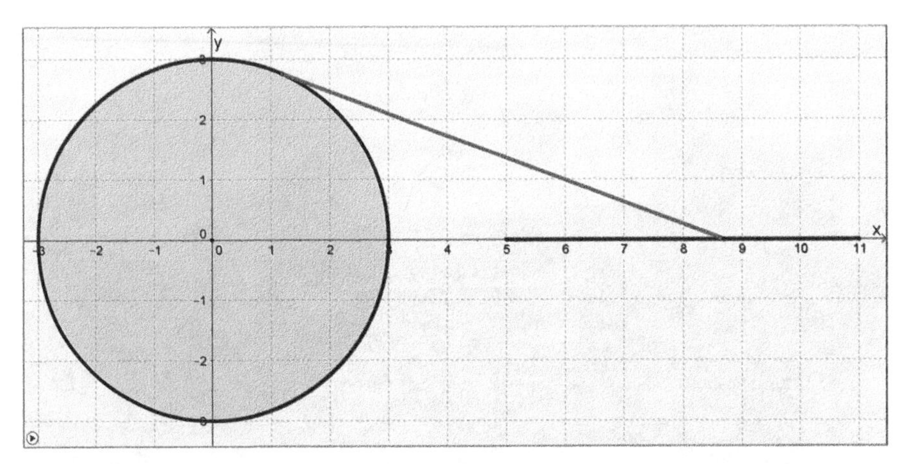

FIGURE 7.50 Simple model connecting circular to linear motion in a locomotive.

Last create the rod as a segment between E and F.

Now hide the help circle d and all labels. Take some time to color the wheel, piston, and rod and make all lines thicker.

Right-click on E and select **Animation on**. The wheel seems to be revolving; the rod connects its circular movement to the piston's linear movement as indicated in Figure 7.50.

By right-clicking on the animated point E in the Algebra window and opening the Algebra tab, you can increase the animation speed if you want to, and if you have already defined a number or slider, you may enter that numbers name to control the animations speed with this number or slider.

You can analyze the properties of this model. Start by showing Graphics 2 with **Ctrl-Shift-2**. If necessary, dock it and drag it to a suitable position in the window. **Ctrl-drag** the axis so that the x-axis shows 0 – 360 degrees and the y-axis shows –3 to 3. Thereafter measure the phase angle of E by clicking on the angle measuring tool and then click on some a point on the x-axis, the center of the wheel and on point E. This will create an angle α.

The formula $= 180/\pi$ will convert an angle from radians to degrees.

The animated point E on the circle and the point that represents the piston are both moving horizontally and in phase with each other. By first clicking in Graphics Area 2 to make it active and then entering **(α/°,x(E))**[1], you create a new point G in Graphics Area 2 that shows how E's x-coordinate changes with the phase angle. Right-click on the point in the Algebra window and select **Trace on**, and you will see the curve it produces. It is similar to a cosine curve.

[1]We defined the angle with a degree symbol. This might seem odd, but it is related to the difference between radians and degrees. In GeoGebra all trigonometric input is considered to be in radians. An angle without a degree symbol is in radians and the degree symbol is handled as a multiplicative constant, having the value $\pi/180$. The notation 36° looks as if it is in degrees, but GeoGebra handles it as $36 \cdot \pi/180$, an angle in radians. Thus, to plot an angle in degrees, we have to "divide by the degree symbol."

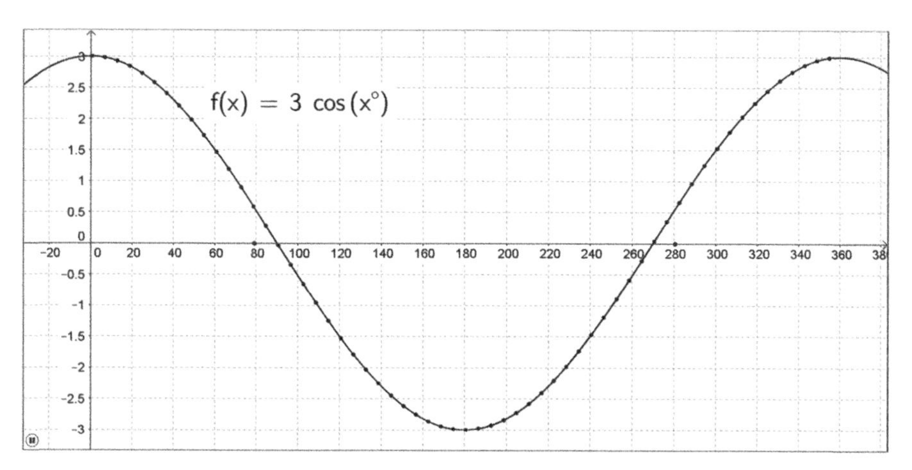

FIGURE 7.51 E's x-coordinate generating a cosine function.

You can also create G's locus by clicking in the Graphics Area 2 to make it active, click on the Locus tool, then click on G and finally click on E. But can you be sure that this is a cosine function? The radius of the wheel is 3. Enter the function **f(x) = 3cos(x°)**, and notice that the result, shown in Figure 7.51, is exactly the same as your earlier results.

In the same way you can create another point with the command

```
(α/°,x(F)-8)
```

that will show the corresponding result for the point representing the piston. Trace the point and create its locus in the same way, and notice that it does not describe a cosine function, even though it resembles such a curve. It is not so difficult to derive a function for the piston's x-coordinate with the assistance of the Pythagoras theorem, obtaining the result shown in Figure 7.52.

Typing

```
Roots[g,0,360]
```

locates the intersection points with the x-axis, and labeled I and J, and you can then calculate

```
fraction = (x(J)-x(I))/360
```

You find that the fraction = 0.56, which means that the piston is placed in the outer half of the cylinder 56% of the time. The piston is moving fastest when the derivative $g'(x)$ has its extreme values. You can find the derivative by typing

```
g'(x)
```

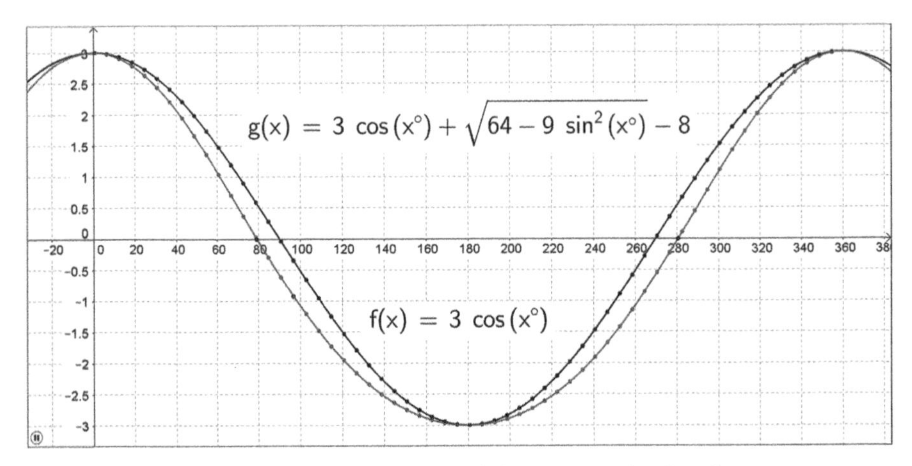

FIGURE 7.52 Piston not describing a pure cosine function.

and then find its extreme values from

```
Extremum[g', 10, 120]        (US English), or
TurningPoint[g', 10, 120]    (UK & AU English)
```

At the highest speed the phase angle is 71.6°.

7.10 MAXIMUM VOLUME

No book on modeling would be complete without this classic exercise: From a sheet of paper or carton with sides a, and b, you can build a box having height c by folding up the sides and folding and taping the corners together. Build a geometrical model of this in GeoGebra and investigate what value of c gives the maximum volume, for different values of a and b. Then try to derive a theoretical expression and show that this expression gives you the results you have already obtained in measuring the model.

As an extension, try to find out how real removal boxes are constructed from a rectangular piece of cardboard and recreate that construction in GeoGebra.

7.11 PASCAL'S SNAIL OR LIMAÇON

In geometry, a *limaçon*, also known as *Pascal's snail*, is defined as the shape formed when a circle rolls around the outside of a circle of equal radius. It can also be defined as the shape formed when a circle rolls around a circle with half its radius so that the smaller circle is inside the larger circle. The cardioid is the special case

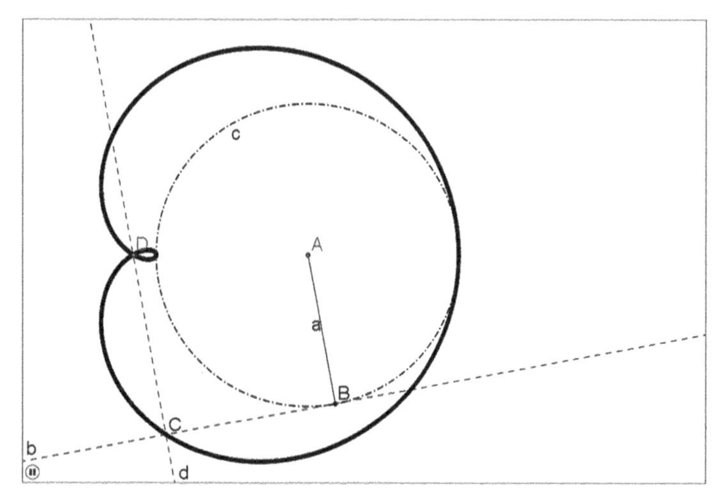

FIGURE 7.53 Pascal's snail.

in which the point generating the shape lies on the rolling circle; the resulting curve has a cusp.

In Figure 7.53 you see one construction. This way of using fixed artifacts to create curves has been around for many years, for instance, in ship building. Investigate the Pascal's snail for different values of a.

7.12 EQUILATERAL TRIANGLE DISSECTION

Create an equilateral triangle ABC and place a point P within it. Through P, create three lines parallel to each of the triangle's sides. These three lines dissect the triangle in six parts, three small equilateral triangles and three parallellograms.

For certain positions of P, the combined area of the three small triangles equals the combined area of the three parallelograms. Can you find these positions? What shape do they form? Can you verify this? Can you prove it? What happens if the original triangle isn't equilateral?

7.13 DIVIDING THE SIDES OF A TRIANGLE

For any triangle ABC, select an arbitrary point M. Join AM, BM, and CM, and extend them to cut BC, AC, and AB at D, E, and F, respectively. Join DE. Label the intersection of CF and DE as N. Join AN, and extend it to meet CB at G.

If M is selected so that D is the midpoint of BC, what can then be said about GC? If you then redo this construction by moving M so that D now falls on the point where G used to be, what can now be said about GC?

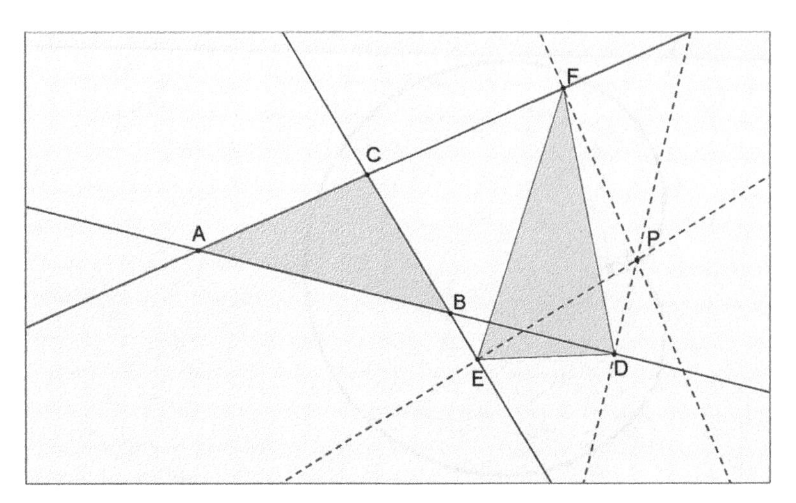

FIGURE 7.54 Creating the pedal triangle.

7.14 THE PEDAL TRIANGLE

By using GeoGebra, we can explore various relationships of the pedal triangle. Given a point P and a triangle ABC, a pedal triangle is defined as the triangle whose vertices are the feet of the perpendiculars from P to the extended sides of triangle ABC. P is called the "pedal point."

To construct the pedal triangle, first construct the points A, B, C, and P using the Point tool. Next use the Line tool to create lines through AB, BC, and CA. Then construct the perpendiculars to these lines through P with the Perpendicular line tool. Now you can use the Polygon tool to create triangle ABC and the pedal triangle DEF; see Figure 7.54.

The pedal triangle exhibits some interesting behavior. First play around a little by moving P around in the plane, in and around triangle ABC. Make notes of everything that you observe.

Here are a few investigations you could try:

- The pedal triangle DEF has different behavior in different regions of the plane. Try to map these regions and how they relate to the original triangle ABC.
- For some positions of P, the pedal triangle has one side parallel and overlapping one of the sides in the original triangle ABC. What is special about these positions? Can you trace them? What curve do they lie on?
- Trace the positions of P where DE is parallel to AB. Then trace the positions where EF is parallel to AB and finally where FD is parallel to AB. What can you observe?
- Similarly, trace the positions where DE is parallel to AB, DE is parallel to BC, and DE is parallel to CA. What can you observe now?
- Sometimes the pedal triangle DEF has a right angle. When? Can you trace these positions? What curve do they lie on?

Once you have a hypothesis, try testing it by creating a curve and placing the pedal point on it by double-clicking P and typing

```
Point[a]
```

as the definition of P, where a is the name of the line, circle, or curve you have created.

As a further confirmation of your hypothesis, try to figure out what would happen if triangle ABC had a different configuration and then check that your predictions are correct.

7.15 THE INFINITY DIAGRAM

Imagine a coordinate system where the distance from 0 to 1 is the same as the distance from 1 to infinity, say 10 cm, as in Figure 7.55. You might argue, in a simplistic way, that $1/1 = 1$, $1/0 = \infty$, and $1/\infty = 0$, and that therefore the function $y = 1/x$ seems to lie on a straight line. You don't actually *know* this, since you don't know the rest of the scale. But what would the scale look like if $y = 1/x$ actually *was* a straight line in this diagram? How many centimeters from the origin would the number 10 lie? What number would lie exactly 5 cm away from the origin?

If you solve this problem, an obvious extension is to ask what the scale would look like if the function $y = x^{-n}$ is a straight line.

Make a dynamic scale where n is controlled by a slider that shows the correct position for a set of numbers, such as {1/10, 1/9, 1/8, …, 1/2, 1, 2, 3, …, 9, 10} for different values of n.

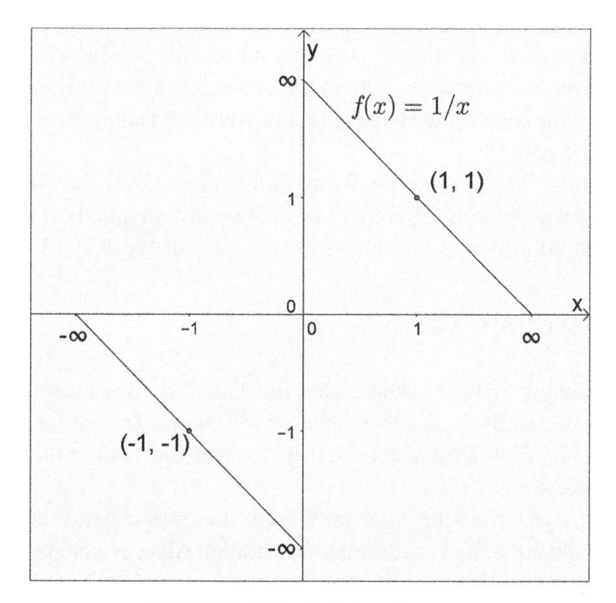

FIGURE 7.55 Infinity diagram.

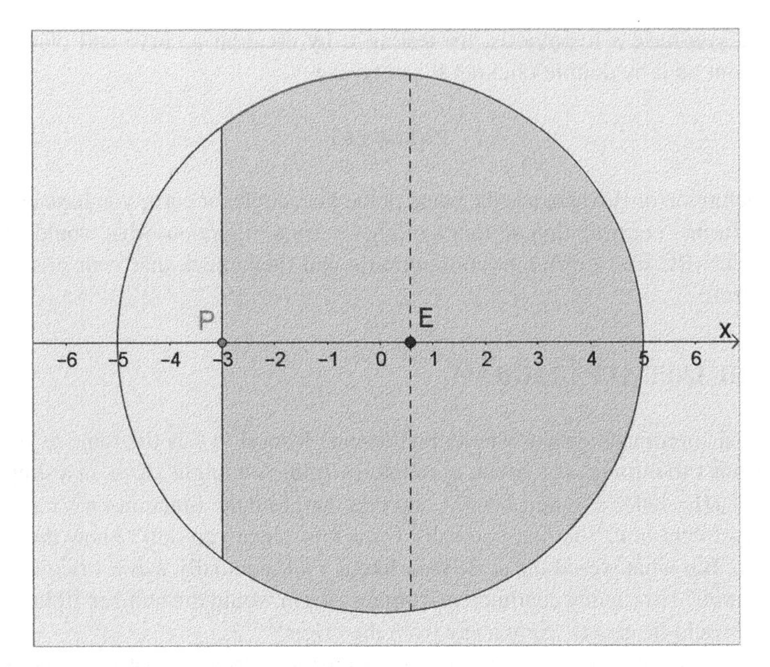

FIGURE 7.56 Dividing a circular segment in two equal areas with a line.

7.16 DISSECTING A CIRCULAR SEGMENT

In Figure 7.56 a circle is divided in two circular segments by a line parallel to the y-axis through P. The larger, right-hand segment is divided again in two parts by another line parallel to the y-axis through E, such that it dissects the segment in two parts of equal area. Plot the x-coordinates of points P and E against each other in a diagram. What does the graph look like? What is the maximum deviation from a straight line, and where does it occur?

As an extension, do the same for P and G, the *centroid* of the circular segment. What are the similarities and differences of the two investigations? Formulas can be found at Wolfram Mathworld: http://mathworld.wolfram.com/CircularSegment.html

7.17 NEUBERG CUBIC ART

Create a quadrilateral ABCD. Then create the *Euler line* for each of the triangles ABC, ABD, ACD, and BCD. See Problem 7.4, *Points in a Triangle,* on how to create Euler lines quickly. Play around a little to get a feel for what is happening as you move the points around.

You will notice that the four Euler lines are sometimes concurrent, meaning they intersect at a single point, and are sometimes parallel. Also, as you move point D, the Euler line belonging to triangle ABC does not move around but the other lines swing around like crazy.

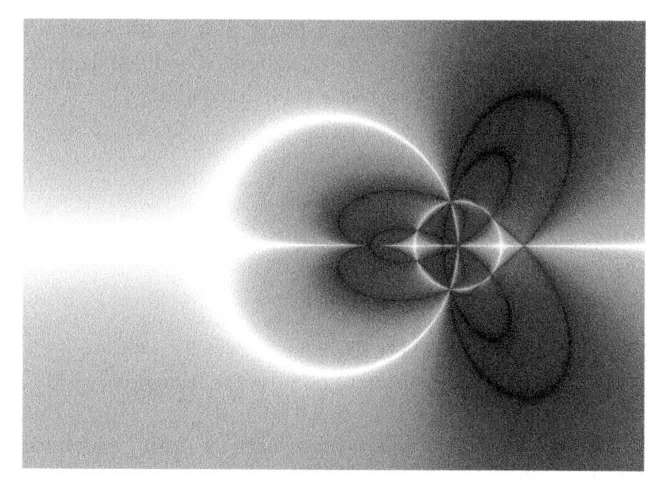

FIGURE 7.57 Art from the Euler line. Photo: Jonas Hall.

Now create a new triangle S, from the intersection points of the three Euler lines not belonging to triangle ABC. As you move D around, the area of S will vary from 0 to infinity.

Next you will need to create a *color model* in order to convert the area of S area to a color. In GeoGebra a color is measured with a number from 0 to 1. You therefore need to find a function that has $[0, \infty]$ as its domain and $[0, 1]$ as its range. There are several such functions, you may want to investigate the following common color models

$$c(x) = \frac{2}{\pi}\arctan(\mathrm{k}x), \quad c(x) = \frac{2}{1+e^{-\mathrm{k}x}} - 1, \quad \text{and} \quad c(x) = \frac{\mathrm{k}x}{\mathrm{k}x+1}$$

to start with.

Make sure that you set a dynamic color for D according to the area of S through your selected color model. Activate the trace of D and move it around systematically to create a *color map*. See Problem 7.4 again for details on how to create a color map. The patterns are quite intriguing, as you can see in Figure 7.57. There are more examples of these images along with more information about them that can be seen at the Swedish GeoGebra Institute's Art section, http://www.geogebrainstitut.se/art/art.asp. To create the same color effects, you need to have similar, but slightly different color models for each of the red, green, and blue components. Can you work out exactly which ones?

7.18 PHASE PLOTS FOR TRIANGLES

First create a geometrical model of a triangle that is determined by two angles α and β and a fixed perimeter. This may prove a challenge in itself. It may be easier to set one side $=1$ first and then later add the scaling that is needed to keep the perimeter constant.

You can then use this model to make a phase plot of all possible triangles. In Graphics Area 2, let the axes represent the two angles α and β and create a point that represents the current triangle. In this phase plot, determine:

- The outline of all possible triangles
- The positions of all right-angled triangles
- The positions of all isosceles triangles
- The positions of all equilateral triangles
- The positions of all acute triangles
- The positions of all obtuse triangles
- The position of the triangle outlining all possible triangles

You can make some interesting *color maps* here by setting the dynamic color of the point representing the current triangle in the phase plot. See Problem 7.4, *Points in a Triangle,* and Problem 7.17, *Neuberg Cubic Art,* for more information on color maps. You could set the dynamic color to reflect the "same-ness" of the angles, or the "right-angled-ness" of the triangle to help you in identifying the positions of these triangles.

What will happen with the outline of all possible triangles if you require that $\alpha<\beta<\gamma$ are the three angles of the triangle? How will this restrict the outline, and where now will the outline triangle lie within itself?

You can also create this diagram as a ternary plot. Figure 7.58 shows an example of how to read a ternary diagram. Ternary diagrams are ideal for showing how three

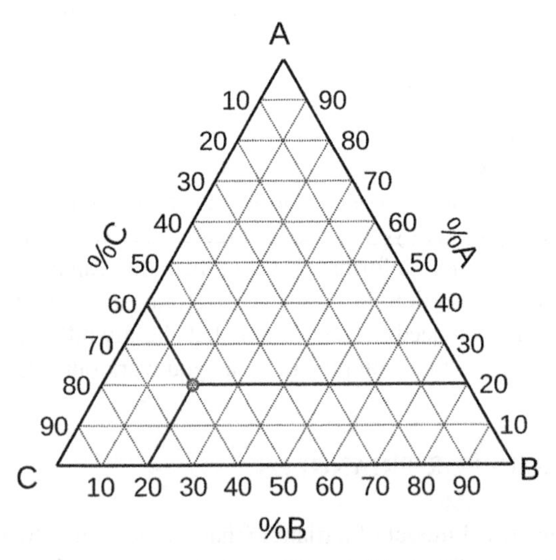

FIGURE 7.58 Reading a ternary diagram. Photo: Cdang (own work) [Public domain], via Wikimedia Commons. https://upload.wikimedia.org/wikipedia/commons/1/16/Lecture_diagramme_ternaire.svg.

variables interact when their sum is constant. If we set $\alpha = 180°$ at the origin and $\beta = 180°$ at (180, 0), then $\gamma = 180°$ lies at

$$\left(90,\ 90\sqrt{3}\right)$$

and a triangle with angles α, β, and γ lies at

$$\left(\beta + \gamma/2,\ \sqrt{3}\gamma/2\right)$$

See the Wikipedia article https://en.wikipedia.org/wiki/Ternary_plot for more on ternary diagrams. What shape now will the outline of all possible triangles take if $\alpha < \beta < \gamma$?

7.19 THE JOUKOWSKI AIRFOIL

This problem involves imaginary numbers. Create a circle with its center A, near the origin and a radius r of approximately 1. Let P be a point on the circle. Then

```
a = x(P)
b = y(P)
z_1 = a + b i
```

will create the complex number z_1. Now let

```
w = z_1 + 1/z_1
```

and activate the trace of w. As z_1 moves around the circle, w will trace out a shape. Use the locus of w with respect to P to generate the shape without resorting to tracing.

If you move the center of the circle and vary the radius, you will find that the shape sometimes looks like the cross section of a wing. In the original theory, the circle has to pass through (1, 0) and contain (-1, 0). Can you find the radius and the coordinates of A that creates the aerofoil?

If instead you define

```
w = z_1 + 1/z_1^n
```

then you may simply investigate the shapes that occur for different A, r, and n. There are plenty of interesting shapes to find. Write a report of what you find and include screen shots of the shapes along with values for A, r, and n.

8

DISCRETE MODELS

Imagine planning a trip around France where your tour is based on a map showing distances between various cities and other points of interest. The map can be viewed as a graph with a point for each city and connections to nearby cities labeled by the distance between the two cities. You might ask yourself a question like "If I drive, how do I visit every city on the map?" This is an example of *the traveling salesman problem* from graph theory. Basically, a graph is a collection of vertices or nodes that are connected by edges. Graph theory is a large domain of so-called discrete mathematics. Discrete models are often modeled with the help of spreadsheets instead of with the help of function curves.

But discrete mathematics is more than graph theory. In the previous chapters you have seen examples of many different types of functions, but almost all have been continuous. The discrete counterpart of continuous functions consists in number sequences such as the Fibonacci sequence or the population of deer from one year to the next. In this latter example we often speak of *discrete time*.

In this chapter then, you will find exaples of linear programming, matrices representing Markov chains, spreadsheet calculations, difference equations and difference schemes, and some programming examples of situations using discrete time. In the discrete world time is not a continuous river but a clock, going tic-toc, tic-toc, tic-toc, ...

Mathematical Modeling: Applications with GeoGebra™, First Edition. Jonas Hall and Thomas Lingefjärd.

Companion website: www.wiley.com/go/Hall/MathematicalModeling

8.1 THE CABINETMAKER

A cabinetmaker in the United States in the Edwardian 1910s might specialize in making chairs and tables, perhaps such a table as shown in Figure 8.1. The cabinetmaker was likely selling finely wrought tables at a profit of 30 USD and chairs at a profit of 10 USD. Because tables and chairs are often bought together, the cabinetmaker had to produce at least three times as many chairs as tables. In this era, it took six days to produce a fine table and three days to produce a simple chair. The cabinetmaker worked 40 days in each two month eriod, spending the remainder on travelling, selling and bying materials. All furniture were then sold at the end of each period. Also, all the furniture had to be stored in the small storeroom where there was only room for four tables or sixteen chairs, or some combination of these furniture pieces. The demand for furniture was high in urban areas at this time, and the storeroom was always emptied at the end of each 2 month period.

How many chairs and tables should the cabinetmaker produce to maximize the profit? How could the profit grow?

Proposed Solution 1

Many students may want to solve this problem by plugging in the numbers right away, but it is also possible to see this as a problem where variables are gradually introduced.

The problem is in fact suitable for *linear programming*. Start by defining the variables x for the number of chairs and y for the number of tables. After that define the *profit function* $P(x, y) = 30x + 10y$, where P is the number of earned dollars each

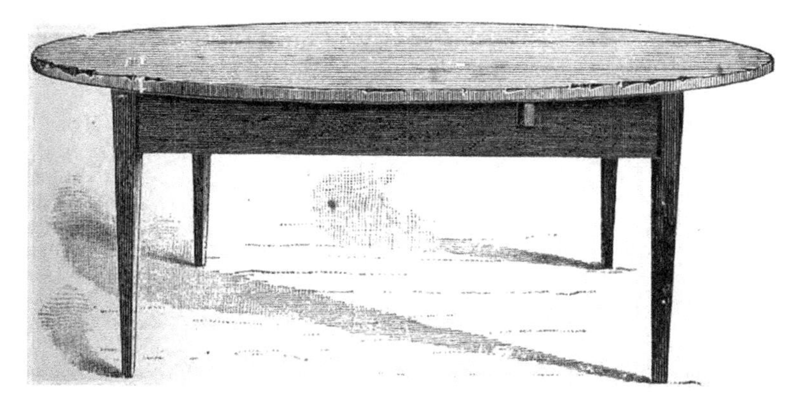

FIGURE 8.1 Table upon which Ohio's first constitution was signed. The table is exhibited in Chillicothe's (Ohio's first capital town) Town Hall. Photo: Henry Howe [public domain], via the Historical Collections of Ohio, the Ohio Centennial Edition / Wikimedia Commons. https://upload.wikimedia.org/wikipedia/commons/9/94/Ohio_constitution_table.png.

two month period. This is the function to maximize. But selecting x or y to be too large will violate the *constraints*. In this problem there are three constraints:

1. The customers' demands for at least three times as many chairs as tables can be expressed as $x \geq 3y$.
2. The working days are limited to 40 days per period, indicating that $3x + 6y \leq 40$.
3. The storage room can store maximally 4 tables or 16 chairs. Assume that you can freely exchange 4 chairs for one table, ignoring fractional tables, writing this as $x/4 + y \leq 4$.

Now type in these inequalities in GeoGebra, as they stand. The $\leq$ sign can be typed as <=, and the inequalities are shown as regions in GeoGebra.

In Figure 8.2 the names of the inequalities have been changed to show more clearly what they are linked to. You can see that there is a region where all inequalities are met, including the unstated requirements that the variables are to be positive numbers, this being the darkest region. The solution must be inside this region. The problem has now been formulated in a graphical representation.

In order to find the optimal solution, first create a profit line:

```
30x + 10y = 40
```

where the number 40 is arbitrary. As you grab the line and move it, this number, and the equation for the line, will be changed. It may be difficult to grab the line with the mouse since the line is overlapping the inequalities, but you can select the line's equation in the Algebra window and use the arrow keys to move it. You can see that

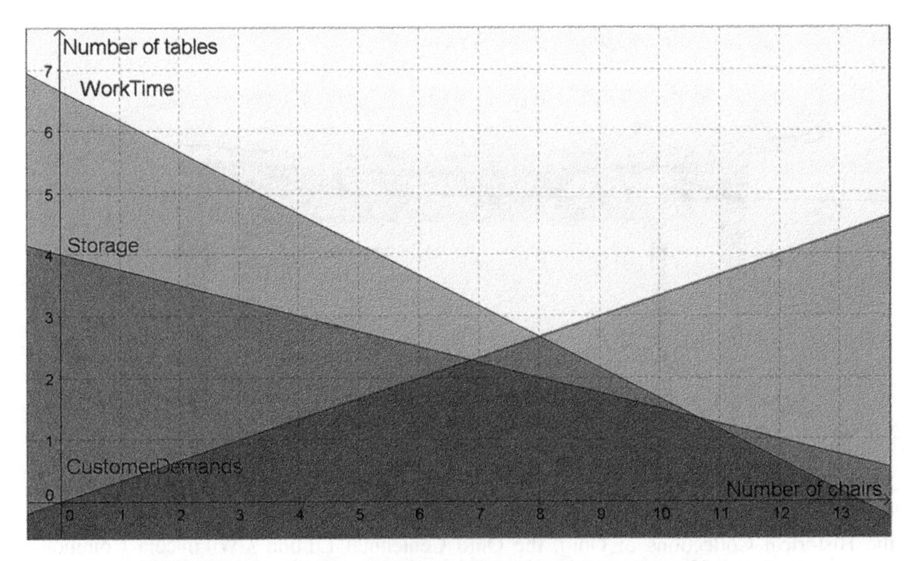

FIGURE 8.2 Overlapping inequalities forming an area of interest.

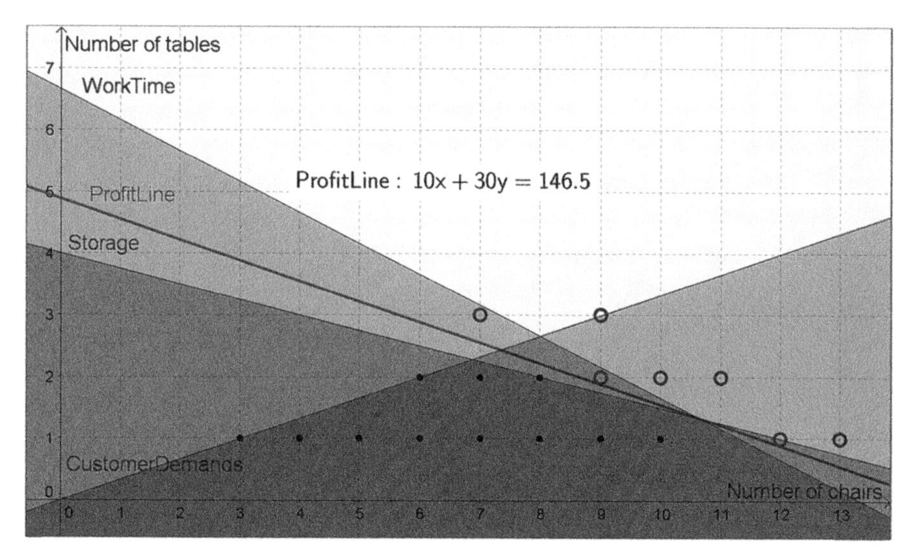

FIGURE 8.3 Profit line through the optimal vertex with interesting points highlighted.

the number is increasing when you move the line to the upper right. The optimal solution should be found at the corner of the WorkTime and Storage inequalities.

Depending on what situation you are modeling, you can sometimes accept decimal numbers in the answers and then the intersection would be optimal. In this case that method would be doubtful. The cabinetmaker could, of course, produce half a chair, but then he will not be able to sell it during this period. So start instead to investigate the integer points where a solution might be found. In Figure 8.3 some interesting integer points have been marked.

The optimal integer solution is marked with a cross. According to this solution the cabinetmaker should produce one table and eleven chairs every period, which will generate an income of 140 USD. This will take 39 days and use almost all the space in the storage room; only one more chair will fit. The small points are solutions that are not optimal, and the circled ones are solutions that are possible if you alter some of the constraints.

Now move the profit line so that it passes through this integer solution. You can see that it also passes through another, equally profitable solution where the production is 2 tables and 8 chairs each period. This fully uses the storage room and only takes 36 days. It is probably best to vary these two strategies in order to better meet the demands from customers.

If the cabinetmaker can imagine loosening the constraints, there could be other possibilities. If everything is sold, then the customer's demands may not be so difficult to handle. In reality though, he will not make much profit by removing this constraint since the profit line is “more parallel” to the other constraints.

But if it is possible to squeeze an extra chair into the storage room, then 2 tables and 9 chairs could be produced each period, which would give an extra 10 USD without exceeding 40 working days each period.

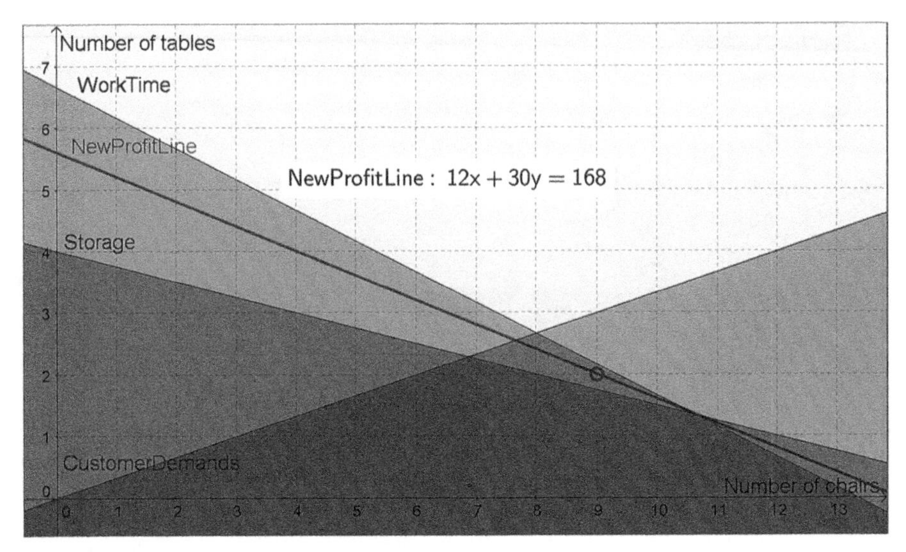

FIGURE 8.4 Raising the price.

Now suppose that the cabinet maker wants to work a bit more than 40 days per period, reducing the time spent on other tasks. Then it may be possible to fill the storage room and produce 12 chairs and 1 table, also giving an extra 10 USD. That would require working 42 days each period, and that everything would fit into the storage room.

Of course, the usual way to get a better income, if the demand is good, is to raise the price. If the cabinetmaker raises the price on the chairs from 10 USD to 12 USD, the profit line will turn a little bit and become more parallel with the working time constraint. The original solution with 1 table and 11 chairs will then be the only optimal solution and give 162 USD. If, in addition, the carpenter finds the storage space for another chair, then 2 tables and 9 chairs will give an income of 168 USD. These solutions are presented in Figure 8.4.

Proposed Solution 2

Integer problems are easy to investigate with simple programming. We have chosen to use Python in this book because, in this programming language, small and simple programs are both easy to write and control. It is easiest to use a program code interpreter on Internet, for example one of these:

http://www.trypython.org/

http://www.tutorialspoint.com/execute_python_online.php

We will use Codingground's tutorial Execute Python Online, shown in Figure 8.5. You can write the code in the main central window and press Execute to get the results in the terminal window below. If you would like to learn more, there are exercises in the right-hand pane.

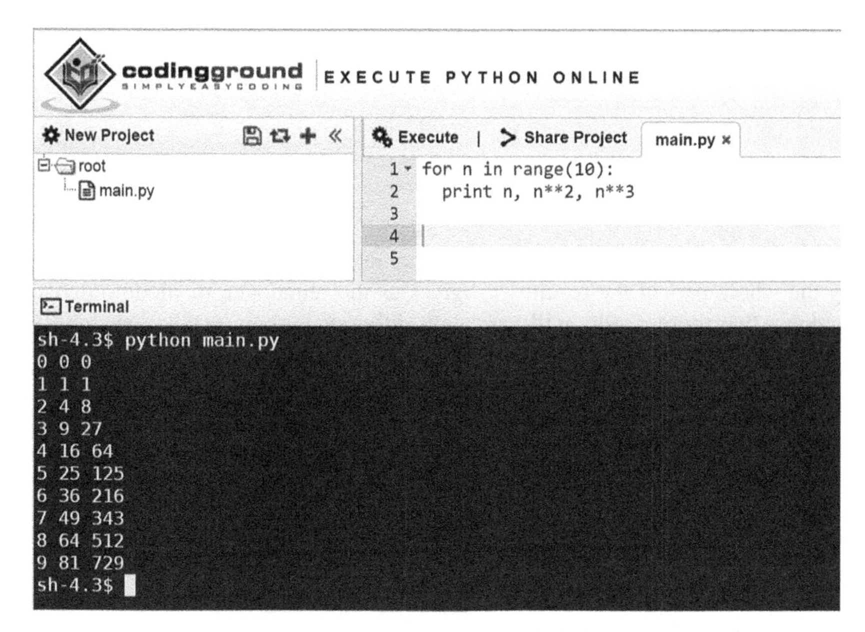

FIGURE 8.5 Coding Python online at Codingground.

Enter the following simple program code, line by line:

```
for n in range(10):
  print n, n**2, n**3
```

Do not forget the initial spaces on the second line! They are important as they define the **for** *loop block*, making the **end** statement unnecessary in Python. Now try:

```
a = 3
b = 5
print "a is not ", b
```

Unfortunately, different versions of Python handle division differently. There are two division signs, "/" and "%." Whether only integers are used or if some decimal numbers are used in a calculation will also affect the outcome. Running the following short program on Codingground

```
print 7/4
print 7%4
print 7/4.
print 7%4.
```

gives the following results

```
1
3
1.75
3.0
```

From which we conclude that "/" gives us division, but dividing integers gives only the whole part of the result. "%" gives us the remainder, or modulo division. We repeat that these results will vary with different Python installations, and you need to make sure you understand how the Python environment you are using works.

Finding numbers that are divisible with 7 and 13 at the same time, type (try it):

```
for n in range(1000):
  if (n%7 == 0) and (n%13 == 0):
    print n
```

These examples illustrate some essential programming points:

- Numbers can be stored in variables.
- You can tell the program to repeat something (using **for** loops).
- You can interrogate the program and control its behavior depending on the answers (using **if** statements).

In this case the program should investigate all the integer combinations of x and y that are constrained by the three conditions $3x+6y \leq 40$, $x \geq 3y$, and $x/4+y \leq 4$. If these conditions are all met, then the program will print what the profit will be for exactly this combination of x and y. In Python the program will be four lines long:

```
for x in range(20):
  for y in range(20):
    if (3*x+6*y <= 40) and (x >= 3*y) and (x/4.+y <= 4):
      print "x =",x,"y =",y,"Profit =",10*x+30*y
```

Notice the decimal point when dividing by 4 to make sure we're in "decimal mode." The result from the run is a list of values where, close to the end of the list, you find the optimal solution:

```
x = 6 y = 2 Profit = 120
x = 7 y = 0 Profit = 70
x = 7 y = 1 Profit = 100
x = 7 y = 2 Profit = 130
x = 8 y = 0 Profit = 80
x = 8 y = 1 Profit = 110
```

```
x = 8 y = 2 Profit = 140
x = 9 y = 0 Profit = 90
x = 9 y = 1 Profit = 120
x = 10 y = 0 Profit = 100
x = 10 y = 1 Profit = 130
x = 11 y = 0 Profit = 110
x = 11 y = 1 Profit = 140
x = 12 y = 0 Profit = 120
x = 13 y = 0 Profit = 130
```

Some issues to think about when writing code:

- Use a text editor, for instance, Notepad or Notepad2, when building your program code so that you can save it. Then you can just copy and paste your code into the Python interpreter you are using.
- The **for** command is based on lists. **range(20)** is a function that generates a list from 0 to 19, and so x and y will therefore not be able to take the value 20.
- **for** and **if** must have a colon last on the line to show that everything that is indented to the right is a single program block. Everything in this block will be carried out if the **if** statement is true or if the **for** statement is yet not finished.
- There is no **end** statement as in many other programming languages. The program block is over when the indented block is finished.
- Tests for equality require the double equal sign "=="
- Multiplication is written as * and $\leq$ as <=. 3**5 means 3^5

Often it is a good idea to let the program itself find the optimum value. This is normally accomplished by storing the current "record" in a variable during the search. In this case you also want the record generating x and y values to be stored for later retrieval. These variables should be initialized at the start of the program. This makes the program a bit longer:

```
# Reset variables for storage of records
xOpt = 0
yOpt = 0
Profitmax = 0
# Looping through all combinations of x and y
for x in range(20):
  for y in range(20):
    # Testing if conditions are met
    if (3*x+6*y <= 40) and (x >= 3*y) and (x/4.+y <= 4):
      # Then test if there is a new maximum
      if Profitmax < 10*x+30*y:
```

```
        # Then save the information
        xOpt = x
        yOpt = y
        Profitmax = 10*x+30*y
# When the looping has finished, print the record
print "x =",xOpt,"y =",yOpt," Profit =",Profitmax
```

As you can see, there are comments in the program code. All code lines starting with # are interpreted as comments, which make it possible to write code that is easy to read. The result is this time just one line:

```
x = 8 y = 2 Profit = 140
```

As a small surprise, the program gives another solution this time. The program is written so that it saves and writes the first solution if there are several of them. Two small changes in the code will take care of that problem. You need to move the **print**-command into the double loop so that it prints every new record, and you need to change the < to <= in the second **if**-command. The program code will look like this after the changes:

```
# Reset the variables for storage of records
xOpt = 0
yOpt = 0
Profitmax = 0
# Going through all combinations of x and y
for x in range(20):
  for y in range(20):
    # Testing if conditions are met
    if (3*x+6*y <= 40) and (x >= 3*y) and (x/4.+y <= 4):
      # Then test if there is a new maximum
      if Profitmax <= 10*x+30*y:
        # Then save the information
        xOpt = x
        yOpt = y
        Profitmax = 10*x+30*y
        # Print the largest profit so far
        print "x =",xOpt,"y =",yOpt,"Profit =",
        Profitmax
```

This small change will get a list of all record profits. Both the optimal solutions are there, at the end of the list:

```
x = 0 y = 0 Profit = 0
x = 1 y = 0 Profit = 10
x = 2 y = 0 Profit = 20
x = 3 y = 0 Profit = 30
x = 3 y = 1 Profit = 60
x = 4 y = 1 Profit = 70
x = 5 y = 1 Profit = 80
x = 6 y = 1 Profit = 90
x = 6 y = 2 Profit = 120
x = 7 y = 2 Profit = 130
x = 8 y = 2 Profit = 140
x = 11 y = 1 Profit = 140
```

You could have done this without the variables **xOpt** and **yOpt,** since you are printing the maximum values as you find them. It you omit them, the code will be four lines shorter. If you choose to leave out the comments too, the code will be an additional six lines shorter, 7 lines in all:

```
Profitmax = 0
for x in range(20):
  for y in range(20):
    if (3*x+6*y <= 40) and (x >= 3*y) and (x/4.+y <= 4):
      if Profitmax <= 10*x+30*y:
        Profitmax = 10*x+30*y
      print "x =",x,"y =",y,"Profit =",Profitmax
```

Comment

To do numerical investigations with programming, going through every single option like you have done in this problem is sometimes called the *brute force method.* In this case you could have worked the problem by hand, but for larger problems with thousands of combinations, programming is a very powerful mathematical problem solving method. In 1977 a famous problem known as the four-color problem was solved by a computer program that was instructed to check all possibilities in a complicated mathematical proof. Often the structures of these programs are similar:

```
# Reset/initiate all needed variables (s = 0)
# Investigate all necessary values (for)
            # Test if conditions are met (if)
                  # If so save the information (s = ...)
# Print the results (print)
```

8.2 WEATHER

Summer in the temperate zones tends to gravitate toward predominantly low-pressure or high-pressure weather. Here we simplify this drastically to the following:

- High pressure gives mostly sunny weather.
- Low pressure will give either sunny or rainy weather.

Let us use a table to form the transition probabilities that weather of a certain type will shift to weather of a different type. These probabilies are summarized in Table 8.1 where the sum of the probabilities for each line = 1.

You can see that, for example, if there is a low pressure giving rainy weather on one day when you have a 70% probability that there will be sunny, but still low-pressure controlled weather on the next day. The transition probabilities given in Table 8.1 are all invented but could, in principle, be determined by observation.

Suppose a situation where the transition probabilities for the different weather types for June 1 have been calculated to 10% for sunny, high-pressure controlled weather, 70% for sunny but low-pressure controlled weather and 20% for rain. These are very unscientifically estimates of the weather during summer in the Stockholm archipelago, shown in Figure 8.6. Calculate the probability that it will rain on June 6. What is the probability that it will rain on July 1? What happens if the transition probabilities are changed? What happens if new measurements give new transition probabilities for the weather on June 1?

Proposed Solution

On June 1 the probability for predominantly high-pressure weather readings is 10%. The probability for high-pressure weather readings for the following day is therefore $10\% \cdot 90\% = 9\%$, where the 90% is taken from Table 8.1. Now this is not the only possibility. You could have a low-pressure controlled readings but still sunny weather on June 1, which would give an additional probability of $70\% \cdot 5\% = 3.5\%$ that it would be high-pressure controlled weather on June 2. Finally you could also have rain at June 1 but still very sunny weather the following day, which gives a contribution of $20\% \cdot 5\% = 1\%$. The total probability for high-pressure controlled weather on June 2 will be

$$p(H) = 10\% \cdot 90\% + 70\% \cdot 5\% + 20\% \cdot 5\% = 13.5\%$$

TABLE 8.1 Transition Probabilities of Three Weather Conditions

Today/Tomorrow	H-sunshine	L-sunshine	L-rain
H-sunshine	0.9	0	0.1
L-sunshine	0.05	0.5	0.45
L-rain	0.05	0.7	0.25

FIGURE 8.6 Beautiful weather at the outer coast line in the archipelago. Photo: Arild Vågen (own work) [CC BY-SA 3.0 (http://creativecommons.org/licenses/by-sa/3.0)], via Wikimedia Commons. https://upload.wikimedia.org/wikipedia/commons/c/ca/Kors%C3%B6_Kroks%C3%B6_Sand%C3%B6n_February_2013.jpg.

In the same way the probability for low-pressure controlled—but sunny—weather and rainy weather will be

$$p(L)=10\% \cdot 0\% + 70\% \cdot 50\% + 20\% \cdot 70\% = 49\%, \text{and}$$

$$p(R)=10\% \cdot 10\% + 70\% \cdot 45\% + 20\% \cdot 25\% = 37.5\%$$

Here H mean high-pressure controlled weather, L stand for low-pressure controlled but sunny weather, and R stand for rainy weather. You have to do quite a lot of calculations just to get from one day to another. In order to calculate the weather forecast for June 6, you need to do all the calculations for another four days. You could do it, of course, but then to make a forecast for July 1 will need another 24 days of calculations, which may be a bit tedious. You need a better and faster method.

This type of situation is not uncommon, and of course, mathematicians have developed special methods and even concepts to deal with it. You could call it calculating with tables, but the name normally used in the mathematics community is matrix calculation.

A matrix is a kind of table with values, but written with large parenthesis around the whole table. The transition probabilities table can be written as a matrix like this:

$$\mathbf{M} = \begin{pmatrix} 0.9 & 0 & 0.1 \\ 0.05 & 0.5 & 0.45 \\ 0.05 & 0.7 & 0.25 \end{pmatrix}$$

The values for weather probabilities for June 1 and 2 will be written as

$$\mathbf{O} = \begin{pmatrix} 0.1 & 0.7 & 0.2 \end{pmatrix}$$

$$\mathbf{N} = \begin{pmatrix} 0.135 & 0.49 & 0.375 \end{pmatrix}$$

where **O** and **N** are labels for the old and the new weather. The complicated calculations may now be summarized through matrix multiplication, giving $\mathbf{O} \cdot \mathbf{M} = \mathbf{N}$. It is like multiplying one table with another table to get a new, third table, as a result. In fact, matrix multiplication is partly defined the way it is because of it's use in these types of situations.

Note that **M** is just like any rate factor in that a single multiplication takes you from the old value to the new one. To calculate the probability for rain on June 6, you need to do the calculation $V = G \cdot M^5$, which is similar to percentage calculations. The multiplication could be carried out with a modern calculator or even a smart phone but there is a profit in knowing how to do it in GeoGebra as well, since this will provide you with interactive possibilities. Matrices are entered into Geogebra as lists of lists with braces. You type them like this:

```
G = {{0.1, 0.7, 0.2}}
M = {{0.9, 0, 0.1}, {0.05, 0.5, 0.45}, {0.05, 0.7, 0.25}}
```

The input is row by row. The calculation from one day to another is the multiplication

```
V = G M
```

To calculate the weather for a day 5 days ahead, use

```
V6june = G M⁵
```

which gives you the resulting weather matrix seen in Figure 8.7, and you can see that the probability for rain on June 6 is 31%.

There are 30 days in June: to calculate the weather for July 1, you have to calculate

```
V1juli=G M^30
```

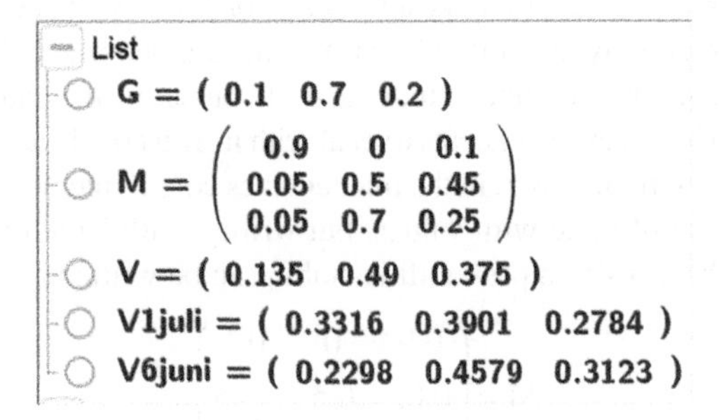

FIGURE 8.7 GeoGebra's matrix calculations.

which gives (0.3316, 0.3901, 0.2784). According to this model there is a 28% probability for rain.

What happens if you change the values in the transition matrix? Create variables for the individual values in Table 8.1:

```
HH = 0.9
Hf = 0
Hr = 0.1
fH = 0.05
ff = 0.5
fr = 0.45
rH = 0.05
rf = 0.7
rr = 0.25
```

Then you have to redefine the transition matric to

```
M = {{HH, Hf, Hr}, {fH, ff, fr}, {rH, rf, rr}}
```

You can now select a number in the Algebra window and change its value with the arrow keys. You can also click on the visibility button next to the number to activate a slider for that number. Before using the arrow keys to change the numbers, it is advisable to first set a smaller increment. Click on the word **Number** in the Algebra window. That will select all numbers. Right-click on them and select **Properties**, the **Sliders** tab, and change the increment to 0.01.

If the numbers are representing probabilities, then the sum of all the numbers should be = 1. Thus you might want to define

```
Hr = 1 - HH - Hf
```

so that Hr changes automatically when either HH or Hf is changed. You can do this in several different ways depending on what variable you want to investigate.

In the same way you can introduce and implement the parameters H0, f0, and r0 and redefine

```
G = {H0, f0, r0}
```

or

```
G = {H0, f0, 1 - H0 - f0}
```

It is easy to discover that almost any weather on June 1 will lead to a 28% probability for rain on July 1 given the current values of the transition matrix **M**.

The theoretical background and how to use matrices is usually studied at university courses, but basic properties of matrices may be beneficial for students in upper secondary school. This can give an excellent training in dealing with concepts and procedures related to percentages, priority rules, and algebra. It will later be possible to reconnect to matrices when teaching equation systems.

The procedure behind the notion of matrices as change factors that can themselves be multiplied many times was developed by Andrej Markov toward the end of the nineteenth century and is usually called *Markov chains*.

8.3 SQUIRRELS

Grey squirrels, shown in Figure 8.8b, were introduced from North America to England from the end of the nineteenth century to the 1920s, when it was discovered that the grey squirrels started to outcompete the domestic red European squirrels, shown in Figure 8.8a. Between 1973 and 1988 every square kilometer in Great Britain was investigated and in every such box the type of squirrels that was found in the area, red (R), grey (G), both (B), or none (N), was recorded. The researchers attempted to investigate how many areas of one type were transformed to another type the next year. The results are in Table 8.2.

Try to determine from these data whether the red squirrel should be considered an endangered species.

(a)

FIGURE 8.8a Red Squirrel. Photo: Pawel Ryszawa (own work) [GFDL (http://www.gnu.org/copyleft/fdl.html), CC-BY-SA-3.0 (http://creativecommons.org/licenses/by-sa/3.0/) or CC BY-SA 2.5-2.0-1.0 (http://creativecommons.org/licenses/by-sa/2.5-2.0-1.0)], via Wikimedia Commons. https://upload.wikimedia.org/wikipedia/commons/e/e1/Red_Squirrel_-_Lazienki.JPG.

(b)

FIGURE 8.8b Gray squirrel. Photo: BirdPhotos.com (BirdPhotos.com) [CC BY 3.0 (http://creativecommons.org/licenses/by/3.0)], via Wikimedia Commons. https://upload.wikimedia.org/wikipedia/commons/0/0a/Eastern_Grey_Squirrel.jpg.

TABLE 8.2 Number of Areas Overrun by Different Squirrel Types from Year to Year

From / To	G	G	B	I
R	2529	61	282	3
G	35	733	25	123
B	257	20	4311	310
I	5	91	335	5930

Proposed Solution

Table 8.2 contains enough information to serve as a *transition matrix*, see Problem 8.2, *Weather,* for another example of this. Enter the table's contents into GeoGebra's spreadsheet, which you open with **Ctrl-Shift-S**. You can hide the graphics area with **Ctrl-Shift-1,** since you will not need it. You can also use the menu option **View**... to open or close windows.

In column F you calculate the sum of every row. In F2 type

```
= B2 + C2 + D2 + E2
```

and copy this formula down by dragging the fill handle. Below this table, in rows 7 to 10, create the probabilities, that is, each value divided by it row sum. The sum of all such rows will be = 1. These numbers express the probability that the area will change or maintain its population of squirrels the next year. You can see that, for example,

▼ Spreadsheet

B7 =B2 / $F2

	A	B	C	D	E	F
1		R	G	B	N	Tot
2	R	2529	61	282	3	**2875**
3	G	35	733	25	123	**916**
4	B	257	20	4311	310	**4898**
5	N	5	91	335	5930	**6361**
6						
7	R	0.87965	0.02122	0.09809	0.00104	
8	G	0.03821	0.80022	0.02729	0.13428	
9	B	0.05247	0.00408	0.88016	0.06329	
10	N	0.00079	0.01431	0.05266	0.93224	

FIGURE 8.9 Spreadsheet transforms a table to a matrix.

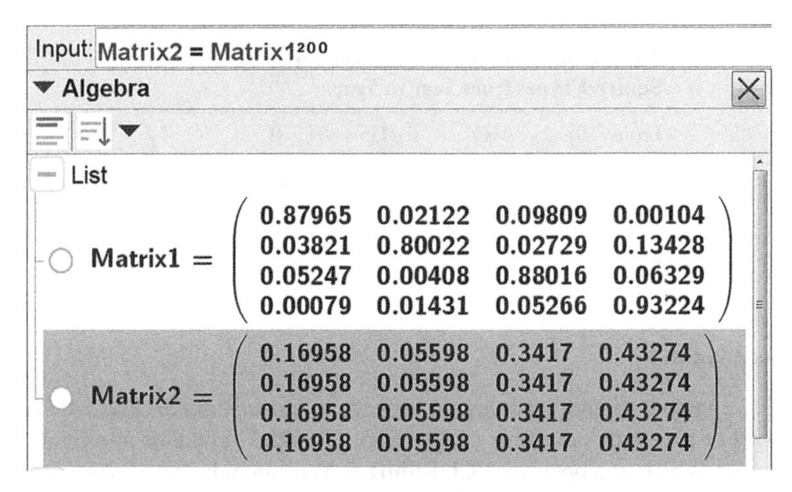

FIGURE 8.10 Matrix2 shows the long-term distribution of squirrel region types.

there is only a 2.73% probability that an area with only grey squirrels this year will change to having only red squirrels the next year. In B7 type:

```
=B2 / $F2
```

Then copy this first to the right and then down, as shown in Figure 8.9. The dollar sign indicates that the column label F won't be replaced when you copy the formula.

Select all these 16 probabilities, right-click on them and select **Create... > Matrix** to set up a matrix in the Algebra window. This matrix can be used for calculations. It represents the transition from one condition in the British fauna a certain year to another condition the next year. If you would like to see the long-term development, you could easily raise this to a power of some large number: 200 will do nicely. Doing this will create a new matrix where each row is the same, which means that regardless of what the fauna distribution looks like today, its distribution will one day end up looking like this.

You can see from the resulting matrix in Figure 8.10 that about 17% of all areas will be populated by red squirrels. According to this model there is no large risk that the red squirrels will completely disappear someday. It might, however, be a good idea to keep track of the populations regularly so that you know that the predicted distribution is stable.

8.4 CHLORINE

An outdoor pool like the one in Figure 8.11 is about to be prepared for the summer season with a daily cleaning program, using chlorine. Initially, on day one, a 15 liter starting dose of chlorine is poured into the swimming pool.

After 24 hours 15% of the chlorine content has disappeared. An extra liter of chlorine is poured into the pool every morning for the rest of the season. How much chlorine will there be in the pool after one day, after you have added the extra daily liter? How much chlorine will there be after two days? After three days?

FIGURE 8.11 Tooting Bec Lido, south London. Nick Cooper at en.wikipedia [CC BY-SA 3.0 (http://creativecommons.org/licenses/by-sa/3.0) or GFDL (http://www.gnu.org/copyleft/fdl.html)], from Wikimedia Commons. https://upload.wikimedia.org/wikipedia/commons/5/5b/Tooting_Bec_Lido_20080724.JPG.

Formulate a recursive relation that will describe the amount of chlorine in the pool. Use this relation to produce a table with values and show with a diagram what will happen with the chlorine in the pool over time. Fit a suitable model to your generated data.

Proposed Solution 1

Start with adding 15 liters of chlorine into the pool at day 0, and record this in the spreadsheet. After one day = 24 hours, there is 85% left of the 15 liters = 12.75 liters. After you have poured in the daily liter, there will be 13.75 liters of chlorine in the pool after which the chlorine content starts to decrease again.

Another 24 hours later the chlorine content is down to 11.69 liters, and you add the daily liter of chlorine to the pool again. The same process continues, and the chlorine content decreases relatively quickly at first. The limiting value for the chlorine content seems to be around 6.67 liters.

The calculation of the chlorine content can be seen as a recursive relation. The same fraction of chlorine is taken away every 24 hours and the same amount, 1 liter, is poured into the pool every morning. The relation can be written as $C_{n+1} = 0.85C_n + 1$, with $C_0 = 15$.

This problem can be modeled into a spreadsheet in GeoGebra. Show the spreadsheet by selecting **Show>Spreadsheet** in the menu. In B3 type

```
= 0.85 B2 + 1
```

and copy this down. In column A you enter 0 and 1, then selecting both these values, copy them down. You can then select all these values in both columns, right-click on the selection and select **Create...>list with points** in order to see them graphically. The points in Figure 8.12 strongly resemble an exponentially decreasing function with an added constant term. Create a model by typing the following commands:

```
a=1
b=1
c=1
m(x)=ab^x+c        (can be hidden)
Fit[list1, m]
```

By right-clicking on B2, having the initial value 15, and then selecting **Show object**, you get a *slider* with which you can adjust the initial value. When you do that, you will immediately see that over time the amount of chlorine does not change but does go to the limit 6.67. Using a recursive relation in a spreadsheet and a general regression tool is very powerful combination.

If you find that the function no longer fits well to the points, the probable reason is that the algorithm used to generate the curve now needs the starting

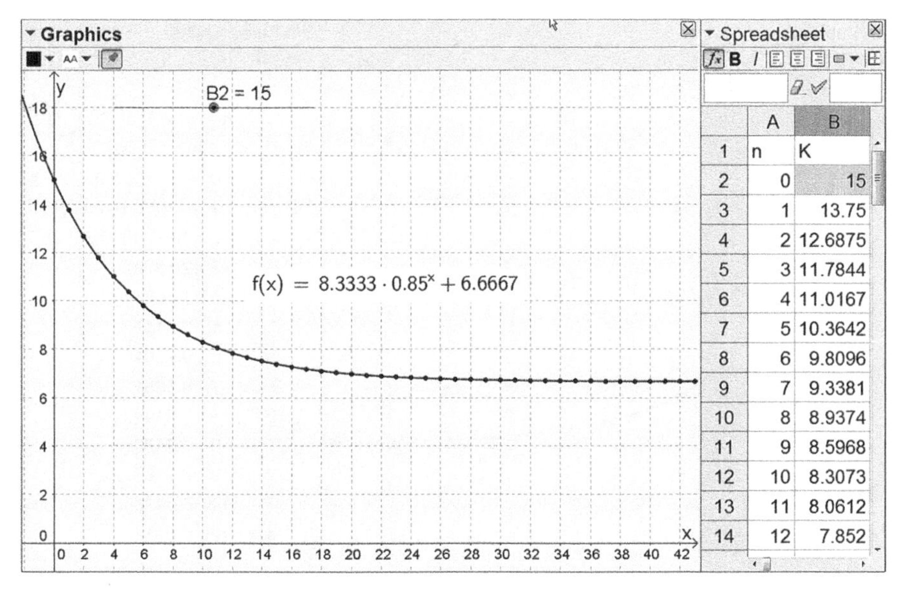

FIGURE 8.12 Decreasing exponential functionmodeled on the data.

values a, b, and c, to be nearer to the actual values. You could then try new starting values, such as b = −1 and c = 6.

You could even try to make your model more general, so that you can adjust the values of the amount of chlorine that remains after 24 hours and the amount of chlorine that is poured into the pool every morning. You need to change the cell values somewhat. Say you decide to set C2 = 0.85 and C4 = 1. Cells C1 and C3 can then be used for the labels.

After that redefine B3 to

```
= $C$2 B2 + $C$4.
```

The extra $ signs mean that these references are *absolute* and that they will not be changed when you copy the formula. The reference to cell B2 is nevertheless still *relative*: it will change when you copy the formula down.

After the formula in B3 has been copied down to all relevant cells, you need to create some sliders for C2 and C4. You can do that by right-clicking on these cells and selecting **Show object**. You also need to adjust the increment, the step length, on the **Slider** tab, in the **Properties** dialogue. In this case 0.01 for C2 and 0.1 for C4 could be suitable values. Figure 8.13 shows these updates to the model.

You can now experiment with this model and investigate if you can adjust the remaining amount of chlorine to your desired level by changing the amount of chlorine that you add every day.

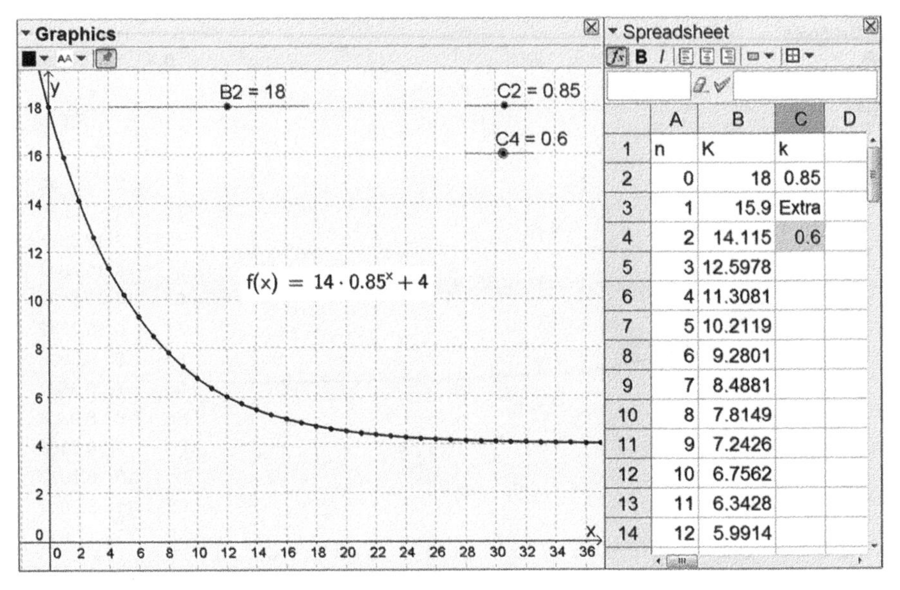

FIGURE 8.13 Model where all parameters can be varied.

Proposed Solution 2

You can also solve the problem algebraically. Start with a description of what happens on the first few days with the amount of chlorine in the pool.

Day 1: $15 \cdot 0.85 + 1 = 13.75$ liters
Day 2: $(15 \cdot 0.85 + 1) \cdot 0.85 + 1 = 15 \cdot 0.85^2 + 0.85 + 1 \approx 12.69$ liters
Day 3: $15 \cdot 0.85^3 + 0.85^2 + 0.85 + 1 \approx 11.78$ liters
Day 4: $15 \cdot 0.85^4 + 0.85^3 + 0.85^2 + 0.85 + 1$

From this you may conclude that the following relation holds:

Remaining amount of chlorine $= 15 \cdot 0.85^n + 0.85^{(n-1)} + 0.85^{(n-2)} + \ldots + 0.85^2 + 0.85 + 1$, where n = the number of days

This is a geometrical series: it is just the term $15 \cdot 0.85^n$ that separates it from the standard form. But you can ignore this term because it goes to zero as n goes to infinity (after very many days). Since this term can be ignored when n goes to infinity, it is not relevant how much chlorine you have as the starting dose: you can have 100 liter or 1 liter and the limit will still be the same because it is the same geometrical series.

The sum of an infinite geometrical series is usually written as

$$S = \frac{a}{1-k}$$

where in this case $a = 1$ and $k = 0.85$, yielding $S = 1/0.15 \approx 6.67$ liters.

If you want to calculate the exact amount of chloride after an exact number of days, you must use the formula for the sum of a finite number of terms with the start term $a = 1$ and also take the term that holds the factor 15 (= 14 liters + 1 liter) into account. The general formula is usually written as

$$S_n = \frac{a \cdot \left(1 - k^{n+1}\right)}{1 - k}$$

Here

$$S_n = 14 \cdot 0.85^n + \frac{\left(1 - 0.85^{n+1}\right)}{1 - 0.85}$$

and you can validate this result by comparing it with the previously calculated values.

8.5 THE DEER FARM

Deer, such as the fallow deer shown in Figure 8.14, have always fascinated us humans, both for their beauty and for the meat they bring to hunters. Suppose that one year a farmer cordons off 20 deer on his land. A female deer will deliver 1 to 3 kids a year. Assume that the every female only delivers 1.5 kids a year in a fenced area, and that about 10% of the population dies every year.

FIGURE 8.14 Fallow deer in field. Av Johann-Nikolaus Andreae (originally posted to Flickr as p9036717.jpg) [CC BY-SA 2.0 (http://creativecommons.org/licenses/by-sa/2.0)], via Wikimedia Commons. https://upload.wikimedia.org/wikipedia/commons/f/f3/Fallow_deer_in_field.jpg.

Construct a mathematical model based on these observations. Calculate the size of the deer population 20 years hence into the future. How many deer can be shot each year if the farmer wants to have a remaining stable herd of at least 200 deer? For how long must the farmer wait until the hunting can start?

A better model will take into account that there are limited resources in the area, in that the most the area can handle is about 500 deer. You can further assume that the number of kids that every female deer gives birth to will goes to zero when the population increases. In what way will the hunt be affected according to this refined model?

Proposed Solution 1

This problem is suitable for modeling into the spreadsheet. Open GeoGebra's spreadsheet by selecting **Show > Spreadsheet** in the menu. Then build the model according to Figure 8.15.

This is quite typical of how you can build a model in a spreadsheet. Use clear labels for the parameters and build the columns from left to right. Create general formulas that can be copied down.

Start by defining a number of parameters according to Figure 8.15:

```
Deaths = 0.1
Hunt = 0
HuntingStarts = 6
N_0 = 20
Newborn = 1.5
```

The year numbers are created manually by entering 1 and 2. Then you select these two cells and copy them down to row 21. In cells B2 to F2 type the following formulas:

In **B2**: `= N_0`
In **C2**: `= round(B2/2 Newborn)`
In **D2**: `= round(Deaths(B2+C2))`
In **E2**: `= If[A2 >= HuntingStarts, Hunt, 0]`
In **F2**: `= (A2, B2)`

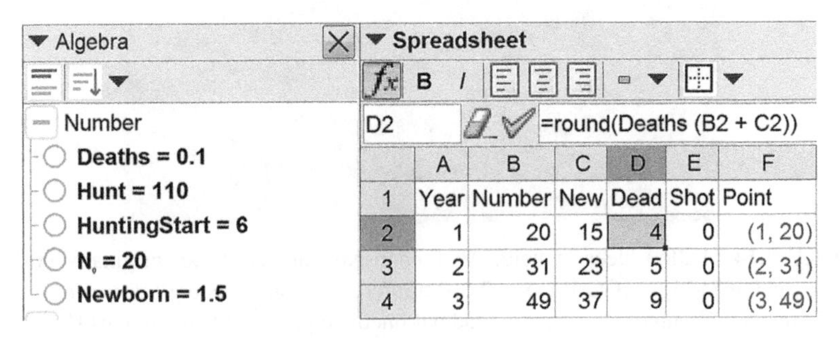

FIGURE 8.15 Parameters and the spreadsheet model.

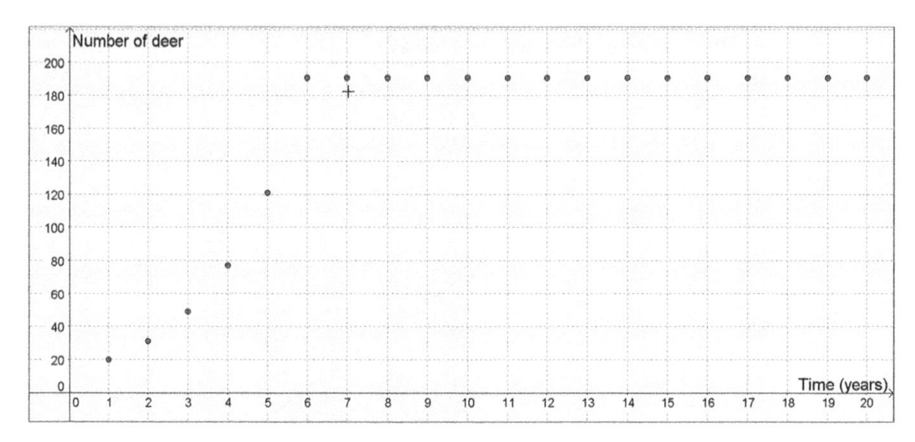

FIGURE 8.16 Hunt starts after six years.

Now select all these cells and copy them down one row, to row 3. After that you need to change cell B3 to

```
= B2 + C2 - D2 - E2
```

Finally select B3:F3 and copy these cells down to row 21.

Observe the function **round()**, which rounds off to whole numbers, and the command **If[<Condition>, <Value if true>, <Value if false>],** which makes it possible to say that *if* you have waited long enough, *then* you can hunt, *but otherwise* not.

The points created in column F are shown automatically in the graphics view. Start by setting **HuntingStarts** to 6 years. After that you can select **Hunt** and change its value with the arrow keys. In Figure 8.16 you can see that a stable deer herd is achieved if you shoot about 110 animals per year. Observe that this solution is unstable, meaning that a small change of the parameters up or down will give a large change in the number of deer after 20 years.

Proposed Solution 2

In order to get a better model and take into account decreasing birthrates when the resources run short, you could create a linear function that gives the number of new kids as a function of the number of animals. You can also introduce the parameter **K = 500**. The function, *New*(*N*), should have the properties that *New*(0) = Newborn and *New*(M) = 0. If linear, then

```
New(N) = Newborn - N Newborn/K
```

Draw this new function in Graphics Area 2, shown in Figure 8.17.

You also need to make some modifications in your calculation. Replace cell C2 with

```
=round(B2 New(B2) / 2)
```

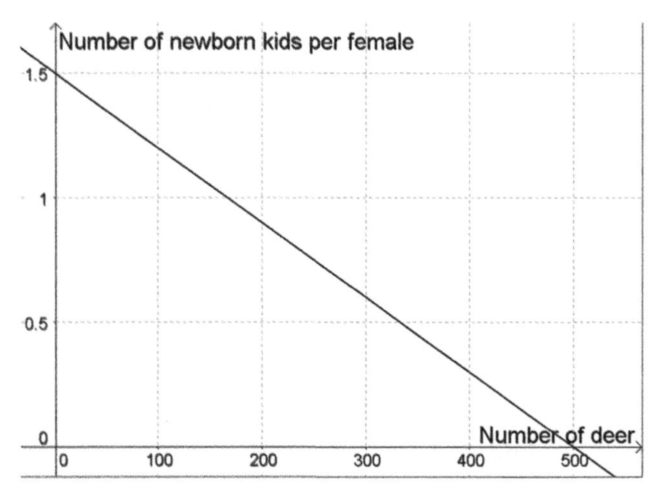

FIGURE 8.17 Birthrate will decrease with the number of animals when the resources are limited.

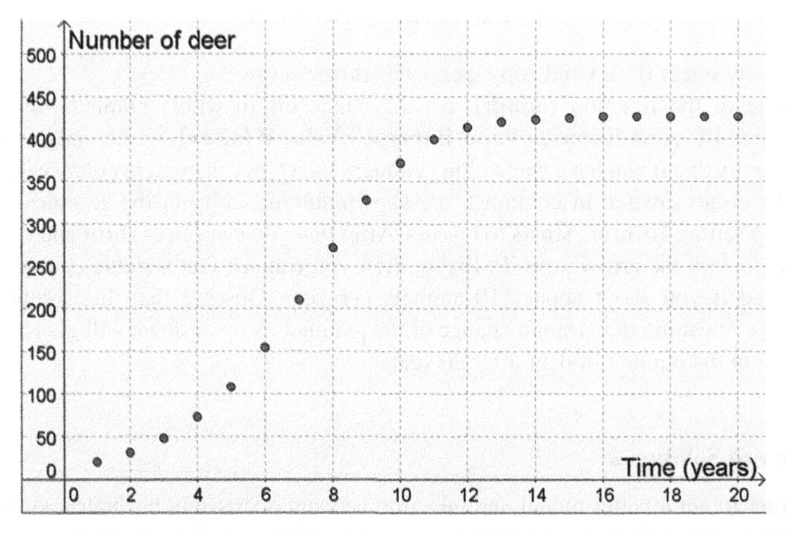

FIGURE 8.18 Deer population living with limited resources but without being hunted.

and copy this down. The results will be different now. You have to wait another year for the hunt to start, and you have to lower the number of deer shot to about 60 animals in order to get a stable herd of about 200 animals. This solution, shown in Figure 8.18, is still very sensitive for changes in the original birthrate, which is why all hunting must be preceded by a careful inventory of the population in terms of numbers and health.

If you never start hunting the deer, the population will stabilize at around 425 animals, a little below the 500 that was assumed to be a reasonable maximum.

Comments

If you let one group of students do this modeling exercise one year, you could instruct next year's group to expand on it. There are quite a few possible expansions for this scenario:

What population is required for different numbers of deer to be shot and still maintain a stable population? Make a list of the different values and see if you can find a function that models this relationship.

How does the number of newborn per female deer affect the hunt? Make a list of the different values and see if you can find a function that models this relationship.

Assume that the deer population needs a certain amount of deployed food in order to grow. Furthermore assume that the food costs money and that every deer that is shot provides a certain income. Is it profitable to deploy as much food as possible or is there a limit to feeding the deer?

Scarcity of resources can result in death rates rather than birthrates. Change the model to reflect this and compare what death rates and birthrates give the same results. Can you find a relationship between the two?

8.6 ANALYZING A NUMBER SEQUENCE

Which is the next number in the sequence 8, 8, 8, 32, 128, 368, …? Is there more than one possible answer?

There is really nothing special about this sequence of numbers. Allow your students to make up their own number sequences. We just want to show you some different ways of tackling this problem, especially building a mathematical model.

Proposed Solution 1

Through any six points there passes exactly one fifth-degree polynomial. So fit a fifth-degree polynomial through the points (1, 8), (2, 8), (3, 8), (4, 32), (5, 128), and (6, 368) and then calculate the same polynomial for the value when $x=7$.

You can enter the points directly into the input field as **(1, 8)**, and so on. Another option is to enter the points into the spreadsheet that you open with **Ctrl-Shift-S**. With the x-coordinates in one column and the y-coordinates in the next column, select the values, right-click on the selected area, and in the menu that appears select the option **Create…>list with points**.

Fitting a fifth-degree polynomial to the points is done by typing

```
FitPoly[A,B,C,D,E,F,5]
```

unless you have already created a list and are using the syntax

```
FitPoly[list1, 5]
```

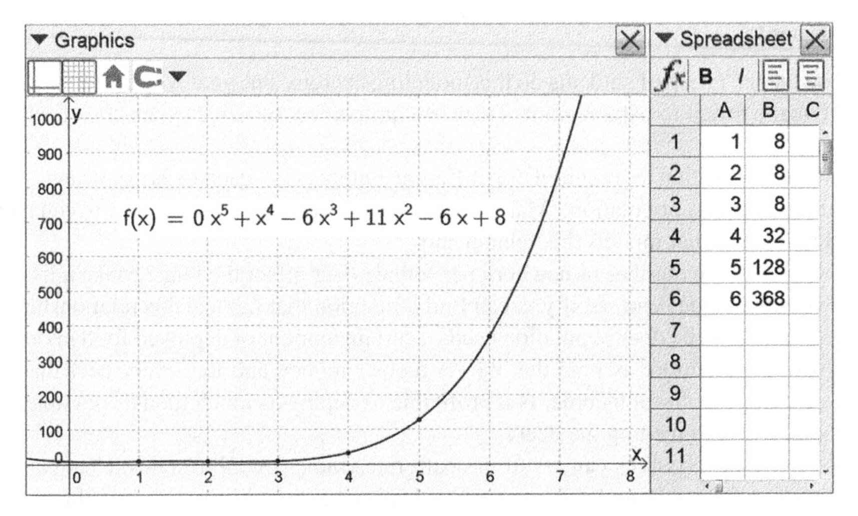

FIGURE 8.19 Polynomial generating a number sequence.

In either case you can get the answer to the original question by evaluating

```
f(7)
```

in the input bar, so you can see that the next number in the number sequence is 848. You can also see in Figure 8.19 that the fifth-degree coefficient is 0, and that the polynomial really is a fourth-degree polynomial: $y = x^4 - 6x^3 + 11x^2 - 6x + 8$.

Proposed Solution 2

If you do not have access to software that can fit a polynomial to a sequence of points but have access to a spreadsheet, or maybe not even that, it is still possible to find a generating polynomial for a number sequence by learning the theory behind all polynomial number sequences. You will use a spreadsheet, but technichally you could use pencil and paper.

The key is to set up a difference schema, where you write down the successive differences in all ways until they become constant. This is sometimes referred to as the *finite difference method*, even though this really addresses much harder topics. In cell C3 type:

```
= B3 - B2
```

and copy down. In cell D4 type

```
= C4 - D3
```

and copy *that* down, and so on. You will see that you get a constant difference of 24 in step 4. If a sequence of successive differences get the constant value a at step i,

▼ Spreadsheet

K2 =G2 + 6A2³

	A	B	C	D	E	F	G	H	I	J	K	L	M
1	x-values	y-values	d1	d2	d3	d4	minus x⁴	d1	d2	d3	minus (-6)x³	d1	d2
2	1	8					7				13		
3	2	8	0				-8	-15			40	27	
4	3	8	0	0			-73	-65	-50		89	49	22
5	4	32	24	24	24		-224	-151	-86	-36	160	71	22
6	5	128	96	72	48	24	-497	-273	-122	-36	253	93	22
7	6	368	240	144	72	24	-928	-431	-158	-36	368	115	22

FIGURE 8.20 Difference schema for analyzing a number sequence.

the number sequence can be generated by a polynomial of degree i where the coefficient for the highest degree term is $a/i!$ In this case, the fourth-degree term is $24/4! = 1$.

From the generating polynomial $f(x) = a \cdot x^4 + b \cdot x^3 + c \cdot x^2 + d \cdot x + e$ where $a = 1$, you now create the difference $f(x) - a \cdot x^4 = b \cdot x^3 + c \cdot x^2 + d \cdot x + e$, which is a number sequence generating a polynomial of degree three. In cell G2 you can create this difference by typing

```
=B2 - A2^4
```

and copying down as usual. Once again you can produce a difference schema, and in Figure 8.20 you can see that it shows a constant difference of –36 after three steps.

Now you can just continue with the same method as before. The third-degree coefficient $b = -36/3! = -6$. Create a new number sequence by subtracting this term too.

$f(x) - a \cdot x^4 - b \cdot x^3 = c \cdot x^2 + d \cdot x + e$, so in K2 type:

```
=G2 - (-6)A2^4
```

to get a new series of terms where the second difference comes to 22. The second-degree coefficient $c = 22/2! = 11$, and when you have subtracted this too, you can identify the final number generating polynomial as $-6x + 8$. The whole original given number sequence can therefore be generated by the fourth-degree polynomial $f(x) = x^4 - 6x^3 + 11x^2 - 6x + 8$.

A difference schema is the discrete equivalent to repeating derivatives, with the difference that you do not need to know the original function to begin with. In a way the method also has some similarities with the integration of solutions for directly integrable differential equations.

Proposed Solution 3

If you have access to the Internet, then you have more options. Send

```
8, 8, 8, 32, 128, 368
```

to Wolfram Alpha and you will get a difference schema, a generating function, and the next few terms, shown in Figure 8.21. The function is not the same, however, and

Possible sequence identification:

Closed form:

$$a_n = \frac{8\left(15\,n^3 - 89\,n^2 + 150\,n - 16\right)}{n^2 - 15\,n + 74} \quad \text{(for all terms given)}$$

Continuation:

$$8,\ 8,\ 8,\ 32,\ 128,\ 368,\ 808,\ 1408,\ 2024,\ 2528,\ 2888,\ \frac{59552}{19},\ 3308,\ 3440,\ \ldots$$

FIGURE 8.21 Wolfram Alpha suggests a rational function.

Search: **seq:8,8,8,32,128,368**

Sorry, but the terms do not match anything in the table.

If your sequence is of general interest, please submit it using the form provided and it will (probably) be added to the OEIS! Include a brief description and if possible enough terms to fill 3 lines on the screen. We need at least 4 terms.

FIGURE 8.22 OEIS only finds well-known series—why not try with 1, 3, 4, 7, 11....

the next few terms are different from what you found earlier. Wolfram Alpha gives a rational function, a quotient of a third- and a second-degree polynomial, which also gives a perfect fit to the given terms.

Using this generating function the next term comes to 808. Which one of these answers is correct? Without more information is it hard to determine that. In some sense *all answers* are correct, since you can always find a function that generates the given terms and one more. We can, for instance, create a seventh-degree polynomial going through the six known points and another point of our choice.

Proposed Solution 4

A search through OEIS, *Online Encyclopedia of Integer Sequences*, http://oeis.org/ will unfortunately not give you an answer this time, since this number sequence is not famous or well known. But OEIS, shown in Figure 8.22, can sometimes be a very powerful tool if you meet number sequences that are difficult to analyze but are previously known.

8.7 INNER AREAS IN A SQUARE

In a square with side length s you construct an inner square by connecting segments from the square's vertices to the opposite edges' midpoints. This produces an inner square, shown in Figure 8.23. Determine the ratio of the inner square's area and the area of the larger original square.

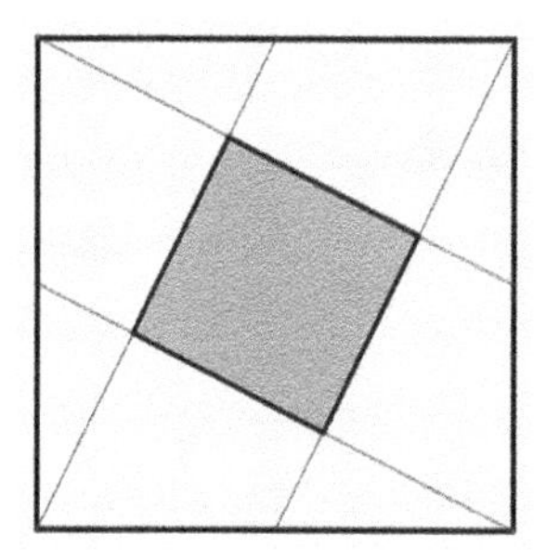 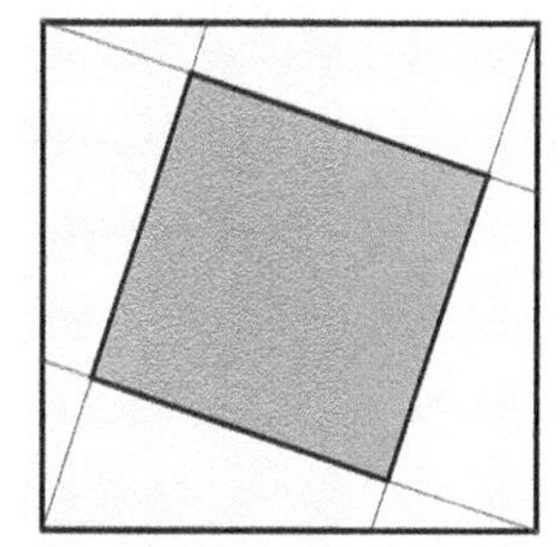 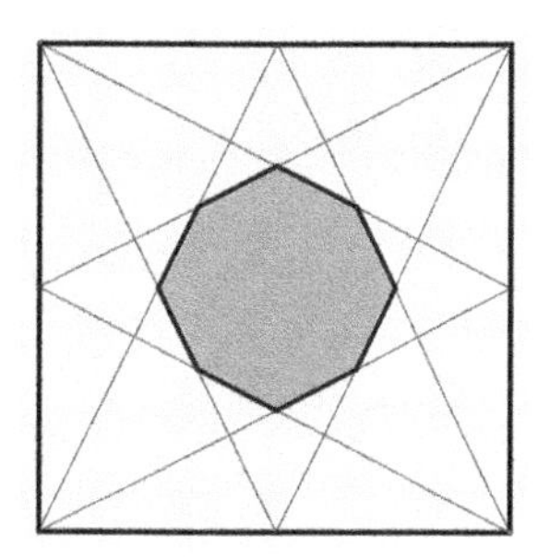

FIGURE 8.23 Inner shapes of a square.

How is the ratio of the two areas changed if instead you construct the inner square by splitting the sides of the large square into three equally long subsections? Four subdivisions? Five? ... Is there a pattern?

If you draw segments from the outer square's vertices to both the opposite edges' midpoints, an inner octagon is constructed. What happens to the ratio of the areas now?

What would happen if you divide the sides of the outer square in other ways? You could draw segments from 2/3 instead of from 1/3, or perhaps 1/4 and 3/4, or 1/5 and 3/5? Investigate!

Proposed Solution

To trisect a segment AB, you could construct three circles with the same arbitrary radius on another arbitrary segment from A, shown in Figure 8.24. You then draw lines though these new equidistant points, parallel to the line from the last point on the arbitrary line to B. These lines intersect AB so that it is divided in three equal parts.

But you can also construct points from other points by direct calculations. The midpoint formula can be written

$$\mathrm{C} = \frac{\mathrm{A}+\mathrm{B}}{2}$$

where A, B, and C, are points and the calculations have to be done on x- and y-coordinates separately. In GeoGebra is it possible to do these calculations directly on the points themelves. If you type

```
C = (A + B)/2
```

the midpoint C will be constructed between A and B. In the same manner you can construct other points. The commands

```
D = (A+2B)/3
E = (2A+B)/3
```

will create two points that trisect the distance AB, as shown in Figure 8.25.

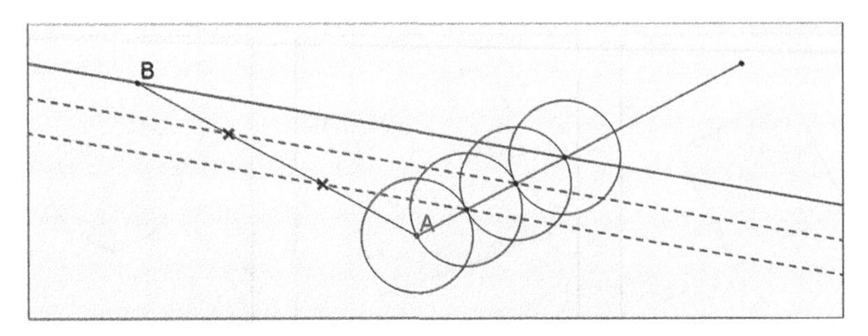

FIGURE 8.24 How to divide a segment in equal parts.

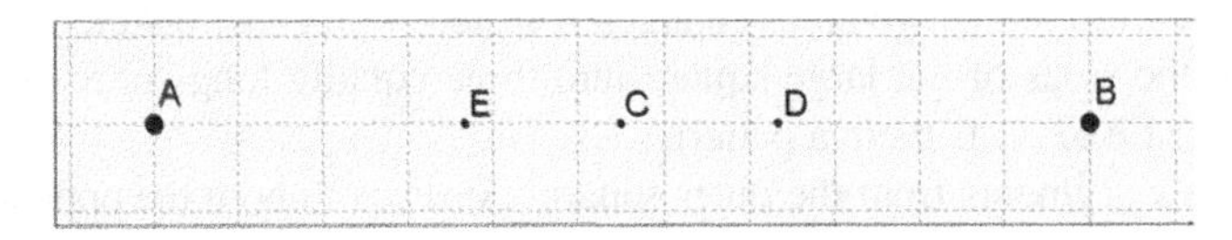

FIGURE 8.25 How to construct points, dividing a segment in different parts.

To create a fully dynamic construction with all the possibilities this entails, you should now start a new window with **Ctrl-N** and construct two integer sliders, n and m, to help you create the points. After that you create a square with the Regular Polygon tool. Assuming that the verticies are named A, B, C, and D, create four points with the following commands:

```
(n A + (m - n) B)/m
(n B + (m - n) C)/m
(n C + (m - n) D)/m
(n D + (m - n) A)/m
```

The inner square is now constructed using the Segment, Intersection, and Polygon tools; see Figure 8.26. Vary n and m and see how the area for the inner polygon is changing. Since the area of the first square is arbitrary, it is the ratio between the areas that is interesting, something you can compute by typing

```
Ratio = poly2/poly1
```

The answer to the first question seems to be exactly 1/5=0.2. You should use a table to record your results, depending of the values for n and m. The simplest way is to fill the table manually, but you could also instruct GeoGebra to do it automatically.

You should start by creating a new tool. In the menu, select **Tools>Create New Tool**. In the list over object, chose **poly2** and click **Next**. GeoGebra suggest the numbers n and m and the points A and B as input objects, which you can accept. Label the tool **InnerSquare** as in Figure 8.27, and click **Finish**.

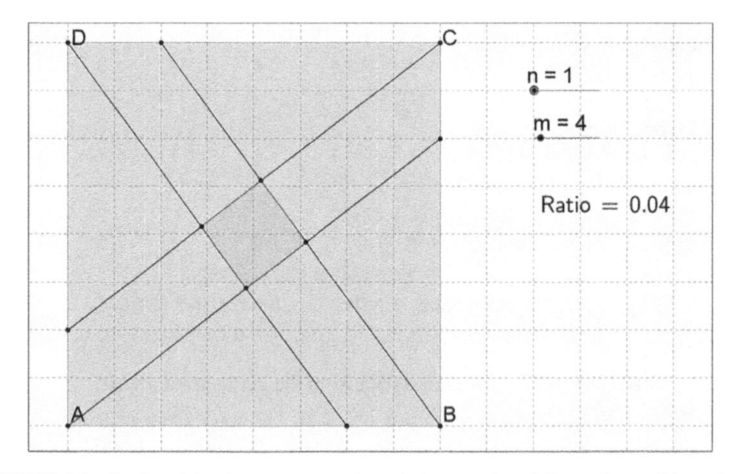

FIGURE 8.26 Ratio of the inner area to the whole area for differently positioned points.

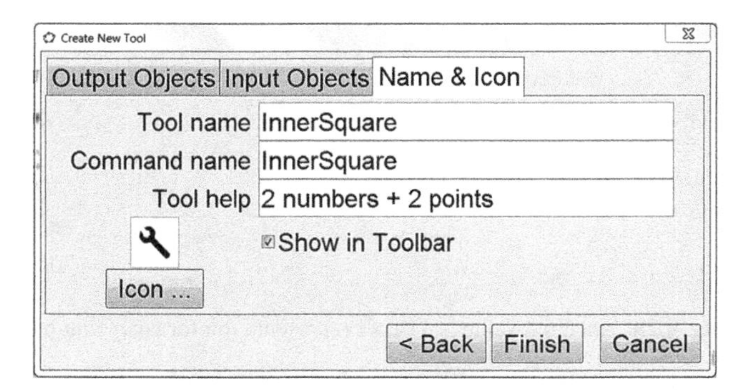

FIGURE 8.27 Creating a new tool.

Now it is time to open the spreadsheet with **Ctrl-Shift-S** or through the menu option **View > Spreadsheet**. Enter values for *n* horizontally in the first row and values for *m* vertically, in column A, and then type

In B2: `InnerSquare[B$1, $A2, A, B] / Polygon1`

This formula can be copied, first down and then to the right as in Figure 8.28. The dollar signs in the formula locks the reference, so make sure that the formulas refer to row 1 and column A all the time.

If for a moment you stop and look at what you are doing, you could argue that you are in the middle of a specialization process where you create special cases to get more information that you will try to generalize later on. An efficient problem-solving process is often a successful interplay between specialization and generalization.

The reason why you first had to create a new tool was that every cell in the spreadsheet only has room for just one command or one object. In this way all objects that you create can be mass-produced.

	A	B	C	D	E	F	G	H	I	J	K
1		1	2	3	4	5	6	7	8	9	10
2	2	0.2	1	?	?	?	?	?	?	?	?
3	3	0.07692	0.4	1	?	?	?	?	?	?	?
4	4	0.04	0.2	0.52941	1	?	?	?	?	?	?
5	5	0.02439	0.11765	0.31034	0.61538	1	?	?	?	?	?
6	6	0.01639	0.07692	0.2	0.4	0.67568	1	?	?	?	?
7	7	0.01176	0.05405	0.13846	0.27586	0.4717	0.72	1	?	?	?
8	8	0.00885	0.04	0.10112	0.2	0.34247	0.52941	0.75385	1	?	?
9	9	0.0069	0.03077	0.07692	0.15094	0.25773	0.4	0.57647	0.78049	1	?
10	10	0.00552	0.02439	0.0604	0.11765	0.2	0.31034	0.44954	0.61538	0.80198	1

FIGURE 8.28 Measured data from the geometrical model.

	A	B	C	D	E	F	G	H	I	J	K
1		1	2	3	4	5	6	7	8	9	10
2	2	5	1	?	?	?	?	?	?	?	?
3	3	13	2.5	1	?	?	?	?	?	?	?
4	4	25	5	1.88889	1	?	?	?	?	?	?
5	5	41	8.5	3.22222	1.625	1	?	?	?	?	?
6	6	61	13	5	2.5	1.48	1	?	?	?	?
7	7	85	18.5	7.22222	3.625	2.12	1.38889	1	?	?	?
8	8	113	25	9.88889	5	2.92	1.88889	1.32653	1	?	?
9	9	145	32.5	13	6.625	3.88	2.5	1.73469	1.28125	1	?
10	10	181	41	16.55...	8.5	5	3.22222	2.22449	1.625	1.24691	1

FIGURE 8.29 Inverted value is a better representation for analyzing patterns.

Now try to analyze the new data. Start by just looking at it and notice that the value $1/5=0.2$ is visible here and there and that it clearly represent different appearances of the fraction $1/5=2/10=3/15=4/20, \ldots$. Other repetitions confirm this.

The values of $0.2=1/5$ and $0.04=1/25$ are exact. Perhaps other values are also inverted values or whole numbers? It seems so. A small and quick change in the spreadsheet formula to

```
= Polygon1/InnerSquare[B$1, $A2, A, B]
```

gives the results in Figure 8.29 instead.

Concentrate on the first two columns and copy these values into a new spreadsheet in a new window of GeoGebra. In cell C3 type:

```
= B3 - B2
```

and then copy this down to see the differences in the number sequence. The differences are clearly factors of four, increasing by four every step.

C3 =B3 - B2

	A	B	C	D	E	F	G
1			d1	d2			d1
2	1	1				-1	
3	2	5	4			-3	-2
4	3	13	8	4		-5	-2
5	4	25	12	4		-7	-2
6	5	41	16	4		-9	-2
7	6	61	20	4		-11	-2

FIGURE 8.30 Difference analysis of a number sequence generated by $2m^2 - 2m+1$ (column B).

If a number sequence's successive differences get a constant value a at step i, the number sequence is generated by a polynomial of degree i where the highest degree coefficient is $a/i!$ In this case you got the constant value 4 after step 2; see Figure 8.30 This means that the ratio between the outer and inner areas can be generated by a second-degree polynomial with the highest degree term $4/2!=2$. Compare this to Problem 8.6, *Analyzing a Number Sequence*.

Now subtract this term $2m^2$ from the values in column B and put the new values in column E. In practice, type:

```
= B1-2A1²
```

in E1 and copy it down. You will find that this new number sequence is linear and can be represented by the relationship $1 - 2m$. This in turn tells you that the original number sequence can be represented by the relationship $2m^2 - 2m+1$ and that the inner area can be calculated as

$$\text{Inner area} = \frac{\text{Original area}}{2m^2 - 2m + 1} \qquad \text{for } n=1$$

In the same way you can analyze the other columns to get a relationship that holds for $n=2, 3, \ldots$ The generating polynomials become

For $n=1$: $2m^2 - 2m+1$
For $n=2$: $m^2/2 - m+1$
For $n=3$: $2m^2/9 - 2m/3+1$
For $n=4$: $m^2/8 - m/2+1$
...
For n: $\dfrac{2m^2}{n^2} - \dfrac{2m}{n} + 1 = \dfrac{2m^2 - 2mn + n^2}{n^2}$

As you can see, algebra is a powerful pattern recognition tool. A general proof for different values of m and n may be difficult to work out, but if you start with a special case, it should be possible to sketch a proof. We will start by proving the first case.

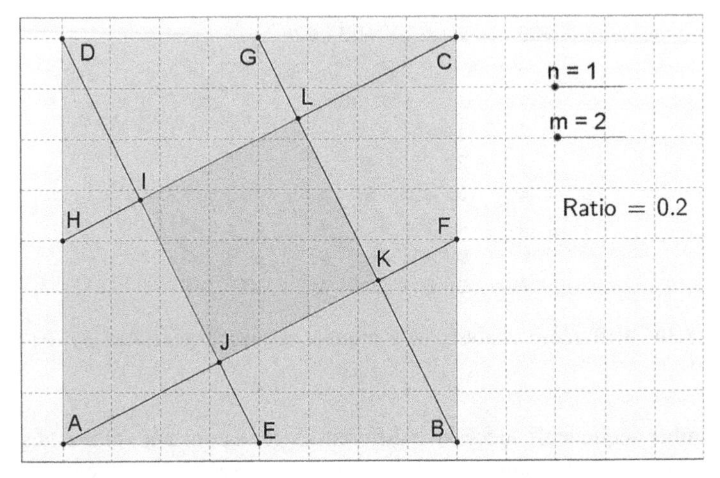

FIGURE 8.31 $n=1$ and $m=2$ gives an inner area that is one-fifth of the outerarea.

In Figure 8.31 we see the simplest case. Assuming AB = 1, without loss of generality, we have the combined area of ΔABF + ΔCDH = 0.5. ΔAEJ ~ ΔABK ~ ΔABF, since they all have a right angle and the angle at A in common. That gives the area of ΔAEJ as 1/5 of ΔABF and 1/4 of ΔABK, since AE is half of AB; now ΔABK, together with ΔBFK, has the same area as ΔAEJ and thus equals the area of ΔABF. ΔAEJ is consequently 1/5 of 1/4 = 1/20 of the whole area. The quadrilateral EBKJ has an area equal to 3/20, by which we obtain the inner, target area of the square IJK L = $1 - 4 \cdot (1/20) - 4 \cdot (3/20) = 4/20 = 1/5$. Q.E.D.

With carefully drawn figures, a structured method and some patience, it should now be possible to construct proofs in the same manner for other configurations until you see how a general proof should be constructed for general values of both n and m. As we did above, starting with special cases is often the best approach when studying complex mathematical situations. The case with the inner octagon is left for the reader—or for the reader's students—to try.

8.8 INNER AREAS IN A TRIANGLE

In an arbitrary triangle ΔABC you can define points to trisect every side. Construct a segment from the opposite vertex to the first trisection point, counterclockwise. These segments define a new triangle, shown in Figure 8.32. How is the area of this triangle related to the area of the outer triangle?

Connect both trisection points with the opposite vertex. The result of this construction becomes an inner hexagon. Investigate the relationship between the inner hexagon's area and the area of the triangle ΔABC.

Subsect all sides of the triangle in five equally large segments. Connect the two central fifth sect points with the opposite vertex (see the figure). The result of this

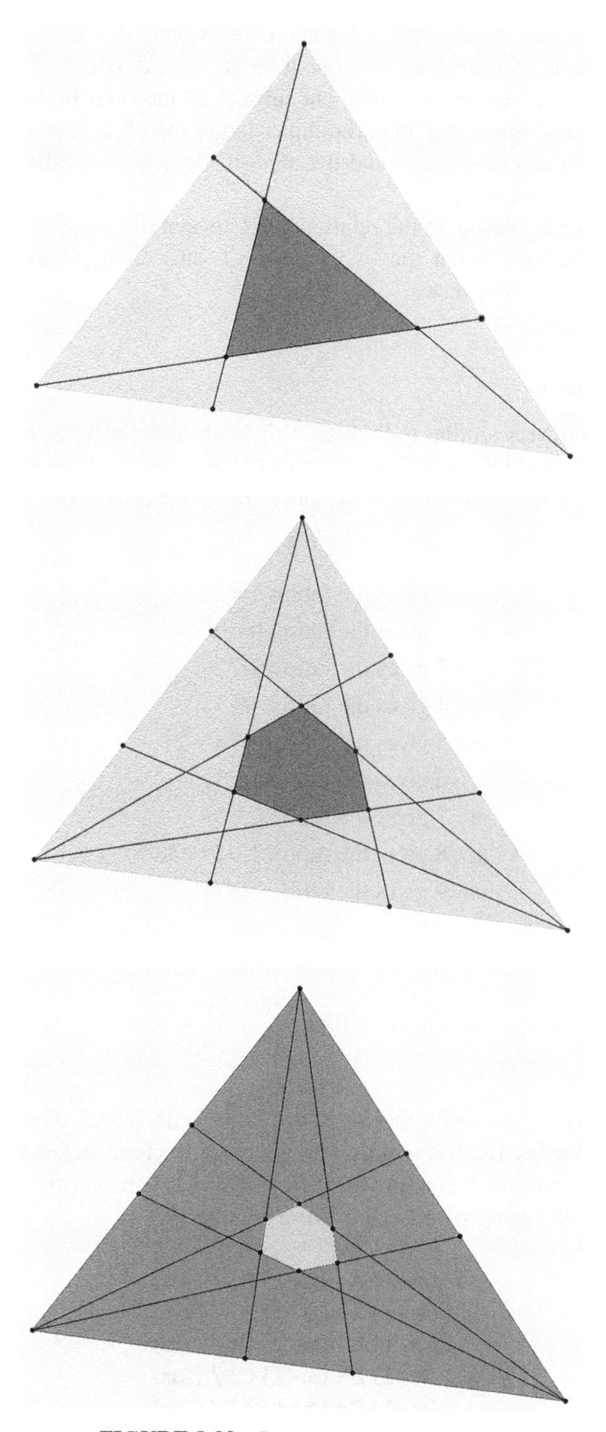

FIGURE 8.32 Inner shapes of a triangle.

construction becomes another inner hexagon. Investigate the relationship between the new inner hexagon's area and the area of the triangle ΔABC.

Do the same construction as above, but subsect all the sides in the outer triangle ΔABC into n parts, where n is an odd number larger than 5. Investigate the relation between the central region's area and the area of the triangle ΔABC in relation to your selection of value of n.

Repeat this, and investigate the relationship between the triangle ΔABC and the area that you will get when the two outermost subsection points are used and connected to the opposite vertex.

Proposed Solution

This problem is in very similar to Problem 8.7, *Inner Areas in a Square*, and you can use similar methods to solve it. The first part of the problem deals with case where $n=1$ and you get the following results once the model, shown in Figure 8.33, is constructed:

$m=1$ gives the ratio$=1=1/1$
$m=2$ gives the ratio$=0=0/3$
$m=3$ gives the ratio$=1/7$
$m=4$ gives the ratio$=4/13$
$m=5$ gives the ratio$=3/7=9/21$
$m=6$ gives the ratio$=16/31$
$m=7$ gives the ratio$=25/43$
$m=8$ gives the ratio$=12/19=36/57$
$m=9$ gives the ratio$=49/73$

...

$$\text{In general: ratio} = \frac{(m-2)^2}{m^2-m+1}$$

This is, of course, just a hypothesis that you now need to justify. Here the command **RationalNumberText[Ratio]** was used to get exact fractions in GeoGebra.

In order to create the hexagons for different subsections m (m odd), create two points at every side using with the following commands:

```
((m-1)A+(m+1)B)/(2m)
((m+1)A+(m-1)B)/(2m)
((m-1)B+(m+1)C)/(2m)
((m+1)B+(m-1)C)/(2m)
((m-1)C+(m+1)A)/(2m)
((m+1)C+(m-1)A)/(2m)
```

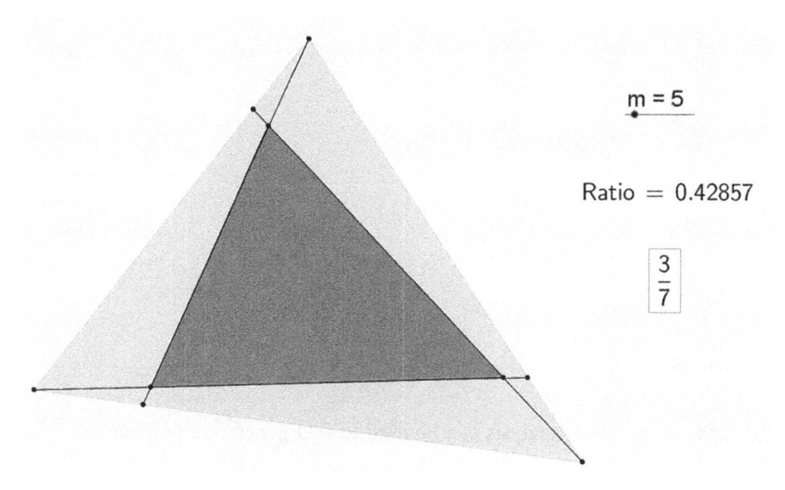

FIGURE 8.33 Inner area of a triangle.

Then create segments and the central hexagon. Measure the area as before.

$m = 3$ gives the ratio $= 1/10$
$m = 5$ gives the ratio $= 1/28$
$m = 7$ gives the ratio $= 1/55$
$m = 9$ gives the ratio $= 1/91$
$m = 11$ gives the ratio $= 1/136$
$m = 13$ gives the ratio $= 1/190$
...

Since you are now jumping two steps at a time, it is a little more difficult to analyze the differences in the number sequence 10, 28, 55, ... You migh,t however, suspect that you once again are looking for a second-degree polynomial. In a new GeoGebra file you can enter the points (3, 10), (5, 28), (7, 55), ..., and then ask GeoGebra for a seconddegree polynomial regression to these points by typing

```
FitPoly[A,B,C...,2]
```

Looking at the result in Figure 8.34, it seems that the relation of the area of the inner area when using the inner subsection points and the original triangle can be written as

$$\frac{8}{9m^2-1} = \frac{8}{(3m+1)(3m-1)}$$

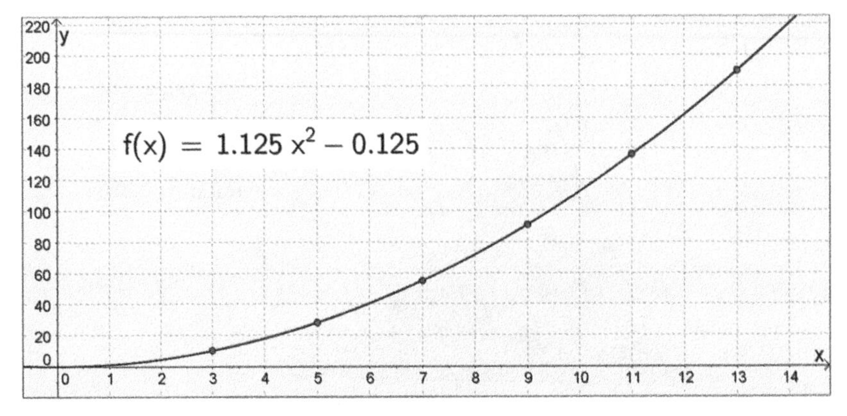

FIGURE 8.34 Regression analysis can be used to find generating polynomials to number sequences.

If you instead use the outer subsection points for the inner area, these can be written as (A+(m − 1)B)/m and ((m − 1)A+B)/m, and so forth. You will get the following result:

m=3 gives the ratio=1/10 (as before)
m=5 gives the ratio=1/3=9/27
m=7 gives the ratio=25/52
m=9 gives the ratio=49/85
m=11 gives the ratio=9/14=81/126
m=13 gives the ratio=121/175
...

The numerators are square numbers, and the denominators can be found by doing a square regression, the same as above. Then this relationship can be written as

$$\frac{2(m-2)^2}{2m^2+m-1}$$

Comment

If you search at *The On-line Encyclopedia of Integer Sequences*, http://oeis.org, for the number sequence 10, 28, 55, 91, 136, 190, …, you will find that these numbers are *nonagonal numbers*, comparable to triangular numbers, square numbers, pentagonal number, and so forth. They are created, as in Figure 8.35, with a center point in the middle and then the next nine points in a ring around the first, 18 (=2·9) points in the next ring, and so forth. But are the nonagonal number a result from the fact that you have a triangle and then create a hexagon from this? Three vertices plus six vertices? There is clearly room for more investigations here!

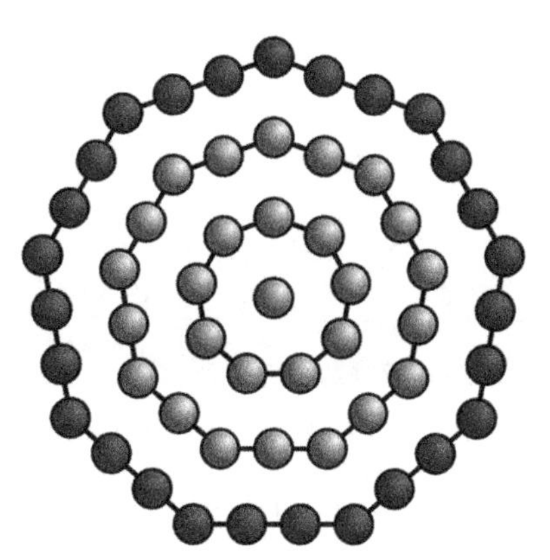

FIGURE 8.35 Nonagon numbers. Photo: Claudio Rocchini (Own work) [GFDL (http://www.gnu.org/copyleft/fdl.html), CC-BY-SA-3.0 (http://creativecommons.org/licenses/by-sa/3.0/) or CC BY 2.5 (http://creativecommons.org/licenses/by/2.5)], via Wikimedia Commons. https://upload.wikimedia.org/wikipedia/commons/7/70/Centered_nonagonal_number.svg.

8.9 A CLIMATE MODEL BASED ON ALBEDO

Figure 8.36 shows the Sun shining above a bank of clouds. The *solar constant*, the amount of power beaming in from the Sun to the Earth, has been measured to about 1,370 W/m^2 perpendicular to the line between the two. Earth's *albedo* (from latin *alba* = whiteness) is close to 0.3, meaning that about 30% of the radiation is reflected out to space again. But Earth is in thermal equilibrium, which means that all the energy thus absorbed is also radiated away from the entire surface of the Earth. The thermal equilibrium can, with the help of the radiation law, be formulated as

$$(1-a)\mathrm{S}\cdot\pi r^2 = 4\pi r^2\cdot\sigma T^4, \text{or} (1-a)\mathrm{S} = 4\sigma T^4$$

The solar radiation falls in perpendicular to Earth's cross section while the heat from the Earth is radiated out from a spherical surface, thus the factor 4. Here a is Earth's albedo, S is the solar constant, σ is Stefan–Boltzmanns constant, and T is the average temperature of Earth in Kelvin. From this relation is it possible to write the average temperature as a function of the albedo, $T = T(a)$.

The average temperature comes to about −18°C, assuming an albedo of about 0.3. It is the greenhouse effect that results in a higher average temperature. We can introduce the Earth's *emissivity*, ε, to explain this. Emissivity (from the Latin word *emi´tto* = to radiate, to beam) is a body's ability to transmit electromagnetic radiationsuch as heat and light. The emissivity ε is smaller than one, signaling the fact that the Earth is radiating less than it would without an atmosphere. We can now write

$$(1-a)\mathrm{S} = 4\varepsilon\sigma T^4$$

FIGURE 8.36 Sunrise over Haleakala. Photo: Ewen Roberts [CC BY 2.0 (http://creativecommons.org/licenses/by/2.0)], via Wikimedia Commons. https://upload.wikimedia.org/wikipedia/commons/7/70/Sunrise_over_Haleakala.jpg.

By measuring Earth's average temperature to about 15°C, it is possible to calculate ε.

Yet at the same time there may be an interesting feedback loop going on here. The albedo determines the average temperature, but the temperature is also affecting the albedo back again. How so? If the planet becomes colder, it will grow large ice caps and the albedo will increase, which will make the temperature fall further. If instead the planet becomes warmer, the icecaps will melt and the albedo will sink. We will then receive more radiation and the temperature will rise even more. We have no analytical model or formula for this, but broadly speaking, you can assume that the following:

- If the average temperature goes below 0°C, most of the planet will become covered with ice and snow, and the albedo will reach its maximum value of about 0.8.
- Earth's current average temperature is about 15°C, and the albedo is about 0.3.
- The albedo will never sink below about 0.15, which will happen to a planet free of both clouds and ice.
- If the average temperature goes up, clouds will form, resulting in an albedo of 0.5.
- If the average temperature goes up a lot, the whole planet will be covered with clouds, giving the entire planet the albedo of 0.5.

Construct a mathematical model that describes how the albedo is determined by the temperature, $a = a(T)$. Calculate Earth's temperature without emissivity, and then use this value to calculate the emissivity. Investigate what will happen in the feedback loop as you repeatedly calculate the temperature from the albedo and the albedo from the temperature.

How come the Earth is not covered by ice or the seas aren't boiling? The feedback loop indicates that our current status should be unstable, but is it unstable in reality? What will it take for us to get into an instability leading us to a catastrophe?

Proposed Solution

Start by defining constants, and then calculate Earth's average temperature from an albedo of 0.3:

```
a_0 = 0.3
amin = 0.15
amax = 0.8
t_0 = 15
T_0 = t_0 + 273.2
SolarConst = 1370
SBConst = 5.67E-8
AvTempK = (((1 - a_0) SolarConst) / (4SBConst))^(1/4)
AvTempC = AvTempK - 273.2
```

You will get an avarege temperature equal to –18°C. It is advisable to set **Options>Rounding>3 Significant Figures** (or more); otherwise, the Stefan–Boltzmanns constant will be represented as 0. Then calculate the emissivity by typing

```
em = (1 - a_0) SolarConst / (4SBConst T_0⁴)
```

which gives em ≈ 0.61.

You now want to create a function that will transform temperature to albedo according to the guidelines you have. You can do this in different ways in GeoGebra:

You can sketch how the function should behave with points. After that you can right-drag a rectangle around the points and use the **Create List** {1,2} tool. Once you have this list, you can fit a function to that list, for instance, an eighth-degree polynomial by typing

```
FitPoly[list1, 8]
```

Or, you could use the Freehand Shape tool to sketch a suitable function using the mouse. This might look like the curve in Figure 8.37.

Or, you could define a function passing through a list of values. The first two values in the list should be the x-values that define the domain of the function, such as

```
f(x)= Function[-80,100,0.8,0.8,0.8,0.4,0.25,0.15,
0.25,0,4,0.5]
```

Regardless of how you create this function (and, in practice, you will probably use more values), create it in Graphics Area 2 because you will use Graphics Area 1 later on to show the development of temperature over time. Show Graphics 2 by pressing **Ctrl-Shift-2**.

Once the feedback function has been created, you show the spreadsheet by pressing **Ctrl-Shift-S** and build the feedback loop model there, shown in Figure 8.38.

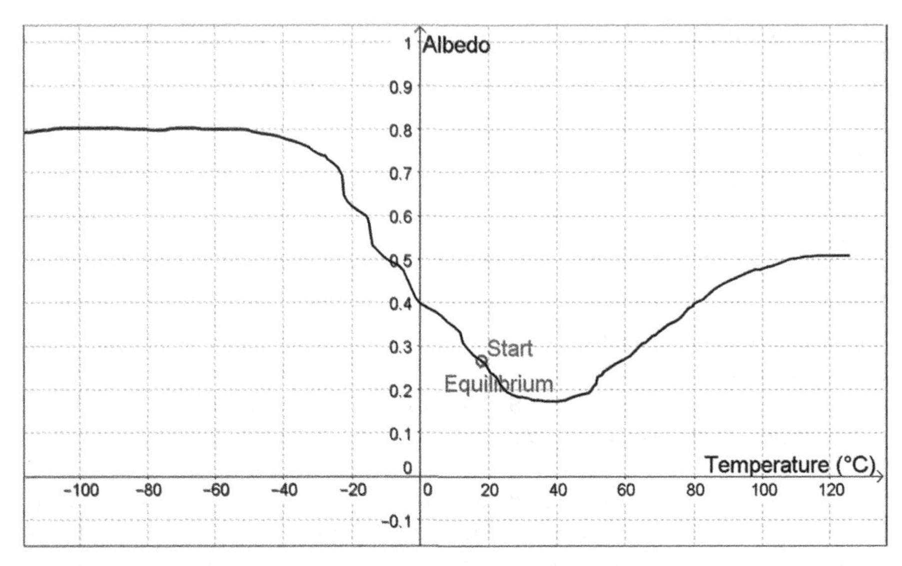

FIGURE 8.37 Freehand function that defines the relation between the average temperature and albedo.

▼ Spreadsheet

D2 =Max[Min[f(C2), amax], amin]

	A	B	C	D	E	F
1	Time step	Albedo in	Temperature	Albedo out	Albedo Point	Temperature point
2	0	0.302	14.9	0.302	(0, 0.302)	(0, 0.149)
3	1	0.302	14.9	0.302	(1, 0.302)	(1, 0.149)
4	2	0.302	14.9	0.302	(2, 0.302)	(2, 0.149)
5	3	0.302	14.9	0.302	(3, 0.302)	(3, 0.149)
6	4	0.302	14.9	0.302	(4, 0.302)	(4, 0.149)

FIGURE 8.38 Spreadsheet model for the feedback loop.

The first two rows in the spreadsheet are calculated like this:

```
A2 = 0
B2 = a_0
C2 = ((SolarConst (1 - B2)) / (4em SBConst))^(1/4) - 273.2
D2 = Max[Min[f(C2), amax], amin]
E2 = (A2, B2)
F2 = (A2, C2/100)
A3 = 1
B2 = D2
```

```
C3 = ((SolarConst (1-B3)) / (4emSBConst))^(1/4) - 273.2
D3 = Max[Min[f(C3), amax], amin]
E3 = (A3, B3)
F3 = (A3, C3/100)
```

Make sure that the points E2, E3, F2, and F3 are created in Graphics 1 and give the points suitable colors. The construction

```
E2=(A2,C2/100)
```

means that you can let the y-axis measure both albedo and temperature divided by 100 at the same time.

The cells A2 and A3 are selected at the same time and copied down until you have a number sequence from 0 to 50. Then you can select the cells B3 to F3 and copy them down as well.

Next create two new points in Graphics Area 2:

```
Start = (A2, B2)
Equilibrium = (A50, B50)
```

It is now easier see what happens over time. Redefine

```
a_0 = f(t_0)
```

and show t_0 as s slider as in Figure 8.39.

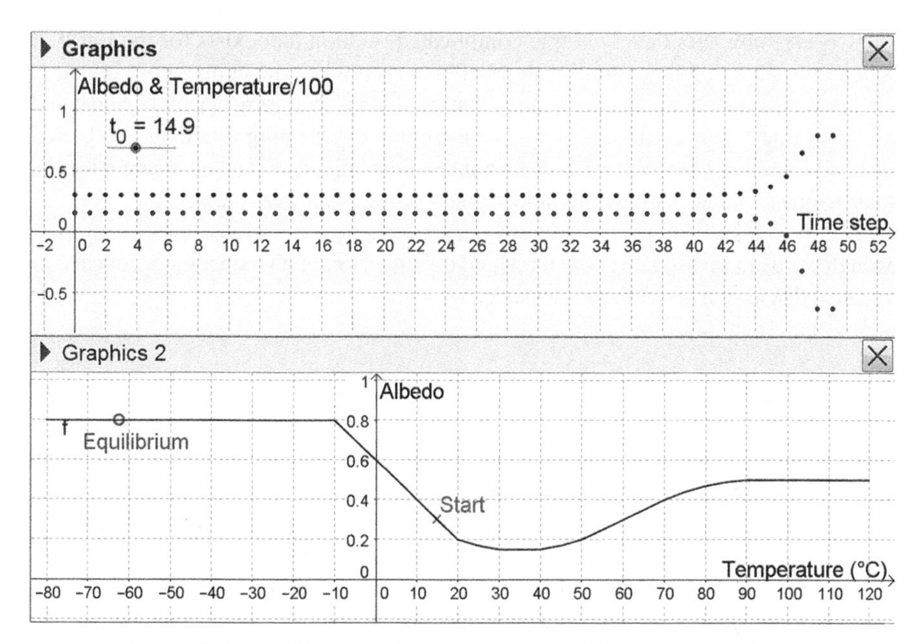

FIGURE 8.39 Climate model exhibiting a global temperature collapse.

Depending on how you have built your feedback function *f*, different things may happen when you adjust the value for t_0. If the slope of *f* is small near the starting point, the equilibrium, the long-term result, is stable and small changes in temperature will only give small changes in albedo. If instead the slope of *f* is too large, then small changes in temperature will grow over time, and the equilibrium position will settle in either of the +30°C or the –60°C regions. Try to change t_0 and see what happens.

You can easily create several different feedback functions and investigate this phenomenon. The most convenient way is to create different lists, **Alt1**, **Alt2**... with values according to the third method above, and then to define

```
f(x) = Function[Alt1]
```

where you can change one list for another to change the function *f*.

You can ignore the actual values and concentrate on the instability. That we now seem to have a reasonably stable climate could be due to the fact that f' is not very large. It is reasonable to assume that f' will decrease when the temperature slowly rises, which indicates that we are fairly safe and not in any obvious danger of a catastrophic overheating that would boil off the oceans. An increase in the Earth's average temperature by several degrees could be considered catastrophic for other reasons, but if would be still possible for humans to survive as a species.

A temperature drop nevertheless could lead to sliding into the steeper part of the graph, resulting in the planet becoming icier, something we could hardly survive if taken to the extremes that this model predicts. These types of unbalanced effects are often called threshold effects, and are typical for chaotic systems. Since we do not know everything about every single complicated relation that exists for the Earth, it is quite possible that there are threshold effects even when the temperature is rising. We know that the planet has been much warmer than now during earlier periods, and we know that Venus could have had a catastrophic overheating that long ago boiled away all its water. Once the water was in the atmosphere, it was an easy target for the solar winds as Venus's magnetic field collapsed due to its freezing core.

This model is too simple to make any deeper realistic conclusions, but still complex enough for us to investigate some threshold effects of chaotic feedback systems and to initiate a discussion about their relevance.

8.10 TRAFFIC JAM

Traffic jams can sometimes occur without any particular causes. See, for instance, the results from an experiment with real cars: http://www.newscientist.com/article/dn13402-shockwave-traffic-jam-recreated-for-first-time.html#.UponhMTTuhM or the computer simulation: http://blogs.kqed.org/lowdown/2013/11/12/traffic-waves or this three-lane simulation: http://www.mtreiber.de/MicroApplet_html5. Finite reaction times lead to domino effects, resulting in *density waves*, congestions that move very slowly or even not at all.

Create a model with the capacity to visualize this type of density waves. Then use the model to answer some of the following questions:

In what direction is the congestion moving? Can your make it stand still?

How fast is the traffic jam moving in relation to the speed of the cars or the value of other parameters?

In what way do your parameters affect your model?

Can you improve your model so that it works better?

Proposed Solution

Often problems of this kind are solved by programming, not caring much about visualization or graphical interfaces. In Geogebra you get the graphics for free, and you only need to program or organize the mathematical structure.

From the links given in the problem, you may be inspired to experiement and do your own simulations. We suggest that you try a number of cars moving around a circular path. We chose to have a relatively small number of cars, nine, initially placed 40° from each other and moving at a pace of 10° per time unit.

A driver that will fall behind will speed up, and a driver that will catch up will slow down. You can model this in the following way: The average time to the next car can be $\Delta t = \Delta s / v_0 = 360/(n \cdot v_0)$ where the distance is measured in degrees. First assume that the driver adjusts the velocity of the car as if the car would cover the actual distance in this time, which gives a car speed of $v = \Delta s / \Delta t = n \cdot v_0 \cdot \Delta s / 360$.

After some initial testing of this structure, we could see that we needed to adjust the model somewhat: one option was to add a normal distribution in order to deal with the impossibility of having an exact velocity. The other option was that we add some reaction time for the "drivers," so we decided not to define the new velocity to exactly the "correct" value but to the old velocity + some of the difference up to the "correct" velocity. When the car is approaching the next car, the speed control must be more precise and detailed to avoid a smash up, and when the car is far away from other cars the speed control can be more relaxed. The final formula for the velocity then became

$$v_{new} = v_{old} + \left(\frac{n \cdot v_0 \cdot \Delta s}{360} - v_{old} \right) \cdot \frac{\tau}{\Delta s + \tau} + N_{\Delta s \cdot \sigma}$$

The factor $\tau/(\Delta s + \tau)$ is close to 1 when $\Delta s \ll \tau$, and close to $\tau/\Delta s$ when $\Delta s \gg \tau$, which gives the "correct" velocity when the cars are close and a slow change of the velocity when the distance is large. The term $N\Delta_{s \cdot}\sigma$ stands for the normal distributed and randomized addition to the velocity where you can set the standard deviation to increase with the distance. Note that this is not the same model as the one given in the link to the computer simulation.

Since you will need to study the system over a long time duration, it will be impractical to let the time run over different rows in a spreadsheet. You will instead use the spreadsheet to describe the current value of the variables, namely the position and velocity of each of the nine cars, and you can use a script to update these while the time is running. The time will be represented by an animated slider that switches between 0 and 1 and then runs the update instructions at every switch.

Start by creating the speed control slider:

```
Speed = 0
```

and set **Min=0**, **Max=100** and **Increment=0.1** in the **Properties** dialogue. Make sure that the slider is visible in the graphics area. Then create

```
Time=0
```

setting **Min=0**, **Max=1**, **Increment=1** and **Speed=Speed** (type "Speed" in the Speed field). Now, by changing the Speed slider, you can change the speed of the clock.

You can then create some more variables, such as

```
n = 9
sigma = 0.01        (set Increment to 0.001)
tau = 3
v0 = 10
dt = 0.1
```

Now is the time to fill the spreadsheet as in Figure 8.40. Note that since you have a variable named *speed*, you cannot simply type this name in cell C1. GeoGebra will then exchange this for the current value of the variable. Instead, you have to type

```
"speed"
```

with quotation marks to be sure that you end up with a text.

	A	B	C	D	E
1	Car	Position	Speed	Point	Distance
2	1	0	10	(1, 0)	40
3	2	40	10	(0.766, 0.6428)	40
4	3	80	10	(0.1736, 0.9848)	40
5	4	120	10	(-0.5, 0.866)	40
6	5	160	10	(-0.9397, 0.342)	40
7	6	200	10	(-0.9397, -0.342)	40
8	7	240	10	(-0.5, -0.866)	40
9	8	280	10	(0.1736, -0.9848)	40
10	9	320	10	(0.766, -0.6428)	40
11					
12				Min:	40

FIGURE 8.40 Spreadsheet part of the simulation of traffic jam.

The positions and speeds are just the numbers you have entered. The distances are defined so that in cell E2 you have

```
= B3 - B2
```

and so forth, all the way down to cell E10, where you find

```
= 360 + B2 - B10
```

The points are created in the D column, where in D2 you find

```
=(cos(B2°), sin(B2°))
```

You also want to make the colors of the points dynamic so that you can show the car with the shortest distance to the next car ahead as red. After you have created the point in cell D2, you right-click on it and select **Properties**, the **Advanced** tab, and fill in the **Dynamical color**:

Red: `If[E2=Min[E$2:E$10],1,0]`
Green: `0`
Blue: `0`

This means that if this particular car happens to have a distance to the next car that is equal to the minimum distance, then the color will be set to red $(1, 0, 0)$; otherwise, it is black $(0, 0, 0)$. You should also set the size and style to something suitable. It can now be copied down to D10.

You are now ready to fine-tune the graphics some more, changing D2 again, setting the blue component of the dynamic color to 1 and the shape to something else for just this car. These changes will make it so much easier to count laps if you have one blue car among all the black ones.

Cell E12 is set to show the least distance with the formula

```
= min[E2:E10]
```

The points are shown in the graphics area where you can now hide both axes and grid lines. Also create two circles by typing

```
Circle[(0,0), 0.95]
Circle[(0,0), 1.05]
```

to sketch the track where the cars are driving. You can see the initial starting positions of the cars in Figure 8.41.

Now you need to do some programming. Start by creating a button for the animation. Click on the Button tool and then click in the graphics area where you want the button to be created. Type "Start" in the **Caption** field and the command

```
Speed =100
```

in the GeoGebra Script field, as shown in Figure 8.42.

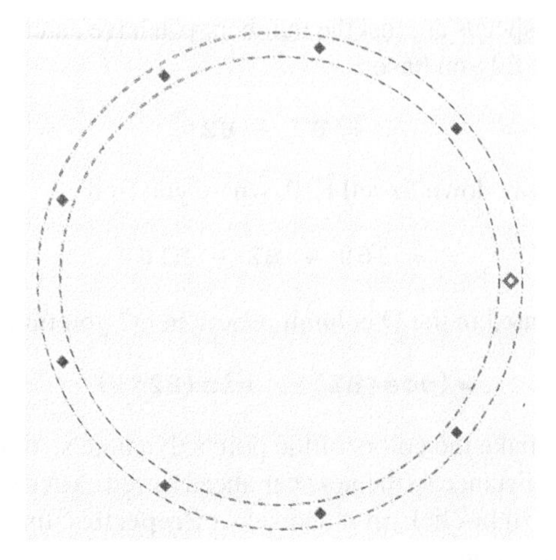

FIGURE 8.41 Cars ready to go.

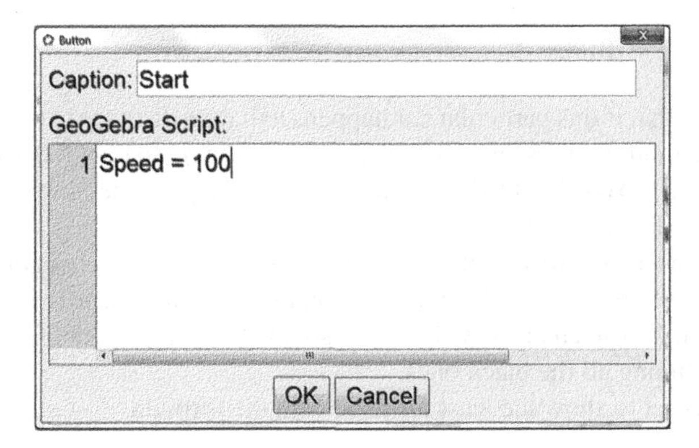

FIGURE 8.42 Creating a command button is easy.

Click on **OK** to create the button. The size of the text can be set in the properties of the button. Every time you click the button, Speed will be set to 100. Also create another similar button in the same manner with the caption "Stop" and with the GeoGebra Script

```
Speed = 0
```

A button can do several things at the same time: Create a third button called "Reset" that will reset all variables to their start values according to Figure 8.43.

All that now remains is the heart of the project and model, namely updating the positions and velocity from one time moment to the next. This script should be executed every time the variable Time is updated. Right-click on Time and select

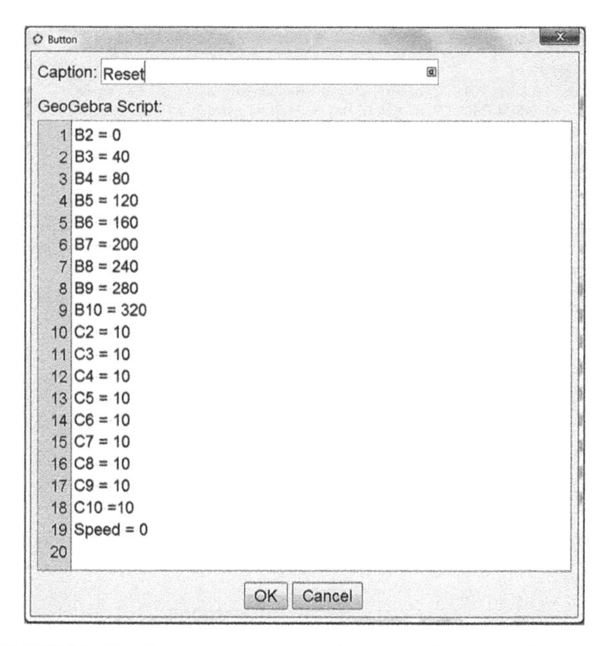

FIGURE 8.43 Longer script executing several commands at once.

Properties, **Scripting**, **On Update**. To be able to read and write the value of a cell in the same scripting command, use the command **SetValue[]**, and in C2 type:

```
SetValue[C2, C2 + (n v0 (B3 - B2)/360 - C2) tau/(B3 - B2 +
tau) + RandomNormalDistribution[mu, (B3 - B2) sigma]]
```

Some spaces are needed as multiplication signs. Complete with the rest of the commands as in Figure 8.44.

Update the positions by typing

```
SetValue[B2, B2 + dt C2]
```

and so forth; this corresponds to letting the position angle to be updated with the value $v \cdot \mathrm{dt}$.

The update command for C10 will be

```
SetValue[C10, C10 + (n v0 (360 + B2 - B10) / 360 - C10)
tau/(360 + B2 - B10 + tau) + RandomNormalDistribution
[mu, (360 + B2 - B10) sigma]]
```

which is different than the other updates of velocity, but the update of B10 is

```
SetValue[B10, B10 + dt C10]
```

precisely the same as the others.

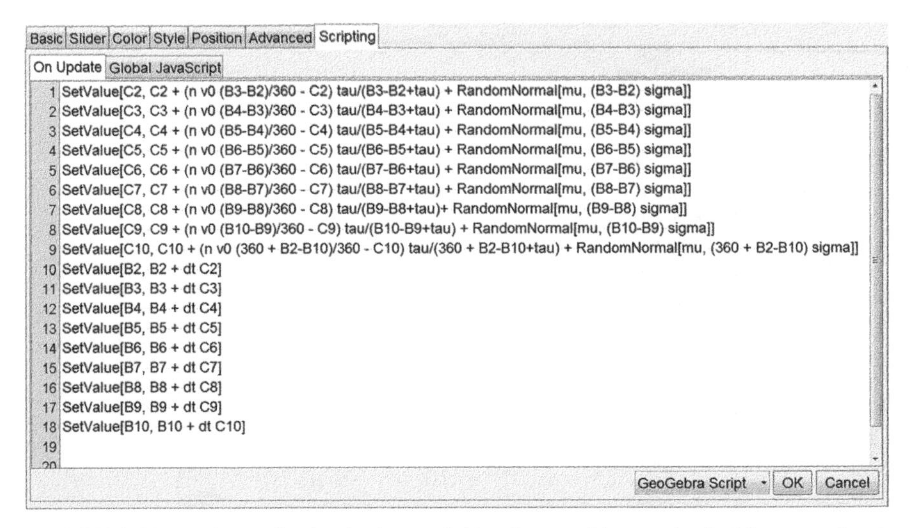

FIGURE 8.44 Update script for the time variable where positions and velocities are updated.

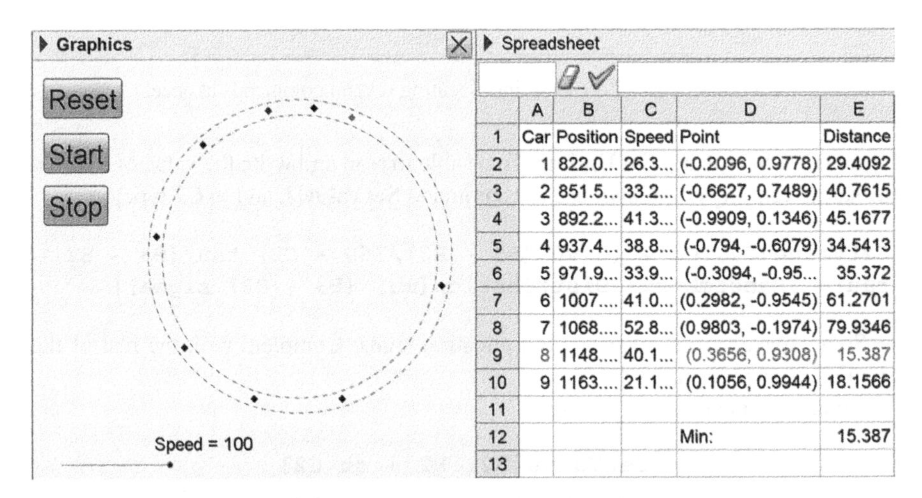

	A	B	C	D	E
1	Car	Position	Speed	Point	Distance
2	1	822.0...	26.3...	(-0.2096, 0.9778)	29.4092
3	2	851.5...	33.2...	(-0.6627, 0.7489)	40.7615
4	3	892.2...	41.3...	(-0.9909, 0.1346)	45.1677
5	4	937.4...	38.8...	(-0.794, -0.6079)	34.5413
6	5	971.9...	33.9...	(-0.3094, -0.95...	35.372
7	6	1007....	41.0...	(0.2982, -0.9545)	61.2701
8	7	1068....	52.8...	(0.9803, -0.1974)	79.9346
9	8	1148....	40.1...	(0.3656, 0.9308)	15.387
10	9	1163....	21.1...	(0.1056, 0.9944)	18.1566
11					
12				Min:	15.387
13					

FIGURE 8.45 Center of the congestion at the gray car.

Last, with this script in place, test if everything is working. Increase the slider's animation speed just a little bit and wait until Time switchs value. Now all positions and velocities should be updated. If everything seems to be all right: click on the Start button. You can try increasing v0 to 20 to get a clearer result; see Figure 8.45.

After having run the model a couple of times, you might conclude that in this model the density wave's center always moves forward in the same direction as the traffic, but with a velocity that is about 20% to25% of the traffic's velocity when v0=20. It should be clear that this is the result of adding better speed control while breaking. In real-world traffic, most drivers are much too aggressive and eager to get up to speed again after passing through a density wave; this model evades that kind of behavior. According to the Wikipedia article on *congestion shockwaves* (at http://en.wikipedia.org/wiki/Traffic_flow#Congestion_Shockwave) density waves usually move in the opposite direction to traffic, at a speed of about 30 km/h or 20 mph.

The parameter τ controls for aggressive steering while driving over long distances. Low values relate to cautious steering and maintaining larger distances between cars, whereas high values relate to reckless steering and less distance to the car right up ahead. Larger values for sigma mean that the distance changes are quicker because the fluctuations are greater, but it can mean that the results are less noticeable because the drivers are more unpredictable. The variable dt is the time interval: decreasing its value gives the same effect as if you increase the initial velocity v0—whereby the cars go faster, and that will show more obvious density waves, and quicker, in the model. See Figure 8.46 for some sample screenshots of a run with $\tau=4$ and $v_0=20$.

8.11 WILDFIRE

A wildfire or a wildland fire such as the Bugaboo scrub fire in 2007, seen in Figure 8.47, is an initially uncontrollable fire in an open terrain. A wildfire's possibility to spread in an uncontrolled fashion is mainly affected by the dryness of the burnable vegetation and of the wind speed. If the area is dry and windy, a wildfire can devastate large areas and be a threat to human life.

In order to be able to make better predictions of weather conditions that can raise the risk of fire, your national weather institute is performing simulations of dryness and wind speeds using adjustable parameters. The national weather institute is looking for individuals who can construct the necessary computer models. Use GeoGebra to construct a model that visualizes how a wildfire will spread based on dryness, wind speed, and other parameters.

Proposed Solution

To show how a wildfire can spread, the model has to be organized like a map and be divided into different zones. Obviously, if a wildfire is burning in one zone, then there is a high probability that the wildfire will spread to adjacent zones. This probability will depend on the wind speed and humidity/dryness in the zone. You also need to identify the different development stages of a wildfire in a particular zone. Each wildfire is small and harmless to begin with, let us say that the intensity i is 0.5. But it will quickly flare up and become fierce at intensity $i=1$. After a while as large amounts of combustible vegetation get consumed, the wildfire will calm down: let us say to $i=0.3$. This is when the open part of the wildfire stops, though there is a low probability that some underlying vegetation will still burn for several days, so let us set $i=0.01$ to indicate that the fire might start to spread again.

To be able to realize this model, you need more components. You need a way to measure time, and you need way to update the map from one time interval to the next. You also need a *Graphical User Interface*, a GUI, with controls to start, stop, and

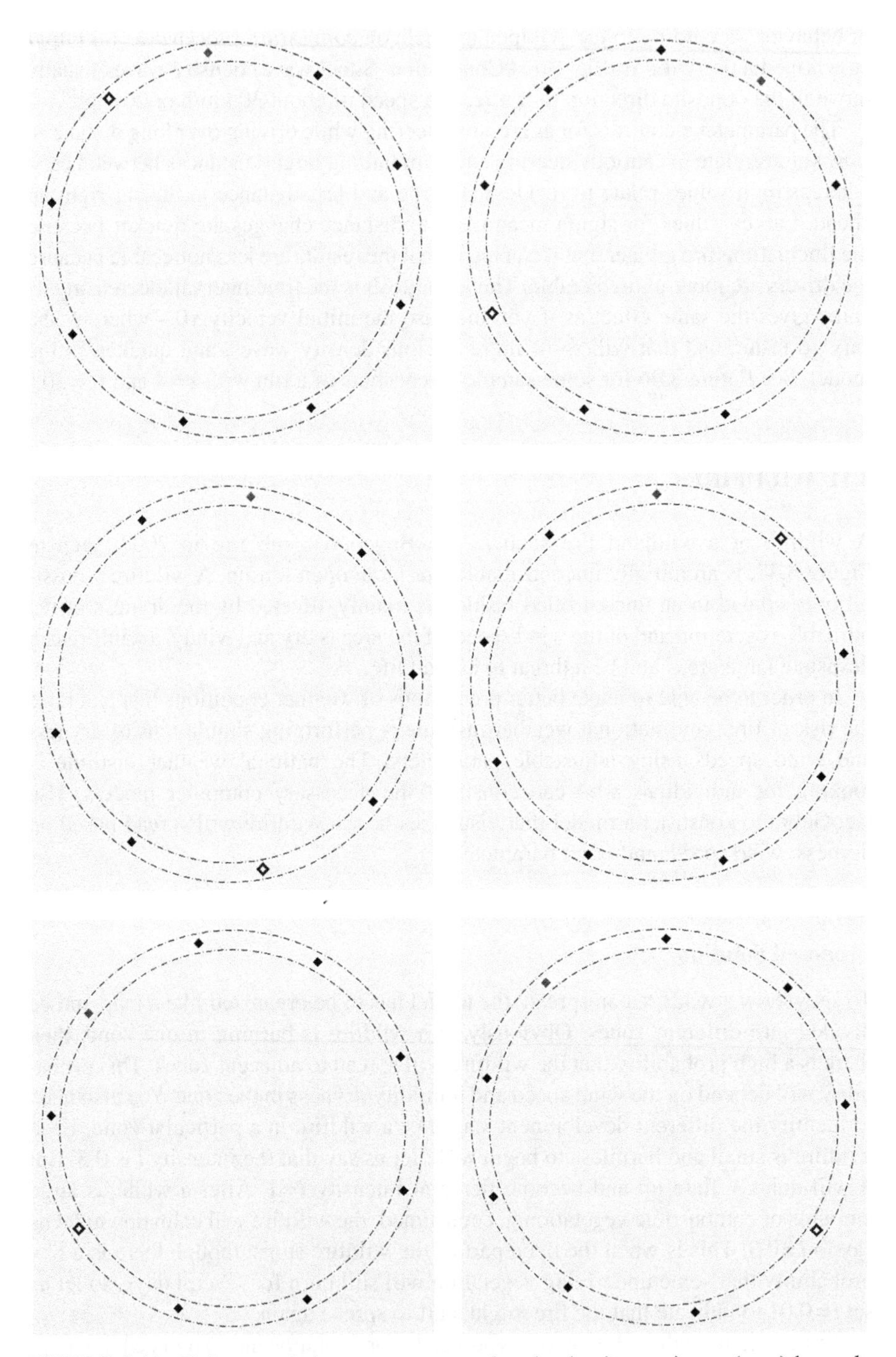

FIGURE 8.46 Hollow symbol car runs one turn but the density maximum (gray) has only moved a little.

FIGURE 8.47 Bugaboo forest fire. Photo: Mark Wolfe/FEMA (image from the FEMA Photo Library) [Public domain], via Wikimedia Commons. https://upload.wikimedia.org/wikipedia/commons/2/23/Bugaboo_forest_fire.jpg.

reset the animation. You will find that the update is the most complicated component and the one that will need most resources. The construction protocol of this model will contain thousands of rows, and you will use programs with hundreds of rows of code. It is possible to do this with simpler, easier, and better programming by simply coding, but we want to show you a somewhat different method to do this with GeoGebra, which contain some powerful ideas.

Before you start the construction, you need to think about the rules for an update. Every region or cell is surrounded by eight other cells. Depending on the wind direction, the values of the variables E, W, N, S, SE, SW, NE, and NW will give the probability for the cell to catch fire from that particular direction. For each cell that is not burning, you can assign a random probability value between 0 and 1. If the sum of the products of intensities and probabilities exceed this number, this means that the wildfire has spread to this cell and the intensity of this cell will be changed from 0 to 0.5. Once a cell catches fire, the fire will not be affected by surrounding factors any more, each cell will burn through the change of intensity from 0 to 0.5; to 1; to 0.3; to 0.01.

Now we can start the construction, Begin by creating three versions of the map in the spreadsheet shown in Figure 8.48. The maps are the 20 by 20 cells in the grid, and you can assume that the area is restricted by roads, rivers, and other barriers that the fire cannot pass. These barriers are represented by a grid of zeros outside the map. Outmost are the row and column numbers. It is important that they be

	A	B	C	D	E	F	G	H	I	J	K	L	M	N	O	P	Q	R	S	T	U	V	W
1		0	1	2	3	4	5	6	7	8	9	10	11	12	13	14	15	16	17	18	19	20	21
2	21	0	0	0	0	0	0	0	0	0	0	0	0	0	0	0	0	0	0	0	0	0	0
3	20	0	0	0	0	0	0	0	0	0	0	0	0	0	0	0	0	0	0	0	0	0	0
4	19	0	0	0	0	0	0	0	0	0	0	0	0	0	0	0	0	0	0	0	0	0	0
5	18	0	0	0	0	0	0	0	0	0	0	0	0	0	0	0	0	0	0	0	0	0	0
6	17	0	0	0	0	0	0	0	0	0	0	0	0	0	0	0	0	0	0	0	0	0	0
7	16	0	0	0	0	0	0	0	0	0	0	0	0	0	0	0	0	0	0	0	0	0	0
8	15	0	0	0	0	0	0	0	0	0	0	0	0	0	0	0	0	0	0	0	0	0	0
9	14	0	0	0	0	0	0	0	0	0	0	0	0	0	0	0	0	0	0	0	0	0	0
10	13	0	0	0	0	0	0	0	0	0	0	0	0	0	0	0	0	0	0	0	0	0	0
11	12	0	0	0	0	0	0	0	0	0	0	0	0	0	0	0	0	0	0	0	0	0	0
12	11	0	0	0	0	0	0	0	0	0	0	0	0	0	0	0	0	0	0	0	0	0	0
13	10	0	0	0	0	0	0	0	0	0	0	0	0	0	0	0	0	0	0	0	0	0	0
14	9	0	0	0	0	0	0	0	0	0	0	0	0	0	0	0	0	0	0	0	0	0	0
15	8	0	0	0	0	0	0	0	0	0	0	0	0	0	0	0	0	0	0	0	0	0	0
16	7	0	0	0	0	0	0	0	0	0	0	0	0	0	0	0	0	0	0	0	0	0	0
17	6	0	0	0	0	0	0	0	0	0	0	0	0	0	0	0	0	0	0	0	0	0	0
18	5	0	0	0	0	0	0	0	0	0	0	0	0	0	0	0	0	0	0	0	0	0	0
19	4	0	0	0	0	0	0	0	0	0	0	0	0	0	0	0	0	0	0	0	0	0	0
20	3	0	0	0	0	0	0	0	0	0	0	0	0	0	0	0	0	0	0	0	0	0	0
21	2	0	0	0	0	0	0	0	0	0	0	0	0	0	0	0	0	0	0	0	0	0	0
22	1	0	0	0	0	0	0	0	0	0	0	0	0	0	0	0	0	0	0	0	0	0	0
23	0	0	0	0	0	0	0	0	0	0	0	0	0	0	0	0	0	0	0	0	0	0	0

FIGURE 8.48 Representation of the map in a spreadsheet.

positioned exactly as they are because you will be using them when you generate the graphical map. The whole area can be copied so that there will be three identical areas in the spreadsheet: A1:W23, A25:W77, and A49:W71. The upper area will show the current situation, the middle area will be used to do the update and the lowest area will contain squares that will be shown graphically. Initially, all cells can be filled with a 0.

It is now time to create the graphical representation of the map. In cell V51 type

```
=Polygon[(U$1,$A52), (V$1,$A52), (V$1,$A51), (U$1,$A51)]
```

This command will create a square in the graphics area and the coordinates will be fetched from the spreadsheet, from the third area's row and column numbers. Right-click on the square in the graphics area and select **Properties**, the **Advanced** tab and define the dynamical colors for

Red: `If[V3 > 0.01,1,0]`
Green: `If[V3 == 0.01,0,1 - V3]`
Blue: `0`

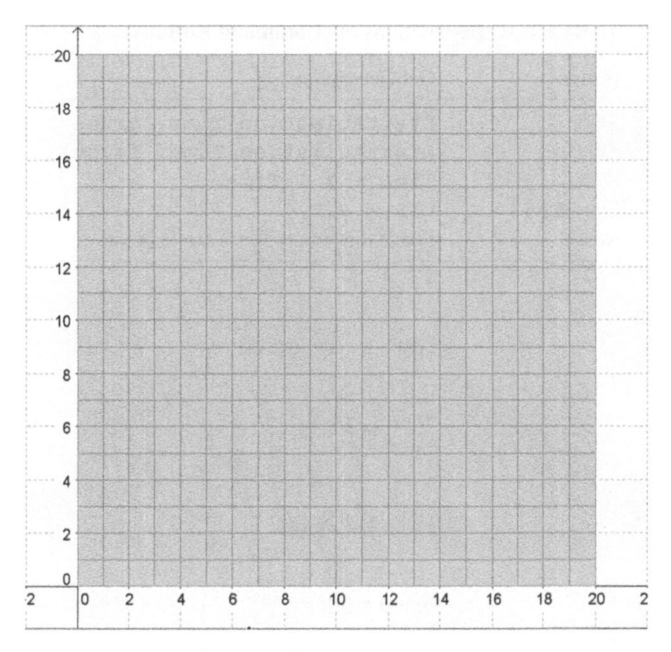

FIGURE 8.49 So far no wildfire.

This will give you reasonable color effects. Once this is done, copy cell V51 down, and then copy the whole column to the left. You should wind up with a green landscape, yet untouched by fire, as in Figure 8.49.

You now need to add some more parameters. Start with the time control and type

```
Speed = 5
Time = 0
```

in the input field. Right-click on Time and select **Object Properties**, the **Slider** tab, and set

```
Min = 0
Max = 1
Step = 1
Speed = Speed
```

Also set Repeat to **Oscillating**. When Time will be animated, it will now toggle its value as a digital clock: 0, 1, 0, 1, 0, ...

Then introduce a variable for the dryness level:

```
R = 0.5
```

TABLE 8.3 Instructions for Command Buttons

Button	GeoGebrascipt
Start	`StartAnimation[Time, true]`
Stop	`StartAnimation[Time, false]`
Step	`Time = 1 - Time`
Start @L12	`L12 = 0.5`
Reset	Leave this empty for the time being.

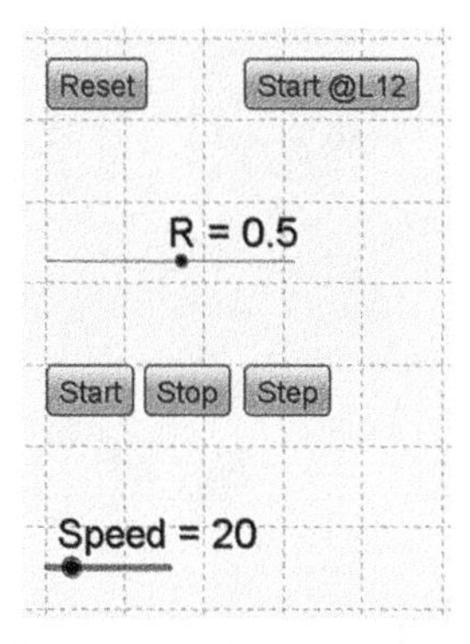

FIGURE 8.50 Simple GUI.

Next create probabilities for different wind directions:

```
E = 0.2R
NE = 0.3R
N = 0.4R
NW = 0.3R
W = 0.2R
SW = 0.1R
S = 0.05R
SE = 0.1R
```

If you would like to, you could make the control of wind values more sophisticated later on, but this will do just fine for now.

The next step is to create the graphical user Interface. Show the numbers R and Speed as sliders and create five command buttons with Comman Button tool from the information in Table 8.3. Figure 8.50 shows what the simple GUI looks like.

You have only two things left to do, but these are the most complicated to do and most difficult to succeed with. You have to create a script that will reset everything to zero, which means that every single cell in the area C3:V22 must be set to 0. Since this area contain 400 cells, it means that it you need 400 rows of code of the type

```
C3 = 0
```

something you would probably rather not type by hand. You further need to create a script that updates the area according to the rules you have set up every time the Time variable changes its value, something that will need 800 rows of complicated code.

The scripts are fortunately uncomplicated enough in their structure, and each row resembles the previous row. It is therefore possible to write a *meta-script* to create these scripts. Start with the easier, update script.

The following eight rows, written in Python, will do the trick:

```
from __future__ import print_function
cols1 = ["", "A", "B", "C", "D", "E", "F", "G", "H"]
cols2 = ["I", "J", "K", "L", "M", "N", "O", "P", "Q"]
cols3 = ["R", "S", "T", "U", "V", "W", "X", "Y", "Z"]
cols = cols1+cols2+cols3
for cv in range(3, 23):
  for row in range(3, 23):
    print (cols[cv], row, ' = 0', sep = '')
```

In **row 1** a more modern form of the print function is imported so that the output will be easier to handle.

In **rows 2–5** lists of letters are created, without exceeding the row length.

In **rows 6–7** a double loop that goes through each cell in every row and every column is set up so that row 8 will be executed once for each cell.

In **row 8,** finally, a specific command is written to the update script for that specific cell. The final argument suppresses extra spaces.

To avoid losing the program by mistake, write it in Notepad or some other text editor. You can then copy the program to an online python compiler. This time we suggest that you use use repl.it at https://repl.it/languages/python3 simply because it is easier to copy multiple lines of output from repl.it than it is, for example, from Codingground. Paste the code into the left window, click on **Run,** and your 400 rows of code will be generated in the right window, as indicated in Figure 8.51.

Once this is done, copy all the 400 rows of output from the Python meta-script to the command button Reset. Right-click on the button, select **Object Properties**, the **Scripting** tab, and the **On Click** tab and paste the script into the big text box. Do not forget to click the OK-button to save the script before you close the dialogue box. You can test the script by changing some of the values in any of the cells in the area C3:V22 and then click on the Reset button.

The update script of the Time variable will be created in the same way, although in several steps. First you must allow area 2 to be a future update

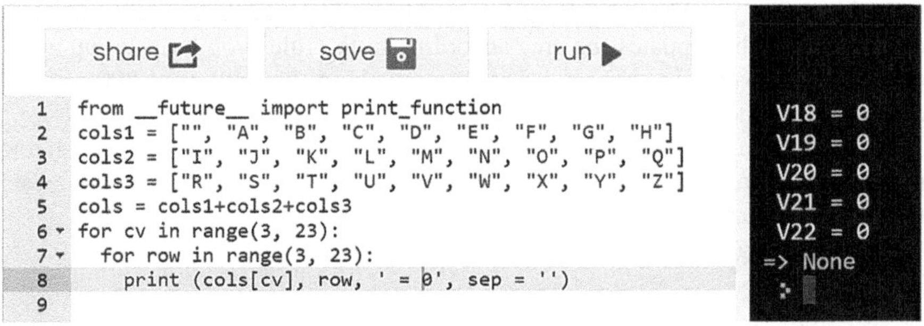

```
share save run
1 from __future__ import print_function
2 cols1 = ["", "A", "B", "C", "D", "E", "F", "G", "H"]
3 cols2 = ["I", "J", "K", "L", "M", "N", "O", "P", "Q"]
4 cols3 = ["R", "S", "T", "U", "V", "W", "X", "Y", "Z"]
5 cols = cols1+cols2+cols3
6 for cv in range(3, 23):
7   for row in range(3, 23):
8     print (cols[cv], row, ' = 0', sep = '')
9

V18 = 0
V19 = 0
V20 = 0
V21 = 0
V22 = 0
=> None
```

FIGURE 8.51 Writing a program to write another program ....

from area 1, according to the rules we have defined; after that you must copy area 2 to area 1. Each of these two parts will need 400 rows of code that will be generated by two different meta-scripts. The first meta-script will implement the update rules and will look like this:

```
from __future__ import print_function
cols1 = ["", "A", "B", "C", "D", "E", "F", "G", "H"]
cols2 = ["I", "J", "K", "L", "M", "N", "O", "P", "Q"]
cols3 = ["R", "S", "T", "U", "V", "W", "X", "Y", "Z"]
c = cols1+cols2+cols3
for v in range(3, 23):
  for r in range(3, 23):
    print ('ggbApplet.evalCommand("', end='', sep = '')
    print (c[v],r+24, '=CopyFreeObject[', end='', sep = '')
    print ("If[", c[v], r, "==0.01,0.01", end='', sep = '')
    print (",If[", c[v], r, "==0.3,0.01", end='', sep = '')
    print (",If[", c[v], r, "==1,0.3", end='', sep = '')
    print (",If[", c[v], r, "==0.5,1", end='', sep = '')
    print (",If[", c[v], r, end='', sep = '')
    print ("==0,If[RandomUniform[0,1]<", end='', sep = '')
    print (c[v-1], r-1, "*NW+", end='', sep = '')
    print (c[v], r-1, "*N+", end='', sep = '')
    print (c[v+1], r-1, "*NE+", end='', sep = '')
    print (c[v-1], r, "*W+", end='', sep = '')
    print (c[v+1], r, "*(E)+", end='', sep = '')
    print (c[v-1], r+1, "*SW+", end='', sep = '')
    print (c[v], r+1, "*S+", end='', sep = '')
    print (c[v+1], r+1, '*SE, 0.5, 0]]]]]]]")', sep = '')
```

When you execute this script at repl.it, it will (somewhat slowly) generate 400 rows of code where every row has the same structure as this line:

```
ggbApplet.evalCommand("C27=CopyFreeObject[If[C3==0.01,
0.01,If[C3==0.3,0.01,If[C3==1,0.3,If[C3==0.5,1,If[C3==
0,If[RandomUniform[0,1]<B2*NW+C2*N+D2*NE+B3*W+D3*(E)+B
4*SW+C4*S+D4*SE,0.5,0]]]]]]]")
```

You see that the rules for the spread of the wildfire are expressed as a series of nested If commands with a product of probabilities at the end. Everything is encapsulated in a JavaScript command because JavaScript is often better at handling complicated expressions than GeoGebraScript is. The Eastern wind's probability is contained in brackets to avoid problems where GeoGebra interprets it as a part of a power of ten, as in 3E+7.

Save the script in text editor and wait for the last part, the copying of area 2 to area 1, which will be generated by the last script:

```
from __future__ import print_function
cols1 = ["", "A", "B", "C", "D", "E", "F", "G", "H"]
cols2 = ["I", "J", "K", "L", "M", "N", "O", "P", "Q"]
cols3 = ["R", "S", "T", "U", "V", "W", "X", "Y", "Z"]
c = cols1+cols2+cols3
for v in range(3, 23):
  for r in range(3, 23):
    print ('ggbApplet.evalCommand("', end='', sep = '')
    print (c[v], r, '=CopyFreeObject[', end='', sep = '')
    print (c[v], r+24, ']")', sep = '')
```

This will generate 400 rows of the following structure:

```
ggbApplet.evalCommand("V18=CopyFreeObject[V42]")
```

Copy these new 400 rows to the text editor and save it. Finally copy all these 800 rows from the text editor, right-click on the variable Time, and select **Object Properties**, the **Scripting** tab, the **On Update** tab and paste it there. Save the script by clicking on OK before you close the dialogue box.

The model is now finished. The start and stop buttons will start and stop the animation. When the animation is stopped, you can step one step at the time ahead with the Step button. You start a wildfire at in the center of the forest by clicking the button Start @L12 and you reset it with the Reset button. You can try to give different values for the wind variables or you can increase or decrease the dryness R. An example is shown in Figure 8.52.

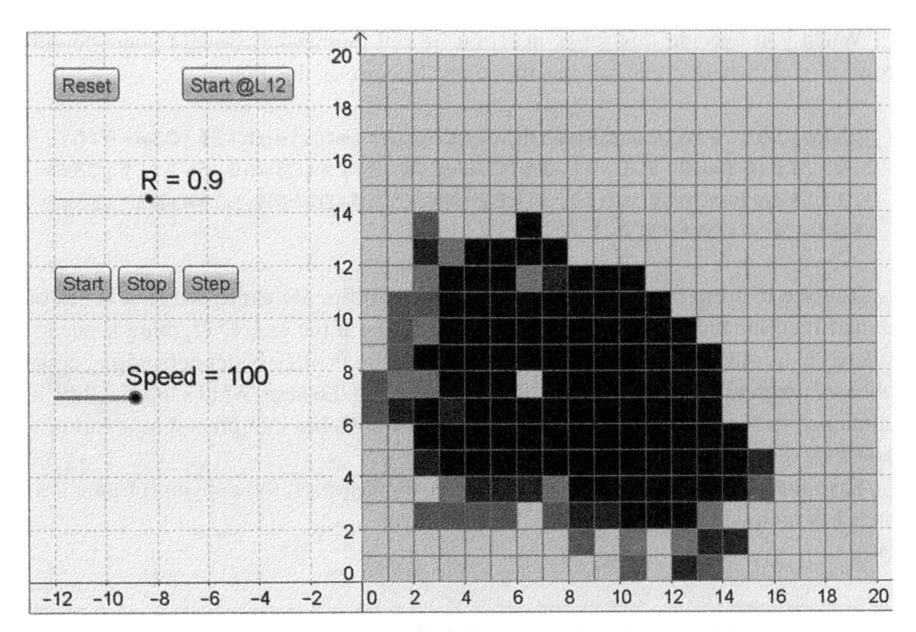

FIGURE 8.52 Wildfire when the wind is coming from the north.

Comments

If you would like to do the simulation faster, just model a smaller surface. Perhaps 15 by 15 or 10 by 10 cells would be enough to do interesting observations or calculations with the model. For instance, you could measure how many wildfires die off by themselves inside a radius of 5 cells for different levels of humidity/dryness in air or land (the parameter R).

The traditional way of doing this type of simulations is to code the situation in Python, JavaScript, or some other high-level program language from the beginning, ignoring GeoGebra's own commands, which are relatively slow. The map can then be represented by two matrices. The disadvantage is that you need to program the graphical map too, which you avoid by letting GeoGebra do what GeoGebra does best, handle geometrical objects with dynamic color in the spreadsheet.

We hope that this example will inspire a student's project, especially a project that includes programming.

8.12 A MODERN CARPENTER

A modern carpenter has a small company where completed wood parts in pinewood are bought on the Internet. The carpenter only produces two different products, bureaus and stools. The only operations involved are assembling and varnishing the wooden parts into bureaus and stools. Varnishing takes more time, but because of the noxious fumes, only a maximum of 25 h of varnishing out of the 40 h of total work per week can be used. The following schema is a description of his production. A bureau takes 30 min to assemble and an hour to varnish and can be sold for a profit of 40 USD. A stool takes only 15 min to assemble and 25 min to varnish. A stool can be sold for a profit of 20 USD.

Can the carpenter complete an order of 40 stools and 10 bureaus in a single week?
Determine the maximum profit that the carpenters's company can make in one week.
Decide on a strategy to increase the carpenter's income.

8.13 CONWAY'S GAME OF LIFE

A well-known simulation similar to the one in Problem 8.11, *Wildfire*, is Conway's *Game of Life* where each cell can be either dead (0) or alive (1). A cell dies if it has fewer than two living cells as neighbors, a cell remains unchanged if it has two living cells as neighbors, a cell is born or lives on if it has three living cells as neighbors, and a cell dies from overpopulation and lack of food if it has more than three living cells as neighbors. This simulation has been the subject of a lot of research and it has been shown, among other things, that the system is able to function as a (very slow) computer. You can find an implementation on http://www.ozwebsoft.com/webwaylife.html.

Create your own simulation in GeoGebra.

8.14 MATRIX TAXIS

The owner of a taxi company keeps data on where the taxis are needed. The city is divided into two sections, the downtown section and the airport section. Eighty percent of the downtown customers ask to be taken to a different downtown location so the taxi stays downtown. Twenty percent of the downtown customers ask to be taken to the airport. Sixty percent of the airport customers ask to be taken downtown, while forty percent go to a different airport location, such as another terminal or a nearby hotel.

Construct a transition matrix. If a fleet of 100 taxis starts out 50% at the airport and 50% downtown, what is their distribution after one run? Two runs? At the end of the day? Is a 50–50 distribution the best plan for starting the day? Explain.

Suppose that the taxi company collects new data that shows 90% of the downtown taxis stay downtown and 10% go to the airport, while 50% of the airport taxis stay at the airport and 50% go downtown. What starting distribution should the company adopt now?

8.15 THE CAR PARK

The owner of a large shopping center has contracted you to design the car park. If ordinary "rectangular" parking is used, more space is needed to reverse out into the driving lane, so while packing the cars close, this plan wastes space in using to wide driving lanes. With angle parking, the driving lane width can be reduced, down to a minimum required to drive cars safely among the parked cars. It also packs cars closer in one direction, while spreading the out in the other direction.

Try to find a relation between the required drive lane width and the angle of parking. Then, assuming the car park is a neat 100×200 m, make a plan of the car park and calculate the number of cars in the car park as a function of the angle. Remember that there is no such thing as a decimal number of cars.

8.16 SELECTING A COLLAGE

Select 3 to 5 promising collages or universities that you may want to go to. Also select three (no more) criteria that are important to you, such as academic reputation, location, etc. Make a table and grade each institution against each criteria according to a fixed scale, say from 0 to 10.

Assume that the relative importance of your critera are numbers x, y, and z, such that $x+y+z=1$. Then obviously $z=1-x-y$.

Your table can be called a *decision matrix*. Work out the scores for the different collages. These scores will be functions of x and y, and for each point (x, y) in a coordinate system, a particular collage will have the highest score and win.

Mark the regions that correspond to a particular collage winning in the coordinate system for your particular choise of collages and criteria. What is the *feasible region*, where you can place points (x, y)?

8.17 APPORTIONMENT

Read (at least part of) the Wikipedia article on apportionment:

https://en.wikipedia.org/wiki/Apportionment_(politics).

Select at least two different methods of apportionment and create an app in GeoGebra or by programming that illustrates this for a sample country with five districts. The populations of each district and the number of representative seats should be easily adjusted parameters. Each different method's result should be visible at the same time as all others so that it is possivle to find discrepancies simply by varying sliders. If you solve the problem by programming, make a search for discrepancies and display these.

As an extension, create a map of the country with five district capitals and use the command **Voronoi** to create a Voronoi diagram, suggesting the boundaries between districts.

8.18 STEINER TREES FOR REGULAR POLYGONS

Read the Wikipedia article on Steiner Trees at https://en.wikipedia.org/wiki/Steiner_tree_problem.

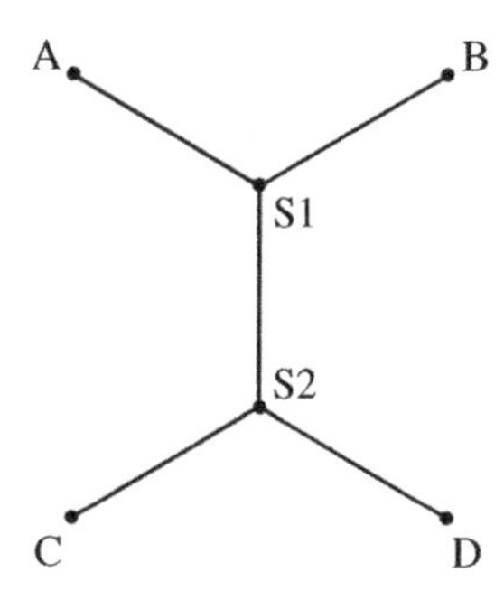

FIGURE 8.53 Steiner Tree for a square. Photo: Bryan Derksen via Wikimedia Commons. Public Domain. https://en.wikipedia.org/wiki/File:Steiner_4_points.svg.

The Steinder for a network of nodes, or vertices, is a set of edges such that there is a path from every node to every other node, and the sum of the length (or weights) of the edges is as small as possible. You could add vertices if this reduces the total length.

In Figure 8.53 you can see the Steiner Tree for a square ABCD. Two extra vertices have been added.

Find the ratio of s, the length of the minimal spanning tree, and a, the side of the square. This can be done experimentally in GeoGebra by creating the vertices, including extra ones, and then using the command **MinimumSpanningTree** to find the layout of the tree. Also you can create the corresponding segments and add them. When you move the extra points around, you can monitor the sum for the minimum.

Then do the same for the Steiner trees of a regular pentagon, hexagon, heptagon, and so on. You might have to think a bit or search to find the minimum layouts. Can you find a function that describes the relationship between the ratio s/a and the number of sides n, in the polygon? Does it approach a limit?

8.19 HUGS AND HIGH FIVES

Returning from the school holidays, all the boys in the class did high fives as they greeted each other and all the girls in the class hugged each other. There were 15 more hugs than high fives. How many students are there in the class?

How does the result vary with the number 15?

This problem is suitable for use in analyzing with either a spreadsheet, having the number of boys along the rows and the number of girls down the columns, or a brute force search program.

8.20 PYTHAGOREAN TRIPLES

Given two integers, p and q, a *Pythagorean triple* (a, b, c) can be constructed from the relations

$$\begin{cases} a = p^2 - q^2 \\ b = 2pq \\ c = p^2 + q^2 \end{cases}$$

The three numbers a, b, and c satisfy the Pythagorean theorem as being possible integer sides of a right-angled triangle.

Letting p and q run along the rows and columns, you can generate these triples as lists in the spreadsheet. You will notice that the same triple, or multiple of a triple, will be generated by several different combinations of p and q. How do these groups of triples position themselves in the spreadsheet? Why?

If you generate the triples as 3D-points instead, you will be able to look at them in the 3D window. Try to color the points of a row, or a column, or all similar triangles and see where they lay in the 3D-pattern you have generated.

You could also generate the 2D-points (a, b) for all c < 1,000. Zoom in so that you see each point and try to explain the patterns you see.

8.21 CREDITS

So you've found a house to your liking and need to borrow 100,000 USD. But what loan to get? You can afford to pay 700 USD/month and you know that the total cost of the loan will be lower, the faster you can pay it back. You get the following offers from the banks you visit:

- You get an interest rate equal to the base rate, p%.
- You get an interest rate of 1% the first year, 2% the second year, and p% all remaining years.
- You get an interest rate equal to the base rate, p%, but a maximum of 7% and a minimum of 2%.
- You take a foreign loan in Swiss franc and get an interest rate equal to $(p - 1)$% but subject to the swings in currency rates.

The base rate p will fluctuate over the years, as will the exchange rate. Find reasonable maximum and minimum values for these and devise a stochastic model where these values are changed from year to year based on normally distributed incremental changes, subject to the maximum and minimum values you have decided on.

Then, using these rates, work out the size of the remaining loan over the years and determine, for each of the alternative credits, the total cost of the loan and the total repayment period.

By pressing F9, all calculations are refreshed with new random components. Do a run of at least 20 samples and find the average results.

This exersize is based on a GeoGebra conference presentation in July 2015 by Christian Dorner and can be found at https://tube.geogebra.org/b/1479777.

8.22 THE PIANO

Imagine a composer in Heaven, sitting at an infinitely wide piano with only white keys. She can only compose melodies that are *scales*, meaning after one note must follow either the note above it or the note below it. How many melodies can she compose that consists of 30 notes if she has to stop on the note 6 steps to the right of the starting note? Work this out before continuing with the rest of the problem.

- - - -

It should perhaps not come as such a big surprise that the answer is connected to Pascal's triangle. There are

$$\binom{29}{17} = 51\ 895\ 935$$

different ways to play such melodies. Check low numbers to convince yourself that this works, that is, how many melodies with 3 notes can be played that finish one note to the right?

Now imagine a composer stuck in Purgatory, sitting at *half* an infinitely wide piano with only white keys. It extends infinitely to the right but has a single leftmost key. How many melodies can he compose that consists of 30 notes if he has to stop on the note 6 steps to the right of the starting, leftmost, note? Try it!

- - - -

It turns out there is an interesting geometrical twist to this problem. Fold Pascal's triangle according to Figure 8.54 so that the circles overlap. The left part should be folded back, under the right part. Subtract the upside-down numbers from the top numbers, and you have the solution. Algebraically, the number of melodies are

$$\binom{29}{17} - \binom{29}{11} = 17{,}298{,}645$$

As you look down the leftmost column in your folded triangle, you will see a series of numbers beginning with 1, 1, 2, 5, ... Look this series up at The Online Encyclopedia of Integer Sequences, OEIS, at http://oeis.org/ and see what you discover.

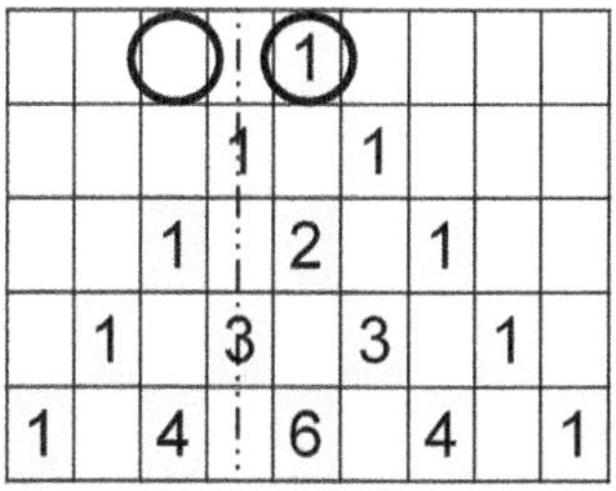

		○		(1)				
			1		1			
		1		2		1		
	1		3		3		1	
1		4		6		4		1

FIGURE 8.54 Folding Pascal's triangle.

For the final question, imagine a piano in Hell, where a composer sits at a piano with only six white keys. How many melodies are possible to compose consisting of 30 notes that start on the leftmost note and finish on the rightmost note? Try to find the answer in at least two ways, by creating a spreadsheet model from the start and by attempting the folding trick. Templates of Pascal's triangle suitable for folding can be found on the book's website.

9

MODELING IN THE CLASSROOM

A teacher who resolves to work on mathematical modeling is embarking on a process consisting of several phases spanning more than just one mathematics course. In the process both the teacher and students grow from being beginners to becoming experienced modelers. The strength and complexity of digital tools such as GeoGebra, and Wolfram Alpha might seem overwhelming at first and almost frightening to tackle. As a teacher, you might even justly question the rationality of teaching students tools that can do so much mathematics before the students know how to do this same mathematics by hand.

Looking at the modeling process from another angle, we might discuss the calculation of square roots. It is quite possible to calculate square roots by hand with an algorithm related to long division. Most would today find it an overly detailed and ceremonious method. As far as we are aware, no one is complaining about the fact that this method is long gone from both curriculum and classroom, replaced with the use of computers, calculators, or smart phones that can calculate square roots to the precision we require. Indeed, today we focus on the concept of square roots and how these are related to squares and use calculations of square roots as a small part of the greater problem-solving processes.

Today, you may learn the symbolical rules for calculating square roots so that you can handle simple expressions, just like the way you learn basic multiplication so that you understand the number system better and are able to calculate simple products. Few people, however, attempt to calculate products like $34.567 \cdot 39.8273$ by hand,

Mathematical Modeling: Applications with GeoGebra™, First Edition. Jonas Hall and Thomas Lingefjärd.

Companion website: www.wiley.com/go/Hall/MathematicalModeling

even though they know the multiplication table. In the same way you are likely to only simplify small and simple symbolic expressions by hand because you know that there exists powerful software that can deal with even quite complex algebraic expressions that occur in professional situations. To be sure, someone has to do theoretical mathematics and create these programs, but in each field there are more tool users than tool makers, and clearly, you don't train everyone to be a tool maker in upper secondary school.

Therefore you should not be discouraged by the available free software that can factorize large numbers to 15 digits, solve differential equations and compute derivatives, calculate integrals, calculate probability, and all more or less automatically. In fact this should enable you to change your focus from algorithms to concepts and more powerful problem solving with more realistic problems. With more realistic problems you might just be able to dispense with the question so frequently heard from students: Why do we have to do this?

Another benefit to bringing computing power, visualization, and explorative methods into the classroom is that with these new tools the subject of mathematics, that for so long has been regarded by students as *antiquated*, now has the potential to be regarded as *modern*.

In this chapter you will find a description of the process that teachers and students typically go through when working with modeling in the classroom—from novice to experts. Depending on the teacher, the students, the course, and personal ambitions, you have an inbuilt freedom to choose how far you would like to go. Nevertheless, you must aim at developing the student's ability to perform mathematical modeling just as you must foster all other mathematical abilities.

9.1 THE TEACHER CREATING DIAGRAMS

For a teacher unaccustomed to GeoGebra and other mathematical modeling tools, it might be a good idea to play around with GeoGebra alone first. GeoGebra, Wolfram Alpha and Excel are excellent programs for creating proper and professional diagrams of different kinds that you can use in your own instructions to the students. In GeoGebra you copy the graphics area by pressing **Ctrl-Shift-C**, after which you switch to the word processor and paste your screenshot with **Ctrl-V**. Learning the basics, such as how to graph a function, how to pan and zoom, how to change the color and style of graphs, lines and points, and so on, can be done in the classroom, but most teachers prefer to do it before confronting the students.

9.2 STUDENT'S LAB REPORTS

One convenient way for teachers to introduce GeoGebra for the students may be to demonstrate how useful GeoGebra can be for writing lab reports in physics or chemistry, and even writing up analyses of correlations between statistical data in mathematics or in social science. It is not difficult for students to enter their measurements into the spreadsheet and then produce diagrams, calculating a correlation, regression, or slope of a linear fit and then to copy these results into a word processor. Compared with making painstakingly hand-drawn diagrams, we have found that this immediately appeals to students.

9.3 MAKING SCREENCAST INSTRUCTIONS

Trying to teach your students exactly how a program works in a full class setting can be frustrating. When you are explaining step number three, someone will ask you what you were doing in step one, and then you will have lost half of the students regardless of what you choose to do or say.

Try instead to make a screencast, describing the process for the students and showing them exactly what is happening on your computer screen with your voice explaining what you are doing. You can put this video on YouTube where you can put all your mathematical screencasts on your own channel, a sort of personal web page on YouTube. The screencasts can be created using Screencast-o-matic (www.screencast-o-matic.com), Community clips which is a Microsoft Office app, or any of many other tools.

The major feature of instructional videos is that students individually can view the parts they think are problematic or especially interesting. They can do this at the time of their own convenience and repeat it as many times as they want. The only thing you as their teacher need to do is to give them the link to your YouTube channel. Naturally, it will work just as well if you want to upload the file to your local network, a Dropbox folder, or your school's web portal.

9.4 DEMONSTRATIONS

GeoGebra is an excellent tool for classroom demonstrations and may equally well serve as a part of a lecture, to prompt a student discussion of a concept or of an observation by a fellow student, or to show how something really works in response to a question. You could introduce theories by showing silent screencasts for the students to discuss or outline a proof. Several examples of both can be found on GeoGebraTube.

In any case, dynamics are important for all demonstrations. Be sure to vary values, point's positions and so on and encourage students to predict the results of the changes. Use sliders, points that are movable on the axis or free points, to vary your constructions. Allow the students to enter points of freehand functions as predictions.

As an example, you may create the slider n and a function $f(x)$. Then type

```
Function[f,0,n]
```

and hide the original function f so you can show the new function little by little. Let the students try to predict where the function will be drawn and place points representing the students' predictions. Then you can draw the function slowly by moving the slider n.

Visibility is especially important here. Use thick lines, strong colors, and large letters to keep the content lively. Make sure you have enlarged the text size and moved the Input bar to the top of the screen. A list of suitable settings can be found in Appendix A, Section A.5, *Customizing GeoGebra.*

You might like to start with already completed constructions, or applets, but after you become more familiar with GeoGebra, you should try to make the constructions on the fly, at the same time as you are describing how to do a certain procedure. When you feel even more relaxed, you will find that you can do a screencast in the classroom while doing the construction so that the students understand how screencasts are done and have the opportunity to check how you made the construction later. Students generally appreciate this and can better follow instructions that are paced and well thought out.

9.5 STUDENTS INVESTIGATING CONSTRUCTIONS

When you and your students have started to get used to working with GeoGebra, you can start working with mathematical modeling. We recommend that you start with problems, exercises, or situations where the construction and model are given and where the students are asked, during a lesson, to investigate, describe, and explain what happens when the values of parameters are changed or when points are moved. It is important that you emphasize that a model is a description of a real situation and that is has a purpose. We recommend that you have at least two other models lined up describing other realistic situations that you can use if the students need more input or if they need to see more modeling examples.

Finished constructions can be uploaded to www.geogebratube.org so that your students only need a web address to access it. You could even put the whole exercise with questions, screencasts, and everything else on GeoGebraTube. In order to be able to put material on GeoGebraTube, you need to create a GeoGebra account with a username and password.

9.6 WORKING IN GROUPS

Encourage or require students to work in pairs or in small groups but emphasize that everyone in the group needs to understand the mathematics used in the modeling process, that all students need to learn how to use GeoGebra, and that everyone needs to spend equal time at the computer, if they share a computer. In some groups it may be best if the student with the lowest confidence with regard to computer literacy is the one who is assigned to work with GeoGebra while the other students in the group help with the instructions.

In group work, students often talk about the assignment among themselves. This is important because this way they develop their language and oral communication skills. When they later present their solutions and/or methods, it can be done in different ways. Some students may be interested in showing their work on the screen, but all groups should be encouraged to do a screencast, and all groups should watch other group's presentations and peer-review them. If you ask students to make screencasts to explain concepts, then limit their length to 30 seconds. This keeps the students' attention focused on what is important and reduces the time you need to review their work.

If there is a certain concept or a specific algorithm that you want to consider in more depth, then you can ask all students to turn in an extra assignment. This will, of course, increase the amount of feedback they will need from you.

One type of problem that usually motivates students to do a little extra work is if you have the facilities to make video recordings of situations that are changing over time: It may be water pouring out of bottles, different ways of candles burning down, cars speeds affecting braking, the arc of a tossed basketball, pendulums with decreasing amplitude, growing plants, and so on. Some such video recordings can be downloaded from the Internet and are usually classified as time-lapse-sequences. By measuring the images in the videos, with an embedded ruler, by counting pixels, or be setting a ruler directly on the screen, students can obtain values to use in constructing a mathematical model. One good collection of video recordings can be found at https://sites.google.com/site/videoeromsammenhaenge/.

9.7 STUDENTS CONSTRUCTING MODELS

The next step is to give your students sufficient time for a larger assignment where they are expected to create their own model. The first time they do this, it would be wise to have a little formal review session on the basics in GeoGebra. Some of the material in Chapter 12, *Introduction to GeoGebra*, is suitable for this review and can be distributed to the students. Students often need clear hands-on instructions the first time: "Do it this way," "Use this tool," "Place the point here," and so on. Use a screen capture tool that enables you to copy suitable images to add to your instructions, such as Gadwin PrintScreen.

The authors of this book often give the students a week to do an assignment and provide detailed instructions for short to medium length modeling tasks. The assignment is introduced in one lesson at which time the groups are also formed. We let the students work on the assignment all through that lesson. During lesson two and lesson three of the same week, the students will work on their regular math schoolwork, but you need to allot some time to answer to any questions and to discuss examples. Finishing the assignment will be the homework for that week and should be done outside regular school hours. You could allow some extra time to complete the homework, but after the first time you try this, it is important that the students understand that they own the problem and it is up to them to find the time for it. Some students will need a lot of time and some will finish quickly. Ideally, a fair portion of the class should be nearly finished after the first lesson and only need to polish off and document their work at home.

Now how should the students present their solutions? We have found that encouraging students to do their own screencasts is an excellent learning experience, if these videos are kept to about 30 to 60 seconds long. This way you can look through them in a reasonable amount of time. To allow longer time could create a situation where you might have to look through as many as 15 video recordings spending 5 minutes on each, something you do not want to do and something you shouldn't spend that much time on in school. Even from a short screencast you can get a reasonably clear picture of what your students understand, and then you will have time to formulate control questions and additional instructions to bring back to each group.

You may want to encourage your students to write simple reports early on. This means that you will need to spend some time teaching them how to write simple mathematical text on the computer with spaces around plus-, minus-, and equal signs, using italics for variables, and raising exponents and lowering indexes. More complicated formulas with fractions and square roots can wait until later on. Remember to make screencasts of your instructions.

It is important that from the beginning students get clear instructions about what you expect from them. Is it just their problem-solving competence that you will evaluate and grade, or do you also intend to evaluate and grade their communication skills? The communication competence consists of both the students' presentation and their authoring of a report. You may want to stress the importance of writing mathematical text correctly in the report, and so forth.

For the simple early assignments you might chose to limit the grading to a few, clearly defined competencies. Further into the course, as the assignments get more complicated, students may be involved in a discussion of what broader competencies might be expected of them and evaluated in their assignments.

9.8 BROADER ASSIGNMENTS

In the more advanced courses in upper secondary school, the assignments may be so much more complex that students will need longer time periods to prepare them. Sometimes they may need a month, spending one lesson per week on the assignment and on questions, answers, and discussions regarding that assignment.

It is important that you the teacher think about the competencies to be developed in the assignments you make and also what you require of the students to get the grades you use in your grading system. Chapter 10, *To Evaluate Mathematical Modeling*, discusses that question more extensively.

There may be students who just study one mathematics course in upper secondary school in your educational system. Your educational system may require mathematical modeling to be taught to such students either as a competence or as part of the central content. If they are expected to be able to reach different grades, you need to provide them with situations where they are able to reach all available grades. For higher grades you would typically require the student to be able to construct or select and change mathematical models to a variety of real situations and to compare the models. Suitable mathematical modeling task for such a course might be to fit linear models to a set of data points, change to exponential models, and then compare the outcomes of these models.

9.9 THE SAME OR DIFFERENT ASSIGNMENTS

As teachers, we are often faced with the choice of having students do the same or different tasks for an assignment. With different assignments is it likely that the presentations will become much more animated and motivate other students to share in the discussion sessions. But this does require that the assignments be about the same in terms of difficulty and depth.

The problems we present in this book are different in that respect. This book only has a certain set of problems, selected so that we could illustrate a variety of techniques. If you are missing some techniques, you should add your own assignment where this technique is used. If you want your group to learn a particular technique, you should create more problems of the same sort.

You the teacher will probably create new problems once you have tried some of ours and have seen how they are constructed. In doing so, you will build your own stock of suitable assignments for your group of students.

Regardless of how you select assignments and how you distribute them to your students, you must always make sure that the assignments are presented as clearly as possible, especially if you present several different assignments at once. Be aware that if you do give out different assignments, then you also have to allot more time for explaining the assignments and for providing guidance and feedback. Your interactions with the students in coaching them can be are very rewarding and learning experience.

9.10 PREVIOUS ASSIGNMENTS

Likely your students are working on a large number of mathematical modeling assignments during the year, and they may even publish their solutions online on the Internet. However, this doesn't that mean that these assignments are spoiled for next year's students. While some assignments will have complete solutions published on the Internet, the way forward is to build on these solutions, solve the modeling situations in different ways, or to find broader explanations or a different angle to the same situation.

There is actually an advantage to seeing previous solutions. Students may both be inspired by the solutions and learn methods from them. In giving students a thorough solution to a previous, complex assignment, you can encourage them to apply the same method for another assignment. In truth, good and well thought out examples that students can learn from are not yet available in textbooks. And we encourage you, the teacher, to send us any useful new tasks that you have created and wish to share; we will then send you a web address where other teachers have supplied their favorite assignments so that you can add to and access this growing common depository of assignments.

9.11 THE CONSULTANCY BUREAU

We would like to end this chapter with a description of a comprehensive mathematical modeling assignment carried out in the *Marina Läroverket* High School in Danderyd, Sweden, during the spring semester of 2012 with students in the MB10b class of the D-course (corresponding roughly to course 4 in the Swedish curriculum of today). In the course curriculum it was clearly stated that students would be required to work on a large model incorporating different areas of mathematics. Previously this project was assigned two weeks before the end of the course and was strictly algebraic. But this time a different approach was tried:

The students in the class were informed at the end of the C-course that they would do a major project in mathematics over their next semester. The project would run in parallel with the more fundamental other courses of mathematics they were required to take in that next semester and their work on this project would have as large an impact on the course grade as the national test result. The scenario was the following:

The students were to be temporarily employed at *The Mad Mathematician's Consultancy Bureau*, a mathematical consultancy agency that accepts external assignments of a mathematical nature. The Bureau's MO was to always, if possible, solve every mathematical assignment in three different ways, compare the results, and then provide the customer with a detailed assessment. The three mathematical solutions could include any of the following:

- Algebraic methods based on the literature and Google searches and explained step by step
- Numerical methods using Excel or GeoGebra
- Simulations of the situation with an analysis in GeoGebra
- Programming to show numerical and graphical results
- Measurements applied to an actual physical model
- Measurements carried out in day-to-day circumstances.

Once the three different solutions to the problem would become available, the results would be explained along with any sources of errors in a written report and presented orally:

- The written report to the customer, on the question at issue, would include the three suggested solutions, \a synthesis, and a discussion. Each solution should document the method used, the data, and the analysis leading to the results. The report would have be typeset with correct typographical treatment of the mathematical text, correctly written formulas, and informative and visually pleasing graphics.
- The oral report to the customer would be short and include computer-generated diagrams, simulations, images, constructions, presentations, and so forth. The presentation would be at most 5 minutes long, including the time for questions.

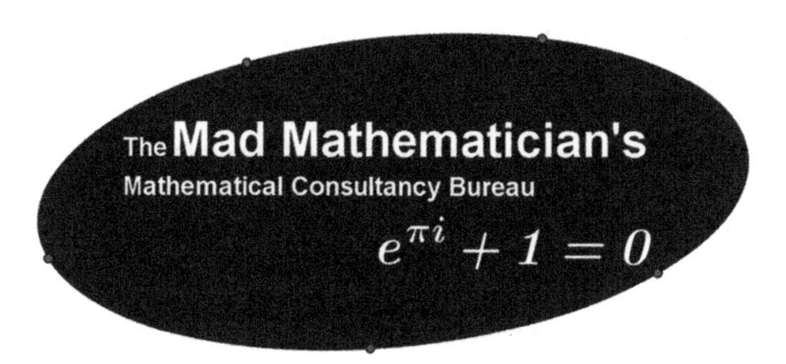

FIGURE 9.1 Logotype of the Consultancy Bureau that all students were told to use in their reports and on the posters.

The presentation, however, could be done as a screen cast should the student suffer from stage fright.

- One informative and artistically pleasing poster from the advertising department should accompany the presentation.
- All written text, both on the poster and the report itself, should carry the company's logotype, as shown in Figure 9.1, which would be designed by the teacher and with several different color options.
- All computer files should be provided together with other applicable materials. Reports and posters should be provided in portable document format (*.pdf) with a given naming scheme.

Each student in the C-course was given a list of 36 different tasks to select from and could also choose to collaborate with another student who selected the same assignment. The maximum was two students per assignment. Students could chose whatever methods to use for the assignments, even if the teacher, in this scenario also the manager of the Consultancy Bureau, was willing to help the students find suitable methods.

The assignments were mainly taken from the currently used textbooks for the course with some additions. The assignments, the student's reports, filmed presentations, posters, and files can, with the students' explicit permissions, be found at the Swedish GeoGebra Institute's website http://www.geogebrainstitut.se/elevmtrl/elevmtrl.asp.

The planning of the project was extensive and detailed, and some additional time was used for drawing up lists with the project names and spaces for students to sign up for the projects, for planning when presentations would be carried out, and for converting different file formats to pdf (hint: let the students do this themselves). In the students own schedules it turned out that they had plenty of time during the semester when they could work on their projects. However, it gradually became clear that students need a lot of support to learn to work like this and that it might be wise to allot more time for instruction in the beginning rather than later. The fact that the teacher told the students that the evaluation of this modeling assignment would do as much for the final grade as the national test would, made for strong student interest and effort.

The work turned out to be fun and exciting for students. Most were able to understand the project and quickly started to work seriously together. Of course, the students did have a lot of questions, and it was important that the teacher could answer the questions and guide the students but also to let them do the work themselves.

Most, if not all, of the students had never before worked on such a large project in mathematics, but most were quick to catch on and became very proud of their work, when finished. Some students did not manage to solve their task in three different ways, but all students managed to solve it in least one way; moreover all students in the class wrote good reports and did excellent presentations or screencast recordings. Altogether, the results were very satisfying, and all students who participated in the project also did well on the national test for this course and all passed the course.

Naturally it would have been more interesting if the students all had gotten real assignments from real customers. Unfortunately, it is difficult to find 36 different tasks with about the same level of difficulty from real external customers. With more time and perhaps with external funding, it might have been possible to approach other persons and companies to serve as customers to provide students with a more authentic experience. That could, of course, lead to a quite another project.

10

ASSESSING MODELING

A mathematical modeling project of the magnitude that we have described in Problem 9.11, *The Consultancy Bureau*, represents an impressive bit of work for a typical student. The students will have developed familiarity with new mathematical objects and concepts normally introduced later in mathematical coursework. They will have solved nonroutine problems they haven't seen before, they will have constructed and evaluated mathematical models, and they will have practiced reasoning and argued why their particular model was the best one possible. Finally they would have communicated their results and the relevance of their results in a way they had probably not yet encountered in mathematics. In short, the students would have developed many useful competencies through their work on the project.

In return the students need to get relevant, fair, and detailed feedback on their contributions, preferably without too much extra work for the teacher. There are indeed several ways to do this, but we have found that clear criteria for grading together will written assessment templates given to students at the beginning of the modeling project to be an important part of the project's success. If students do not know that their project work will be evaluated properly, then there is a danger that they might instead concentrate on solving standard problems in the textbook and on performing well on standardized tests. We encourage you to explain to your students how the modeling project will be valued in relation to other coursework and centralized tests when you do the grading.

Mathematical Modeling: Applications with GeoGebra™, First Edition. Jonas Hall and Thomas Lingefjärd.

Companion website: www.wiley.com/go/Hall/MathematicalModeling

10.1 TO EVALUATE MATHEMATICAL MODELING ASSIGNMENTS

All mathematical modeling assignments are presented and rooted in a context. Often the assignments are distributed by the teacher to a group of students or to an entire class of students. There are always constraints framing the project's context that affect the content and the teaching and learning of mathematics within that mathematical modeling project. A mathematical modeling assignment is often related to a situation in real life (IRL) and therefore also connecting to other subject areas. The exceptions in this book are mainly in Chapter 7, *Geometrical Models*. Regardless of whether your mathematical modeling project is related to real life, we suggest that you frame and formulate it so that students will learn both cognitive and practical competencies.

Cognitive competencies would include the capability of understanding, simplifying, interpreting context, assumptions, formulations, procedural management, and mathematical results, comparisons, critiques, and validations. These competencies correlate relatively well with the following actions that students can perform:

- Simplification and clarification of contexts
- Simplification of assumptions
- Formulation of relevant assumptions
- Use of mathematical concepts to justify findings, interpretation, and argument
- Identification of independent and dependent variables in functional relations
- Identification of mathematical relations and linking them to IRL situations
- Representation of mathematical concepts in algebraic form
- Application of correct formulas
- Use of derivations results to clarify interpretations
- Choice of appropriate technology for calculations
- Choice of appropriate technology for producing a graphical representation of a model
- Choice of appropriate technology for handling arithmetic in tables
- Choice of appropriate general graphical representations
- Close attention to IRL implications from mathematical results

10.2 CONCRETIZING GRADING CRITERIA

Grading criteria that focus on competencies rather than the central content are relatively the same around the globe and also for different courses in mathematics. These can be summarized in a list, table, or matrix, such as the one in Table 10.1. We have used the seven competencies in the Swedish curriculum as a starting point, but we might have used the Big 5 framework, or the 21st century skills framework, or any other national curriculum. In Table 10.1 you can, of course, add or delete any competencies to create your own local framework. We have used a grading scale with

TABLE 10.1 Summary of Criteria for Grading in Mathematics

Mathematics	E	C	A
1. Concepts and representations	Describes **briefly some** C + R Alternates with **some certainty** between R Can solve problems **in familiar situations**	Describes **extensively some** C + R Alternates with **some certainty** between R Can solve problems	Describes **extensively several** C + R Alternates with **certainty** between R Can solve **complex** problems
2. Procedures and standard problems	Handles **some simple** P Solves **S** with **some certainty**	Handles **several P including advanced algebraic expressions** Solves **S** with **certainty**	Handles **several P including advanced algebraic expressions** Solves **S** with **certainty and in an efficient way**
3. Problem solving	Formulates, analyzes, and solves P **of a simple character with few concepts requiring simple interpretations**	Formulates, analyzes, and solves P **with several concepts requiring advanced interpretations**	Formulates, analyzes, and solves P **of complex character, including several concepts requiring advanced interpretations** **Discovers general relationships presented with symbolic algebra**
4. Modeling	Applies **given M** Evaluates with **simple** opinions reasonability of M together with strategies and methods for M	**Selects + applies M** Evaluates with **simple** opinions reasonability of M together with strategies, methods, **and alternatives for M**	**Selects, applies, and changes** M Evaluates with **nuanced** opinions reasonability of M together with strategies, methods, **and alternatives for M**
5. Reasoning and proofs	**Simple R** with **simple** opinions **R**	**Well-formulated R** with **nuanced** opinions **R** **Performs simple P**	**Well-formulated and nuanced** R with **nuanced** opinions **Develops** R **Performs P**
6. Communication, language, and presentation	**with some certainty** in numbers, written language, … **with some** mathematical symbols	**with some certainty** in numbers, written language, … **Uses** mathematical symbols **with some adjustment to purpose and situation**	**with certainty** in numbers, written language, … **Uses** mathematical symbols **with adjustment to purpose and situation**
7. Relevance	**Simple** reasoning about the relevance of **something in the course's context** and its meaning	**Well-formulated** reasoning about the relevance of **some different areas in the course** and their meanings	**Well-formulated and nuanced** reasoning about the relevance of **some different areas in the course** and their meanings

levels A to E, where A is the highest achievable level. **Bold** entries represent the *level distinguishing value indicators*.

If you have never before seen a grading system like this one, it might be difficult to interpret, and for students it may take some time to understand how it may correspond to your grading. It may therefore be useful to make direct links between the levels noted in the table and the expectations you have of the assignments you are using.

Notice that mathematical modeling competency is listed in the center of the other competencies. In one direction you will see that a student needs to be good in reasoning in order to build a mathematical model and to be able to discuss the relevance of the model. In order to reason, students need a working mathematical language. In the other direction you will see that modeling is connected to problem solving and problem solving is connected to routine problems and understanding concepts.

The relationships are not equally clear in the other direction: You can engage your students in mathematical activities, including all six of the other competencies, but without automatically engaging them in mathematical modeling. This is an important point. How do *you* bring all competencies into play?

Looking again at the grading criteria for mathematical modeling, you might note to begin with, that it is only enough on the basic E-level to work with given models. If students construct their own model, then they are working at the C-level. In order to work at the A-level, however, a student is required to change, adjust, or customize a mathematical model or even come up with a model from scratch. In the given examples of modeling tasks in this book, there are sometimes more than one solution in order to show that you can use different models to solve the same problem. You can introduce new variables, change the regression model, include the origin as a measured point, add a variable controlled by a slider and dynamically investigate how a regression model will change when the slider is changing—all are examples of how a student can customize a model and work at the top level.

Second, the plausibility of the solution constituting the mathematical model, together with chosen strategies and methods, should be evaluated by the students working on the task. Examples of what constitutes valid strategies could be to measure data and fit a function, or to figure out a theoretically based idea about what function will work and then adjust some parameters to get the best fit. Examples of methods could be to calculate error limits by systematically excluding one data point at a time from the list of data points that build up the best fit, called *jackknife resampling*. Another example of methodology could be to solve a differential equation numerically in a spreadsheet.

In the grading columns of Table 10.1, you can see that students may give simple opinions (at E- and C-level) or nuanced opinions (at A-level), and that the students working at C- and A-level may evaluate these opinions and the selection they made of model, strategies, and methods. A nuanced opinion may take several different perspectives and may have the form of "*we see... but on the other hand...*," and also show a certain accuracy and richness of details. To reach the top level in problem solving requires the student to think about alternative ways to solve the problem together with nuanced and detailed comments on the choices and to have shown accuracy in the computational work.

The following example of a student evaluation for the fish farm in Problem 5.1, *The Fish Farm II*, shows work at the top level:

> *To begin with, we thought that the quadratic supply function would be the best choice, since it had more parameters to work with than the linear and the cubic function. But then we thought that when the price goes down to zero, it would attract a lot of interested persons and that situation corresponds better to the cubic function.*

This student has solved the problem in several ways and commented on the differences in a nuanced way by giving different perspectives on what may be appropriate in leading to a solution. The comment also shows a certain attention to detail by the student in deliberating a solution. The accuracy, however, is shown more in the actual report than in the comments.

We are now ready to give a more specific explanation of the grading criteria for the mathematical modeling competency. In the text below the label "problem" means a mathematical modeling task.

- **For an E** you should have solved the problem according to the instructions. Furthermore you should have commented on the plausibility of both the answer and the model.
- **For a C** you should have solved the problem in least two ways (for example, by using different supply functions). Furthermore you should have compared these solutions and commented on the plausibility of both the answer and the models with their implied assumptions.
- **For an A** you should have solved the problem in least two ways (for example, by using different supply functions). Furthermore, you should have compared these solutions and commented on the plausibility of both the answer and the models with their implied assumptions. These comment should be both nuanced and detailed. You should be able to argue for both solutions from different standpoints and have enough details to give weight to your arguments.

Here we have just given one specific criteria for the assignment, but it is quite possible to give more. You could also present the criteria differently to make it clearer. The criteria for an A could be written:

For an A you should have

- solved the problem
  - in at least two ways (with different supply functions)
  - with accuracy

- compared these solutions where you have commented on
  - the answers' plausibility
  - the models' plausibility and implied assumptions
- these comments should be
  - nuanced, meaning you should be able to argue for both solutions
  - detailed, to give weight to both arguments

In the criteria for the problem-solving competency, you can see that it is sometimes the problem's intrinsic difficulty that decides what level the student will reach. Students should be given opportunity to work with rich problem situations that can be solved at many levels of difficulty. You should try to avoid limiting your students by giving them very easy problems, unless you are introducing problem solving to them. The criteria for problem solving are the ones you must clarify the most, so the students understand what level of work they must do to achieve a certain grade. Here is another example, again taken from Problem 5.1 *The Fish Farm II*:

For an E you should have:

- Used a supply function of your own choice and followed the instructions to find the maximal profit and what price that should be set to get this profit
- Clearly stated price and maximal profit and commented on the plausibility of the solution and of the mathematical model.

For a C you should in addition have:

- Used both of the supply functions and carried out both calculations correctly and appropriately (suitable names on parameters, units on the axis, colors, etc.) to get the maximal profit and the price that should be set to reach this profit as well as investigated how this profit depends on the cost per fish
- Commented on the different results for the different supply functions. Is the difference large? What does it depend on?

For an A you should in addition have:

- Used a modified supply function and analyzed how this supply function depends on the measured value
- Done an optional, small expansion/observation by varying the values for the parameters and investigating *dProfit/dParameter* and showed that you understand the concept of *stability*

- Commented on what limitations you see in these mathematical models. Under what circumstances can they be valid? Are these circumstances realistic?
- Commented on what economical optimization does not take into consideration from natural resources and environmental perspectives
- Presented thoughts, conclusions, and analysis in a structured and readable way

As you probably have noticed here, the criteria for the problem solving and the mathematical modeling competencies (and partially some other competencies too) have been woven together.

10.3 EVALUATING STUDENTS' WORK

In a written report the student could create, in peace and quiet, a whole presentation that demonstrates the student's knowledge through all levels of competencies. A report may take some time to read, but it is generally not to so hard to evaluate because the student's level of knowledge will often be clear.

Ideally, the report was compiled and written under the best of circumstances. Unfortunately, a lot of reports are written at the last minute, the day before they are to be turned in and often student require an extension. If students get more time to finish the problems early, and get support with their planning, they can learn to start early on the report, and produce a better result.

In a mathematical modeling project the focus is rarely on routine problems or proofs, which is why the table columns do not show these competencies. Table 10.2 shows an example of how the evaluation can both be further shortened and made more specific at the same time.

At the time of assessment and grading, regardless of whether it is a report, an oral presentation, or a poster, video, or screencast recording, or something else being assessed, it is useful to have a copy of the assessment for each student where you can select and point to observed criteria and take notes on references and comments. If students give oral presentations, it is wise to record them because it is easy for your mind to drift in the assessments from the first to the last presentation. This will also help you in discussions over grading, when you do comparisons, or if you want a second opinion from a colleague.

A poster is a good way to allow students to think about another target group than the teacher or the original owner of the problem. A poster addresses potential future buyers of your mathematical competency and should be designed so it attracts attention at the same time as it makes your personal or company logotype well known. Note that we do not talk about a white paper with some notes and pasted diagrams, we are talking about a colorful, professional looking computer-produced product that can illustrate the strength of your company in different ways and in for a long time, such as the poster in Figure 10.1, originally produced by a Swedish student, translated by the authors.

TABLE 10.2 Summary of Grading Criteria Adjusted to Mathematical Modeling

E	C	A
Concepts and representations: You use words and concepts in a correct way. You show understanding of how tables, graphs, formulas, and text connect together.		
Comprehensively, some switches with some certainty Well-known situations	Extensively some switches with some certainty	Extensively several switches with certainty Complex P
Problem solving: You show ability to solve the problem and interpret the answer.		
Simple	Advanced interpretations	Complex and advanced interpretations
Models: You use several different models or methods to solve the same problem. You evaluate the plausibility of the answers and the models and methods. You discuss the models underlying the assumptions and limitations. You can change the model. You can select suitable methods for a specific problem.		
Given models	Selects suitable models	Selects and customizes models
Simple opinions	Simple opinions including alternatives	Nuanced opinions including alternatives
Reasoning and separating guesses from well-founded claims: You describe your way of thinking and explain not only just *how* you reach your results but also *why* you set up equations, applied models, etc. You evaluate your solutions and describe why you think one solution is better than another.		
Simple reasoning	Well-founded reasoning	Well-founded, nuanced reasoning
Simple evaluations	Nuanced evaluations	Nuanced evaluations
Communication language and presentation: You write mathematical text with correctly typed formulas and symbols and you use a correct language. You round off correctly, use suitable units and you present the text so it will be easy to read. You are clear in every step of the presentation and there is a visible chain of thoughts in your report. In your oral presentation you will use a computer in a professional way and you will stick to the time schedule. You will summarize the most important findings in your report and your poster will be aesthetically pleasing.		
Some certainty—can be followed Some symbols, etc.	Some certainty—easy to follow	Secure—clear chain of thought
	Uses symbols, etc., with some customizing	Uses symbols, etc., with good customizing
Relevance: You comment, with arguments, and give examples of new applications of your solution, or think of new problems that can be answered with the same or similar methods.		
Simple reasoning	Well-founded reasoning	Well-founded nuanced reasoning

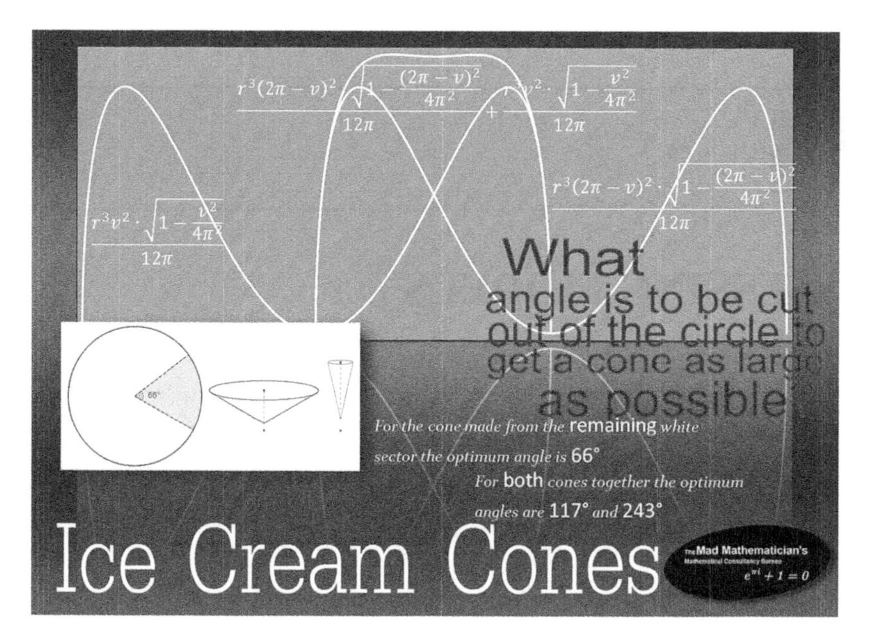

FIGURE 10.1 Poster by Sebastian Genas.

By addressing these issues, you show to your students that mathematics is not an isolated subject, but one that has real-life relevancy. Doing evaluations this way may just be the difference between engaging and motivating students or not.

11

ASSESSING MODELS

Even a relatively simple model such as one that describes the cooling of a cup of coffee or a linear model that describes results in pole vaulting is afflicted with certain limitations. This means that mathematical models are seldom correct in the same sense that routine problems are where there is often a single correct answer that can be checked at the end of the textbook. Mathematical models that try to predict human or other living creatures' behavior are at best probable or reasonable.

When you confront a model's assumptions with reality, you may find that the correlation between even a simple mathematical model and reality can sometimes be very good within certain limits. The population of several animals and plants, globally and locally, are unfortunately decreasing exponentially. If nothing radical is done, these species will die out. As one example consider the peregrines on the Swedish west coast, which were decreasing exponentially from the 1950s to the 1970s, when exhaustive rescue operations were launched. With changed circumstances, the population decline was broken and the original mathematical models were no longer valid. You can also easily find examples of models exhibiting exponential growth. Bacteria may under certain circumstances increase exponentially. This is also true for larger animals. The cormorant has grown exponentially in several countries since the mid-1950s.

Evaluating mathematical models is often about trying to evaluate the size of the errors in the predictions that the models are doing. This error can be expressed in many different ways, some of which are presented here. To explain how to treat error estimates fully lies outside the scope of this book because it is aimed at upper secondary school. Nevertheless, some ground work can be covered.

Mathematical Modeling: Applications with GeoGebra™, First Edition. Jonas Hall and Thomas Lingefjärd.

Companion website: www.wiley.com/go/Hall/MathematicalModeling

11.1 RELATIVE ERROR

A common measure of the size of an error is the *relative error,* defined as the absolute, or actual, error divided by the measurement in question. You can use an estimate, a standard deviation, a standard error, or a confidence interval as the absolute error. You can use one particular measurement or the mean of several measurements as the measurement in question. The relative error is often given in percent.

For example, you measure a slope of a fitted line to be 2.45 and then remove one data point at a time to calculate the slope again. These new values are 2.3, 2.4, 2.5, and 2.6. The errors are –0.15, –0.05, 0.05, and 0.15. The mean of these is, of course, 0, which is why you need to calculate the standard deviation for the slopes instead to get an error estimate of 0.112. The relative error for the slope can then be calculated as 0.112/2.45 = 4.6%. This is an example of the process known as *jackknife resampling*. Another example of this may be found in Problem 4.8, *Concentration*.

If you have a relationship between different quantities, for example, that the velocity equals the distance divided by the time, and you measure different distances and times and calculate the relative errors of these measurements, what will the relative error be in for the speed? The simple answer is: about as large as the largest relative error in your measurements. With this simple method you can find an error estimate for your results. And if several different investigations give different relative errors, you can determine which of the investigations is the most accurate. This is a particularly useful property of the relative error. Learning how to calculate *error propagation* will improve on these simple methods, but for many simple situations, they suffice.

11.2 CORRELATION

Linear models' strengths or weaknesses are often measured in terms of correlation. The *correlation coefficient*, or *Pearson product-moment correlation coefficient*, says something about how close the *linear* relation is between two data sets. The correlation coefficient, r can have a value between –1 and +1. If r is 0, then there is no linear relation at all. But if r is equal to +1 or – 1, it means that the relation is 100% positive or negative. Normally no relations have a 100% correlation. Note that r is only relevant for linear models. The correlation coefficient is calculated in GeoGebra by typing

```
CorrelationCoefficient[list1]
```

where list1 contains the data points.

Sometimes the quantity r^2 is used instead of r. This quantity is often called the *determination coefficient.* As before, the values close to 0 mean that no relationship exists, but now high values mean that there is a relation, regardless of the slope's direction. The determination coefficient, r^2, may be perceived as more intuitive than the correlation coefficient, r, since low value = weak relation and large value = strong relation. The determination coefficient can be more generally defined for nonlinear relations.

In many investigations and mathematical modeling situations we are interested in analyzing the relationships between variables. Imagine, for instance, the vast number of studies on the relationship between smoking and lung cancer or the relationship between industrial exhausts and the quality of seawater. A first step in such analyses is often to measure the level of covariation in the variables being investigated.

In most mathematical modeling situations it is important to be clear about what type of variables you are dealing with. But, when you are analyzing pairs of variables, the situation is more complicated than when you have only a single variable. You might have a pair of variables where one variable is quantitative and the other variable is qualitative, for example, a person's length (quantitative) and favorite color (qualitative). Mixing variables in that way can make mathematical modeling difficult. In this book we have only used quantitative variables.

11.3 SUM OF SQUARED ERRORS

The correlation coefficient measures the *linearity* between variables in a linear model. In a comparison, in order to determine which mathematical model is the better one to use, the SSE, *sum of squared errors*, is applied. SSE, sometimes referred to as SS_{Res}, is defined as

$$\text{SSE} = \Sigma\left(y_i - f\left(x_i\right)\right)^2$$

where (x_i, y_i) are data points and the function f is the fitted function. In practice, this is the basis for the *least squares fitting method*. You measure the vertical error of each data point, square the errors, and sum the squares. The lower the value, the lower is the total error. In GeoGebra you can type

```
SSE[list1,f]
```

where list1 is a list containing your data points and f is your function. You can also find this value in the probability calculator when you show statistical data Σx.

11.4 SIMPLE LINEAR REGRESSION

A central mathematical modeling concept is *regression*. This concept is important when you want to predict how future data will behave, or when you want to interpolate between your data or even when you want to extrapolate outside the domain of your data, something one should never undertake lightly.

A model for a simple linear regression can be described like this: There are n data points $(x_1, y_1), \ldots, (x_n, y_n)$, where $x_1, \ldots, x_n$ and $y_1, \ldots, y_n$ are measured values. You want to determine the equation $y = \mathrm{a}x + \mathrm{b}$ for a line that goes through the set of data points $(x_1, y_1), \ldots, (x_n, y_n)$ in such a way that the total deviation from each point to the line in some way becomes as small as possible.

The points P_1 to P_6 represents the measured values $(x_1, y_1), \ldots, (x_6, y_6)$. A line is drawn through the data point set, and for every point you measure the distance to the line parallel to the y-axis. Square all these distances and sum the squares. Then determine the parameters a and b in the equation for the line so that the sum of the squares becomes as small as possible. The method is called the *method of least squares*, and the line is called a *regression line*. Regression lines are often written $\hat{y} = ax + b$. In GeoGebra this geometrical construction can be made dynamically, as shown in Figure 11.1. Moving the points will also change the regression line.

The parameter b is not always so interesting, while the *regression coefficient* a almost always has a meaningful interpretation.

The accuracy of your regression relation can be estimated through the *correlation coefficient* r, which enables meaningful comparisons of completely different materials. Should $r = 1$, then the points lie on a straight line and there is a perfect linear relation. This is literally never the case in practice. Every linear model calculated like this on real data has an error, so the model should be written

$$y_i = \mathrm{k}x_i + \mathrm{m} + \varepsilon_i$$

where the error in every point

$$\varepsilon_i = (y_i - \hat{y}_i)$$

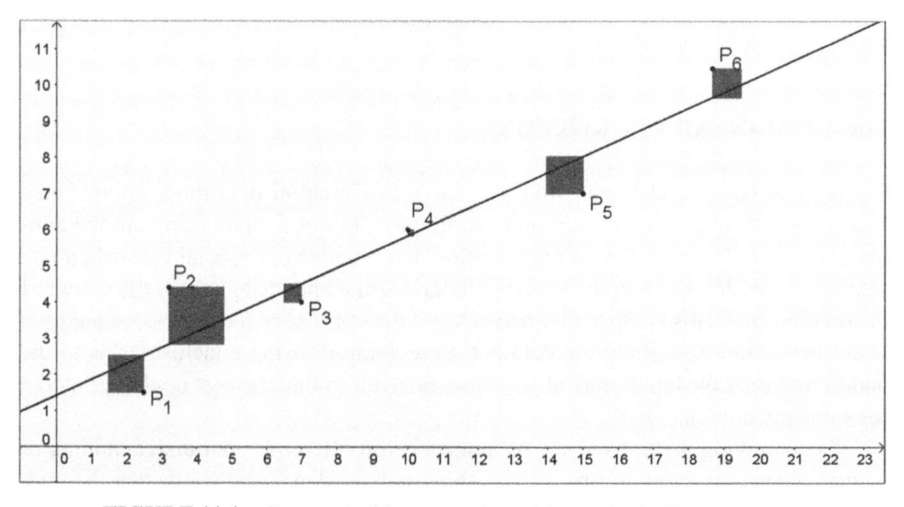

FIGURE 11.1 Geometrical interpretation of the method of least squares.

is called the *residual*. Residual diagrams, where you can plot ε_i against x_i often tell you whether the errors you have are systematic or random. Systematic errors suggest that your model is poorly chosen. See Problems 4.3, 4.7, and especially Problem 4.10, *Tides*, for examples on how to work with residual diagrams in GeoGebra.

Note that while many computer programs and calculators now can perform fast and simple calculations of regression lines and correlation coefficients, it is up to you to interpret and understand the meaning of the possible relationship you think you see.

11.5 MULTIPLE REGRESSION ANALYSIS

Often relationships are so complicated that they are not possible to describe with just a simple linear model. One of the more obvious limitations is that it only accounts for one explanatory independent variable's impact on the y-variable. A possible expansion could be to study mathematical models that are still linear but with several independent variables, that is,

$$y=\alpha+\beta_1 x_1+\beta_2 x_2+\varepsilon$$

$$y=\alpha+\beta_1 x_1+\beta_2 x_2+\beta_3 x_3+\varepsilon$$

$$\ldots$$

Mathematical models where you have more than just one explaining variable are called *multiple regression models*. In many social science studies there are several, different strong, explanatory variables (e.g., study success that could be explained by the level of education of your parents, the level of study-friendliness in your home, the level of literacy, the cultural and social backgrounds) and you will then find yourself in the middle of quite complex models. The study of multiple regression analysis is thus far outside the framework of upper secondary school and far outside the scope for this book.

11.6 NONLINEAR REGRESSION

Linear regression models are often just an approximation of a more complicated reality. Therefore it is sometimes quite natural to look for a relationship among nonlinear models. The diversity of nonlinear models entails very special care when you select your model, and ideally, you should have some knowledge about the nature of the relationship. This is especially necessary if the purpose of the regression analysis is not just about describing the data but more about drawing conclusions as to the underlying relationship. It may also be very difficult to find the best nonlinear model for some phenomena.

When studying natural science phenomena, you often use your understanding of natural science in trying to find relationships and also have some hope that the relationship is possible to describe by a relatively pure mathematical model. You know that some relationships are quadratic whereas others are exponential, logarithmic, and so on. Nevertheless, with too few observations it may still be difficult to be sure of a mathematical relationship without a very deep underlying understanding of the nature of the relationship.

Even if today we have the technology that can compute fits for exponential relationships, the literature on regression analysis often contains methods for linearizing data, meaning to take logarithms of the left, right, or both sides of a model. This procedure is still valid, since it is easier for the human brain to interpret linearity than exponentiality, and the like. See Problem 4.3, *Warm-Blooded Animals*, or Problem 5.10, *Cardiac Output*, for examples.

Modern computer software frequently comes with standard algorithms to fit regression functions to data points that might be *linear, quadratic, cubic, logarithmic, and exponential* or to follow a *power*-relation. Sometimes there are possibilities to fit data to a *logistical* relationship or to a *trigonometrical* relationship. In addition, GeoGebra and more advanced computer software often have the possibility to fit data to models of your own choice. It is important to learn how to use your tools effectively in order to reach what is called *instrumental genesis*. This means that an instrument such as GeoGebra can be used to do more than was thought originally, meaning you can solve harder and different problems than you thought you can do. A detailed description of how GeoGebra fits data to mathematical models of your choice can be found in Appendix B, Section B.4, *GeoGebra's Generic Fitting Commands*, and Section B.5, *Example of a Generic Fit*.

11.7 CONFIDENCE INTERVALS

A simple modeling situation is that which occurs when you are looking at unbound randomized selections at statistical investigations. If you ask 1,000 persons in a country with free elections: "Will you vote for the same party that you voted for in the last election?" and 882 respond yes (ignore the problems of dropouts for now), you know that 88.2 % of the 1,000 persons gave a positive response. You can then assume (as our model does) that the same proportion, 88.2% of the population, will respond yes if asked.

In such situations it is useful to know the error margins of the result. They are often given as a 95% confidence interval. In this case we have $p=88.2\%$ and $n=1{,}000$. The error margin is then calculated as

$$f = 1.96 \cdot \sqrt{\frac{p(1-p)}{n}} = 1.96 \cdot \sqrt{\frac{0.882 \cdot 0.118}{1000}} = 0.019995 \approx 0.020$$

and the result can then be presented as $88.2\% \pm 2.0\%$ (95% confidence, or $p=0.05$). If a later inquiry gives 90.1%, it is within the error margins from the previous measurement. You can then conclude that the measured increase *is not significant* (and should be ignored). This situation is so common in media nowadays that is should be considered general knowledge to know what *statistically significant* means.

TABLE 11.1 Measured Times for 10 Swings of a Pendulum

Time (s)
20.5
20.8
20.6
20.5
20.4
20.6
20.5
20.6
20.7
20.6

The somewhat odd factor 1.96 is calculated from the fact that you want exactly 95% confidence. If you change the factor to 2, you are actually calculating 95.45% confidence. The square root gives the standard error, which corresponds to one standard deviation for normally distributed data. If you replace the 1.96 with the number 3, ±3 standard deviations, you get 99.7% confidence. In several scientific experiments today a confidence interval of ±5 standard deviations is required for claiming a new discovery, corresponding to 99.99994% confidence. One such example is the experiments that lead to the discovery of the Higgs particle.

When it comes to calculating mean values, the process is the same. If you build a pendulum and measure the time it takes for the pendulum to move 10 times back and forth, you may get the results in Table 11.1, from which you can calculate the mean $\overline{x} = 20.58$ and the standard deviation $\sigma = 0.1077$.

Calculating the standard deviation is relevant because the measured times can be assumed to depend on several small, concurrent—but independent—factors, and the times can therefore be considered to be normally distributed. Observe that there are plenty of situations where the errors **not** are normally distributed. Caution must always be taken.

The standard deviation representing the standard error in itself gives only 67% confidence, so the standard deviation 0.1077 is multiplied with the factor 1.96 give a 95% confidence interval for *every separate single* measurement. This can be written as 20.58 ± 0.21. In the long run, 19 out of 20 measurements will end up within this interval and just one out of 20 will end up outside the interval.

The mean value is more accurate, however. The more values you collect, the more stable the mean value becomes. The error limits for the mean value are given by

$$f = 1.96 \cdot \frac{\sigma}{\sqrt{n}} = 1.96 \cdot \frac{0.1077}{\sqrt{10}} = 0.06675 \approx 0.067$$

and the mean value can therefore be presented as $\overline{x} = 20.58 \pm 0.07$ (95% confidence).

But remember that, on average, one out of 20 investigations has a mean value that falls outside this interval, something very well illustrated in the web comic xkcd: http://xkcd.com/882/. The Wikipedia article on Confidence Interval, https://en.wikipedia.org/wiki/Confidence_interval#Misunderstandings, gives even more cautionary notes.

When you are fitting parameters in a mathematical modeling situation, you are probably interested in the error limits that the parameters get. A common situation is that you are interested in a, the slope of a straight line. The standard error for a can be calculated as

$$\varepsilon_a = \sqrt{\frac{\Sigma\left(y_i - f\left(x_i\right)\right)^2}{(n-2)\cdot\Sigma\left(x_i - \bar{x}\right)^2}}$$

In GeoGebra you type

```
ErrorA = sqrt(SumSquaredErrors[list1,f]/
((Length[list1]-2)*SXX[list1]))
```

where the data points are collected in list1 and the fitted straight line is named $f(x)$ and constructed with the command

```
RegressionPoly[list1, 1]
```

If you want the 95% confidence interval for a, you need to calculate

```
a - 1.96 ErrorA
a + 1.96 ErrorA
```

Standard deviations, mean values, and so forth, are nowadays calculated by most calculators and statistically aware computer software. In general, it is harder to find programs that calculate SSEs with relative ease for different models or give error limits for fitted parameters. As shown, GeoGebra can do part of this. Perhaps future versions of GeoGebra will be able to calculate error limits for parameters automatically, independent of the regression model, giving even more computational power to students for free.

11.8 2D CONFIDENCE INTERVAL TOOLS

It is possible to draw an ellipse on top of a scatterplot that represent a 95% 2D confidence region. In Figure 11.2 you see two such ellipses, one corresponding to 95% confidence and the other to 90% confidence. The construction is complex and goes outside the scope of this book, but readymade tools that can be opened in GeoGebra can produce these ellipses from a list of points or, if you want, an arbitrary confidence interval from a list of points and a number.

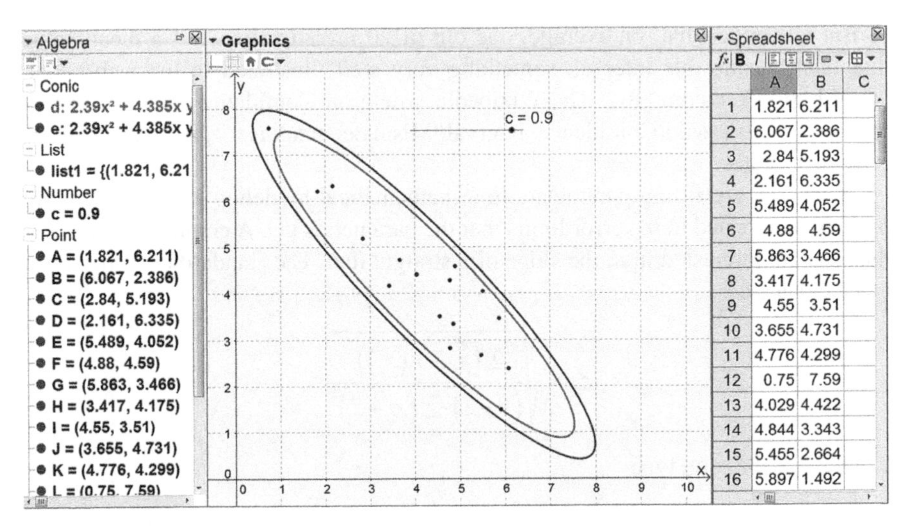

FIGURE 11.2 Two 2D confidence ellipses.

These tools may be downloaded from Book Companion Site or from GeoGebra-Tube at http://tube.geogebra.org/material/simple/id/1496461#material/1496469.

12

INTERPRETING MODELS

Being a teacher of mathematics and dealing with mathematical models on a daily basis, it is easy to fall into the trap of thinking that everyone understands models the way you do. In this chapter you will explore some aspects related to the interpretation of models, such as whether to view the model as an *event* or as a *process*.

12.1 MATHEMATICAL REPRESENTATIONS

Mathematical representations such as diagrams, histograms, functions, graphs, tables, and symbols can make it easier for us to understand abstract mathematical concepts or phenomena described in mathematical terms. We humans today are facing a world that is shaped by increasingly complex, dynamic, and powerful systems that transmit information through various different media. It you Google Ebola you will likely find diagrams showing that Ebola is exponentially growing in some unfortunate countries. How would humans with no education in exponential growth or decrease interpret such a model? An EKG tracing, or EKG diagram, is an iconic piece of information but a very difficult mathematical representation to interpret correctly. Traffic jams in Germany are often presented as 3D diagrams that are hard to understand. Being able to interpret, understand, and work with complex systems calls for knowledge of important mathematical processes that are addressed by information given in graphical representations.

Mathematical Modeling: Applications with GeoGebra™, First Edition. Jonas Hall and Thomas Lingefjärd.

Companion website: www.wiley.com/go/Hall/MathematicalModeling

In most countries, mathematics teachers are using many different representations to help students understand mathematical objects and concepts. Geometrical constructions, graphs of functions, and a variety of diagrams of different kinds and of different origin are used to introduce new concepts and to study relationships, dependencies, and change. Mathematical representations, structures, and constructions are also used when you study other subjects, such as biology, chemistry, economy, physics, and the social sciences.

12.2 GRAPHICAL REPRESENTATIONS

Representations are normally a structural system with strong relationships to theories. They may be seen as constructions that connect abstract and concrete mathematics. A representation can be said to be something that stands for something else. Examples of this are using graphical, numerical, symbolical, and language representations of mathematical objects and situations. Graphical representation may also hide *situated information*, so called as when an object's speed or acceleration may be determined from a graph's slope, something not explicitly obvious from the graphical representation itself.

An understanding of a graph can involve several processes:

- perceptual processes for recognition of graphical patterns
- perceptual processes that operate on graphical representations and create qualitative/quantitative meaning
- conceptual processes that translate the visual attributes at hand, such as quantities, scales, and symbols to relevant concepts

Perception is a psychological definition used to describe how you interpret what you experience and the process of perception changes your experiences to understandable and manageable information.

One could argue for specific ways of activating human intuition by use of interaction with graphical artifacts. Some representations include special attributes that may quickly catch a person's attention. This is sometimes called *compelling* visual attributes. It may be any or all of the visual qualities that are embedded in the graphical representation, such as vertices, edges, and contours.

> *My claim is that, even though what-you-see-is-what-you-get is not cued strongly in all contexts, it is cued strongly with respect to the compelling visual attribute of a representation.* (Elby 2000, p. 484)

Which one of these visual properties one notices first is strongly context dependent. You need to know where a car is before you can connect its path to a graph. If you do not know where a train is, it is much harder to understand a graphical representation showing a train's path from A to B.

12.3 A SAMPLE MODEL INTERPRETED

The graph in Figure 12.1 illustrates a train that moves for 4 hours. The distance is in kilometers and the time t is in hours. The train starts from a station at $t=0$.

Conceptual questions:

1. When does the train run at maximal velocity?
2. When does the train run at minimum velocity?
3. What could be said about the train's direction by interpretation of the graph?
4. What could be said about the velocity of a train according to the time–distance graph?

The students' alternative iconic interpretations of graphical representations are well known from research and will often also affect the interpretation of this imaginary train's movements. Whenever we humans face a graphical representation of a situation, we make an interpretation that is partly personal. Some students may choose an interpretation from daily life activities such as going over a hill, running a cable car in San Francisco, or even taking a rollercoaster ride. This corresponds to the selection of a daily life discourse instead of a mathematical discourse when talking about the train.

The concept of slope is important to understand and recognize when interpreting the train's movements. If not, students might well try to create a daily life context for the train instead of considering the model's explanatory power's validity and reliability. Once they have selected a daily life discourse in their arguments, it may be difficult to explain the train's movement scientifically. That is to say, there is less room for a formal mathematical discourse in their thinking when a daily life explanation dominates their thinking.

These students may select to interpret the graph as a mathematical description of an "event," even though the distance graph is a representation of a "process." Conceptual change occurs when the concept is reassigned from the "event" category to the "process" category.

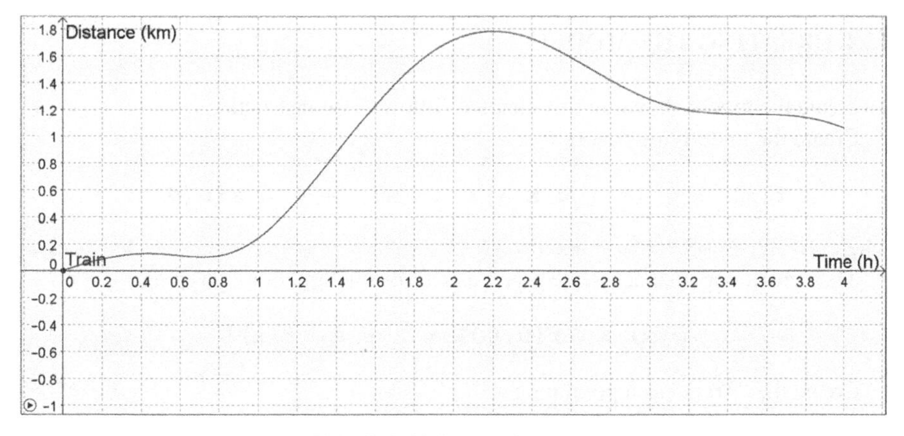

FIGURE 12.1 Moving train.

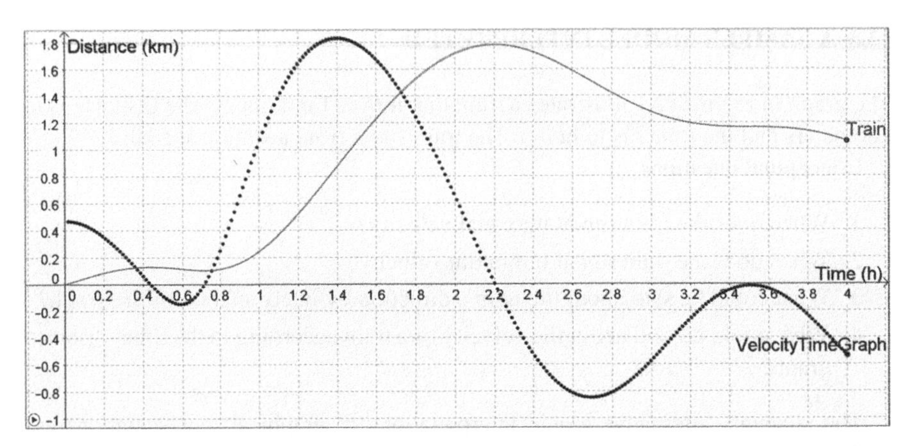

FIGURE 12.2 Train's movement and velocity in the same graphical representation.

In a graphical representation there is always some hidden meaning from the presence of a coordinate system and axes labels. It is quite easy for students to miss the meaning of a slope, so it remains hidden for them. Although most students are able to develop and use perceptual processes for recognition of graphical patterns, the meaning of the perceptual process that operates on this graphical representation and what qualitative/quantitative meaning it creates is difficult to judge, and for some students it is hard to actually translate the visual attributes at hand, such as quantities, scales, and symbols for relevant concepts.

One way to overcome this, and to make students bridge the gap between "event" and "process," could be to use GeoGebra and actually spend time to let the students explain the implication of the slope and how the distance–time graph can be the source for the velocity–time graph, shown in Figure 12.2. Ideally, these should be drawn in different Graphics windows above each other, since the y-axes represent different quantities.

12.4 CREATING THE MODEL

To create the model shown in this chapter, follow these instructions:

Create the underlying function

```
f(x) = e^(a + b/x + c ln(x)) + 1/a sin(d x)
```

where a=7.7, b=–8.9, c=–4 and d=3.6

Then create the distance–time graph as part of this function by typing

```
g(x) = If[0.001 ≤ x ≤ 4, f(x)]
```

Create the "Train" as a point on g

```
Train = Point[g]
```

The "pen" tracing the velocity–time graph is easiest constructed as a point having the same x-coordinate as Train but using the derivative as the y-coordinate:

```
VelocityTimeGraph = (x(Train), f'(x(Train)))
```

Now you enable tracing for VelocityTimeGraph, select suitable colors for the points and the graphs, hide the underlying function f, label and zoom the axes, and animate Train. The animation can be paused by using the pause/play control in the lower left of the Graphics window.

To create VelocityTimeGraph in its own Graphics window, press **Ctrl-Shift-2** to open Graphics Area 2 just before creating VelocityTimeGraph. Drag the area by its title bar so that it is directly above or below Graphic Area 1 and adjust the x-axes so that they match.

You may also directly create the derivative

```
VTgraphAll(x) = f'(x)
VTgraph(x) = Function[0.001 ≤ x ≤ n, VTgraphAll]
```

Accept to create a slider n. By varying n, you can "draw" the velocity–time graph slowly, stopping at interesting points to let the students participate by predicting the next section.

APPENDIX A

INTRODUCTION TO GEOGEBRA

GeoGebra was created in 2001 by Markus Hohenwarter at the University of Salzburg. Since then Markus and a dedicated and continuously growing team of developers have transformed this program to one of the leading pedagogical computer programs for mathematics education around the world. It is a multi-platform, open source project, free for educational purposes.

GeoGebra can best be described as a mathematical laboratory. It handles dynamic geometry, graphing, spread sheets, statistics, regression, algebra, matrices, complex numbers, differential equations, symbolic algebra, programming, and so on. It is an environment particularly suited for investigative work, easy enough to be used in lower grades, yet powerful enough to be a useful tool for university students. Yes, there are other spreadsheet programs, other programs for dynamic geometry, graphing, CAS, and the like, but only in GeoGebra can you find all of these components welded together in a user-friendly package, free for, and aimed at, mathematics education.

The different components work together in a dynamic, interactive, and intuitive way. The spreadsheet can contain both numbers and objects like lines, points, and circles. The symbolic algebra is dynamic and can be connected to both sliders and geometric elements. Enter an algebraic expression and see a geometrical object visualize this. Drag the object and see the algebraic expression change. Build a model of a situation, and then measure and analyze it directly in the model. GeoGebra is, from the very beginning, built to support multiple representations of mathematical objects and concepts.

Mathematical Modeling: Applications with GeoGebra™, First Edition. Jonas Hall and Thomas Lingefjärd.

Companion website: www.wiley.com/go/Hall/MathematicalModeling

In this appendix you will find an overview of the program if you aren't already familiar with this mathematical environment. It is not a complete introduction, focusing on basics and giving some insight into some techniques relevant for modeling work. If you want more information, begin with the resources at GeoGebra's website:

- The forum: http://forum.geogebra.org
- GeoGebraTube: http://tube.geogebra.org
- User manual: http://wiki.geogebra.org/en/Manual
- Tutorials: http://wiki.geogebra.org/en/Tutorials

The keyboard shortcuts page is also a great help:
http://wiki.geogebra.org/en/Keyboard_Shortcuts

There is also a dedicated YouTube channel and a Facebook group:

- https://www.youtube.com/user/GeoGebraChannel
- https://www.facebook.com/geogebra/

A.1 GEOGEBRATUBE AND THE ECOSYSTEM

GeoGebra is part of an ecosystem that, among other things, contains an active user forum and the GeoGebraTube website that acts as a file depository for your constructions and, like YouTube, is a place to search and share constructions with others. You find it at http://tube.geogebra.org. Anyone can download constructions from GeoGebraTube and read posts in the forum. But, to be able to upload constructions and post in the forum, you will need a user account. This is nowadays normally obtained the first time you install GeoGebra, but can also be obtained by going to GeoGebra's website http://www.geogebra.org and clicking through on the **Log in** button in the top right corner and then on **Register**. You may specify a unique username and a password or simply register using one of your Facebook, Twitter, Google, or Microsoft accounts. See Figure A.1.

The same credentials can be used throughout the GeoGebra ecosystem. In GeoGebra you can now upload a construction from the menu by selecting **File > Share**, or **File > Export > Dynamic worksheet as webpage**. The construction will then be uploaded to your GeoGebra account. Add a description and tag it with appropriate search terms. This is important because it enables you and others to find it. Use a unique tag such as your school's name combined with the name of the course, class, or your own name as well as descriptive search terms. Your students now only need the link to the construction on GeoGebraTube or the search tags to be able to access your construction directly from their browser. From there, they could download the construction to their computers in order to make their own copies.

Students and others who find your construction (don't worry, there are privacy settings though you are strongly advised to share your work with the community) can

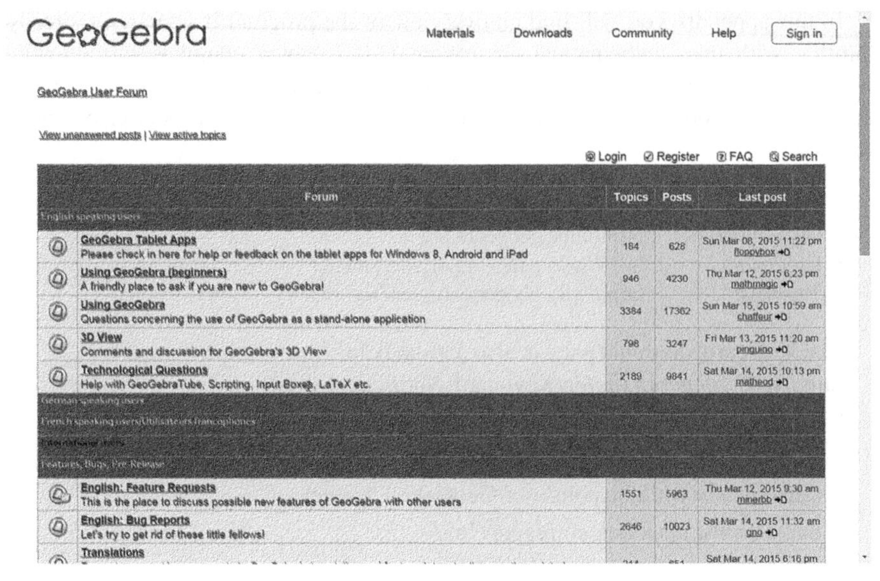

FIGURE A.1 GeoGebra forum.

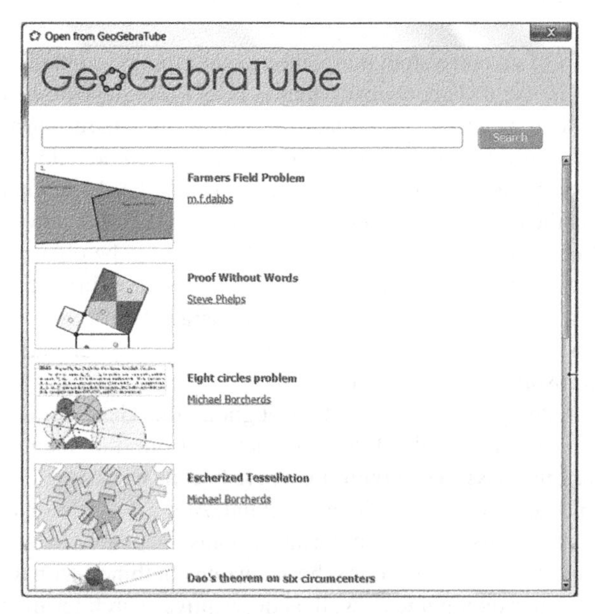

FIGURE A.2 Open from GeoGebraTube... dialogue.

also open it directly from inside GeoGebra. The Menu option **File>Open from GeoGebraTube...** will open a search box where you may search all currently shared constructions. With the construction open in GeoGebra, you can make changes and save it as your own. See Figure A.2.

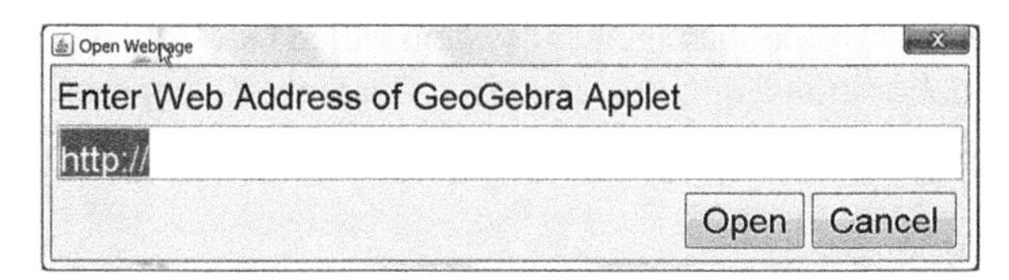

FIGURE A.3 Shift-click Open to open constructions on any website.

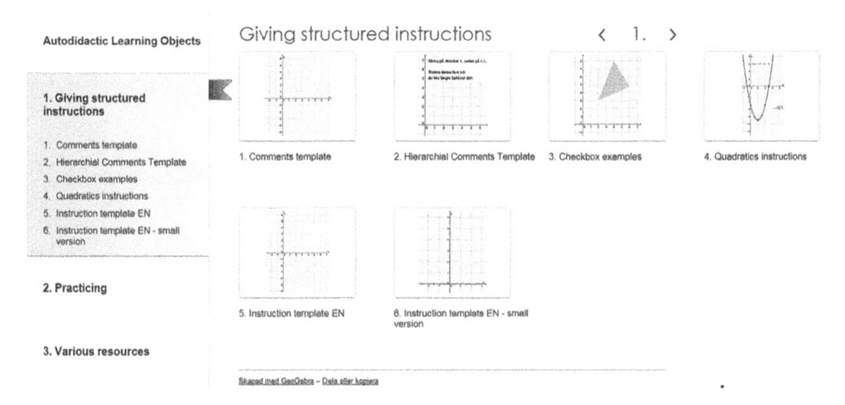

FIGURE A.4 GeoGebraBook with chapters.

Holding down the **Shift** key as you click on **Open from GeoGebraTube...** will open a dialogue where you can open a construction embedded in any website. This alternative can also be used if you know the link to a construction on GeoGebraTube. See Figure A.3.

A.2 GEOGEBRABOOKS

In GeoGebraTube you can also collect different constructions in GeoGebraBooks. A GeoGebraBook can be subdivided in chapters, each chapter having several pages. On each page you may have a mixture of text, images, videos, GeoGebra constructions, links, presentations, and so on. This means that you may collect all constructions, instructions, and links to other resources relevant to a particular course you are teaching in one easy location a single click away. GeoGebraTube is in this sense an easy-to-use web authoring system. See Figure A.4.

If you wish to embed a construction on your, or on your school's website, first upload it to GeoGebraTube, put it in a GeoGebraBook if you wish, and then download the embed code from your construction's page on GeoGebraTube.

By uploading your constructions to GeoGebraTube, you are also ensuring their future compatibility against changes in Java or HTML5. In addition, it is a simple way of distributing searchable constructions to your students. By organizing them into GeoGebraBooks, you can build complete collaborative resources for courses, conferences, and classrooms. You can collect your students work in GeoGebraBooks and publish it to the world.

There is also a classroom management system called GeoGebraGroups where you can organize collaborations, quizzes, and tests. GeoGebraGroups are very useful for creating collaborative GeoGebraBooks.

A.3 GEOGEBRA ON DIFFERENT DEVICES

Although GeoGebra was originally made for the computer, it has since moved on to conquer other platforms as well. GeoGebra currently works for

- Computers
  - Windows
  - MacOS
  - Linux
  - ChromeOS
- Tablets
  - Android
  - iPhone
  - Windows
- Phones
  - Android
  - iPhones (not released at time of writing)

GeoGebra also runs directly from the browser, using the tablet layout where the main menu is located on the right-hand side off the screen. Try it at http://web.geogebra.org.

On phones the functionality is limited. However, a phone app called *GeoGebra Sensors* or *Geomatech Sensors* is planned, which can be run on the phone and connected though a dynamically generated access code to another instance of GeoGebra running in a browser or a tablet. The data from the phones accelerometer and other sensors can then generate a data stream that can be visualized as a function, a list, or a number, in a sense using the phone as a remote data logger.

In this book we have used the traditional Windows user interface.

A.4 THE USER INTERFACE

This book uses settings useful to have in the classroom, giving a slightly different look to the user interface compared to the original "out-of-the-box" settings on a Windows computer. See Figure A.5. Exactly what changes to apply to achieve these effects will be covered in Section A.5. In this section you will cover the different parts of the user interface so that you may familiarize yourself with them:

- The menu
- The toolbar
- "The four little buttons on the right"
- The input bar

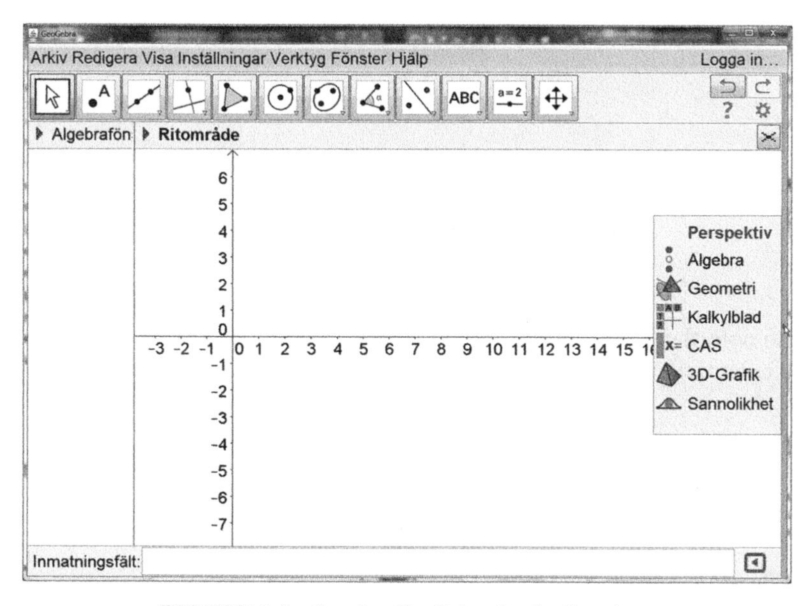

FIGURE A.5 Starting GeoGebra for the first time.

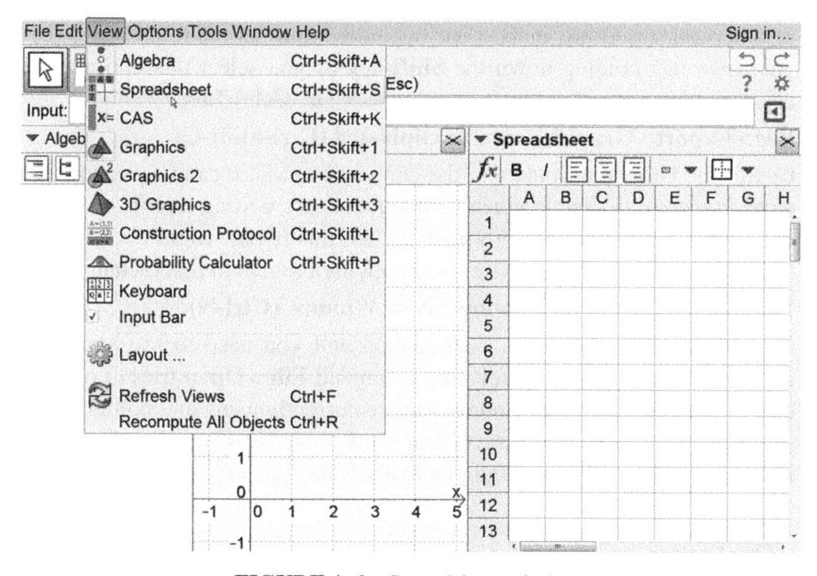

FIGURE A.6 Spreadsheet window.

- The command help button
- The Algebra window
- The Graphics window
- The sidebar

In addition, there are windows not visible from the start, such as the Spreadsheet window. See Figure A.6.

FIGURE A.7 Login for easier uploads.

Also note the login link in the top right corner. Click there and log in with your credentials and you may easily access your uploaded constructions. See Figure A.7.

A.4.1 The Menu

Some menu options deserve special attention:

- **File>Open from GeoGebraTube...**

  If you find a construction on GeoGebraTube and want to open it in GeoGebra, copy the link and use this option. It allows you to draw on hundreds of thousands of other constructions, make your own alterations to them, and save them as your own. By holding down the **Shift** key as you select this option, you can enter an arbitrary web address, even outside GeoGebraTube.
- **File>Export>Graphics view to clipboard (Ctrl-Shift-C)**

  Copies the Graphics window to the clipboard so that it can be pasted into your word processor. Valuable when your students are writing reports or when they need written instructions with images (like this book). Right-click and drag a rectangular selection first to limit the copying process to that rectangle.
- **File>New Window or Window>New Window (Ctrl-N)**

  You are in the middle of a construction and you need to test something in another window. Unfortunately, the command **File>Open** tries to open a new blank construction in the window you are currently using and politely asks you whether you would like to save your work first. Hit **Cancel** and then **Ctrl-N** instead, which is the command you really want.

A.4.2 The Toolbar

GeoGebra can serve as many tools—more than are easily displayed in a toolbar. The solution to this is to make dropdown tool menus. Each tool has a small white arrow in the bottom right corner of the icon that turns red when the mouse hovers above it. If you click on the arrow, rather than on the tool itself, a menu with related tools drops down as in Figure A.8.

Once a tool is selected, it is active until you select another tool. You could, for instance, create several points at once by first selecting the Point tool •A and then

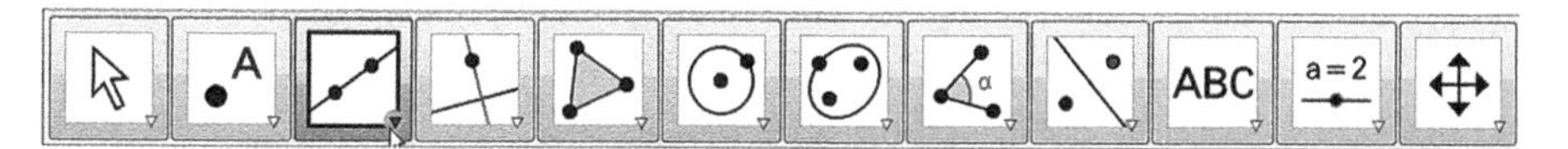

FIGURE A.8 Dropdown toolbar.

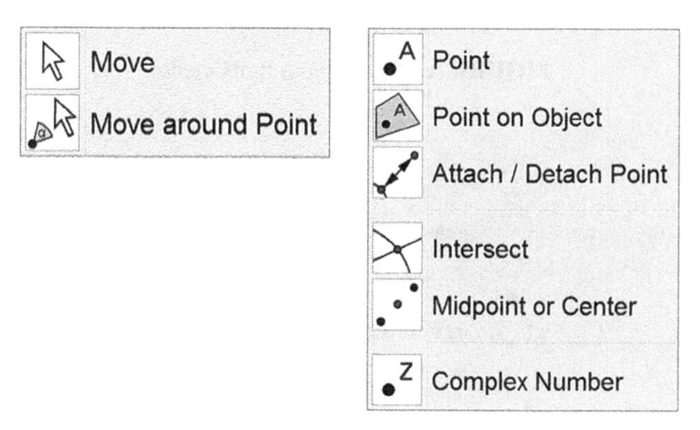

FIGURE A.9 **Selection** and **Point** tools menus.

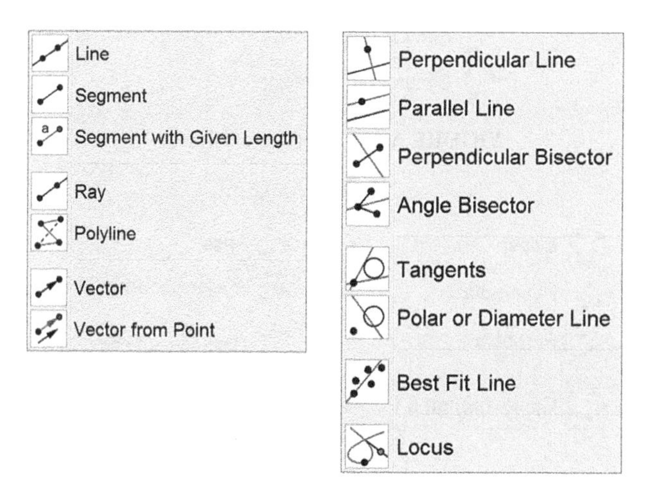

FIGURE A.10 **Line** and **Construction** tools menus.

repeatedly clicking in the Graphics window. You first select the tool, then the objects you wish to apply the tool to.

When you press the **Escape** key, the selection tool **Move** is activated. More important, any other tool and selection is dropped.

From left to right, the tools are grouped in the following dropdown menus you see in Figures A.9, A.10, A.11, A.12, A.13, A.14, and A.15.

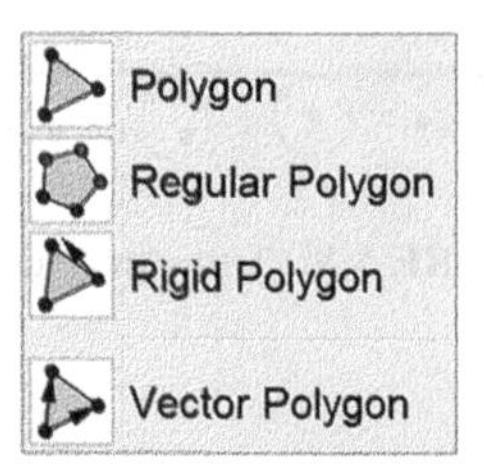

FIGURE A.11 **Polygon** tools menu.

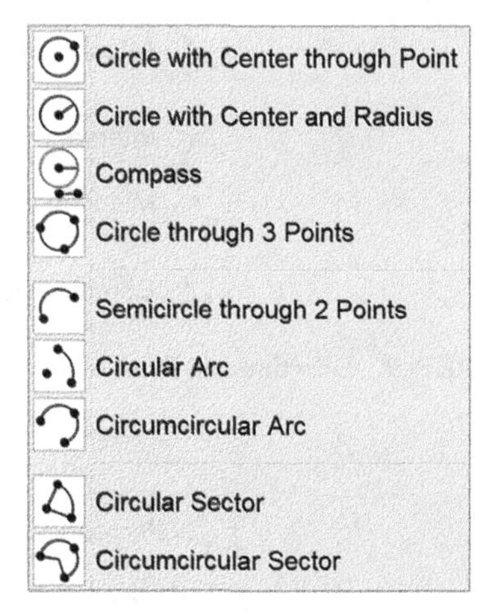

FIGURE A.12 **Circle** tools menu.

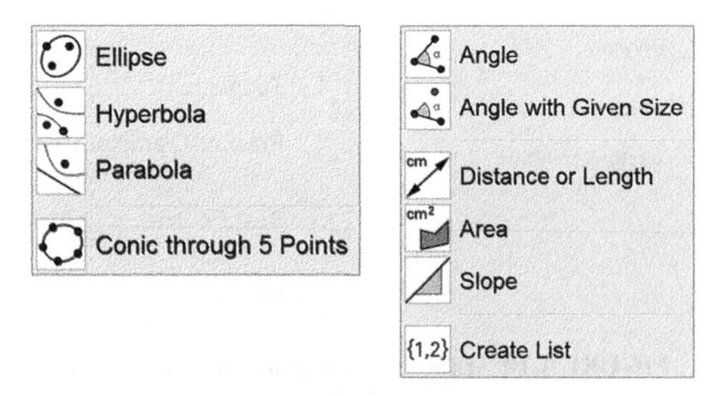

FIGURE A.13 **Conics** and **Measurement** tools menus.

Most of the special tools and all of the action tools lead to dialogues of their own, most of which are activated after you click once in the Graphics window where you want the object to appear. The toolbar itself is customizable though the menu option **Tools > Customize toolbar...** You may for instance remove

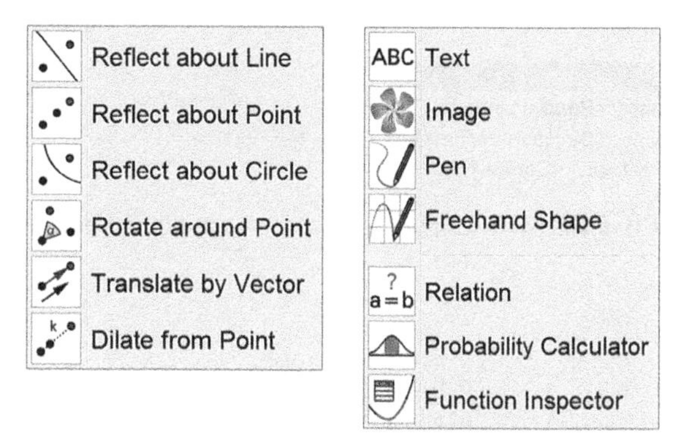

FIGURE A.14 **Transformation** and **Special** tools menus.

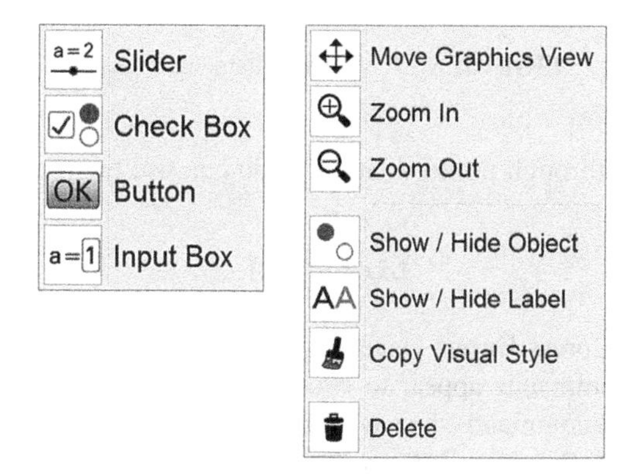

FIGURE A.15 **Action** and **Window** tools menus.

tools for students and have them solve problems with only a limited set of tools.

A.4.3 "The Four Little Buttons on the Right"

To the very right of the toolbar there are four small buttons. At the top you will find buttons for **Undo (Ctrl-Z)** and **Redo (Ctrl-Y)**. Under them you will find a help button giving the same information about the active tool that is also shown in the tool tips balloon bubbles. Finally there is the **Preferences** button leading to the **Properties** dialogue, which will be covered later in Section A.5.5.

A.4.4 The Input Bar/Command Line

All that can be done with tools can also be done with commands. The GeoGebra equivalent of a command line is called the *input bar*, or the algebra input. In the Tablet version this is replaced by an Input bar at the top of the algebra field.

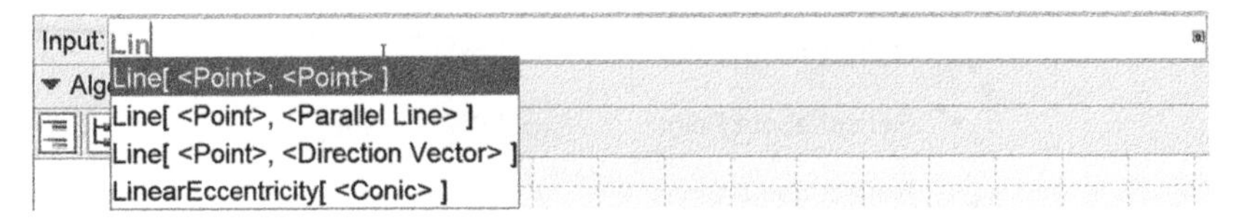

FIGURE A.16 Input bar where you can type commands.

α	β	γ	δ	ε	ζ	η	θ	κ	λ
μ	ξ	ρ	σ	τ	φ	ϕ	χ	ψ	ω
Γ	Δ	Θ	Π	Σ	Φ	Ω	∞	⊗	≟
≠	≤	≥	¬	∧	∨	→	∥	⊥	∈
⊆	⊂	∡	²	³	°	ί	π	e	

FIGURE A.17 Special characters input.

To create a line through points A and C, you can use the line tool, but you can also type

```
Line[A,C]
```

in the input bar and press **Enter**. Once you have typed a few characters, dropdown lists of available commands appear to choose from. See Figure A.16. In addition, there are many more commands than available tools; so many commands do not have corresponding tools. Pressing **Enter** will move the typing cursor to the input bar without the need of using a mouse.

To enter special characters, click the alpha sign that appears to the far right of the input bar when you place the typing cursor in the field. See Figure A.17. A menu of special characters like $^\circ$, α, β, γ, π, and $\leq$ is presented. Several of these can also be inserted using keyboard shortcuts; thus **Alt-o, Alt-a, Alt-b, Alt-g, Alt-g, and Alt-<** will produce the characters mentioned while **Alt-e** will produce the Euler constant $e = 2.718...$, and **Alt-I** will result in the imaginary unit i. **Alt-2** will give you a raised power 2, **Alt-3** a raised power 3, and so on. A complete list of shortcuts can be found at http://wiki.geogebra.org/en/Keyboard_Shortcuts.

A.4.5 The Command Help Button

To the right of the input bar is a small arrow hiding a window that, when open, gives you a list of all GeoGebra's functions, grouped in categories. If you select a command, that command's syntax will be shown, with an additional option of opening the command's user manual entry in a web browser. See Figure A.18.

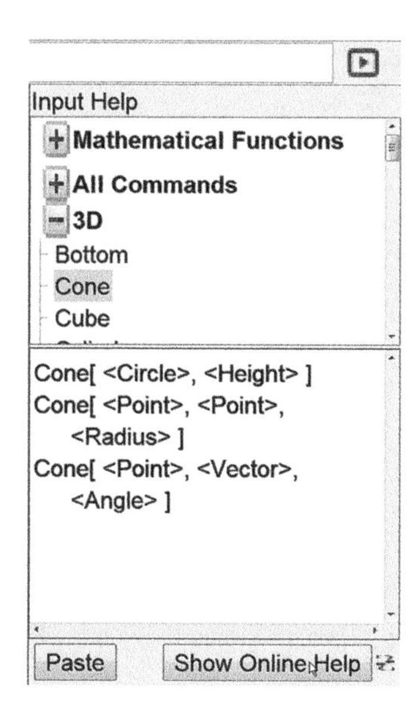

FIGURE A.18 GeoGebra has an extensive help system.

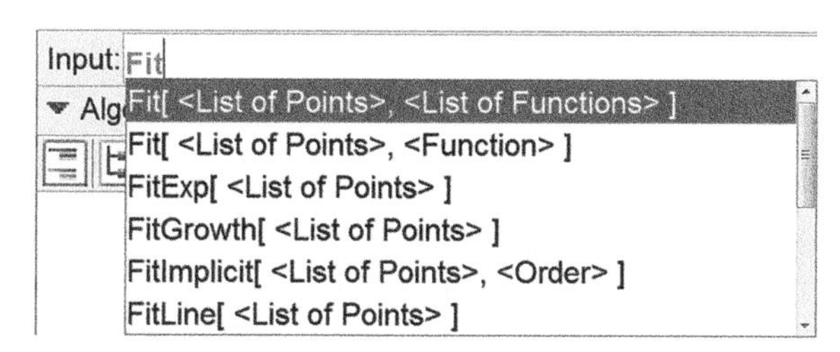

FIGURE A.19 GeoGebra will autocomplete commands for you.

As you type in a command in the input bar, the syntaxes of the possible commands are shown automatically. Type **TAB, comma,** or **right arrow** to go to the next parameter once you have selected a command! See Figure A.19.

A.4.6 The Algebra and Graphics Windows

In the Algebra window the algebraic representations of the objects are shown. A point is shown as a set of coordinates, a function with its expression, and so on. See Figure A.20. When you want to select an object, you can do so by clicking the graphical representation in the Graphics window or the algebraic representation in the Algebra window. Sometimes, such as when an object is animated, it is far easier to use the Algebra window.

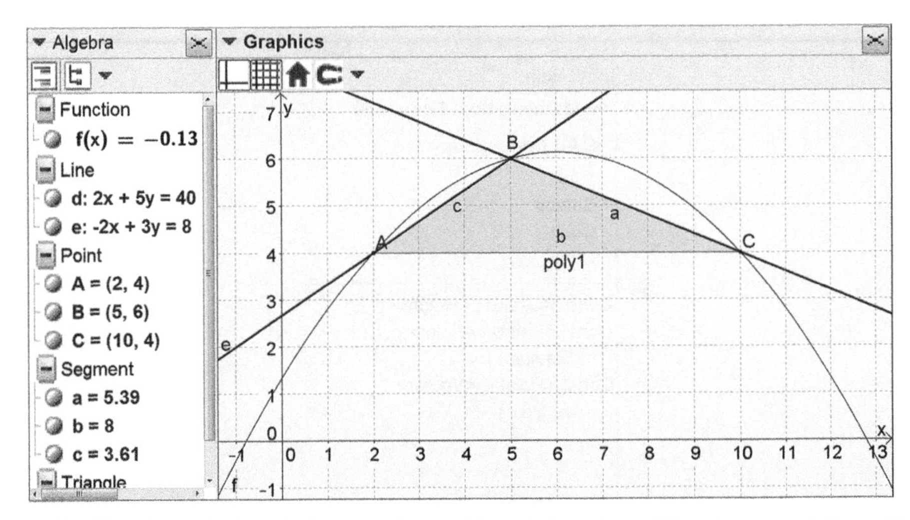

FIGURE A.20 Algebra window and the graphics window show different representations of the objects.

To the left of each object in the Algebra window is a small blue or white circular button controlling the visibility of each object. It functions as a show/hide object toggle.

A.4.7 The Style Bar

Below the words **Algebra** and **Graphics** lies the Style Bar that can be shown or hidden by clicking the small arrow to the left of these words. We suggest that you leave these Style Bars visible. Using them, you can change the sort order in the Algebra window, change the color and style of objects, or show/hide coordinate axes and grid lines. See Figure A.21.

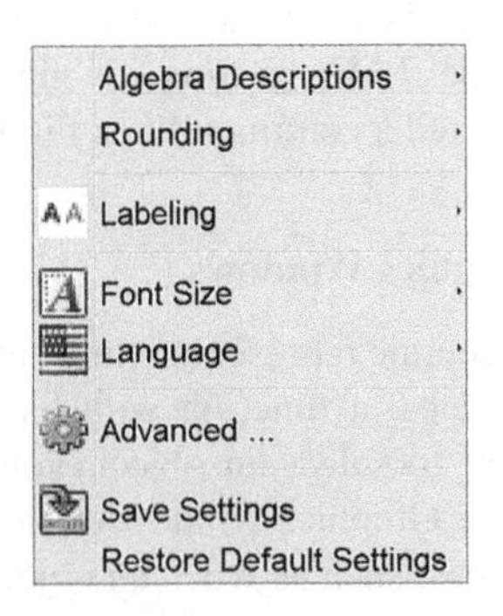

FIGURE A.21 Options menu.

Depending on which objects are currently selected, different style tools are visible in the Graphics window. The Algebra window has two tools. The first show or hides auxiliary objects such as spreadsheet cells and text objects. The other tool allows you to sort the objects in the Algebra window.

A.4.8 The Sidebar

Starting GeoGebra will, with the original settings, display the Sidebar, from which a *perspective* can be chosen. A perspective is a particular arrangement of GeoGebras different windows. Initially the perspective **Algebra** is shown with the Algebra and Graphics windows. If you select the perspective **Spreadsheet** instead, the Algebra window will be swapped out for the spreadsheet window.

Since it is so easy to pick and choose which windows to open from the **View menu** or from keyboard shortcuts like **Ctrl-Shift-S**, the authors normally have the Sidebar turned off.

A.5 CUSTOMIZING GEOGEBRA

In all figures showing screenshots from GeoGebra it is clear that some of the standard settings are adjusted. Most changes are made to create a more appealing interface for the students. This section will outline exactly which settings to change.

A.5.1 Saving Settings

As soon as you change a setting in GeoGebra, it is saved together with your construction. Next time you open that particular construction, all the settings that were active when you saved it, are still active. As soon as you open a new window though, all settings are restored to their saved, default state. If you wish to change this behavior, you need to change the default settings.

Always start with a new, empty file when changing default settings or you may accidentally save changes you did not really mean to save. With the empty file, adjust the settings to your liking, and then select **Options>Save settings** from the menu. See Figure A.22.

If, at any time you discover that your saved settings have been jumbled, you may always use the emergency solution **Options>Restore default settings**.

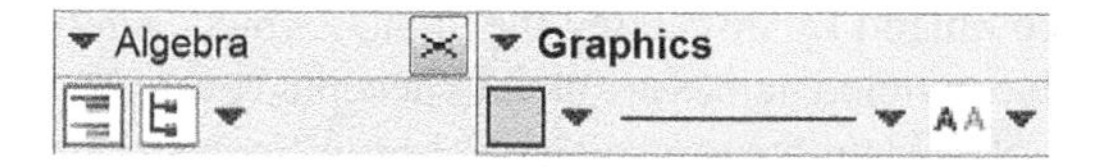

FIGURE A.22 Style Bars allow immediate adjustments to object appearances.

A.5.2 Distributing Settings

Once you have adjusted the settings to your liking, every new file you open will not only display these settings but also keep them when the file is opened by others. The opposite is also true: constructions from others opened by you will display the settings their creators decided.

If you wish your students and colleagues to use the same settings that you have arduously compiled, the simplest way is to distribute an otherwise empty file using these settings. Open GeoGebra and immediately save a file called **MySettings.ggb**. Put the file on the school network, email it, or upload it to GeoGebraTube. Ask everyone to open the file and immediately select **Options>Save Settings** from the menu and then exit GeoGebra. Everyone following this advice will get the same settings as the original **MySettings.ggb**. In case you have a very kind network administrator there are solutions where the students don't need to do anything at all. Search in the Forum for "distribute settings."

A.5.3 The Options Menu

Here follows a list of settings the authors have set as their defaults that can be set from the **Options** menu. Consider this and the following lists as a smorgasbord of possibilities and explore on your own from these examples.

- **Options>Rounding>4 or 5 decimal places**

  The default value of two is rarely enough to adequately show small values, meaning these are rounded to zero, which may be very confusing for students. Ideally, students themselves should learn to change this setting early on.
- **Options>Labeling>No new Objects**

  This is useful when creating lots of data points from values in the spreadsheet. You may still **Show Label** for individual points and objects by right-clicking on them.
- **Options>Font Size>20 pt**

  This is one of the more important settings when you are presenting in front of a class. It makes it possible for students in the back row to read names, values, commands, and so on. This setting also controls the size of axes labels and objects in the Algebra window. The font size for the menu is controlled from the **Advanced** tab in the Properties dialogue.
- **Options>Language - ...**

  Learning a command's name in another language can be done by using the command in the known language and then swapping language. You can see the name of the command by inspecting the created object's definition in its properties. English command names are always accepted even if you work in another language. So, if you prefer to use GeoGebra in Spanish but cannot recall the Spanish command to fit a line to some data points, ask someone for the English equivalent command `FitLine[]`, use it, and then check the definition of the line to find the Spanish command name `AjusteLineal[]`.

A.5.4 The Properties Dialogue

The Properties dialogue contains most of GeoGebras settings. You may access it in a handful of ways:

- By clicking the settings button to the right in the toolbar and selecting which tab you want. See Figure A.23.
- By right-clicking the Graphics window background, and then select **Graphics...** See Figure A.24.
- By selecting **Options>Advanced...** from the menu.
- By selecting **View>Layout...** from the menu.
- By right-clicking any object and selecting **Object properties...**

Once in the dialogue you can switch between different sections by clicking the icons at the top. The sections work like tabs, and different ways of accessing the dialogue simply pre-selects a particular tab. See Figure A.25.

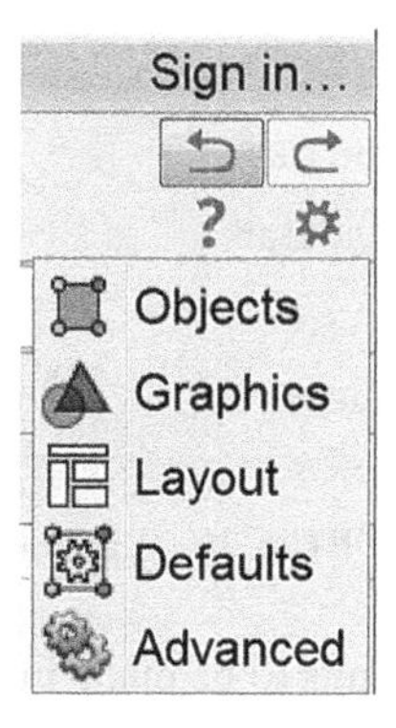

FIGURE A.23 Quick way to the Properties dialogue.

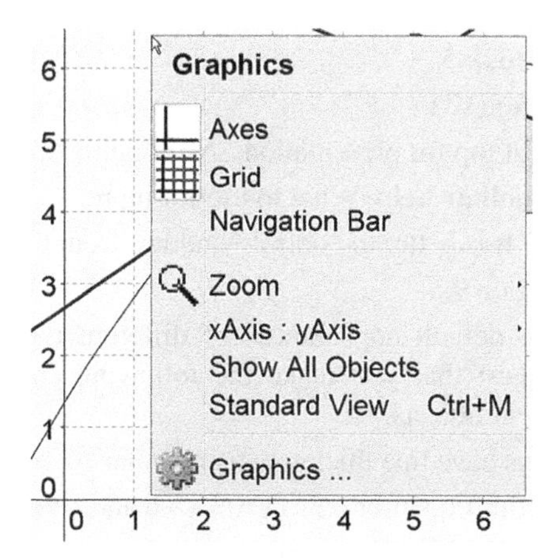

FIGURE A.24 Right-click menu for the graphics window.

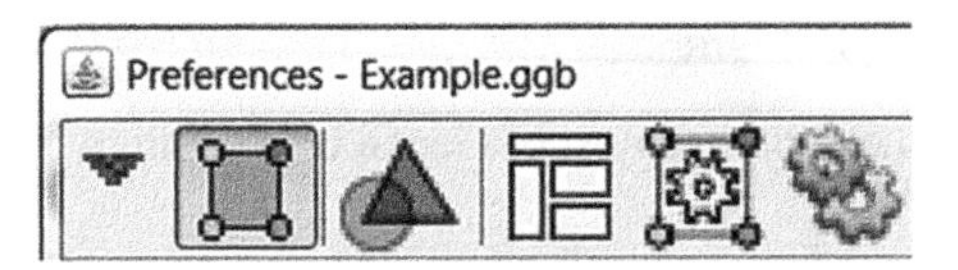

FIGURE A.25 Different tabs are represented as icons.

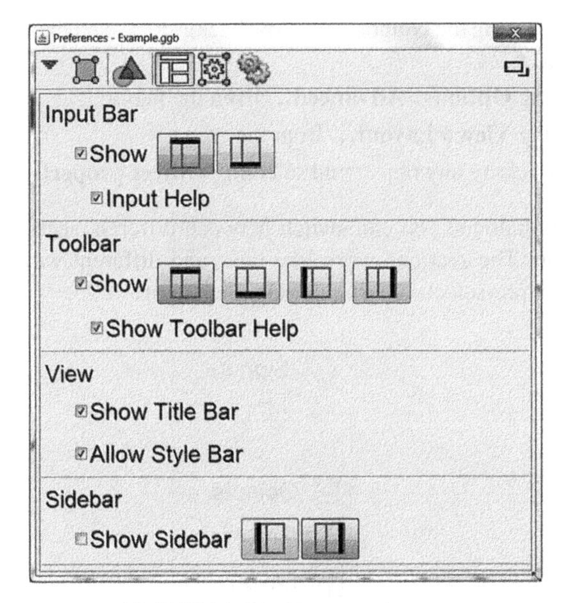

FIGURE A.26 **Layout** tab.

Here is a list of suggested settings to be made in the Properties dialogue.

- In the **Advanced** section:

  Font Size > Menu Font Size: 20 pt: Makes menu more visible**Tooltips > Tooltip Timeout (in seconds)**: 5
- In the **Layout** section :

  Input bar: Show at top for presentations. See Figure A.26.

  Toolbar > Show toolbar help. What to use for input.

  Sidebar. Turn off. It is better to control windows from the menu.
- In the **Default** section:

  This is where the default appearances of different types of objects can be changed. We suggest that you make the following changes to enhance the visibility of different objects:
- Lines and functions have line thickness 5 and color (0, 51, 204)
- Lines are shown with Equation set to $y = \mathrm{m}\, x + \mathrm{b}$ and color (204, 0, 0).
- Polygons have color (255, 173, 0) and opacity 25%
- Texts have white background
- In the **Graphics** section :

Basic xAxis yAxis Grid

FIGURE A.27 Tabs in the Graphics window settings.

Here is another layer of tabs; see Figure A.27.

- **Basic tab > Miscellaneous > Background Color: Light Yellow (255, 250, 205)**

  Makes the screen less glaring, which many report as being easier on the eyes.
- **xAxis and yAxis tabs > Label**: ***x*** and ***y***

  These settings are normally changed depending on problem context, but *x* and *y* are good standard labels and nice to have as a default.
- **Grid tab > Show Grid** (checked)

  The grid allows you to place objects at integer values, line up sliders, reading graphs, and so on. This may also be set from the Style Bar.

A.5.5 Adjusting the Graphing Window

The graphing window, namely the part of the coordinate system currently shown in the Graphics window, can be set in different ways:

- The easiest way may be to **Ctrl-Drag** the individual axes until you are pleased with what you see. Dragging the background will pan it.
- Right-clicking the graphics background and selecting **Graphics...** will open the Properties dialogue where you can enter the minimum and maximum values for the coordinate axes. This corresponds to the WINDOW button on graphing calculators. You may also set axes labels here.
- Right-click the graphics background and select **Zoom...** to zoom in or out a given factor with both axes simultaneously. See Figure A.28.
- Right-click the graphics background and select **xAxis:yAxis** to adjust the *y*-axis to a particular scale relative to the *x*-axis.

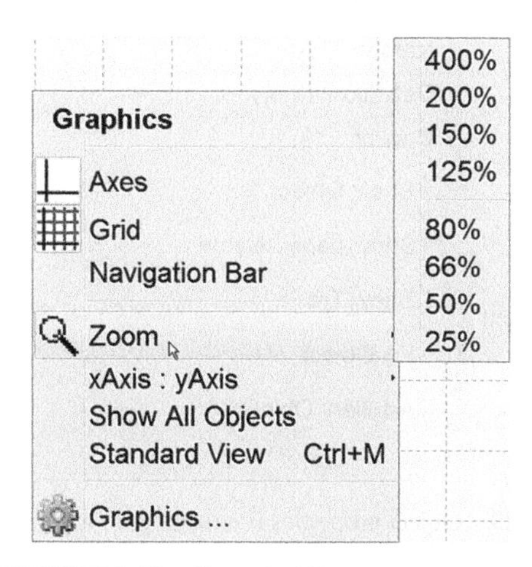

FIGURE A.28 Several different ways to zoom.

A.5.6 Miscellaneous Settings

The same settings you made in the Properties dialogue for the Graphics window may also be done for Graphics Window 2—but you must first show Graphics Window 2 in order for its section icon to become visible. The icon is similar to the icon for Graphics Window 1: . In the same way, a section icon will be shown for CAS and spreadsheet properties when these windows are open.

The positions of different windows in relation to each other, their size, the position of the coordinate axes, and the size and position of the program window itself are other settings that are saved when you select **Options > Save settings**. For this reason it is not wise to save your settings half-way into a problem when these settings appear to be no longer optimal.

A.5.7 Object Properties

There is a section in the Properties dialogue that controls the properties of individual objects. This section is opened when you right-click an object and select **Object Properties...**. All objects can be accessed simultaneously from this dialogue. Settings made here only affect the current construction.

The settings you are most likely to want to change are color, line thickness, line style, point style and sliders minimum, maximum and step values. Apart from the sliders, these settings may also be set in the Style Bar when the object is selected.

Note that all objects can be selected directly in the object tree in the left pane in the dialogue. Several objects can be selected simultaneously by **Ctrl-click** or **Shift-click** or by selecting the heading. In Figure A.29 you can select both points by clicking the word **Point**.

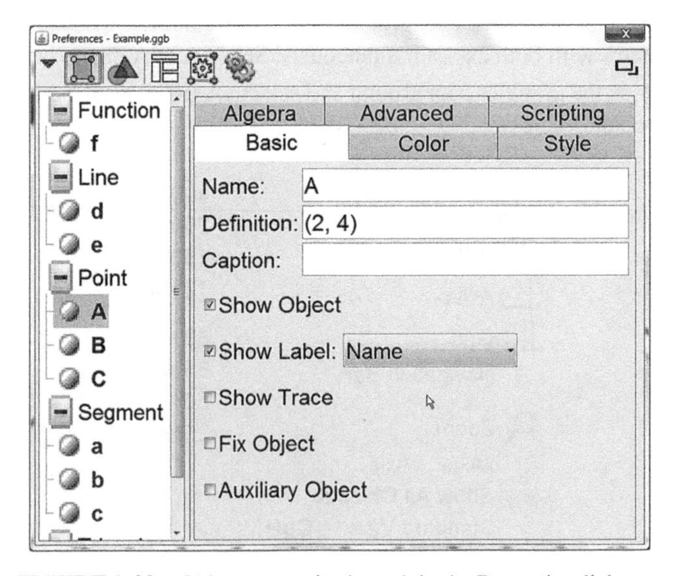

FIGURE A.29 Object properties is a tab in the Properties dialogue.

A.6 SLIDERS

A slider is a graphic representation of a number allowing you to control its value at ease. Sliders are central when working with GeoGebra. See Figure A.30.

Create a slider using the Create slider tool. Click in the Graphics window where you want your slider and adjust the values in the dialogue to your liking. See Figure A.31.

A slider can be created automatically by entering the name and value in the input bar, that is, **s = 300**, and then clicking the white visibility toggle to the left of the number's algebraic representation in the Algebra window. See Figure A.32.

The slider will then be shown with its default settings, that is, **Min**=–5 , **Max**=5 and **Increment**=0.1. The minimum and maximum values are adjusted to include the current value, in this case **Max**=300.

Sliders will be created automatically if you enter a valid expression without first defining your parameters. By entering

```
f(x) = x² + p x + q
```

and then pressing **Enter** (use **Alt-2** or **^2** for the exponent), you are faced with the question **Create slider(s) for p, q?** Press **Enter** or click **Yes**, and they will be created for you. See Figure A.33.

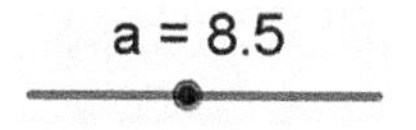

FIGURE A.30 Graphic representation of a number.

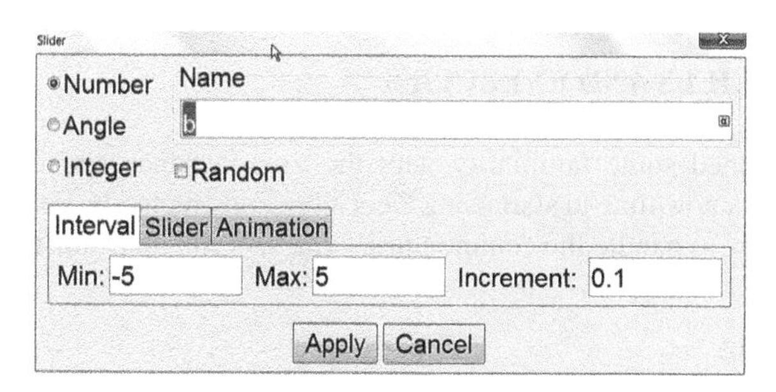

FIGURE A.31 Slider dialogue.

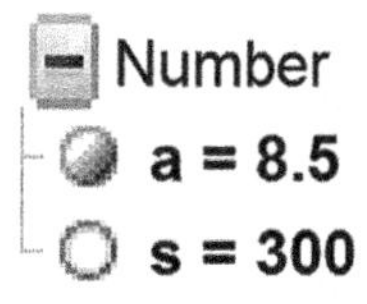

FIGURE A.32 Blue/white buttons control the object's visibility.

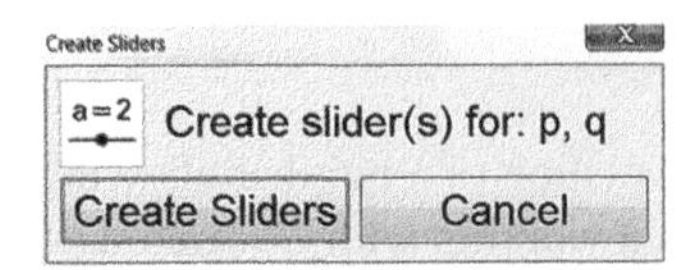

FIGURE A.33 Automatic slider creation.

Sliders may be animated by right-clicking on them and selecting **Animation On**. An animated slider representing time makes it possible to enact realistic processes such as models of the solar system and ballistic motion. While a slider is animated, a point or a line will leave traces or record positions to the spreadsheet for further analysis. All points on other objects can be animated.

Sliders can be created so that they only take integer values. If n is such a slider, you can type

```
f(x) = (x + 2)^n
```

and then change the slider and observe the function change.

A number can at any time be given a new value by entering it in the input bar, for instance,

```
s = 1500
```

If it is a slider, then its **Min** and **Max** values will change to accommodate this new value.

Values like **Min, Max, Step,** and others, can be controlled by other sliders. In this way one slider can control the animation speed of another slider.

A.7 BASIC SKILLS AND EXERCISES

Having established some familiarity with the user interface and learned how to customize it, it is now time to start using GeoGebra constructively. You will see some examples of how to handle the common tasks you will run up against.

A.7.1 Graphing

To graph a function, enter it into the input bar and press **Enter**. Try the following:

- `0.2x²` (press **0.2 x** and press **Alt-2**)
- `y = 3x - 2` (defined as a line, not a function)
- `x < 1` (vertical inequality, can be shown on x-axis)
- `g(x) = A sin(k x)` (with spaces for multiplication, automatic sliders)
- `g'(x)` (derivative)
- `x³+3x²y+x y²-y² = 1` (implicit function, observe space)

You can also create a function by fitting it to a set of data points:

- **`FitLine[(0,2),(3,4),(6,5),(8,6)]`**
- **`FitPoly[A,B,C,D,2]`** (if points A-D already exist)

You can find extrema, roots, inflexion points, and the like, with the appropriate commands. Note that these commands frequently treat polynomials and other functions differently. In the following commands it is assumed that *p* is a polynomial and *g* is some other function.

- `p(x) = x³ - 3x² - 3x + 3`
- `g(x) = 1 + 1/((x - 1)(x - 2)(x - 3))`
- **`Extremum[p]`** (US English), or **`TurningPoint[p]`** (UK & AU English)
- **`Extremum[g,0,4]`** (US English), or **`TurningPoint[g,0,4]`** (UK & AU English)
- **`Root[p]`**
- **`Root[g, 0.5]`**
- **`Roots[g, 0, 4]`**
- **`Inflection[p]`**
- **`Intersect[p, g, 0, 4]`**

A.7.2 A Dynamic Triangle

Create a dynamic geometrical construction to investigate the behavior of a triangle's altitudes. Open a new window by typing **Ctrl-N**.

Start by creating three arbitrary points using the New Point tool . Click on the tool, and then on their intended positions. The points are automatically named A, B, and C.

Then create three lines through these points using the Line tool . They will be named a, b, and c, in the order of creation.

Using the Normal Line tool , you can now create new lines. Click on A and then on the line not running through A, whereupon a new line, normal to the one you used, is constructed. Do this two more times for B and C. The new lines will be named d, e, and f. You may notice that these altitude lines intersect in a common point; see Figure A.34.

Now create the intersections of the triangle's sides a, b, c and their respective altitude lines d, e, f. You can do this using the New Point tool if you click close enough to the intersections for GeoGebra to recognize which lines to intersect. These are automatically selected when the cursor is close enough. If the area is too crowded for this to be efficient, you could use the dedicated Intersect tool and then click anywhere on the two lines of interest. Your new points, the altitude's *foot points*, will be named D, E, and F.

For clarity, dash the altitude lines by selecting d, e, and *f* simultaneously and then selecting the dashed line style from the Style Bar. See Figure A.35.

Now create the triangle object by clicking the polygon tool and successively on points A, B, C, and finally A again, indicating that you are done. This will create the sides and the polygon as shown in Figure A.36.

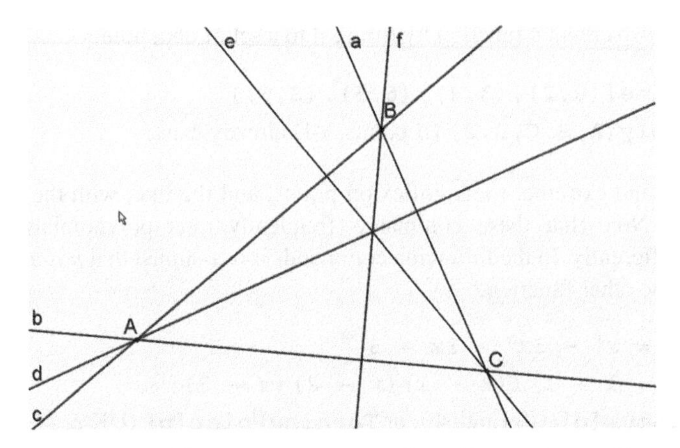

FIGURE A.34 Geometric construction without axes or grid.

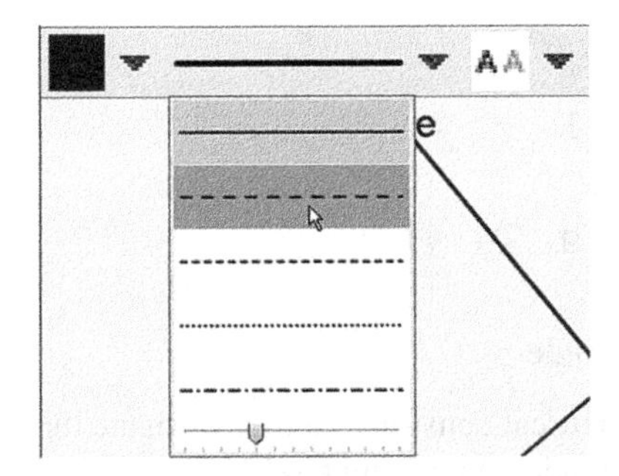

FIGURE A.35 With the lines selected, choose a dashed line in the Style Bar.

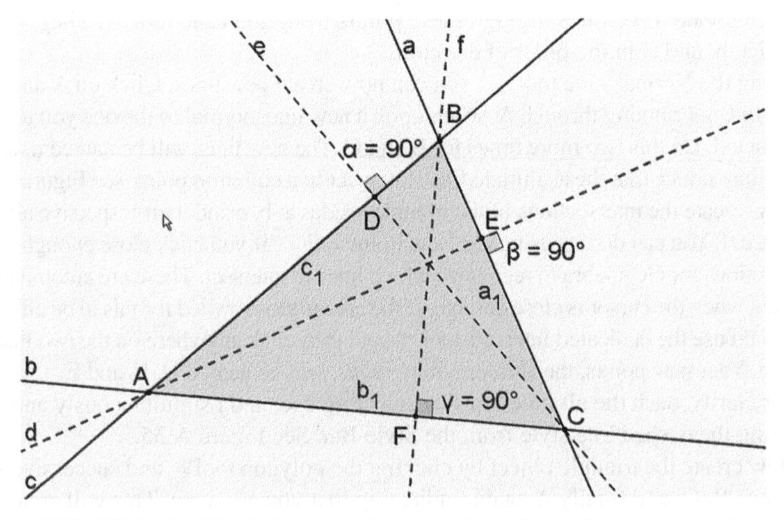

FIGURE A.36 Finished construction, complete with angle measurements.

Press Escape and then drag a vertex of the triangle. If the triangle and its altitudes stay connected, the construction is correct. If they seem to fall apart, it is not. This is known as *the drag test*.

As you drag the vertices you are sure to notice that this intersection sometimes coincides with one of the triangle's vertices. When does this happen. Can you prove it?

A.7.3 Creating Objects in the Spreadsheet

GeoGebra's spreadsheet is in many ways more powerful than other spreadsheets, in that it handles other objects than just numbers and formulas that give rise to new numbers. All objects that GeoGebra can represent can be created and managed in the spreadsheet. The name of the object will be the same as the name of the cell. Conversely, if you create an object some other way and name it "E6," the object will be shown in the spreadsheet cell E6.

Objects in the spreadsheet tend to be quite many and are not by default shown in the Algebra window. They are defined as *auxiliary objects*, and are only shown if you click the button that will toggle the visibility of auxiliary objects. This button sits in the Style Bar of the Algebra window. See Figure A.37.

In the example shown in Figure A.37, a point in cell C2 is created by entering

```
=(A2,B2)
```

You can then copy this formula down to create the rest of the points automatically, by dragging the *fill handle*, the little square, down. In the D column you can create the joining line segments by typing

```
=Segment[C2,C3]
```

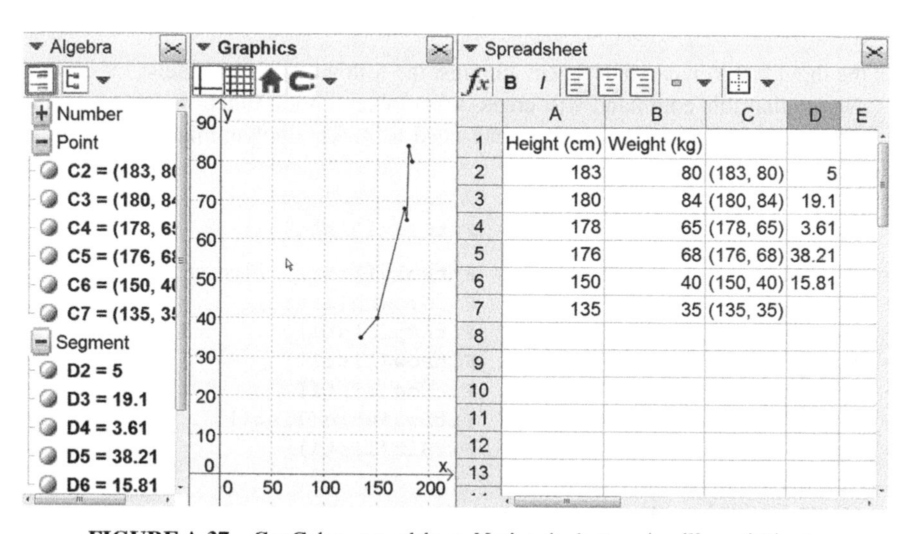

FIGURE A.37 GeoGebra spreadsheet. Notice the button **Auxiliary Objects.**

in cell D2 and then copying this down. By setting the properties of the first object, these will be copied as the object is copied, thus propagating color, style, and other properties. In the same way you can create sets of circles, lines, functions, and so on.

A.7.4 Fitting Functions to Data

To do a regression, that is, finding the function that minimizes the sum of the square of the errors, you need two things: a data set and an idea of what kind of function that is appropriate to fit to the data. You can create a list of data points by any one of several different methods:

- Enter the data in the spreadsheet and select both columns simultaneously, right-click and select **Create>List of points**.
- Create the points as in the example in Section A.7.3 and then select the points in the spreadsheet, right-click, and select **Create>List**.
- If the points are already created, type

  ```
  list1 = {A,B,C}
  ```

  in the input bar.
- If the points are already created, select the Create list tool {1,2}, and then select the point in the Graphics window by dragging a rectangle around them.

Once you have a list of points you can select the function to fit by beginning to type

```
Fit...
```

- Linear — $\mathrm{m} \cdot x + \mathrm{b}$ — **`FitLine[list1]`**

in the input bar. You can then select amongst the following basic models:

Notice that this command will create a line object, not a function. If you would like to make function calls, you instead need to make a polynomial regression of degree 1.

```
FitPoly[list1, 1]
```

- Polynomial — $\mathrm{a} \cdot x^{\mathrm{n}} + \ldots$ — **`FitPoly[list1, 3]`** creates a cubic
- Exponential — $\mathrm{C} \cdot \mathrm{a}^{x}$ — **`FitGrowth[list1]`**
- Exponential — $\mathrm{C} \cdot e^{kx}$ — **`FitExp[list1]`**
- Power — $\mathrm{C} \cdot x^{\mathrm{a}}$ — **`FitPow[list1]`**
- Logarithmic — $\mathrm{C} \cdot \ln x + \mathrm{a}$ — **`FitLog[list1]`**
- Logistic — $\mathrm{c}/(1 + \mathrm{a} \cdot e^{kx})$ — **`FitLogistic[list1]`**
- Trigonometric — $\mathrm{a} \cdot \sin(\mathrm{b}x + \mathrm{c}) + \mathrm{d}$ — **`FitSin[list1]`**

GeoGebra is also capable of fitting *any* function to data. For more information on this, see Appendix B.

A.7.5 Rectangles

Sections A.7.5, A.7.6, A.7.7, A.7.8, A.7.9, A.7.10, and A.7.11 largely build on each other, getting progressively more difficult. You might therefore wish to read these sections in order, building the constructions as you go along.

There already exist tools for creating squares and arbitrary polygons, but rectangles are neither. The simplest way of creating them is to use the coordinate system. Create three points A, B, and C, using the New Point tool •ᴬ such that A lies in the origin, B on the x-axis, and C on the y-axis. Press **Escape** and try to drag the points. A should be immobile, confined to the origin, and the others should be confined to the axes but capable of moving along them.

Then type

```
D = (x(B), y(C))
```

in the input bar and press **Enter**. This will create a point with the same x-coordinate as B and the same y-coordinate as C. The functions x and y extract the coordinates of a point. You can now create the rectangle using the Polygon tool and clicking on the points in the order A, B, D, C, A. This will create the rectangle and its sides. See Figure A.38.

Points B and C, which are sliding on the axes, act as sliders of a sort and may be animated. Also, depending on what you wish to highlight, the individual points, sides, and rectangle surface may be renamed or given new color, size, style, and so on. Save your construction for later use.

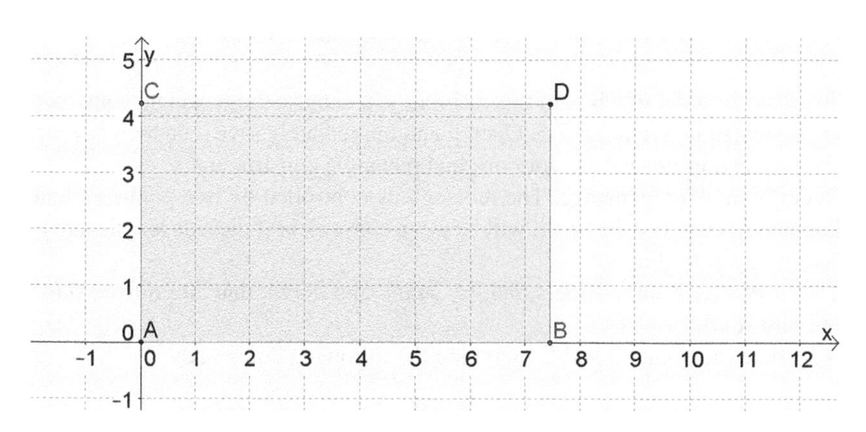

FIGURE A.38 Rectangle controlled by point B and C.

A.7.6 Boxes

This exercise is based on Section A.7.5, where you created a rectangle. The simplest type of perspective drawing often found in mathematics is where you draw all depth-going lines at 45° angle and draw them half as long as their actual length. Hidden lines are dashed. A cube would then look as that in Figure A.39.

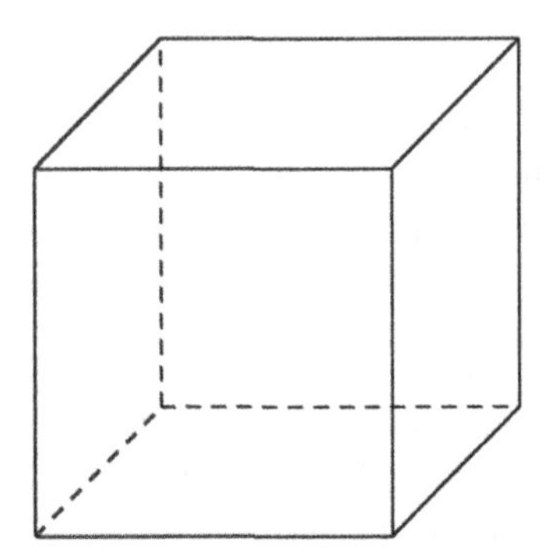

FIGURE A.39 2D drawing of a cube in GeoGebra.

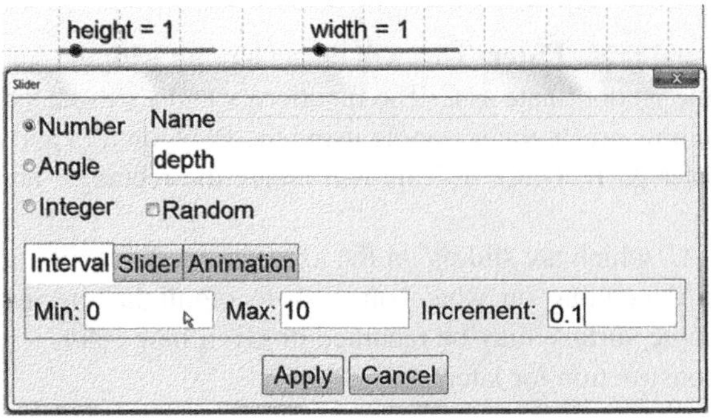

FIGURE A.40 Creating three sliders to control the box.

To create a model of a box in this fashion, you can use the rectangle construction that you created in Section A.7.5. Open it and immediately save it under a new name to ensure you don't overwrite your original rectangle construction.

Begin by making a change. The rectangle is controlled by two points, B and C, sliding on the axes. For the box it will be more convenient if the side lengths are controlled by separate sliders.

Therefore create three sliders, *height*, *width,* and *depth*, that all go from 0 to 10. See Figure A.40.

Next redefine points B and C by typing the following commands:

```
B = (width, 0)
C = (0, height)
```

Check that the rectangle responds as expected when you operate the sliders.

It is now time to now create the points at the back of the box. A point with space coordinates (x, y, z) will be projected onto the surface coordinates

$$\left(x+\frac{z}{2\sqrt{2}}, y+\frac{z}{2\sqrt{2}}\right)$$

The square root of 2 can be written as **sqrt(2)** in GeoGebra, but since exact forms aren't necessary here, you might as well write 1.4. To simplify things, however, first define an auxiliary variable

```
v = depth/(2sqrt(2))
```

and press **Enter**. The four points at the back of the box are then created by typing (see Figure A.41):

```
E = (v, v)
F = (width + v, v)
G = (v, height + v)
H = (width + v, height + v)
```

You can now choose if you want to create polygons for the visible surfaces with the Polygon tool or just create the edges with the Segment tool. It is also possible to use commands:

```
Side = Polygon[B, F, H, D]
Top = Polygon[D, H, G, C]
Segment[E, F]
Segment[E, A]
Segment[E, G]
```

The hidden sides are dashed by selecting them and using the Style Bar. Finish off by making some calculations. In the input bar you can type:

```
TotalArea = 2(width height + height depth + depth width)
Volume = width height depth
```

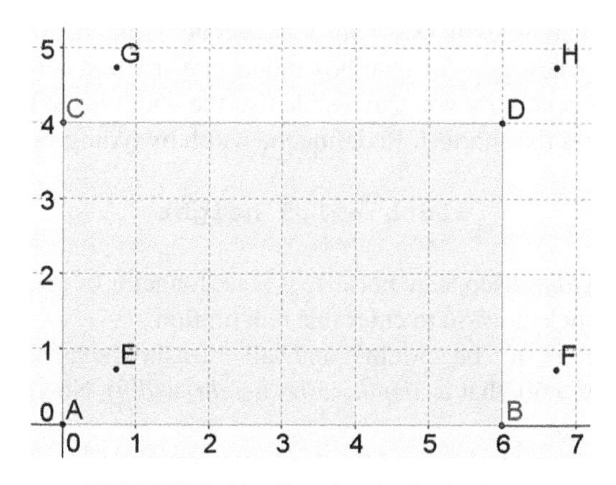

FIGURE A.41 Creating the back plane.

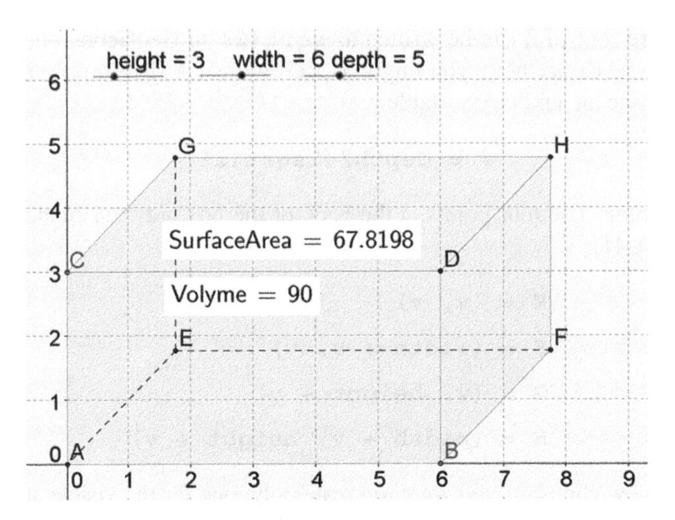

FIGURE A.42 Dynamic box.

Notice the need for suitable variable names. Drag *TotalArea* and *Volume* from the Algebra window and drop them in the Graphics window to create text labels.

Our dynamic box can be seen in Figure A.42. Using this method, you can create simple 3D models in 2D. Of course, your box can be constructed in GeoGebra's native 3D-environment as well.

A.7.7 Container Problems

This exercise is based on Section A.7.6. You can now modify your box model to analyze a typical problem in various ways. Let's say that you have been asked to construct a nearly cubic container with a capacity of 50 ml. The largest dimension is to be exactly 1.5 times the smallest, and the total surface area of the box is to be minimized.

This is a typical optimization problem, normally done using derivatives in the calculus courses, but we will use it to demonstrate some standard techniques in GeoGebra. Create a new copy of your box model construction before you continue.

Without loss of generality you can decide that the width is the longest dimension and that the height is the shortest. Redefine the width by typing

```
width = 1.5 height
```

The slider for *width* disappears because it is no longer a free variable. You could also have double-clicked *width* to enter this redefinition.

If the volume is to be 50 cm^3 and all measurements are in cm, then $height \cdot width \cdot depth = 50$, that is, $depth = 50 / (height \cdot width)$. Now enter

```
depth = 50/(height width)
```

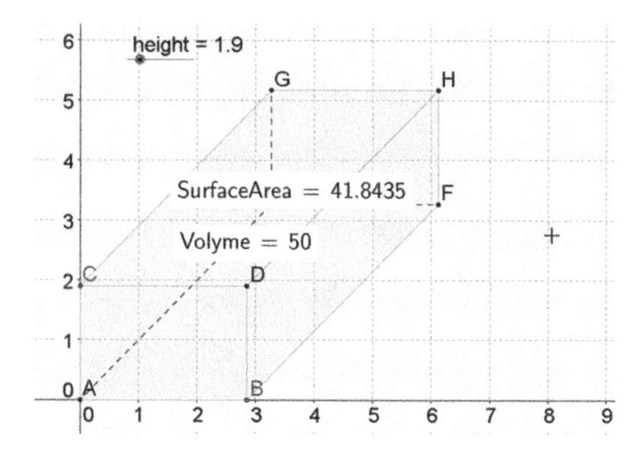

FIGURE A.43 Width and depth now both depend on the height of the box.

and notice the *depth* slider disappearing. You have now formulated the problem in terms of the variable *height*. Vary *height* using its slider and notice how the box behaves. See Figure A.43. Then save your work.

A.7.8 Tracing in Graphics Window 2

This exercise is based on Section A.7.7 where you created a 50 ml box whose largest dimension was exactly 50% larger than the shortest. Now suppose that you wish to investigate what happens to the total surface area as you vary the variable *height*.

Show Graphics Window 2 by selecting **View>Graphics 2** from the menu. Make sure that axes and grid lines are shown, and **Ctrl-drag** the axes so that you see at least 0 – 5 on the x-axis and 0 – 150 on the y-axis. Click once in this new Graphics window to make sure it is active, as indicated by the text "Graphics 2" being presented in bold. Then type

```
P= (height, TotalArea)
```

in the input bar and press **Enter**. This will create a single point in Graphics Window 2. By varying the parameter *height*, you can see how the position of this point changes.

Now right-click this point and select **Trace on**. Then right-click *height* and select **Animation on**. The point will now leave systematic tracks, and in Figure A.44 you can clearly see the approximate value of the minimum. Use the small pause control in the lower left area of the Graphics window to stop and restart the animation. You may clear the trace and refresh the screen with **Ctrl-F**.

Notice that the minimum area is achieved when the height is approximately 3 cm. You can find the complete trace by using the Locus tool or the command

```
Locus[P,height]
```

However, GeoGebra doesn't know enough about the path to find the exact function, so you cannot find the minimum automatically.

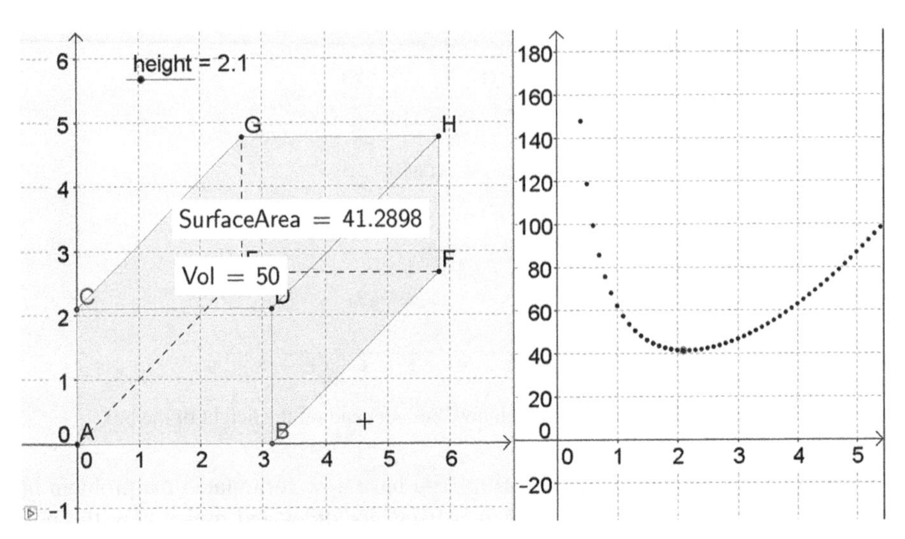

FIGURE A.44 Your model goes in one window, your analysis in the other.

A.7.9 Collecting Data to the Spreadsheet

This exercise builds on Section A.7.8 where you used Graphics Window 2 to view how the total area changes with the height of the box. Unfortunately, you couldn't find the minimum area with any precision to speak of. If you need precision and cannot find the function algebraically, you could collect some data points from this path and fit a polynomial to them. Then you can find the minimum of the polynomial.

First, show the spreadsheet, pressing **Ctrl-Shift-S** or the menu option **View>Spreadsheet**. Adjust its position and size in relation to the other windows if necessary.

To collect data, you right-click the point in Graphics Window 2 and select **Record to spreadsheet**. In the resulting dialogue, just press **Close**, and then restart the animation or change the value of *height* manually. The coordinates of the point will now be recorded to the spreadsheet, as seen in Figure A.45, where arrow keys are used to change the value of *height*. Using the arrow keys to change a number or a slider will change the value with the current value of the number's or slider's increment property, here 0.1. Holding down **Shift** while using the arrow keys will temporarily make the increment 10 times smaller, for increased precision.

Having recorded values on both sides of the minimum, you can now stop the recording of further values by again right-clicking the point and selecting Record to spreadsheet. Since the line representing the point is already selected, press **Remove** and then **Close**. See Figure A.46.

Having accurate values in the spreadsheet, you now want to fit them to a polynomial. First, make sure Graphics Window 2 is selected. After that, select the 5 to 10 points closest to the minimum, right-click, and select **Create...>List of points** as indicated by Figure A.47.

Spreadsheet

	A	B
1	x(P)	y(P)
2	2.11	41.2831
3	2.12	41.2783
4	2.13	41.2753
5	2.14	41.2741
6	2.15	41.2747
7	2.16	41.2772
8	2.17	41.2813
9	2.18	41.2873
10	2.19	41.295
11	2.2	41.3043

FIGURE A.45 Recording P:s coordinates to the spreadsheet.

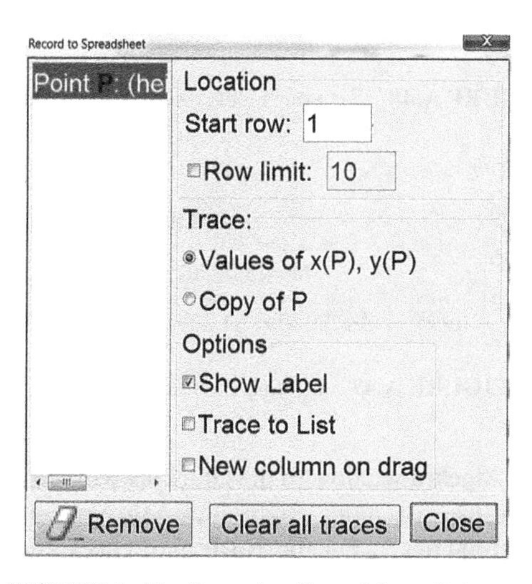

FIGURE A.46 Record to Spreadsheet dialogue.

This list will be named list1 and will be visible in the Algebra window. The points themselves will be visible in Graphics Window 2, near the minimum. To fit a quadratic function to these points, use the command

```
FitPoly[list1,2]
```

This polynomial will be called $f(x)$, and you find its minimum value with the command

```
MinArea=Extremum[f].
```

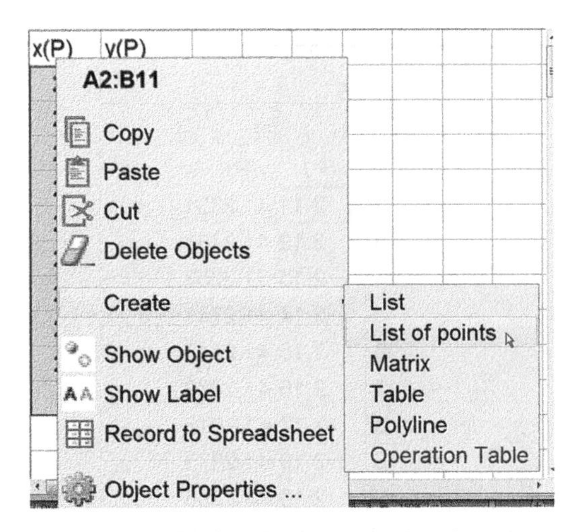

FIGURE A.47 Creating a list of points from data.

Show Label: Name & Value

FIGURE A.48 Set your labels to your preferences.

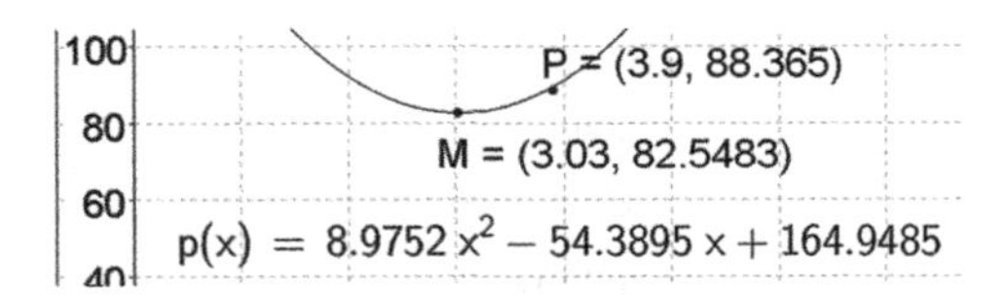

FIGURE A.49 Finding the minimum value.

Drag $f(x)$ from the Algebra window to the Graphics window to create a text label. Also show the label of the extremum. Right-click **MinArea** in the Algebra window and select **Object properties...** On the **Basic** tab, check **Show label** and select **Name & Value** from the list, as shown in Figure A.48. The minimum area is 82.55 cm^2 when the height is 3.022 cm. See Figure A.49.

A.7.10 Using CAS

This exercise builds on the problem in Section A.7.7 where you found the minimum surface area of a box. CAS is short for Computer Algebra System and stands for automatic symbolic algebra calculations. There are several rather special ways to work with the CAS window, so we suggest that you read through the CAS section introduction at https://wiki.geogebra.org/en/CAS_View if you intend to do more work with CAS.

Begin by selecting **View > CAS** from the menu to show the CAS window, and then enter the following command, one on each line, separated by pressing **Enter**:

```
w := 1.5h
d := 50/(w h)
TotAr := 2(w d + d h + h w)
Derivative[TotAr, h]
Solve[$4 = 0, h]
```

Use the *assignment symbol* := to assign values to variables and functions. As you can see in Figure A.50, the $-sign refers to the result in line 4. Pressing **F1** while in the CAS window will present a short list of CAS-specific methods like this. Also use new variable names, since the old ones have values associated with them.

You now have an algebraic result. You can use this to compute the exact value of the minimum area by first extracting the value of the height from the list. After that you work your way through the formulas again like this:

```
OptH := RightSide[Element[$5, 1]]
OptW := 1.5 OptH
OptD := 50/(OptW OptH)
OptAr := 2(OptW OptD + OptD OptH + OptH OptW)
```

Figure A.51 shows the exact form of the minimum area. Press **$ Ctrl-Enter** to get a numerical value ≈ 82.55 cm².

5
Solve[$4 = 0,h]
$\rightarrow \left\{ h = 5 \cdot \frac{\sqrt[3]{6}}{3} \right\}$

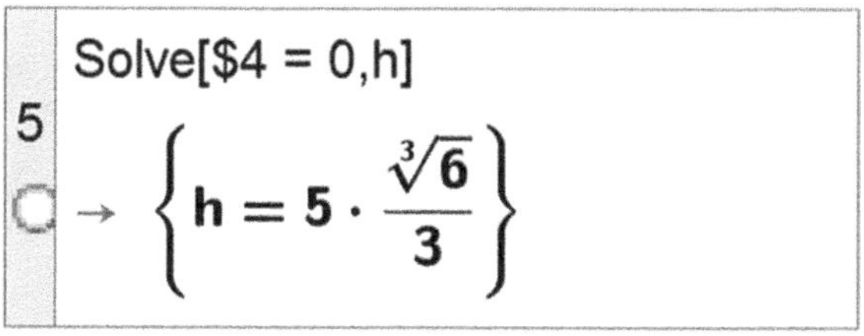

FIGURE A.50 CAS will return exact algebraic solutions whenever possible.

9
OptAr := 2 (OptW OptD + OptD OptH + OptH OptW)
$\rightarrow \sqrt[3]{6}^{2} \cdot 25$

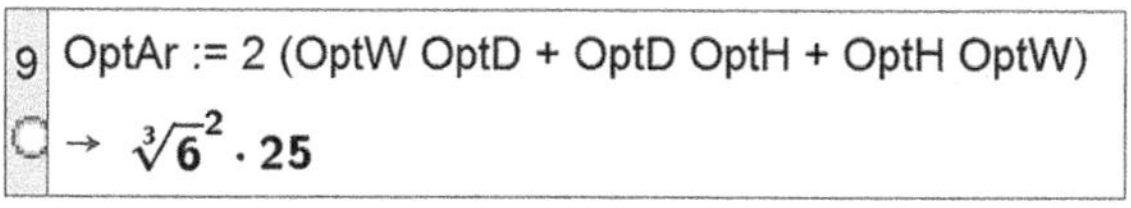

FIGURE A.51 Least surface value in exact form.

A.7.11 Inserting Dynamic Text

As you drag objects from the Algebra window to the Graphics window, you create a text label, but sometimes you may wish to create more flexible and complex text boxes. In the example in Section A.7.7, there is a text label saying Volume=50, no matter the current state of the box. This may be true but not very informative. The text is static and lacks a unit. It would be more interesting to connect the volume to the dynamic sides of the box.

The Insert Text tool ABC allows you to create powerful dynamic text boxes. Click on the tool and then in the graphics area where you want to position the lower left corner of the text box to open the Text dialogue as seen in Figure A.52

If you only type in text, this will become static just as before. But you can at any time insert dynamic objects like the variables *height*, *width,* and *depth* from the list of objects. The multiplication dot can be found in the list of symbols. Should you feel the need to construct more complicated mathematical instructions, check the **LaTeX formula** check box and select ready-made templates from the lists, or type LaTeX commands of your choice in the Edit field. Most mathematical LaTeX commands are implemented.

Dynamic objects are shown in small boxes. You can perform new calculations on these objects inside the boxe, or simply insert a new empty box.

To conclude this example, insert Volume = height $*$ width $*$ depth = 50 cm^3 and click **OK**. Try to change *height* by dragging its slider and see what happens to your text.

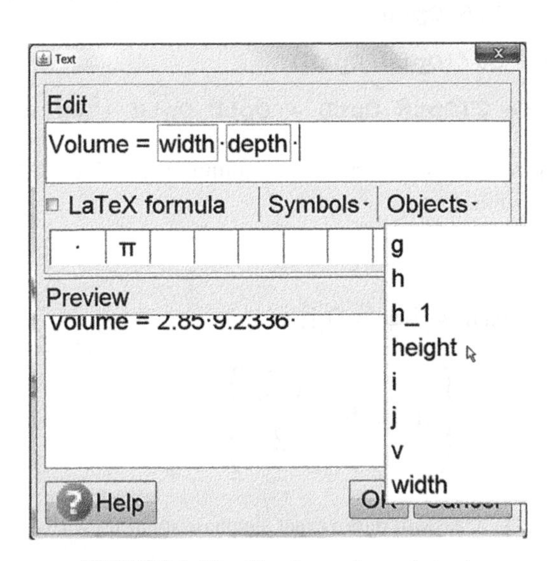

FIGURE A.52 Creating a dynamic text.

A.8 WRITING INSTRUCTIONS

To produce written instructions for your students, you need to do at least two things: create screenshots and quickly change the format, or style, of your text.

GeoGebra can create copies of the graphics area using **Ctrl-Shift-C**. To copy just a section of the Graphics window, first select it by dragging a rectangle round the area of interest. Then go to your word processor and insert from the clipboard as usual, with **Ctrl-V**.

The contents of the spreadsheet may be copied to other spreadsheets or tables but commands specific to GeoGebra will not work there.

The tool icons are available at the following address:

http://wiki.geogebra.org/en/Category:Tools_Icons

To copy other parts of the user interface, you will need to have a good tool for copying screen clips. Nowadays a tool like this is included in your operating system, but depending on how picky you are, you may want to use a third-party program such as Gadwin PrintScreen. With the correct settings, all you need to do is to press **PrntScr**, select a rectangle (which may be fine-tuned before capture), and press **Enter** to both save the selected part of the screen to the clipboard and to an image file for later reference.

It is also useful to be able to quickly format text to mimic **menu commands** and `typed text`. Menu commands are typically in bold, but for typed text you may want to change font, size, and style. Depending on your word processor, you may be able to define a text format for this, and then bind a keyboard shortcut to this format. Doing this will make it much easier to produce good-looking instructions for your students and colleagues.

A.9 REMEMBER

Input:

- Capital letters matter: `X` is not the same as `x`.
- Space can be a multiplication sign: `ax` in `y = ax + b` is interpreted as a new, previously undefined variable. Instead, use `y = a x + b` or `y = a*x + b`.
- Use up and down arrows in the input bar to scroll through previous commands.
- Pressing **Enter** will move the text cursor to the input bar.
- F3 will copy a selected object's definition to the input bar for easy redefinitions.
- The input help is found under the little triangle ◂ to the right of the input bar.

Sliders:

- Use **Arrow** keys to control sliders, numbers and points.
- Set the increment in the **Properties** dialogue, on the **Slider** tab.
- Use **Shift-Arrow** for fine-tuning with 1/10 of the increment.
- Use **Ctrl-Arrow** for larger jumps with 10x the increment.
- Use **Alt-Arrow** for really large jumps with 100x the increment.

Points, lists, and commands:

- Points are written within parentheses: `(3,5), (4.7,3.52), (a,f(a))`...
- `x(A)` and `y(A)` returns the x- and y-coordinates for point A.

- Lists are written within curly braces: `data = {A,B,C,D,E}`.
- Commands have square brackets and capitals: `Intersect[f,g]` but ordinary parentheses are also allowed: `Intersect(f,g)`.

Alt key:

- **Alt-a** gives α, **Alt-b** gives β, etc. **Alt-o** gives the degree sign 360°.
- **Alt-i** gives you the imaginary unit *i*, **Alt-p** gives π, and **Alt-e** gives e=2,718…
- **Alt-2**, **Alt-3**… give you whole number powers. Use **^2.5** for other powers.
- Indexes are typed with an underscore: **S_1 = 3**.
- Special characters are found under the alpha sign α to the right in the input bar

Graphics window:

- The axes have names: `xAxis` and `yAxis`.
- `x(P)` and `y(P)` gives you the coordinates of the point P.
- **Ctrl-Drag** the axes to zoom and **Drag** the background to pan.
- **Escape** de-selects any selected objects and activates the selection tool.

Objects:

- Rename objects *immediately* after creation by just starting to type the name.
- Double-click objects to redefine them.
- Drag from the Algebra window to the Graphics window to create a text label.
- Hide or show objects by clicking the visibility button ● in the Algebra window.

Miscellaneous:

- **Ctrl-F** will erase tracks from tracing objects.
- **Ctrl-Shift-S** opens the spreadsheet. **Ctrl-Shift-2** opens Graphics Window 2.
- **View>Keyboard** to get a virtual keyboard on-screen.
- Make use of the user manuals, guides, and forums online. Links are in the Help-menu.

APPENDIX B

FUNCTION LIBRARY

Modeling often requires you to write down a suitable function to be fitted to a given data set. To be able to do that, you have to obtain some experience with different functions and their properties. Exactly which functions the students need to be exposed to in their education will, of course, vary, but in this appendix an overview is given of the most common functions, and even some less common ones.

Students should definitely know how functions are parametrized, that is, which parameters can be changed, and how that affects the situation being studied. In several instances the same basic type of function can be described and parametrized in different ways, depending on what is suitable for a given situation. Perhaps the basic function needs to be translated sideways or vertically? Perhaps it needs stretching? How do you do these basic transformations?

At some point it is often useful to let the students sum up what they already know about functions. One way of doing this is to let the students compose a catalog of different functions, where they can look up what different functions look like. The teacher may present a list of functions to include and the actual catalogs could be constructed online as Wikis or in OneNote class notebooks. This work could then be followed up with some simple modeling tasks, like having the students fit suitable functions to different shapes of objects in pictures, or to the "lines" on the palms of their hands, which they can photograph.

The appendix also shows how the general fitting command works in GeoGebra. An example is the function fit to data based on a photograph of Sydney Harbor Bridge in Section B.5, Figure B.22.

Mathematical Modeling: Applications with GeoGebra™, First Edition. Jonas Hall and Thomas Lingefjärd.

Companion website: www.wiley.com/go/Hall/MathematicalModeling

B.1 DIFFERENT FUNCTIONS AND THEIR PARAMETRIZATIONS

This section will show the most common functions a student is likely to encounter in upper secondary school, and how to parametrize them.

B.1.1 Linear Functions

A linear function is characterized by the fact that its graph is a straight line. Linear functions thus have a constant slope. For each step you take to the right on the x-axis, the function's value increases or decreases by the same *amount*, no matter where you are on the graph. This is different from exponential functions, which increase or decrease with the same *percentage* for each step to the right that you take on the x-axis.

Ordinary proportionalities are linear functions that pass through the origin.

All (continuous) relationships between two quantities can be approximated by a linear function in a sufficiently small interval. For this reason linear functions are often used as a first approximation in many modeling problems.

Figure B.1a and b show two examples of linear functions. In Figure B.1a, you see a proportionality with a positive slope, and in Figure B.1b, you see a linear function with a negative slope.

The linear function is expressed in many different ways, of which the most common are:

$y = \mathrm{a}x + \mathrm{b}$, $y = \mathrm{a} + \mathrm{b}x$ and $y = \mathrm{m}x + \mathrm{b}$. Some countries use other combinations, for instance, $y = \mathrm{k}x + \mathrm{m}$. In this book, $y = \mathrm{a}x + \mathrm{b}$ is used.

The most basic form of a linear function is the proportionality:

$$y = \mathrm{a}x$$

It is worth noticing that many graphing calculators and other fitting tools are able to fit data to this simple model. To do so in GeoGebra, use the command:

```
Fit[list1, {x}]
```

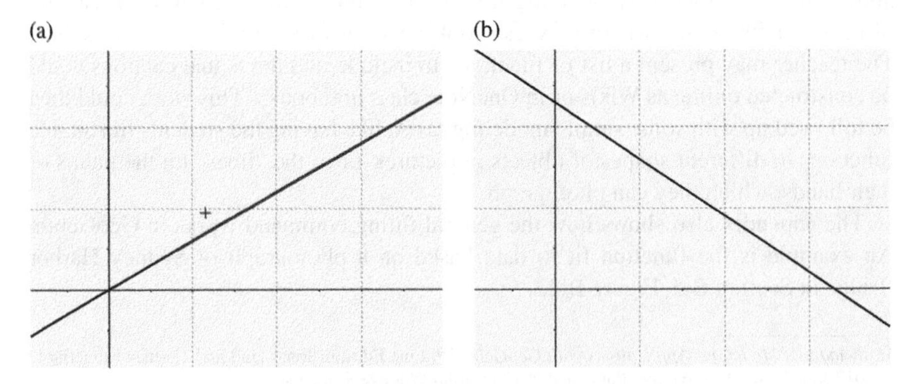

FIGURE B.1a and b Linear functions.

The standard *slope-intercept* form is

$$y=\mathrm{a}x+\mathrm{b}$$

where a is the slope, a = $\Delta y/\Delta x$, and b is where the graphs intersects the y-axis. This point is called the *y-intercept*.

If you need the function to pass through a given point (x_0, y_0), you can translate the function to this point and write:

$$(y-y_0)=\mathrm{a}\cdot(x-x_0), \quad \text{or} \quad y=\mathrm{a}\cdot(x-x_0)+y_0$$

$$(\text{known as the } \textit{one-point formula})$$

If you need the function to pass through two given points, first find the slope $\mathrm{a}=\Delta y/\Delta x$ and then use the one-point formula above.

A line intersecting the axes at $x=x_c$ and $y=y_c$ can be written in *intercept form*, as either of these forms:

$$\frac{x}{\mathrm{x_c}}+\frac{y}{\mathrm{y_c}}=1 \qquad y=\mathrm{y_c}\cdot\left(1-\frac{x}{\mathrm{x_c}}\right)$$

The general form of a linear function is appropriate for systems of linear functions or equations and has the advantage of handling vertical lines. A linear function can be written in general form as

$$\mathrm{a}x+\mathrm{b}y=\mathrm{c}$$

where the parameters a and b are *not* the same as in the slope-intercept form above. If $\mathrm{a}=0$, then the line is horizontal $y=\mathrm{c/b}$, and if $\mathrm{b}=0$, then the line is vertical, $x=\mathrm{c/a}$.

B.1.2 Quadratic Functions

Quadratic graphs have a minimum or maximum value known as the functions *vertex*. It is also called the graphs *turning point*, or *extremum*. On one side of this point, the graph is increasing, and on the other side, it is decreasing. This increase is a linear function, meaning $y'=\mathrm{a}\,x+\mathrm{b}$ and $y''=\mathrm{a}$.

Many physical phenomena can be modeled using quadratic functions. Typical examples include projectile motion, breaking, and kinetic energy.

Figure B.2a and b show two examples of quadratic functions. Figure B.2a shows a function with a=¼, and Figure B.2b shows a function where a=–2.

The basic form of the function is

$$y=\mathrm{a}x^2$$

where a represents the "shape" of the function; a = $\Delta y/(\Delta x)^2$, where Δx and Δy must be measured from the vertex

$$\Delta y=y-\mathrm{V}_y \text{ and } \Delta x=x-\mathrm{V}_x$$

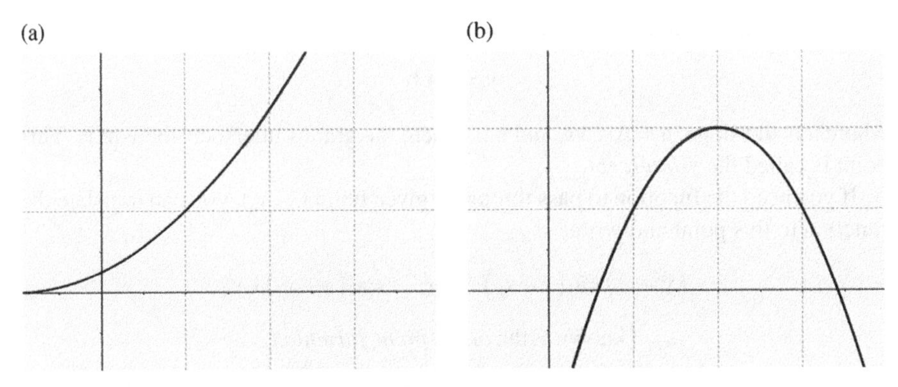

FIGURE B.2a and b Quadratic functions.

If the vertex lies in (V_x, V_y), then you can translate the function to this point and write

$$(y - V_y) = a \cdot (x - V_x)^2 \quad (\textit{vertex form})$$

If the graph intersects the x-axis at $x = x_1$ and $x = x_2$, you can use this to factor the function:

$$y = a \cdot (x - x_1)(x - x_2) \quad (\textit{factor form})$$

In general, though, a quadratic function is represented as a polynomial:

$$y = ax^2 + bx + c \quad (\textit{general form})$$

where a represents the "shape" or "openness,"–b/2a (= V_x) represents the horizontal shift, and c represents the y-intercept.

There are many relationships between these different parameters. For instance, the x-coordinate of the vertex is the average of the two x-intercepts, $V_x = (x_1 + x_2)/2$ and $c = a \cdot x_1 \cdot x_2$.

B.1.3 Exponential Growth

Exponential functions are very common in modeling situations, so common that we have divided the study of them in three distinct sections. In this section we study exponential growth.

The basic exponential function has the property that it increases by the same *percentage* for each step you take on the x-axis, unlike the linear functions, which increase by the same *amount* for every step on the x-axis. Exponential growth is characterized by a very "slow" initial growth followed by a very "fast" growth later on. There is no specific point where the transition between "slow" and "fast" takes place, though the rate of change is exponential in its own right.

Exponential functions are used to model unlimited population growth, compound interest, and other situations where the rate of change can be expressed as being

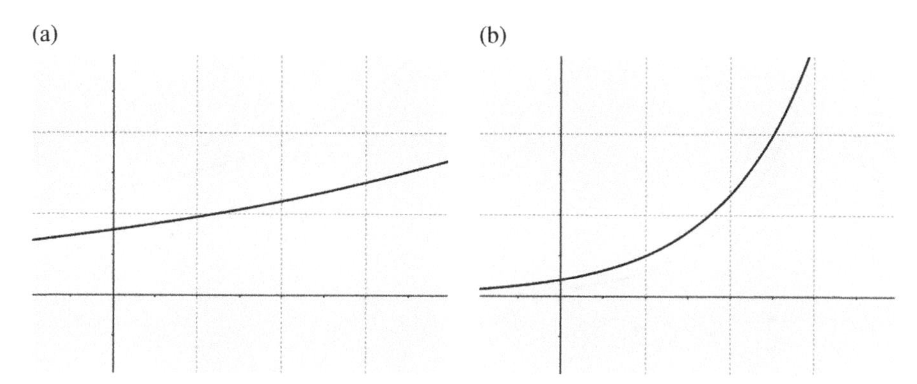

FIGURE B.3a and b Increasing exponential functions with $y'>0$.

proportional to the original function. Thus the basic differential equation $y'=\mathrm{k}\cdot y$ has an exponential function as its solution. The graph is sometimes called a J-curve due to its shape, and to distinguish it from the S-curves of logistic functions.

Figure B.3a and b show two examples of exponential growth. The graph in Figure B.3a can be conceived of as a small part of the graph in Figure B.3b.

The basic exponential function is usually expressed with either an arbitrary base or a fixed base such as 2 or *e*. In the latter case a multiplicative parameter is introduced in the exponent part of the function. Thus typical ways of expressing a basic exponential function can be

$$y=\mathrm{c}\cdot\mathrm{a}^{x} \quad \text{or} \quad y=\mathrm{c}\cdot\mathrm{e}^{\mathrm{k}\cdot x}$$

where c is the y-intercept and $\mathrm{a}=e^{\mathrm{k}}$ is the *change factor*, the number you multiply with for each step you take on the x-axis. For exponential growth this is >1.

If the graph does not tend to 0 to the left, you can translate the function vertically by adding a constant

$$y=\mathrm{c}\cdot\mathrm{a}^{x}+\mathrm{b} \quad \text{or} \quad y=\mathrm{c}\cdot\mathrm{e}^{\mathrm{k}\cdot x}+\mathrm{b}$$

where b is the value the function tends to for large negative x-values.

If your data set has large x-values, such as years 2015, 2016, …, you can translate the function horizontally, either by subtracting a constant number from your data (e.g., 2015) or by specifying the function as

$$y=\mathrm{c}\cdot\mathrm{a}^{(x-\mathrm{d})}+\mathrm{b}\,\text{eller}\; y=\mathrm{c}\cdot\mathrm{e}^{\mathrm{k}\cdot(x-\mathrm{d})}+\mathrm{b}$$

where d is selected to be roughly in the middle of your data sets x-coordinates.

B.1.4 Exponential Decline

A change factor $\mathrm{a}=e^{\mathrm{k}}<1$ indicates that the function is decreasing. Changing the sign of the change factor has the same effect as reflecting the graph in the y-axis (or in the line $x=\mathrm{d}$). The graph exhibits a sharp fall which gradually turns into a slow decrease.

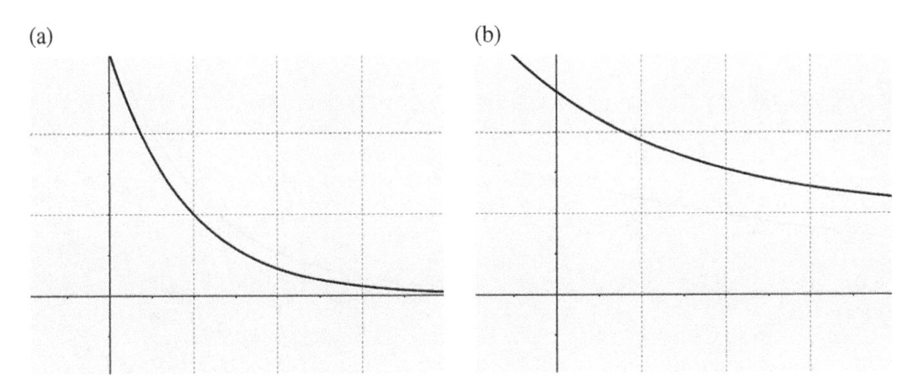

FIGURE B.4a and b Decreasing exponential functions with $y'<0$.

Some situations modeled by exponential decline are radioactive decay, cooling, and effects of medication and pollution.

Figure B.4a and b show the typical shapes of exponential decline or decay.

The basic function can be expressed as

$$y=c\cdot a^{-x}, \text{ or } y=c\cdot e^{-k\cdot x}$$

where c is the y-intercept and $a=e^{k}$ is the *change factor*, the number you multiply with for each step you take on the x-axis. For exponential decline this is <1.

For radioactive decay, the half-life $T_{1/2}$ is introduced and indicates the time it takes for half of a particular sample to decay. Then the functions can be expressed as

$$y=c\cdot 2^{-t/T_{1/2}}=c\cdot 0,5^{t/T_{1/2}}$$

y and c are often named N and N_0: $N=N_0\cdot 2^{-t/T_{1/2}}=N_0\cdot 0,5^{t/T_{1/2}}$

As in the case of exponential growth, the function may be translated vertically by adding a constant $+\,b$, or horizontally by exchanging x or t for $x-d$ or $t-d$.

B.1.5 Exponential Rise

The combination of a negative value of both c and a will give you a function as shown in Figure B.5a and b. It begins with a sharp rise, but because the rate of change is decreasing, the graph tends to an upper limit.

These types of exponential functions occur as you charge a capacitor, measure the temperature of a steak in an oven, or study limited population growth or terminal velocity. The basic function in this case is

$$y=b+c\cdot a^{-x}, \text{ or } y=b+c\cdot e^{-k\cdot x} \qquad \text{(observe that c is negative, often } c=-b)$$

where $b-c$ is the y-intercept and $a=e^{k}<1$ is the *change factor*, the number you multiply with for each step you take on the x-axis. For an exponential rise this is <1.

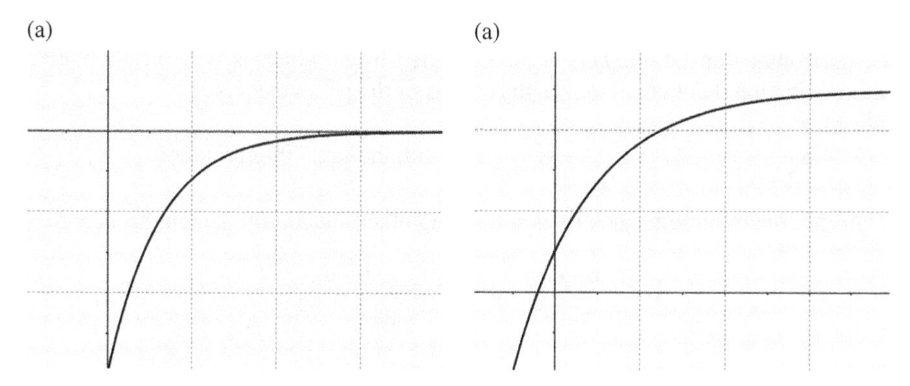

FIGURE B.5a and b Increasing exponential functions with $y' < 0$.

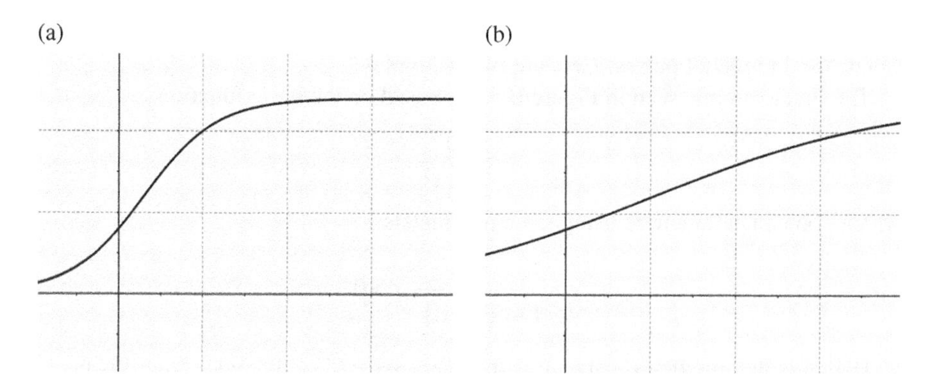

FIGURE B.6a and b Logistic functions.

As in the case of exponential growth, the function may be translated horizontally by exchanging x or t for $x - \text{d}$ or $t - \text{d}$.

B.1.6 Logistic Growth

Exponential growth is all very well, but sooner or later growth must reach some sort of limit that then may be expressed as insufficient foodstuff, lack of natural resources, or a lack of interest to invest. Many such situations can then be modeled with a logistic function, an *S-curve*. These functions typically start as exponential growths but gradually turn into exponential rises, leveling off toward a maximum value often referred to as the *carrying capacity*, K. Two examples of this can be found in Figure B.6a and b.

These functions are excellent applications of basic exponential functions. The basic logistic function is often expressed as either one of

$$y = \frac{\text{b}}{1 + \text{c} \cdot \text{a}^{-x}} \qquad y = \frac{\text{b}}{1 + e^{-\text{k}(x - \text{c})}} \qquad \left(\text{with different interpretations of c}\right)$$

Where $a = e^k$ is the "speed" of the rise, b is the maximum value (= K), and c affects the horizontal translation. This is also affected by a, which is why an alternative parametrization that better isolates the effects of the parameters is

$$y = \frac{b}{1 + a^{-(x-c)}} \qquad y = \frac{b}{1 + e^{-k(x-c)}} \qquad \text{(with the same interpretations of c)}$$

In case the minimum capacity is something other than 0, say d, then this becomes

$$y = \frac{b - d}{1 + a^{-(x-c)}} + d \qquad y = \frac{b - d}{1 + e^{-k(x-c)}} + d$$

B.1.7 Pulses: Symmetric Singular Pulses

A pulse is characterized by a single, distinct maximum with the rest of the function flattening out in both directions from this maximum. Pulses may appear in all sorts of contexts: signals, sound, medication, and so on. There are several functions that can be used to model pulses. Here are two examples:

The first example, seen in Figure B.7a, is based on the basic function

$$y = \frac{1}{1 + x^2}$$

which you can parametrize using three parameters

$$y = \frac{a}{1 + c(x - d)^2}, c > 0$$

Here a is the maximum value, c is the "narrowness" of the pulse, and d is the x-value of the center of the pulse. It is possible to exchange the square for another parameter if the shape of the pulse is wrong.

Our second example, seen in Figure B.7b, is useful if you want a faster, more exponential decline. It is based on the function

$$y = e^{-x^2}$$

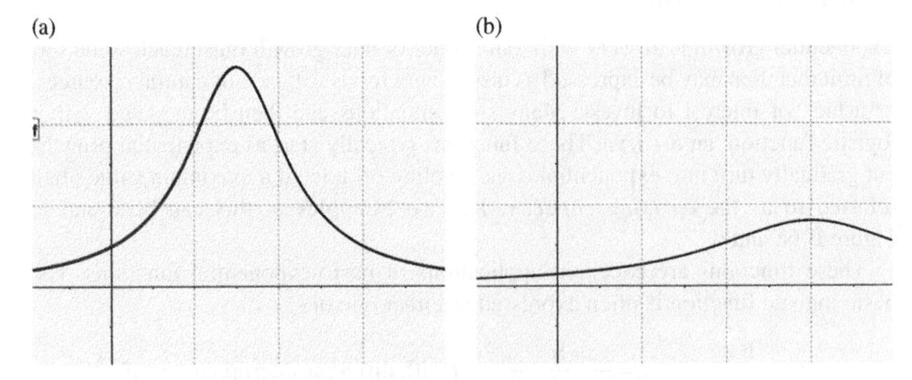

FIGURE B.7a and b Symmetric single pulses.

which you again can parametrize with three parameters, giving

$$y = c \cdot e^{-k \cdot (x-d)^2}$$

where c controls the height of the graph, k its width, and d, as before, the location of its center. This is the well-known normal distribution function, the Gaussian bell curve.

B.1.8 Pulses: Asymmetric Single Pulse

Asymmetric pulses are often more realistic than symmetric pulses and have a clear start for the process, which is normally placed at the origin. Two different main sub-types of asymmetric pulses may be distinguished, depending on what happens initially. Either the graph rises aggressively from the origin, gradually flattening out toward the maximum as in Figure B.8a, or the process starts slowly and only reaches its maximum rate of change after a while as in Figure B.8b. These two main behaviors could be called *aggressive* and *cautious*, respectively.

Aggressive processes can be chemical processes when a compound is added to a solution and immediately begins reacting with it, or a collision of some sort. Cautious processes could be biological processes where a small amount of controlling genes, hormones, or proteins first need to grow, or produce secondary compounds for the process to speed up. An epidemic is another example of a process that can be modeled with a cautious, asymmetric pulse.

An example of an aggressive pulse is

$$y = a \cdot e^{-p \cdot x} - b \cdot e^{-q \cdot x} + c, \qquad \text{where } a, b, p, q > 0, \text{and } p < q$$

This function will tend to c for large values of x and will intersect the y-axis at $y = a - b + c$. To make it tend to 0 and pass through the origin, set $c = 0$ and $a = b$.

An example of a cautious pulse is the product of a power function and a decreasing exponential function:

$$y = c \cdot x^b \cdot a^{-x}, \quad y = c \cdot x^b \cdot e^{-k \cdot x}, \qquad \text{where } a = e^k > 0 \text{ and } k > 0$$

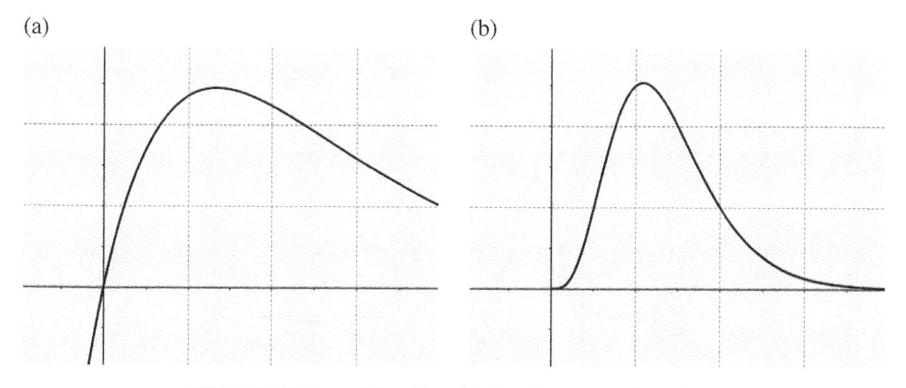

FIGURE B.8a and b Two kinds of asymmetric pulses.

An advanced example of this is the gamma distribution function

$$y = \frac{c \cdot x^{a-1} \cdot e^{-\frac{x}{b}}}{b^a \cdot \Gamma(a)}, \qquad \text{where } \Gamma(x) \text{ is the gamma function}$$

B.1.9 Double Pulses

The derivatives of the functions for symmetric single pulses work well for modeling double pulses where an initial pulse is followed by a similar negative pulse. The parameters mean the same things here as in the previous two sections.

Function 1: $y = -\frac{2 \cdot a \cdot c \cdot (x-d)}{\left(1 + c \cdot (x-d)^2\right)^2}$ (see Figure B.9a)

Function 2: $y = -2 \cdot c \cdot k \cdot (x-d) \cdot e^{-k \cdot (x-d)^2}$ (see Figure B.9b)

For a distinct start, use the derivative of the cautious asymmetric pulse:

Function 3: $y = b \cdot c \cdot x^b \cdot a^{x-1} + c \cdot x^b \cdot a^x \ln a$ (see Figure B.9c)

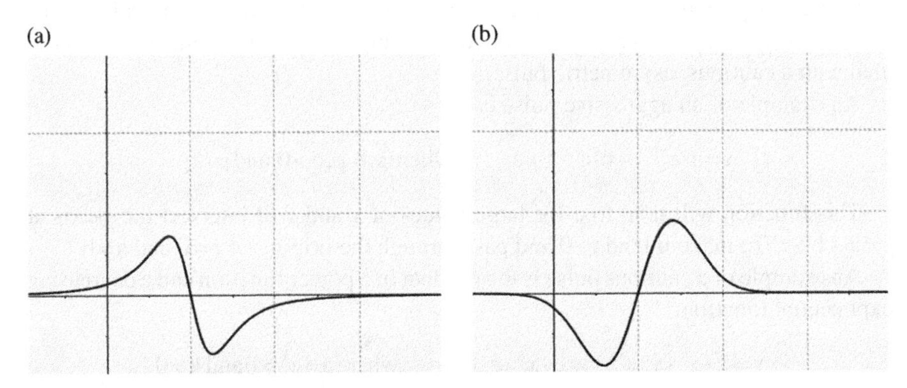

FIGURE B.9a and b Examples of double pulse functions.

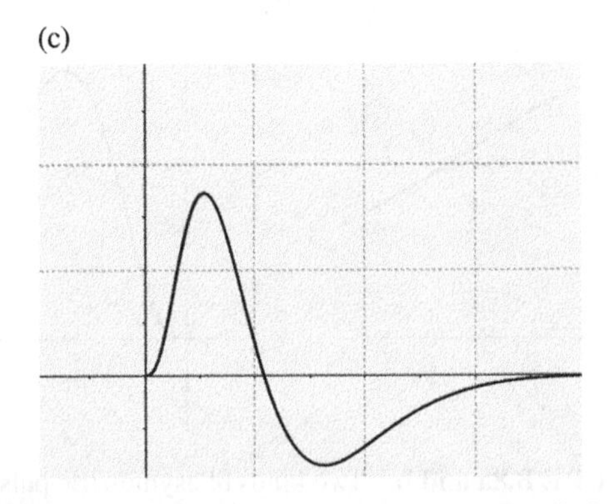

FIGURE B.9c Example of a double pulse with a distinct start.

B.1.10 Power and Root Functions

These functions are useful for modeling growth that starts at the origin. There are many scientific processes and relations that can be modeled using power functions, often where size, lengths, areas, or volumes play a role.

The basic function is

$$y = \mathrm{c} \cdot x^{\mathrm{a}}, \qquad \text{where a} > 0 \text{ and } x > 0$$

C represents the value for $x = 1$. If you want to move this function b steps to the right and d steps up, then use

$$y = \mathrm{c}\left(x - \mathrm{b}\right)^{\mathrm{a}} + \mathrm{d}$$

If $\mathrm{a} > 1$, the function's slope will gradually increase. If $\mathrm{a} < 1$, the function's slope will gradually decrease but never become negative. Figure B.10a and Figure B.10b show examples of this behavior.

If $\mathrm{a} = 0.5$, you get the square root function, $y = c\sqrt{x}$

If $\mathrm{a} = -1$, you get the inverse proportionality $y = \mathrm{c}/x$, which can be used to model electric fields or phenomena that depend on the distance from linear energy sources, such as the illumination from fluorescent lighting or the energy in a tsunami created from a linear fault. See Figure B.10c.

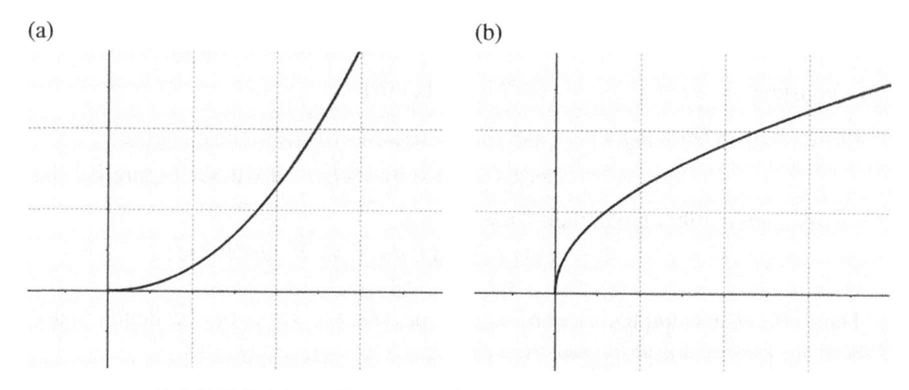

FIGURE B.10a and b Power functions with positive exponents.

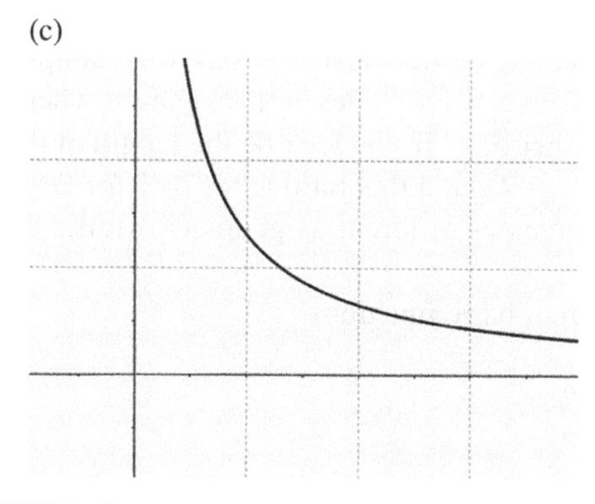

FIGURE B.10c Power function with a negative exponent.

If a = −2, you get the inverse square law $y = c/x^2$, which can be used to model gravitational force, electric force, illumination from a point light source, and so on.

All power functions can be drawn as straight lines if you plot log y versus log x. Doing this will sometimes allow you to find badly measured data easier.

B.1.11 Rational Functions in General

A rational function is a division between two polynomials and is often recognized from its vertical asymptotes. The basic function can be written as

$$y = \frac{a + bx + cx^2 + \ldots}{p + qx + rx^2 + \ldots}$$

Let *Dn* and *Dd* denote the degree of the nominator and the denominator, respectively.

If you have a clear plot of the entire graph, you can determine the minimum value of *Dn* by counting the number of vertical asymptotes. The other clue you get from the graph is by studying the overall behavior for large values of x. If the function then resembles

$y = 0$, then the $Dn < Dd$ (see Figure B.11a)

y=constant, then $Dn = Dd$ (see Figure B.11b)

$y = ax + b$, then $Dn = Dd + 1$ (and the coefficients of the highest degree terms divided by each other = a; see Figure B.12b)

$y = n$th degree polynomial, then $Dn = Dd + n$
($Dn = 3$ and $Dd = 1$; see Figure B.11c)

The vertical asymptotes x-coordinates represent the zeros of the denominator. If there are r vertical asymptotes, then $r \leq Dd$.

B.1.12 Rational Functions: Proportions

Ratios and proportionalities often lead to relatively simple rational functions. What is the relation between two sides in a rectangle where one side always is 2 cm longer than the other one? If you let x be the length of the short side, then the ratio can be written $x/(x+2)$, and this ratio tends to 1 for large values of x Similar ratios often occur in areas as different as geometry, traffic analysis, biology, and chemistry.

There are two different basic functions:

Function 1: $y = c\dfrac{x}{a + x}$

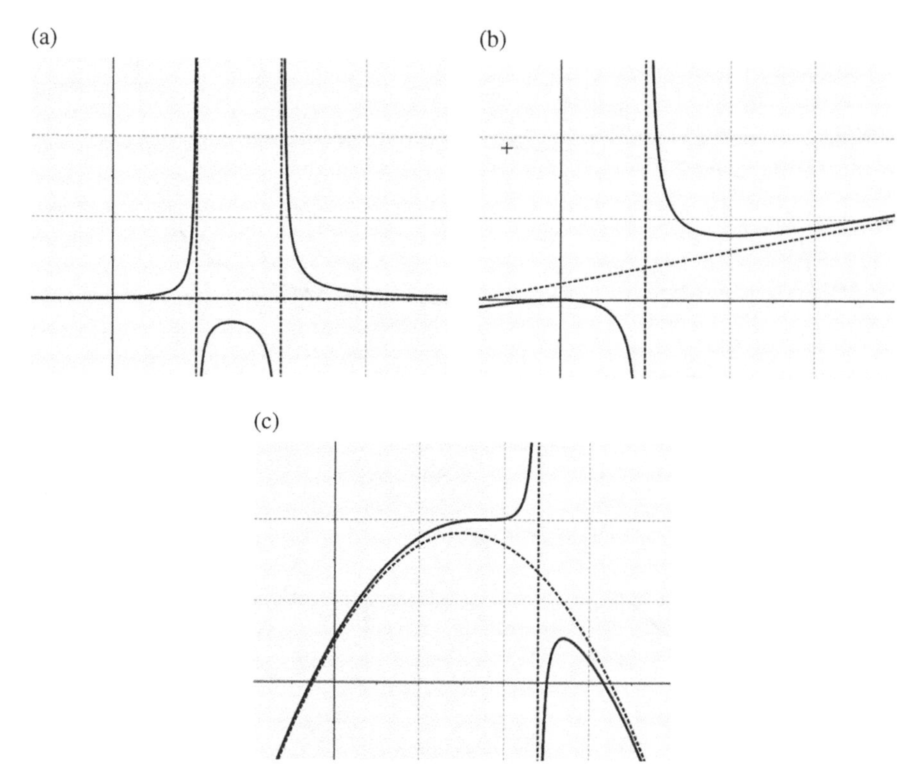

FIGURE B.11a, b, and c Rational functions are characterized by one or several asymptotes.

where a represents the constant part added, and c represents an enlargement factor. This function has a horizontal asymptote $y = \mathrm{c}$. You can determine a as the x-value where $y = \mathrm{c}/2$. In Figure B.12a, a = 0.3 and c = 2.

Function 2: $y = \mathrm{c}\dfrac{x}{\mathrm{a} - x}$

where a represents the constant part from which you can deduct a variable part, and c represents an enlargement factor. In this case you get vertical asymptote $x = \mathrm{a}$. You can determine c as the y-value where $x = \mathrm{a}/2$. In Figure B.12b, a = 3 and c = 0.3.

B.1.13 Logarithmic Functions

Logarithmic functions rise from minus infinity as seen in Figure B.13a. The rate of change is constantly decreasing, but the slope is always positive, unless multiplied by a negative number, as in Figure B.13b.

Logarithm functions can be used to model situations where your senses are involved in interpreting a physical phenomenon such as when sound intensity is translated sound levels using the decibel scale, or when the energy in an earthquake

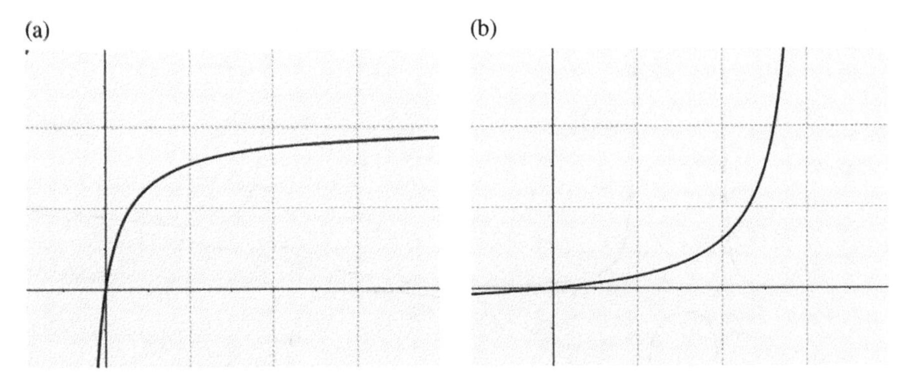

FIGURE B.12a and b Simple rational functions tending to an asymptotic value.

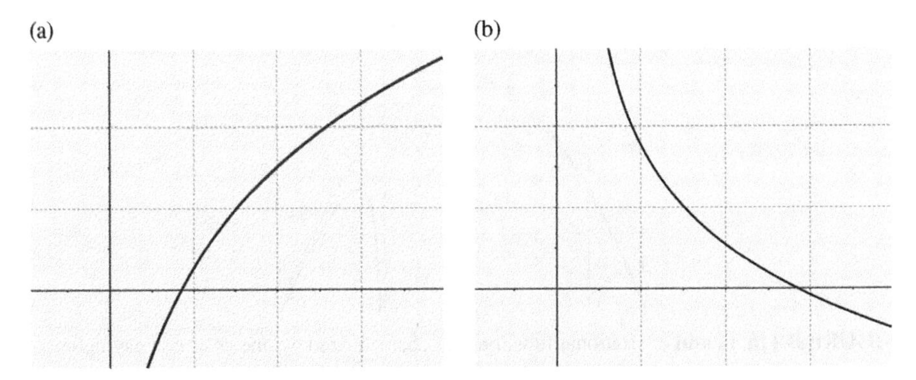

FIGURE B.13a and b Logarithmic functions.

is translated into the perceived effects using the Richter scale, or indeed, when the absolute luminosity of the stars are translated into a perceived, relative magnitude. They can also be used to model sport records over time.

The basic function is

$$y = \log x$$

but this may be translated or stretched in the usual way:

$$y = \mathrm{a} \cdot \log(\mathrm{b}x + \mathrm{c}) + \mathrm{d}$$

It is possible to use logarithms of any base. Using one or another will result in different parameter values but the same function.

B.1.14 Periodic Functions

All you have to do is sit still and look out the window to experience periodic phenomena. The daily rhythms of your body functions, the seasonal changes, a

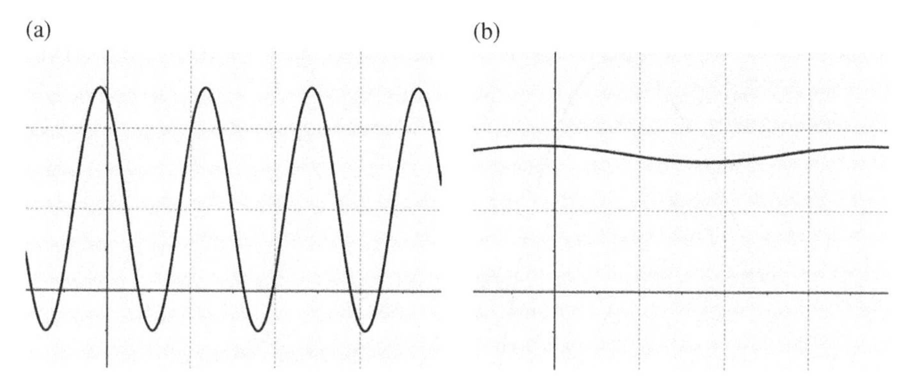

FIGURE B.14a and b Periodic functions.

bouncing ball, a bus shuttle service, are all examples that can be modeled using periodic functions.

The simplest periodic functions are sine and cosine functions. They are characterized by regular variations with the same height on all the peaks and all the valleys and the same distance between peaks and valleys, as can be seen in Figure 14a and b.

The basic sine function is usually parametrized as

$$y = \mathrm{a} \cdot \sin\left(\mathrm{b}\left(x + \mathrm{c}\right)\right) + \mathrm{d}$$

Here a is the amplitude (half the vertical distance between peaks and valleys), b is the angular frequency (how "fast" things are happening, often called ω), c is the phase (horizontal translation), and d is the vertical height of the center line (vertical translation).

B.1.15 Symmetric and Asymmetric Extreme Values

A frequent task is finding the position of a maximum or minimum point with height accuracy. A common approach is to fit a polynomial to the data close to this point. If you have theoretical arguments that show the data to be symmetric around the point, you can fit a quadratic function to these points.

If, however, there are no such theoretical arguments and the data do not show an obvious symmetry, you will get a better fit if you use a polynomial of at least degree three:

$$y = \mathrm{a} + \mathrm{b} \cdot x + \mathrm{c} \cdot x^2 + \mathrm{d} \cdot x^3 + \ldots$$

If the data seem to have a specific number of extreme points, you should try to fit a polynomial of degree = total number of extrema + 1. Figure B.15a and b show examples of third-degree polynomials.

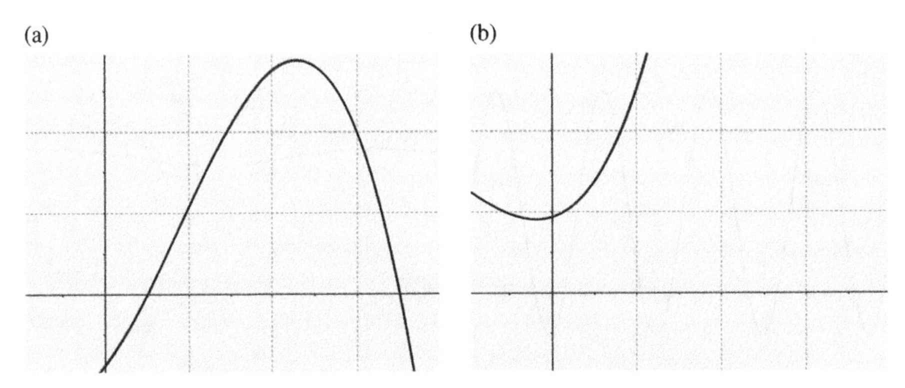

FIGURE B.15a and b Examples of asymmetric extreme points.

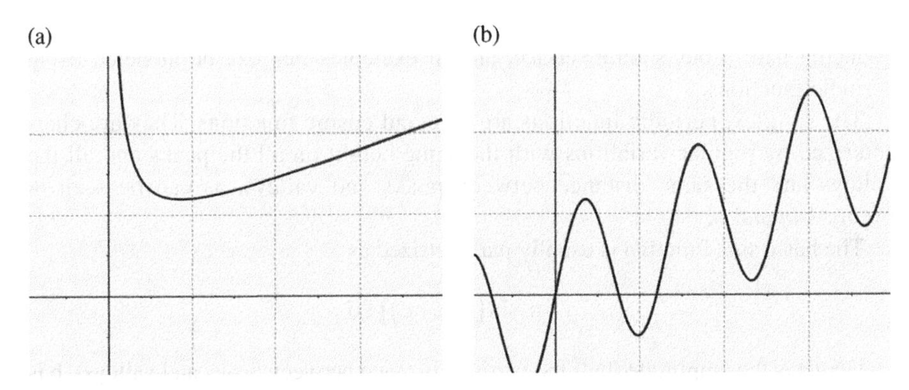

FIGURE B.16a and b Examples of linear combinations of basic functions.

B.1.16 Combinations: Sums (Linear Combinations)

Combining different functions into one vastly enlarges the possibilities of finding suitable functions. A basic linear combination of two functions can be written

$$y=\mathrm{a}\cdot f(x)+\mathrm{b}\cdot g(x)$$

Here a and b are constants and $f(x)$ and $g(x)$ are two different functions.

Sometimes one function can be dominant (=larger) for small values of x and the other function can be dominant for large values of x. In Figure B.16a you see such a combination. For small values of x, the function looks like perhaps $y=1/x$. For larger values of x, it seems more like a straight line, $y=\mathrm{a}x+\mathrm{b}$. You can construct a linear combination of these two basic functions by writing

$$y=\mathrm{c}/x+\mathrm{a}x+\mathrm{b}$$

In the second example in Figure B.16b the behavior for small and large values of x does not seem to differ, but you can still see a periodic variation superimposed on a straight line through the origin, and so you can write

$$y=\mathrm{a}\cdot\sin(\mathrm{b}x)+\mathrm{c}x$$

B.1.17 Combinations: Products

Multiplying two functions can result in many different new functions. One class of these multiplies a function $A(x)$ with a periodic function $f(x)$. The first function $A(x)$ can then be interpreted as a time-dependent amplitude for the periodic function $f(x)$:

$$y = \mathrm{A}(x) \cdot f(x)$$

In the first example in Figure B.17a, $f(x)$ is a sine function and $\mathrm{A}(x)$ is a declining exponential function. There seems to be a constant added as well, so

$$y = \mathrm{C} \cdot \mathrm{e}^{-\mathrm{k}x} \cdot \sin(\mathrm{b}x) + \mathrm{d}$$

In the second example in Figure B.17b, the amplitude of the sine function seems to increase linearly with the distance from the origin, so we assume that

$$y = \mathrm{C} \cdot x \cdot \sin(\mathrm{b}x)$$

Examples of applications for the first example include damped oscillations, bouncing balls, springs. The first phase of a resonance phenomenon where the amplitude is being built up could be modeled using our second example.

B.1.18 Combinations: Gradual Transitions

Should you ever find yourself in the position where you suspect that a phenomenon is being controlled by two different underlying processes where each process is dominant in different intervals but a linear combination does not seem to do the trick, you may want to try a different way to combine two functions. One way of doing this is to add the functions with weights that depend on x. Ideally, you should have a weight function that has a range from 0 to 1 over the entire x-axis. For this, you can use the logistic function

$$L(x) = \frac{1}{1 + e^{-\mathrm{k}(x-\mathrm{c})}}$$

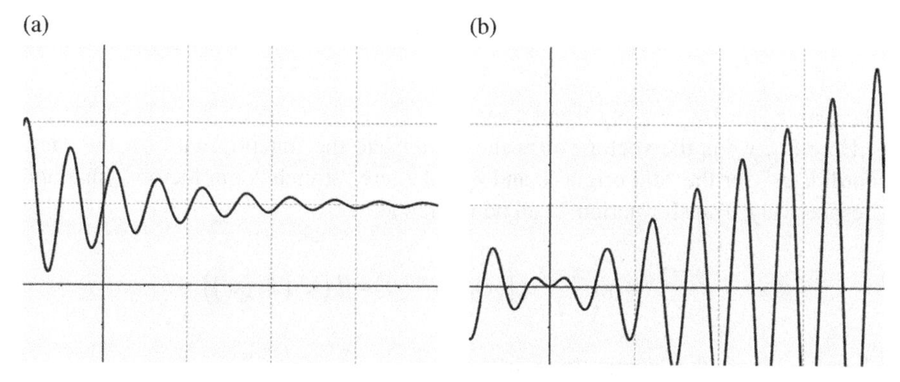

FIGURE B.17a and b Examples of products of functions.

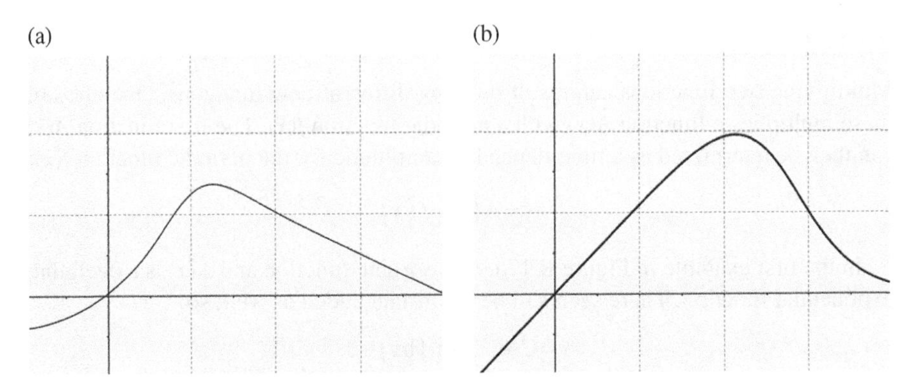

FIGURE B.18a and b Gradual transitions between functions.

and then combine the basic functions f and g like this

$$y = f(x)\cdot L(x) + g(x)\cdot(1 - L(x))$$

The parameter c is the x-coordinate for the midpoint of the transition and k determines the "abruptness" of the transition. In Figure B.18a you see an example of a transition between a quadratic function and a linear function, and in Figure B.18b the transition is from a linear function to an exponential decline.

B.2 LINEAR TRANSFORMATIONS IN GENERAL

In Section B.1 you saw several examples of how different parameters can both translate and stretch the function both vertically and horizontally. A change that only consists of a translation and a stretch is sometimes called a linear transform. In general, any basic function $f(x)$ can undergo a linear transform to a new function $g(x)$ in several different ways:

Transformation 1: $g(x) = \mathrm{a}\cdot f(\mathrm{b}\cdot x + \mathrm{c}) + \mathrm{d}$

Transformation 2: $g(x) = \mathrm{z_y}\cdot f\left(\dfrac{x - \mathrm{x_1}}{\mathrm{z}_x}\right) + \mathrm{y_1}$

where $\mathrm{a} = \mathrm{z_y}$, $\mathrm{b} = 1/\mathrm{z_x}$, $\mathrm{c} = -\mathrm{x_1}/\mathrm{z_x}$ and $\mathrm{d} = \mathrm{y_1}$.

Here (x_1, y_1) is the vector you want to translate the function with (or the "new coordinates" for the "old origin"), and z_x and z_y are "stretch/zoom factors" for x and y respectively. Transformation 2 can be written as

$$\frac{y - \mathrm{y_1}}{\mathrm{z_y}} = f\left(\frac{x - \mathrm{x_1}}{\mathrm{z}_x}\right) \text{or } \mathrm{k_y}(y - \mathrm{y_1}) = f(\mathrm{k}_x(x - \mathrm{x_1}))$$

where $\mathrm{k} = 1/\mathrm{z}$ are the "compression factors."

In other words, a linear transform may be achieved by substituting

- $(x - x_1)$ for x
- $(y - y_1)$ for y
- Then multiplying either or both of these parentheses with a compression factor.

Another way of thinking about this is to imagine an *inner* linear function in x, and an *outer* linear function in y.

In the first example in Figure B.19a, you can see a quadratic equation that has been translated three steps to the right and then compressed by a factor of 2 vertically.

In Figure B.19b, you have an exponential function that has been moved up one step and then compressed a factor 2 vertically as before, but also stretched by a factor of 10 horizontally. Observe that all stretches are relative to the "new origin," here (0, 1). In the final example in Figure B.19c, a sine function has been translated to a "new origin" at (1, 1) and then stretched a factor 1,5 vertically and compressed a factor 4 horizontally. Also see Section B.3.

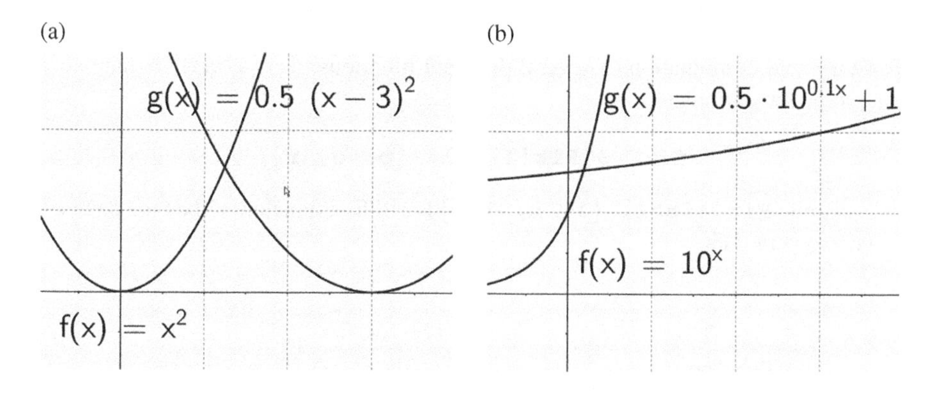

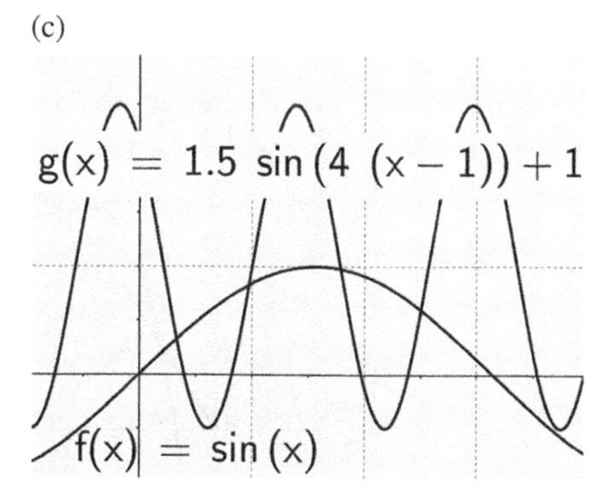

FIGURE B.19a, b, and c Examples of linear transforms.

B.3 DRAGGING A FUNCTION IN GEOGEBRA

Dragging a function in the graphics view in GeoGebra or selecting it and moving it using the arrow keys will update the Algebra window to show the new, translated function $f(x-T_x)+T_y$,where (T_x, T_y) represents the translation vector. This is an example of a simple linear transformation without any stretching. For the function $y=x^2$, the new function is recognized as a quadratic in vertex form. Figure B.20 shows an example of this.

For any quadratic, you may hold down the Alt key while dragging to change the quadratic coefficient.

B.4 GEOGEBRA'S GENERIC FITTING COMMANDS

If none of the standard models seem to fit your data you may specify any model function you wish and tell GeoGebra to fit a function of that particular type to your data. This is called *generic fitting* and can be done using the **`Fit`**`[...]` command in either one of two different ways. In describing these methods, we will assume that your data are contained in **`list1`**.

B.4.1 Linear Combinations

To fit a linear combination of several different functions: $f(x)=\mathrm{a}\cdot p(x)+\mathrm{b}\cdot q(x)+\mathrm{c}\cdot r(x)\ldots$ type the command

```
f(x) = Fit[list1, {p, q, r}]
```

Example: **`Fit[list1, {1, x, 1/x}]`** will fit a function of the type $f(x)=\mathrm{a}+\mathrm{b}x+\mathrm{c}/x$ to your data.

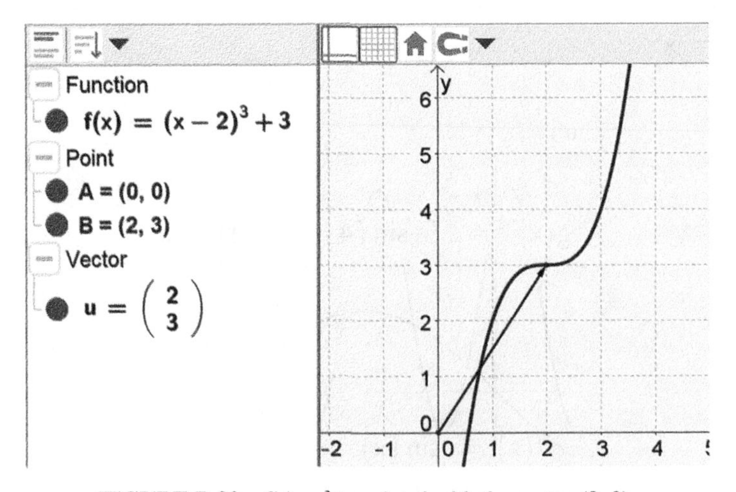

FIGURE B.20 $f(x)=x^3$ translated with the vector (2, 3).

This method does not require you to define parameters or a model function first but can only deal with linear combinations of simple functions.

B.4.2 Arbitrary Model Functions

To fit a completely arbitrary function that depends on a number of parameters *f*(*x*, a, b, c...), you must first construct the model function *m*(*x*, a, b, c...), accepting the creation of new sliders for each parameter. This will be plotted with the current values of the parameters and not show anything real. For this reason it is often customary to hide *m*. Before doing this however, you can first adjust the parameter values to something reasonable as these values are taken as staring values for the algorithm. GeoGebra uses *m* to detect which parameters to optimize. You can use any name for the model function, not just "*m*."

After these preliminary steps you execute the fitting command:

```
f(x) = Fit[list1, m]
```

As an example, type in the following commands, one after the other, and hit **Enter** after each one:

```
a = 2
b = 1
k = -3
m(x) = e^(k x) (a x + b)
```

Observe the spaces that work as invisible multiplication signs. You can adjust the sliders so that *m*(*x*) somewhat resembles your data to give the algorithm some useful starting values. Then hide *m*(*x*) and type:

```
f(x) = Fit[list1, m]
```

B.5 EXAMPLE OF A GENERIC FIT

In the image of Sydney Harbor Bridge, shown in Figure B.22, you find an example of a shape that can be modeled with quadratic functions. The lower part of the bridge valve works better for this than the upper part, which visibly deviates from a quadratic function toward the ends. What would a function look like that could model the upper valve better than a quadratic? It would surely be an even function, but typically, students have yet to discover functions of this shape, so a little help may be needed. The function $y = 1/(1+x^2)$ is an example of such a function. By experimenting with the number 2 in the exponent a more generic function can be found, but for now just accept the function as it is. By introducing suitable parameters in this expression and then fitting these to the measured data, you can come up with a working model.

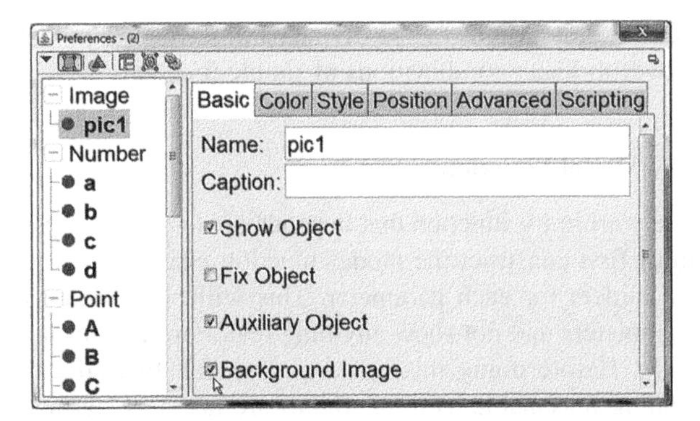

FIGURE B.21 Sending an image to the background.

This is how the modeling process might work in GeoGebra: Start by inserting the picture in GeoGebra using the **Insert Image** tool. Then right-click the image and select **Properties**. Under the **Basic** tab you check **Background Image** as shown in Figure B.21. This means that the image won't block other objects and that the grid will be visible in front of the image. You can still position the image by dragging its control points.

Now use the **New Point** tool to place some points on the upper bridge valve. You could use the stylebar to set a smaller point size before you start and you could select **No New Objects** from the **Options>Labelling** menu. The points' position can be fine-tuned later by setting their **Step** property on the **Slider** tab in the **Object Properties** dialogue to a small value such as 0.01 and using the arrow keys to move the points. The default value for the **Step** property is 0.1.

You must now give some thought to your model function. Using what you have learned about linear transformations in Section B.2, add parameters for translating and stretching the function and write

$$y = \frac{\mathrm{a}}{(\mathrm{b}\cdot x+\mathrm{c})^2+1}+\mathrm{d}$$

In GeoGebra, start by creating the four parameters a, b, c, and d as sliders, as can be seen in Figure B.22. After that you type

```
f(x) = a/((b x + c)^2 + 1) + d
```

(`^2` can be exchanged for **Alt-2**)

You could, of course, have typed the function first, allowing GeoGebra to create the sliders for you. It is quite instructional to let the students try to fit the function manually by adjusting the sliders by hand. This is by no means too difficult, but it is also very clear that the precision is not good. In Figure B.23 you see a frozen moment in the middle of this process.

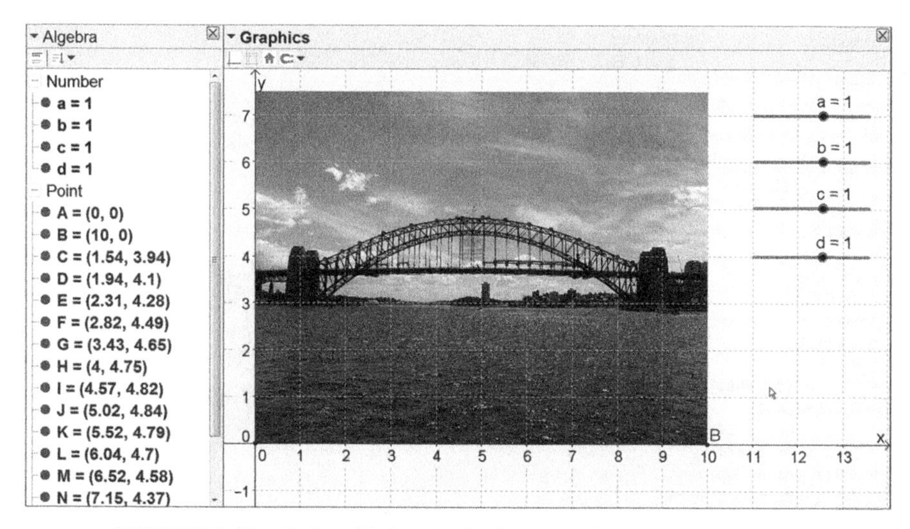

FIGURE B.22 Sydney Harbor Bridge image in GeoGebra with sliders.

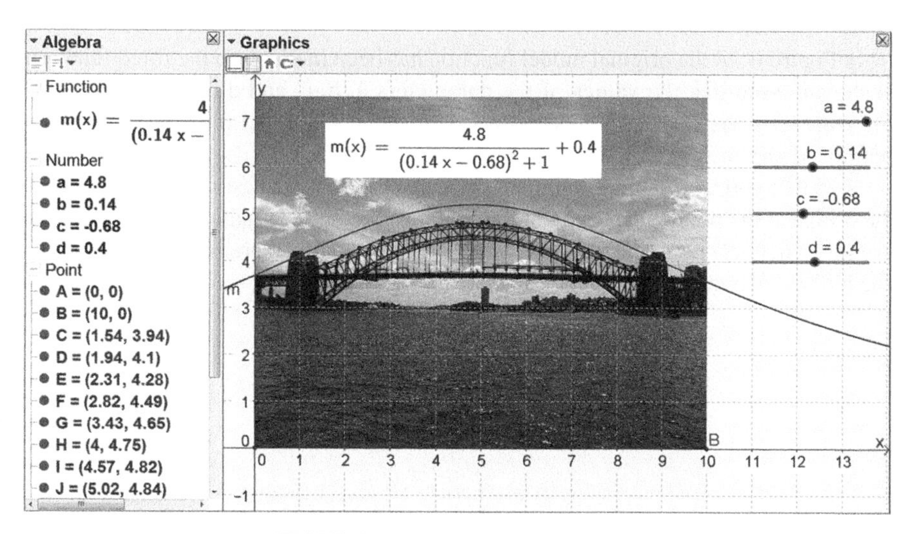

FIGURE B.23 Modeling the bridge.

To fit this model to the data automatically, enter the following command:

```
Fit[{A, B, C, D, E, F, G, H, I, J, K, L}, f]
```

You could have created a list of the points first, but since the list is only needed once, it is just as quick to type the list on the fly.

You might find that you sometimes cannot get a result. Depending on the number of data points and their precision, the algorithms are sometimes unable to find a solution. You can remedy this by reducing the number of data points or by moving them about a little.

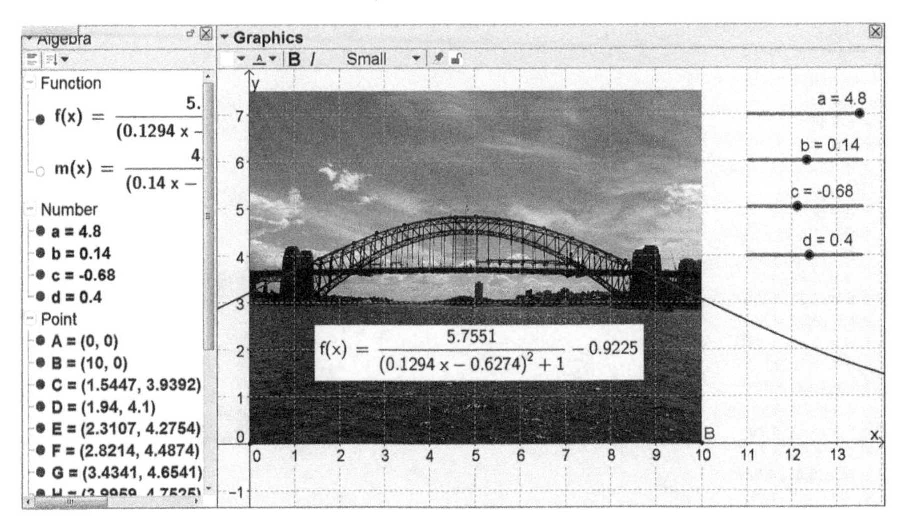

FIGURE B.24 The fitted function.

In Figure B.24 the original model function has been hidden and the fitted function is shown. Note that the values of the parameters a, b, c, and d have not changed. They are only the starting points for the algorithms. The fitted function p show the optimized parameter values.

Now that you have a functioning model, you can answer questions like how tall is the bridge at the point where the lower valve meets with the roadway. You could also use this when teaching integrals and ask the weight and total area of an acrylic wind barrier between the valve and the roadway.

INTEGER PROPERTIES

This table gives some short information on the first 568 integers, one for each page of this book. First the prime factorization is given within braces {...}. If the number itself is the only number occurring within the braces, then it is a prime number.

The various facts displayed are meant both to work as a dictionary of sorts, but also as triggers of curiosity, intended to be investigated further.

The following abbreviations apply:

F	Fibonacci number
T	Triangle number
P	Pentagonal number
C	Catalan Number

For example; C_5 refers to the fifth Catalan number.

Some mathematicians have made a name for themselves for having the numbers as personal friends. The Authors have a truly remarkable story about Ramanujan, beginning with the number 1729, which the margins of this book, alas, are too small to contain.

Mathematical Modeling: Applications with GeoGebra™, First Edition. Jonas Hall and Thomas Lingefjärd.

Companion website: www.wiley.com/go/Hall/MathematicalModeling

#	Information
1	{}, I, Fibonacci number, Square, Benford's law, T_1, P_1..., 0! = 1!, $1^1 = 1^2 = 1^3$...
2	{2}, F_3, C_2, 2!, 2·2 = 2+2, The base of the binary number system
3	{3}, Number of dimensions we live in, 1!+2!, F_4, T_2
4	{2·2}, Square, Tetrahedral number, Each integer is the sum of at most 4 squares
5	{5}, V, There are 5 Platonic solids, F_5, P_2, C_3
6	{2·3}, T_3, 3!, 1+2+ 3 = 1·2·3, First Perfect number
7	{7}, As I was going to S:t Ives..., 1/7 = 0.142857...
8	{2·2·2}, Cube, F_6 Only Fibonacci number being a cube except for 1, 4!!
9	{3·3}, 9-point circle, Square, $9^2 = 81$ and 8+1 = 9, 1!+2!+3!
10	X, {2·5}, T_4, The base of the decimal number system, Tetrahedral number
11	{11}, 11^n gives Pascal's triangle on row n, Palindromic number, Prime twin
12	{2·2·3}, P_3, Number of pentaminos, Number of faces on a Dodecahedron, A dozen, 3·4
13	{13}, There are 13 Archimedean polyhedra, F_7, Prime twin, Baker's dozen
14	{2·7}, C_4, Pyramidal number, Number of faces of a Cuboctahedron
15	{3·5}, The row sum of a magic 3x3-scuare, 5!!, T_5
16	{2·2·2·2}, $4^2 = 2^4$, Square, The base of the hexadecimal number system
17	{17}, 2^3+3^2, Number of plane symmetry groups (wallpaper patterns), Prime twin
18	{2·3·3}, 18 = 9+9 and 81 = 9·9, Octane has 18 isomers.
19	{19}, All integers are the sum of at most 19 4th powers, Prime twin,
20	{2·2·5}, Number of faces on an Icosahedron, Tetrahedral number, A score, (6·5·4)/(3·2·1), 4·5
21	{3·7}, Number of dots on a regular d6, F_8, T_6
22	{2·11}, P_4, 22/7 ≈ π, $1^4+2^3+3^2+4^1$
23	{23}, 23! has 23 digits
24	{2·2·2·3}, 4!, Cannonball problem
25	{5·5}, $25 = 5^2$, $3^2+4^2 = 5^2$, Square, Amorph
26	{2·13}, Only number between a square and a cube
27	{3·3·3}, $27^3 = 19{,}683$ and 27 = 1+9+6+8+3, Cube
28	{2·2·7}, T_7, Second Perfect number, Number of standard dominos
29	{29}, $2^2+3^2+4^2$, Prime twin
30	{2·3·5}, Smallest Sphenic number, Prime factorial 5, 5·6
31	{31}, $1+5+5^2$, Mersenne prime, Prime twin
32	{2·2·2·2·2}, 2^4+4^2, $1^1+2^2+3^3$
33	{3·11}, $1^1+2^2+3^3$, 1!+2!+3!+4!
34	{2·17}, F_9, The row sum of a magic 4x4-square
35	{5·7}, C_9H_{20} has 35 isomers, Number of hexaminos, Tetrahedral number, P_5
36	{2·2·3·3}, Square, T_8, 1+2+...+36 = 666
37	{37}, All integers are the sum of at most 37 5th-powers
38	{2·19}, The row sum of a magic hexagon, $2^2+3^2+5^2$
39	{3·13}, $3^1+3^2+3^3$
40	{2·2·2·5}, Often used as an arbitrarily large number, $3^0+3^1+3^2+3^3$
41	{41}, x^2+x+41 gives primes for $0 \le x \le 39$, Prime twin
42	{2·3·7}, C_5, the row sum of a magic 3x3x3-cube, 6·7
43	{43}, $1+6+6^2$, 4th number in the Sylvester series, Prime twin
44	{2·2·11}, Number of ways that 5 letters can be put in 5 envelopes so that no letters end up in the correct envelope.

#	Information
45	{3·3·5}, $45^2=2025$ and $20+25=45$, T_9
46	{2·23}, Number of chromosomes in human cells
47	{47}, $47+2=49$ and $47\cdot 2=94$
48	{2·2·2·2·3}, Smallest number with exactly 10 divisors, 4 dozen, 6!!
49	{7·7}, Square whose digits are squares
50	{2·5·5}, $1^2+7^2=5^2+5^2=3^2+4^2+5^2$, L
51	{3·17}, P_6, The shortest side in the smallest Heronian tetrahedron
52	{2·2·13}, Number of weeks per year, Number of cards in a typical deck,
53	{53}, 53=35h
54	{2·3·3·3}, Can be written as the sum of three squares on three different ways
55	{5·11}, $55^2=3025$ and $30+25=55$, Pyramidal number, F_{10}, T_{10}
56	{2·2·2·7}, Tetrahedral number, 7·8
57	{3·19}, $1+7+7^2$, 2^5+5^2
58	{2·29}, 2+3+5+7+11+13+17
59	{59}, Number of ways to stellate an Icosahedron, Prime twin
60	{2·2·3·5}, Three score=5 dozen, The first number with 12 divisors
61	{61}, 5^2+6^2, Prime twin
62	{2·31}, Sigmund Freud was reputedly afraid of the number 62
63	{3·3·7}, 111111b
64	{2·2·2·2·2·2}, Square, Cube, $4^3=8^2=2^{2\cdot 3}$, Grahams number$=g_{64}$
65	{5·13}, $8^2+1^2=7^2+4^2$, The row sum of a magic 5x5-square
66	{2·3·11}, T_{11}, Number of books in the bible, LXVI
67	{67}, The second largest Heegner number
68	{2·2·17}, The largest number equal to the sum of two prime numbers in exactly two different ways: 7+61=31+37
69	{3·23}, 69! is the largest factorial$<10^{100}$
70	{2·5·7}, P_7, Cannonball problem, (8·7·6·5)/(4·3·2·1), $70^2=1^2+2^2+\ldots+23^2+24^2$
71	{71}, $71^2=7!+1$ (Brocard's problem), Prime twin
72	{2·2·2·3·3}, $72^5=19^5+43^5+46^5+47^5+67^5$, 8·9
73	{73}, Sheldon Coopers number: $P_{21}=73$ and $P_{12}=37$, Prime twin
74	{2·37}, The density for packed spheres is 74%
75	{3·5·5}, $C_{10}H_{22}$ has 75 isomers
76	{2·2·19}, Automorphic: $76^2=5776$
77	{7·11}, $4^2+5^2+6^2$
78	{2·3·13}, T_{12}, Number of cards in a Tarot deck
79	{79}, $x^2+79x+1601$ gives prime numbers for $1\le x\le 79$
80	{2·2·2·2·5}, Four score, The Pareto principle: 20 % of the causes represent 80 % of the effect
81	{3·3·3·3}, $8+1=9$ and $9^2=81$, Square
82	{2·41}, The (8·2):e prime+the (8+2):e prime
83	{83}, $3^2+5^2+7^2$
84	{2·2·3·7}, Tetrahedral number, Diophantus age, $2^5+3^3+5^2$
85	{5·17}, $1+4+4^2+4^3$, $9^2+2^2=7^2+6^2$
86	{2·43}, $3^2+4^2+5^2+6^2$
87	{3·29}, $2^2+3^2+5^2+7^2$

(Continued)

#	Information
88	{2·2·2·11}, Number of keys on a piano
89	{89}, 8^1+9^2, F_{11}
90	{2·3·3·5}, Number of degrees in a right angle, $2^2+3^2+4^2+5^2+6^2$, 9·10
91	{7·13}, $1^2+2^2+3^2+4^2+5^2+6^2$, Pyramidal number, T_{13}
92	{2·2·23}, P_8, Number of Johnson polyhedra, Number of solutions to the 8 queen problem
93	{3·31}, 333 in base 5
94	{2·47}, The sum of the prime factors are 49 = 94 backwards
95	{5·19}, Hexagonal Pyramidal number, 5! – 4!
96	{2·2·2·2·2·3}, 2^6+2^5
97	{97}, $2\cdot2^3+3\cdot3^3$
98	{2·7·7}, $1^4+2^4+3^4$
99	{3·3·11}, $2^3+3^3+4^3$, $99^2=9801$ and 98+01=99
100	{2·2·5·5}, Square, $1^3+2^3+3^3+4^3$, 2^6+6^2, C
101	{101}, Palindromic prime number, Prime twin
102	{2·3·17}, $102^7=12^7+35^7+53^7+58^7+64^7+83^7+85^7+90^7$
103	{103}, Prime twin
104	{2·2·2·13}, The first boring number in this list, and therefore not boring
105	{3·5·7}, 7!!, T_{14}
106	{2·53}, The second boring number in this list
107	{107}, Prime twin
108	{2·2·3·3·3}, Hyper factorial $3=1^1\cdot2^2\cdot3^3$
109	{109}, Prime twin
110	{2·5·11}, $5^2+6^2+7^2$, 10·11
111	{3·37}, The row sum in the smallest magical prime square
112	{2·2·2·2·7}, 1·2+2·3+3·4+4·5+5·6+6·7
113	{113}, $355/113 \approx \pi$
114	{2·3·19}, The number of suras in the Quran
115	{5·23}
116	{2·2·29}
117	{3·3·13}, The longest side in the smallest Heronian tetrahedron, P_9
118	{2·59}
119	{7·17},
120	{2·2·2·3·5}, T_{15}, Tetrahedral number, The sum of the divisors = 2·120, 5!
121	{11·11}, Number of holes in Chinese chess, $121=11^2$, $1+3+3^2+3^3+3^4$, Square
122	{2·61}, 122·122 = 14884 and 48841 = 221·221
123	{3·41}, The 10th number in the Lucas series
124	{2·2·31}
125	{5·5·5}, $125=5^{1+2}$, $10^2+5^2=11^2+2^2$, Cube
126	{2·3·3·7}, 126=6·21
127	{127}, $127=-1+2^7$, Mersenne prime
128	{2·2·2·2·2·2·2}, $128=2^{8-1}$
129	{3·43}, The sum of the 10 first primes
130	{2·5·13}, 1!+1!+2!+3!+5!
131	{131}, Palindromic prime
132	{2·2·3·11}, C_6, 11·12

#	Information
133	{7·19}, 2^3+5^3
134	{2·67}, $134^2-67^2=13467$
135	{3·3·3·5}, $1^1+3^2+5^3$
136	{2·2·2·17}, T_{16}, $1^3+3^3+6^3=244$ and $2^3+4^3+4^3=136$
137	{137}, Remove any digit to leave a prime, Prime twin
138	{2·3·23}
139	{139}, Prime twin
140	{2·2·5·7}, Pyramidal number, $1^2+2^2+3^2+4^2+5^2+6^2+7^2$
141	{3·47}
142	{2·71}
143	{11·13}, Number of three digit primes
144	{2·2·2·2·3·3}, $F_{12}=12^2=144$ and $21^2=441$, A dozen dozen, Highly totient
145	{5·29}, $12^2+1^2=8^2+9^2$, 3^4+4^3, 1!+4!+5!, P_{10}
146	{2·73}
147	{3·7·7}
148	{2·2·37}
149	{149}, Tribonacci prime, Prime twin
150	{2·3·5·5}, Dunbar's number
151	{151}, Palindromic prime, Prime twin
152	{2·2·2·19}, 3^3+5^3
153	{3·3·17}, 153=3·51, $1^3+5^3+3^3$, 1!+2!+3!+4!+5!, T_{17}
154	{2·7·11}, 0!+1!+2!+3!+4!+5!
155	{5·31}, $5^1+5^2+5^3$
156	{2·2·3·13}, $1+5+5^2+5^3$, 12·13
157	{157}, $1+12+12^2$
158	{2·79}, Number of digits in 100!
159	{3·53}, $C_{11}H_{24}$ has 159 isomers
160	{2·2·2·2·2·5}, $2^3+3^3+5^3$, The sum of the 11 first primes
161	{7·23}, Hexagonal Pyramidal number
162	{2·3·3·3·3}
163	{163}, $1+2\cdot3^4$, The largest Heegner number
164	{2·2·41}
165	{3·5·11}, Tetrahedral number
166	{2·83}, CLXVI
167	{167}, 1·6·7=3·(1+6+7)
168	{2·2·2·3·7}, Number of hours per week
169	{13·13}, $13^2=169$ and $961=31^2$, Square
170	{2·5·17}, $2\cdot(1+4+4^2+4^3)$
171	{3·3·19}, T_{18}
172	{2·2·43}
173	{173}, $1^3+7^3+3^3=371$
174	{2·3·29}, $5^2+6^2+7^2+8^2$
175	{5·5·7}, $1^1+7^2+5^3$
176	{2·2·2·2·11}, P_{11}
177	{3·59}, 2^7+7^2
178	{2·89}, 13^2+3^2

(*Continued*)

#	Information
179	{179}, Prime twin
180	{2·2·3·3·5}, $180^3 = 6^3 + 7^3 + 8^3 + \ldots + 68^3 + 69^3$
181	{181}, Palindromic prime, Prime twin
182	{2·7·13}, 13·14
183	{3·61}, $2^2 + 3^2 + 7^2 + 11^2$
184	{2·2·2·23}, $2^3 \cdot 23$
185	{5·37}, $13^2 + 4^2 = 11^2 + 8^2$
186	{2·3·31}
187	{11·17}
188	{2·2·47}
189	{3·3·3·7}, Number of digits in numbers 1-99
190	{2·5·19}, T_{10}, $4^2 + 5^2 + 6^2 + 7^2 + 8^2$
191	{191}, Prime twin
192	{2·2·2·2·2·2·3}, Has 14 divisors
193	{193}, Prime twin
194	{2·97}, $7^2 + 8^2 + 9^2$, 193, 194 and 195 are the sides of an…
195	{3·5·13}, …almost equilateral, Heronian triangle
196	{2·2·7·7}, Square
197	{197}, The sum of the 12 first primes, $2^3 + 4^3 + 5^3$, Prime twin
198	{2·3·3·11}, 11 + 99 + 88
199	{199}, $3^2 + 4^2 + 5^2 + 6^2 + 7^2 + 8^2$, Prime twin
200	{2·2·2·5·5}, $1 + 7 + 7^2 + 7^3$
201	{3·67},
202	{2·101}, $(2 + 3 + 5 + 7)^2 - (2^2 + 3^2 + 5^2 + 7^2)$
203	{7·29}
204	{2·2·3·17}, Pyramidal number, $204^2 = 23^3 + 24^3 + 25^3$
205	{5·41}, Number of prime twins less than 10000
206	{2·103}, $1^3 + 2^3 + 2^3 + 4^3 + 5^3$
207	{3·3·23},
208	{2·2·2·2·13}, $2^2 + 3^2 + 5^2 + 7^2 + 11^2$
209	{11·19}, $1^6 + 2^5 + 3^4 + 4^3 + 5^2 + 6^1$
210	{2·3·5·7}, T_{20}, P_{12}, Prime factorial 7, 14·15
211	{211}, Number of primes a clock can show (0000-2359)
212	{2·2·53}
213	{3·71}
214	{2·107}
215	{5·43}
216	{2·2·2·3·3·3}, Cube, $6^3 = 5^3 + 4^3 + 3^3$, $216 = 6^{2+1}$
217	{7·31}
218	{2·109}
219	{3·73}, Number of symmetry groups in space (crystal structures)
220	{2·2·5·11}, Tetrahedral number, Amicable with 284
221	{13·17}, 2·3·5·7 + 11 = 13·17, 122·122 = 14884 and 48841 = 221·221
222	{2·3·37}
223	{223}
224	{2·2·2·2·2·7}, $2^3 + 3^3 + 4^3 + 5^3$

#	Information
225	$\{3\cdot3\cdot5\cdot5\}$, $(1+2+3+4+5)^2=1^3+2^3+3^3+4^3+5^3$, Square
226	$\{2\cdot113\}$, $\pi^{226}=226\ldots$
227	$\{227\}$, Prime twin
228	$\{2\cdot2\cdot3\cdot19\}$
229	$\{229\}$, Prime twin
230	$\{2\cdot5\cdot23\}$, $6^2+7^2+8^2+9^2$
231	$\{3\cdot7\cdot11\}$, T_{21}
232	$\{2\cdot2\cdot2\cdot29\}$
233	$\{233\}$, F_{13}
234	$\{2\cdot3\cdot3\cdot13\}$
235	$\{5\cdot47\}$, $2+3=5$ and all digits are primes
236	$\{2\cdot2\cdot59\}$
237	$\{3\cdot79\}$
238	$\{2\cdot7\cdot17\}$, The sum of the 13 first primes
239	$\{239\}$, Prime twin
240	$\{2\cdot2\cdot2\cdot2\cdot3\cdot5\}$, Has 20 divisors, $15\cdot16$, Number of solutions for the Soma cube
241	$\{241\}$, Prime twin
242	$\{2\cdot11\cdot11\}$
243	$\{3\cdot3\cdot3\cdot3\cdot3\}$, 3^5, $1/243=0{,}004115226337448559\ldots$
244	$\{2\cdot2\cdot61\}$, $1^3+3^3+6^3$
245	$\{5\cdot7\cdot7\}$, $8^2+9^2+10^2$
246	$\{2\cdot3\cdot41\}$
247	$\{13\cdot19\}$, P_{13}
248	$\{2\cdot2\cdot2\cdot31\}$
249	$\{3\cdot83\}$
250	$\{2\cdot5\cdot5\cdot5\}$
251	$\{251\}$, $1^3+5^3+5^3=2^3+3^3+6^3$
252	$\{2\cdot2\cdot3\cdot3\cdot7\}$, $(10\cdot9\cdot8\cdot7\cdot6)/(5\cdot4\cdot3\cdot2\cdot1)$, Hexagonal Pyramidal number
253	$\{11\cdot23\}$, Triangular star number, T_{22}
254	$\{2\cdot127\}$
255	$\{3\cdot5\cdot17\}$, 11111111b, FFh
256	$\{2\cdot2\cdot2\cdot2\cdot2\cdot2\cdot2\cdot2\}$, Square, $2^8=4^4=16^2$
257	$\{257\}$, Fermat prime
258	$\{2\cdot3\cdot43\}$, $6^1+6^2+6^3$
259	$\{7\cdot37\}$, $1+6+6^2+6^3$
260	$\{2\cdot2\cdot5\cdot13\}$, The row sum of a magic 8x8-square
261	$\{3\cdot3\cdot29\}$
262	$\{2\cdot131\}$
263	$\{263\}$
264	$\{2\cdot2\cdot2\cdot3\cdot11\}$
265	$\{5\cdot53\}$, Number of derangements of 6 items
266	$\{2\cdot7\cdot19\}$
267	$\{3\cdot89\}$
268	$\{2\cdot2\cdot67\}$
269	$\{269\}$, Prime twin
270	$\{2\cdot3\cdot3\cdot3\cdot5\}$

(*Continued*)

#	Information
271	{271}, Prime twin
272	$\{2\cdot2\cdot2\cdot2\cdot17\}$, $16\cdot17$
273	$\{3\cdot7\cdot13\}$
274	$\{2\cdot137\}$
275	$\{5\cdot5\cdot11\}$
276	$\{2\cdot2\cdot3\cdot23\}$, T_{23}, $1^5+2^5+3^5$
277	{277}
278	$\{2\cdot139\}$
279	$\{3\cdot3\cdot31\}$, $3^2+3^3+3^5$
280	$\{2\cdot2\cdot2\cdot5\cdot7\}$
281	{281}, The sum of the 14 first primes, Prime twin
282	$\{2\cdot3\cdot47\}$
283	{283}, $(6!-5!-4!-3!-2!-1!-0!)/2$, 2^5+8+3^5, Prime twin
284	$\{2\cdot2\cdot71\}$, Amicable with 220
285	$\{3\cdot5\cdot19\}$, $1^2+2^2+3^2+\ldots+8^2+9^2$
286	$\{2\cdot11\cdot13\}$, Tetrahedral number
287	$\{7\cdot41\}$, P_{14}
288	$\{2\cdot2\cdot2\cdot2\cdot2\cdot3\cdot3\}$,
289	$\{17\cdot17\}$, $289=(8+9)^2=(2+3+5+7)^2$, Square
290	$\{2\cdot5\cdot29\}$, $10\cdot$the 10th prime
291	$\{3\cdot97\}$
292	$\{2\cdot2\cdot73\}$
293	{293}, $293^{202}=293\ldots$ and $202^{293}=202\ldots$
294	$\{2\cdot3\cdot7\cdot7\}$
295	$\{5\cdot59\}$
296	$\{2\cdot2\cdot2\cdot37\}$
297	$\{3\cdot3\cdot3\cdot11\}$, $297^2=88209$ and $88+209=297$
298	$\{2\cdot149\}$
299	$\{13\cdot23\}$
300	$\{2\cdot2\cdot3\cdot5\cdot5\}$, T_{24}
301	$\{7\cdot43\}$
302	$\{2\cdot151\}$, $9^2+10^2+11^2$
303	$\{3\cdot101\}$
304	$\{2\cdot2\cdot2\cdot2\cdot19\}$
305	$\{5\cdot61\}$
306	$\{2\cdot3\cdot3\cdot17\}$, $17\cdot18$
307	{307},
308	$\{2\cdot2\cdot7\cdot11\}$
309	$\{3\cdot103\}$,
310	$\{2\cdot5\cdot31\}$
311	{311}, Prime twin
312	$\{2\cdot2\cdot2\cdot3\cdot13\}$, $312^2=14^3+15^3+16^3+\ldots+24^3+25^3$
313	{313}, Prime twin, Palindromic prime, Donald Duck's car's license plate number
314	$\{2\cdot157\}$
315	$\{3\cdot3\cdot5\cdot7\}$
316	$\{2\cdot2\cdot79\}$

#	Information
317	{317}
318	{2·3·53}
319	{11·29}
320	{2·2·2·2·2·2·5}
321	{3·107}
322	{2·7·23}
323	{17·19}
324	{2·2·3·3·3·3}, Square
325	{5·5·13}, $1^2+18^2=6^2+17^2=10^2+15^2$, T_{25}
326	{2·163}
327	{3·109}
328	{2·2·2·41}, The sum of the first 15 primes
329	{7·47}
330	{2·3·5·11}, P_{15}
331	{331}
332	{2·2·83}
333	{3·3·37}
334	{2·167}
335	{5·67}, 355/113 ≈ π
336	{2·2·2·2·3·7}
337	{337}, Number of possible resistances values in a circuit with 8 resistors
338	{2·13·13}
339	{3·113}
340	{2·2·5·17}, $4+4^2+4^3+4^4$
341	{11·31}
342	{2·3·3·19}, 18·19
343	{7·7·7}, $343=(3+4)^3$, Cube
344	{2·2·2·43}
345	{3·5·23}
346	{2·173}, The 13th Smith number and the sum of its digits = 13
347	{347}, $347=7^3+4$, Prime twin
348	{2·2·3·29}
349	{349}, Prime twin
350	{2·5·5·7}
351	{3·3·3·13}, T_{26}
352	{2·2·2·2·2·11}
353	{353}, $353^4=30^4+120^4+272^4+315^4$
354	{2·3·59}, $1^4+2^4+3^4+4^4$
355	{5·71}, 355/113 ≈ π, $C_{12}H_{26}$ has 355 isomers
356	{2·2·89}
357	{3·7·17}
358	{2·179}
359	{359}
360	{2·2·2·3·3·5}, Number of degrees in a circle,
361	{19·19}, Number of intersections on a Go board, Square
362	{2·181}

(Continued)

#	Information
363	$\{3 \cdot 11 \cdot 11\}$, $3^1+3^2+3^3+3^4+3^5$
364	$\{2 \cdot 2 \cdot 7 \cdot 13\}$, Tetrahedral number
365	$\{5 \cdot 73\}$, Number of days in a year, $10^2+11^2+12^2=13^2+14^2$
366	$\{2 \cdot 3 \cdot 61\}$, $8^2+9^2+10^2+11^2$, Number of days in a leap year
367	$\{367\}$
368	$\{2 \cdot 2 \cdot 2 \cdot 2 \cdot 23\}$
369	$\{3 \cdot 3 \cdot 41\}$
370	$\{2 \cdot 5 \cdot 37\}$, $3^3+7^3+0^3$
371	$\{7 \cdot 53\}$, $3^3+7^3+1^3$
372	$\{2 \cdot 2 \cdot 3 \cdot 31\}$, Hexagonal Pyramidal number
373	$\{373\}$, $3^2+5^2+7^2+11^2+13^2$, Palindromic prime
374	$\{2 \cdot 11 \cdot 17\}$
375	$\{3 \cdot 5 \cdot 5 \cdot 5\}$
376	$\{2 \cdot 2 \cdot 2 \cdot 47\}$, P_{16}, Automorphic: $376^2=141376$
377	$\{13 \cdot 29\}$, $2^2+3^2+5^2+7^2+11^2+13^2$, F_{14}, T_{27}
378	$\{2 \cdot 3 \cdot 3 \cdot 3 \cdot 7\}$
379	$\{379\}$
380	$\{2 \cdot 2 \cdot 5 \cdot 19\}$, $19 \cdot 20$
381	$\{3 \cdot 127\}$, The sum of the first 16 primes
382	$\{2 \cdot 191\}$
383	$\{383\}$, Palindromic prime
384	$\{2 \cdot 2 \cdot 2 \cdot 2 \cdot 2 \cdot 2 \cdot 2 \cdot 3\}$, 8!!
385	$\{5 \cdot 7 \cdot 11\}$, $1^2+2^2+3^2+\ldots+8^2+9^2+10^2$
386	$\{2 \cdot 193\}$
387	$\{3 \cdot 3 \cdot 43\}$
388	$\{2 \cdot 2 \cdot 97\}$
389	$\{389\}$
390	$\{2 \cdot 3 \cdot 5 \cdot 13\}$
391	$\{17 \cdot 23\}$, $(20+3) \cdot (20-3)$
392	$\{2 \cdot 2 \cdot 2 \cdot 7 \cdot 7\}$
393	$\{3 \cdot 131\}$
394	$\{2 \cdot 197\}$
395	$\{5 \cdot 79\}$
396	$\{2 \cdot 2 \cdot 3 \cdot 3 \cdot 11\}$
397	$\{397\}$, $3^2+8^2+18^2$
398	$\{2 \cdot 199\}$
399	$\{3 \cdot 7 \cdot 19\}$, $7+7^2+7^3$
400	$\{2 \cdot 2 \cdot 2 \cdot 2 \cdot 5 \cdot 5\}$, Square, $1+7+7^2+7^3$
401	$\{401\}$
402	$\{2 \cdot 3 \cdot 67\}$
403	$\{13 \cdot 31\}$
404	$\{2 \cdot 2 \cdot 101\}$, 404 – Not found
405	$\{3 \cdot 3 \cdot 3 \cdot 3 \cdot 5\}$, $4^3+5^3+6^3$
406	$\{2 \cdot 7 \cdot 29\}$, T_{28}
407	$\{11 \cdot 37\}$, $4^3+0^3+7^3$
408	$\{2 \cdot 2 \cdot 2 \cdot 3 \cdot 17\}$

#	Information
409	{409
410	{2·5·41}
411	{3·137}
412	{2·2·103}
413	{7·59}
414	{2·3·3·23}
415	{5·83}, $7^2+8^2+9^2+10^2+11^2$, $(4^5+1^5+5^5)/(4+1+5)$
416	{2·2·2·2·2·13}
417	{3·139},
418	{2·11·19}
419	{419}, Prime twin
420	{2·2·3·5·7}, 20·21
421	{421}, 14^2+15^2, Prime twin
422	{2·211}
423	{3·3·47}
424	{2·2·2·53}
425	{5·5·17}, P_{17}
426	{2·3·71}
427	{7·61}
428	{2·2·107}
429	{3·11·13}, C_7
430	{2·5·43}
431	{431}, Prime twin
432	{2·2·2·2·3·3·3}, $4\cdot 3^3\cdot 2^2$
433	{433}, Prime twin
434	{2·7·31}, $11^2+12^2+13^2$
435	{3·5·29}, T_{29}
436	{2·2·109}
437	{19·23}, There are 4·3·7 primes less than 437
438	{2·3·73}
439	{439}
440	{2·2·2·5·11}, $2^3+3^3+4^3+5^3+6^3$, The sum of the 17 first primes
441	{3·3·7·7}, Square, $1^3+2^3+3^3+4^3+5^3+6^3$
442	{2·13·17}
443	{443}
444	{2·2·3·37}
445	{5·89}
446	{2·223}, $9^2+10^2+11^2+12^2$
447	{3·149}
448	{2·2·2·2·2·2·7}
449	{449}, 449! is the largest factorial $<10^{1000}$
450	{2·3·3·5·5}
451	{11·41}
452	{2·2·113}
453	{3·151}
454	{2·227}

(*Continued*)

#	Information
455	{5·7·13}, Tetrahedral number
456	{2·2·2·3·19}
457	{457}
458	{2·229}
459	{3·3·3·17}
460	{2·2·5·23}
461	{461}, Prime twin
462	{2·3·7·11}, 21·22
463	{463}, Prime twin
464	{2·2·2·2·29}
465	{3·5·31}, T_{30}
466	{2·233}
467	{467}
468	{2·2·3·3·13}
469	{7·67}
470	{2·5·47}
471	{3·157}
472	{2·2·2·59}
473	{11·43}
474	{2·3·79}
475	{5·5·19}
476	{2·2·7·17}
477	{3·3·53}, P_{18}
478	{2·239}
479	{479}
480	{2·2·2·2·2·3·5}
481	{13·37}, 15^2+16^2
482	{2·241}
483	{3·7·23}
484	{2·2·11·11}, Square
485	{5·97}
486	{2·3·3·3·3·3}
487	{487}
488	{2·2·2·61}
489	{3·163}
490	{2·5·7·7}
491	{491}
492	{2·2·3·41}
493	{17·29}
494	{2·13·19}
495	{3·3·5·11}
496	{2·2·2·2·31}, T_{31}, $1^3+3^3+5^3+7^3$, The third Perfect number
497	{7·71}
498	{2·3·83}
499	{499}, 497+2=499 and 497·2=994
500	{2·2·5·5·5}, D

#	Information
501	{3·167}, The sum of the first 18 primes
502	{2·251}
503	{503}, $2^3+3^3+5^3+7^3$
504	{2·2·2·3·3·7}, 12·42=21·24
505	{5·101}, The row sum of a magic 10x10-square
506	{2·11·23}, $1^2+2^2+3^2+\ldots+11^2$, 22·23
507	{3·13·13}
508	{2·2·127}
509	{509}, $12^2+13^2+14^2$
510	{2·3·5·17}
511	{7·73}, 111111111b
512	{2·2·2·2·2·2·2·2·2}, Cube, $(5+1+2)^3=8^3$,
513	{3·3·3·19}
514	{2·257}
515	{5·103}
516	{2·2·3·43}
517	{11·47}
518	{2·7·37}, $5^1+1^2+8^3$
519	{3·173}
520	{2·2·2·5·13}
521	{521}, Prime twin
522	{2·3·3·29}
523	{523}, Prime twin
524	{2·2·131}
525	{3·5·5·7}, Hexagonal Pyramidal number
526	{2·263}
527	{17·31}
528	{2·2·2·2·3·11}, T_{32}
529	{23·23}, Square
530	{2·5·53}, The sum of the three first Perfect numbers 6+28+496
531	{3·3·59}
532	{2·2·7·19}, P_{19}
533	{13·41}
534	{2·3·89}
535	{5·107}
536	{2·2·2·67}, The number of solutions to the Stomachion puzzle, The number of pages in the original edition of this book where these Integer facts were displayed next to the pagination
537	{3.179}
538	{2.269}, the total number of votes in the Electoral College of the United States
539	{7.7.11}
540	{2.2.3.3.5}, untouchable
541	{541}, 100th prime number
542	{2.271}
543	{3.181}
544	{2.2.2.2.2.17}

(*Continued*)

#	Information
545	{5.109}
546	{2.3.7.13}, sum of 8 consecutive primes
547	{547}
548	{2.2.137}
549	{3.3.61}, every positive integer is the sum of at most 548 ninth powers
550	{2.5.5.11}
550	{2, 5, 5, 11}
551	{19, 29}, The sum of three consecutive primes: 179 + 181 + 191 = 551, number of trees with 12 vertices
552	{2, 2, 2, 3, 23}
553	{7, 79}
554	{2, 277}
555	{3, 5, 37}
556	{2, 2, 139}, The first three digits of 4556
557	{557}
558	{2, 3, 3, 31}, Divides the sum of the largest prime factors of the first 558 positive integers
559	{13, 43}
560	{2, 2, 2, 2, 5, 7}, 42 + 122 + 202 = 560, The number of different moves a bishop can make on a chessboard
561	{3, 11, 17}, The smallest Charmichael number: a560 – 1 is divisible by 561, far all values of a
562	{2, 281}, 112 + 212 = 562
563	{563}, The largest known Wilson prime: 5632 divides (563 – 1)! + 1 (the only other Wilson primes are 3 and 13)
564	{2, 2, 3, 47}, The largest number that can be written with all the numbers from 1 to 5 without repetition as a sum of two primes in two ways: 564 = 541 + 23 = 523 + 41
565	{5, 113}, 62 + 232 = 92 + 222 = 565. Also the hypothenuse of two Pythagorean triples: 5652 = 2762 + 4932 = 3962 + 4032
566	{2, 283}, The number of ways to place three points on a 12 × 12 grid so that no three points lie on a line
567	{3, 3, 3, 3, 7}, The smallest possible sum of primes that are formed using all 10 digits: 567 = 401 + 89 + 67 + 5 + 3 + 2, also: 5672 = 321489 (the only other number with this property is 854)
568	{2, 2, 2, 71}, 102 + 122 + 182 = 568

INDEX

Entries in *italics* refer to figures, entries in **boldface** refer to tables.

Mathematical Modeling: Applications with GeoGebra™, First Edition. Jonas Hall and Thomas Lingefjärd.

Companion website: www.wiley.com/go/Hall/MathematicalModeling

LIST OF PROBLEMS BY NAME

Mathematical Modeling: Applications with GeoGebra™, First Edition. Jonas Hall and Thomas Lingefjärd.

Companion website: www.wiley.com/go/Hall/MathematicalModeling

Printed and bound by CPI Group (UK) Ltd, Croydon, CR0 4YY

16/04/2025

14658520-0005